Agricultural Acarology

Introduction to Integrated Mite Management

Agricultural Acarology

Introduction to Integrated Mite Management

Marjorie A. Hoy
University of Florida
Gainesville, USA

CRC Press
Taylor & Francis Group
Boca Raton London New York

CRC Press is an imprint of the
Taylor & Francis Group, an **informa** business

CRC Press
Taylor & Francis Group
6000 Broken Sound Parkway NW, Suite 300
Boca Raton, FL 33487-2742

Printed in the United States of America on acid-free paper
Version Date: 20110701

International Standard Book Number: 978-1-4398-1751-3 (Hardback)

Library of Congress Cataloging-in-Publication Data

Hoy, Marjorie A.
Agricultural acarology : introduction to integrated mite management / author, Marjorie A. Hoy.
p. cm. -- (Contemporary topics in entomology series ; 7)
"A CRC title."
Includes bibliographical references and index.
ISBN 978-1-4398-1751-3 (hardcover : acid-free paper)
1. Soil mites. 2. Acarology. 3. Soil mites--Biological control 4. Soil mites--Integrated control. 5. Agricultural pests. 6. Beneficial insects. 7. Soil management. I. Title. II. Series.

QL458.H69 2011
632'.6542--dc23 2011022999

Visit the Taylor & Francis Web site at
http://www.taylorandfrancis.com

and the CRC Press Web site at
http://www.crcpress.com

Contents

Supplementary Resources Disclaimer

Additional resources were previously made available for this title on CD. However, as CD has become a less accessible format, all resources have been moved to a more convenient online download option.

You can find these resources available here: www.routledge.com/9781439817513

Please note: Where this title mentions the associated disc, please use the downloadable resources instead.

Preface

Many biologists are intimidated by the thought of working with mites or ticks, in part because they are so small relative to insects that they require specialized monitoring and handling methods. Furthermore, identifying mites can be a daunting task, requiring specialized clearing and slide-mounting methods, new terminology, and complex keys. Unfortunately, compared to insects, mites have relatively few morphological traits that allow easy identification; mites lack antennae, wings, and other easy-to-see morphological traits for discriminating among families, genera, and species. Yet, the ability to work with pest mites in agriculture is essential, although it is becoming increasingly difficult because fewer and fewer courses in acarology are being taught in universities, and fewer and fewer taxonomists are available to assist in mite identifications.

The goal of this book is to provide pest-control workers and students with the tools to manage mite pests of agriculture using the concepts of integrated pest management (IPM), so mite management becomes an integrated effort rather than one based largely on chemical control. Another goal is to emphasize how knowing the biology, ecology, and behavior of pest and beneficial mites (and some beneficial insects) in agricultural systems will allow IPM-based methods to be developed. An emphasis is placed on using biological control and other management tools that are compatible with biological control whenever possible.

A look at the Contents reveals the diversity of information provided. Readers will learn to discriminate between mites and other arthropods and will be introduced to the tactics used in agricultural pest management programs for mites, in addition to basic information on the biology, behavior, and ecology of key mite species in crop-based agriculture. The focus is on biological control, including a review of the biology, behavior, and ecology of important natural enemies (predatory insects, predatory mites, spiders, and pathogens) of pest mites in crops. The various approaches used to manage pest mites are discussed, including their advantages and limitations. The photographs in the text and on the accompanying CD will aid in the use of a dissecting microscope to identify the life stages of plant-pest mites and phytoseiids. Also provided are summaries of several model cropping systems and the reasons for deploying specific IPM tactics in each. The biology and impact of pest mites on honey bees and their management are discussed, in addition to the mites and ticks that affect farm animals and management of mite pests in stored products and households.

Selected references are cited that provide an entry into the literature for each topic and, when feasible, review articles are cited. I apologize to those acarologists whose excellent papers are not cited. A full list of relevant references would be very long; for example, the number of references published between 1970 and 1992 just on the predatory mite *Metaseiulus occidentalis* includes at least 450 papers.

The accompanying CD contains supplementary information, including numerous color photographs of mites and the damage they cause and PDFs of key publications. Websites change rapidly, so only a few are listed, despite the fact that many useful sites offer accurate information and excellent photographs.

The overview of agricultural acarology focuses on integrated mite management (IMM). Different pest problems require different combinations of tactics in IMM, as demonstrated by the model systems presented. Ideally, these models will be useful for the reader who needs to develop and implement IMM programs in another crop or geographic region. Chemical control and the types of pesticides useful for managing mites and ticks are discussed, although specific recommendations cannot be made, with the exceptions of oil and sulfur, because registrations, pesticide resistances, and legal issues change rapidly. Terms that are in **bold** are defined in the glossary, and an index provides access to topics and species.

Acknowledgments

I thank the students in the Doctor of Plant Medicine program at the University of Florida for their inputs into this textbook, which was developed as the result of a required course in the Doctor of Plant Medicine curriculum. The mission of this program, established in 1999, "is to provide all segments of agriculture with rapid, accurate, and scientifically sound diagnoses and management strategies for all types of plant health problems through the activities of broadly trained Plant Doctors. It is our belief that Plant Doctors will have a positive impact on worldwide agriculture by increasing the productivity, usefulness, and profitability of plants while lessening non-target effects of management practices to humans and the environment." As a result, this book has an applied, problem-solving orientation that should be useful to entomologists and others attempting to manage mites in agricultural systems.

This book was written with support from the Department of Entomology and Nematology, Institute of Food and Agricultural Sciences of the University of Florida, and the Davies, Fischer, and Eckes Endowment in Biological Control. Where possible, photographs of the mites and their natural enemies are provided in the text and on the CD to facilitate identifications; these photographs and illustrations are incredibly important because most readers will want to identify their mites using a hand lens or a dissecting microscope. Specialized equipment and skill are required to produce high-quality photographs of these tiny arthropods, so I thank the colleagues who kindly provided photographs or other materials for this book and the accompanying CD.

I especially acknowledge faculty, staff, and students in the Department of Entomology and Nematology, University of Florida (Lyle Buss, Jim Castner, Jim Price, Skip Choate, Patricia Toth, Jerry Butler, Howard Frank, and Harvey Cromroy). I thank the University of California Statewide IPM Program (especially Jack Kelly Clark) for permission to use their superb photographs. In addition, I thank Elizabeth H. Beers (Washington State University, Wenatchee), Valerie Behan-Pelletier (Canadian National Collection of Insects and Arachnids, Ottawa, Canada), Robert Cating (University of Florida, Gainesville), Ross P. Field (formerly at the University of California–Berkeley), Beth Grafton-Cardwell (University of California in Parlier), Hugh Gourlay (Landcare Research, New Zealand), Nancy Fike Knop (formerly at the University of California–Berkeley), Roy Norton (State University of New York, Syracuse), Heather Proctor (University of Alberta, Edmonton, Canada), David Walter (currently at the Royal Alberta Museum, Edmonton, Canada), and Steve Yaninek (International Institute of Tropical Agriculture, currently at Purdue University, Lafayette, Indiana). All sources of photographs are acknowledged in the figure legends.

The accompanying CD includes some publications. I thank the Regents of the University of California for permission to provide the seminal paper on integrated pest management published in *Hilgardia* by Stern et al. (1959). The U.S. Department of Agriculture provided permission to include *An Illustrated Guide to the Plant Abnormalities Caused by Eriophyid Mites in North America* by Keifer et al. Dr. Kirby Stafford, Connecticut Agricultural Experiment Station, New Haven, Connecticut, granted permission to include a PDF of his *Tick Management Handbook.* I thank Tuomas Kostiainen and the Institute of Food and Agricultural Sciences of the University of Florida for allowing me to include a PDF of *The Phytoseiidae as Biological Control Agents of Pest Mites and Insects: A Bibliography.* I am exceedingly grateful to Niklaus Hostettler for his excellent assistance in organizing and editing the many illustrations. Jennifer Gillett-Kaufman, Ke Wu, and Nicole Giannacco provided expert editorial assistance.

I acknowledge the acarologists and entomologists who had a significant influence on my understanding of integrated mite management, especially Bill Allen, Robert Beer, Joseph Camin, Donald Flaherty, Stan Hoyt, Carl Huffaker, and Robert van den Bosch. I thank my husband, Jim, for his patience and encouragement during the production of this book.

Marjorie A. Hoy
Gainesville, Florida

Author

Marjorie A. Hoy, PhD, received her BA degree in zoology and entomology at the University of Kansas at Lawrence, and an MS and a PhD in entomology at the University of California–Berkeley, where she specialized in acarology, biological control, insect ecology, genetics, and evolution. After obtaining her doctorate, she took an assistant scientist position at the Connecticut Agricultural Experiment Station in New Haven and a research scientist position with the U.S. Forest Service in Hamden, Connecticut, where she worked on genetic improvement of natural enemies of the gypsy moth. She moved to the University of California–Berkeley in 1976, where she was an assistant, associate, and full professor and is now a professor emerita. In California, she conducted research on integrated pest management of insects and mites in almonds, grapes, apples, pears, and citrus. In 1992, she took a position as an Eminent Scholar in biological control at the University of Florida, Gainesville, where she has conducted classical biological control of citrus pests and the red palm mite. Currently, she is working on an analysis of the transcriptome and of the genome of a predatory mite. She has published over 350 scientific papers in refereed journals and written two editions of *Insect Molecular Genetics*. She has served as major professor for 15 PhD and 10 MS students and has supervised multiple postdoctoral scientists. She teaches a course in agricultural acarology to students in entomology and the Doctor of Plant Medicine program at the University of Florida, as well as a course in insect molecular genetics.

PART I

Introduction to Acarology

Part I introduces you to the world of mites and ticks, providing an overview of their taxonomic status as arthropods and a discussion of their evolution, basic anatomy, and biology. Much of the information provided on basic physiology and anatomy is relevant to the tactics that can be deployed in integrated mite management (IMM) programs discussed later. On the CD provided, you will see photographs illustrating some of the diversity of mite morphology and biology.

CHAPTER 1

General Introduction to Acarology

Acarology is the study of mites, or Acari or Acarina. It is a specialized field of study in the larger topics of invertebrate zoology and entomology. Because some mites are economically important pests of agriculture (crops, honey bees, stored food products, and livestock), are household pests, or are vectors of diseases of humans and livestock, acarology is often studied in entomology departments.

Mites have been studied for centuries. A tick-transmitted fever was mentioned in an ancient Egyptian papyrus in 1550 BC. Homer mentioned ticks in 850 BC, and Aristotle described mites parasitic on locusts. Hippocrates, Plutarch, Aristophanes, and Pliny all mentioned mites. Until about 1660, mites were referred to as 'lice,' 'beesties,' or 'little insects.' The terms 'Akari' and 'mite' began to be used about 1650. Linnaeus used the generic name *Acarus* in the first edition of the *Systema Naturae* and named the type of the genus *Acarus siro* (a grain mite) in 1758.

Acarology was an active field in Europe in the late 19th and early 20th centuries. Beginning with World War II, an increased awareness of the economic importance of mites and ticks arose in the United States and elsewhere due to a concern about diseases transmitted by ticks and chiggers (which transmit scrub typhus). After the war, with the increased use of synthetic organic pesticides such as DDT to control a wide array of insect and mite pests, spider mites became much more serious agricultural pests around the world, and applied agricultural acarology developed (Baker and Wharton 1952, Jeppson et al. 1975).

Mites are not just agricultural pests, however. They are of intrinsic interest to zoologists and ecologists because they rival the insects in their numbers and diversity. Admittedly, mites usually are less obvious due to their small size. Most mites are 0.08 to 1.0 millimeters (mm) in length; however, some ticks and red velvet mites may reach lengths of 10 to 20 mm. We do not know just how diverse or abundant mites are due to the fact that the taxonomy of mites is 50 to 100 years behind that of the insects. This is because mites are so small that they require rather specialized methods and equipment to study them, and if they are not of economic importance they often are overlooked. It is not unusual for new species, genera, and even families of mites to be discovered, and collecting in distant or exotic regions is not required to discover new mite taxa. By 1950, about 30,000 species of mites in 1700 genera had been described. As of 1999, approximately 40,000 species of mites had been described, and estimates of unnamed species range from 0.5 to 1 million species (Walter and Proctor 1999).

Some taxonomists have suggested there may be more mite species than insect species because mites are smaller and able to occupy smaller niches than insects. Certainly, they outdo the insects in the diversity of habitats they lay claim to because of their small size. Mites have colonized all of the habitats that insects have colonized; some are terrestrial and some aquatic in freshwater, and some are even found in the oceans (Evans 1992, Houck 1994, Proctor 2006, Krantz and Walter 2009). Mites are freeliving or are parasitic on plants and on vertebrate and invertebrate animals; for example, some mites live within the ears of moths (Treat 1975), and a single bird or mammal may serve as a host to several mite species because mites are so small that they can colonize different portions of their host. Different mite species may occur on the outside of feathers, within the quills of feathers, in the nasal passages, or on the legs of their bird hosts. Likewise, plant species typically have three to seven species of mites—some feeding on the plant, some feeding on pollen or fungi, and some that are predators of the plant-feeding mites (Huffaker et al. 1970, McMurtry et al. 1970, Keifer et al. 1982, Hoy et al. 1983, Helle and Sabelis 1985, Gerson and Smiley 1990, Ochoa et al. 1994, Kostiainen and Hoy 1996, Lindquist et al. 1996, Meyer 1996, Zhang and Liang 1997, Bolland et al. 1998, Gerson et al. 2003, Zhang 2003).

Assuming that there are 400,000 species of plants and at least one species-specific, plant-feeding mite on each plant species, then there are 400,000 species of plant-feeding mites alone. If each of the approximately 60,000 vertebrate animals contains at least two species-specific mites, then that would account for another 120,000 mite species. Some mites can be found in surprising places; for example, mites (Cloacaridae) with a highly modified morphology are found in the cloacae of turtles (Camin et al. 1967). These mites may be transmitted sexually, and it appears that different turtle species have different mite species. Mites also are parasitic on or predators of other invertebrates.

Mites are able to exploit very small habitats; for example, mites have colonized the nasal cavities, genital organs, feather shafts, skin pores, and lungs of their vertebrate hosts (Evans 1992). Mites have colonized the soil, where they serve as predators and fungivores, and they are important in recycling minerals in the soil (Balogh 1972). Mites are found in diverse soil, aquatic, aerial, and cave habitats throughout the world, including the Antarctic (Balogh 1972, Dindal 1977, Colloff and Halliday 1998). See Figures S1.1 to S1.12 on the CD for photographs illustrating the diversity of mite morphology and biology.

One of the reasons mites are so diverse is that they have had a long evolutionary history comparable to that of the insects (Figure 1.1). Spiders, scorpions, mites, and ticks (chelicerate arthropods) form one of the most diverse arthropod groups on land. Using molecular data, Jeyaprakash and Hoy (2009) estimated that the orders and classes of spiders, scorpions, mites, and ticks diversified in the late Paleozoic. Dunlop and Selden (2009) confirmed that the oldest known occurrences of chelicerate orders in the fossil record are indeed ancient; for example, the oldest mite fossil is from the Devonian (410 million years ago), while scorpion fossils (428 million years ago) and spider fossils (312 million years ago) are of similar age.

Acarology is the study of mites *and* ticks. *All ticks are mites, but not all mites are ticks!* Ticks are pests of agriculture when they transmit diseases to livestock and humans or cause irritation or damage to them by their feeding. Mites are pests of agriculture when they feed on plants or transmit diseases to crops; however, many mites remain undescribed, or their role in the ecosystem is unknown. This text focuses on a small subset of economically important mites. For a broader overview of acarology, see general texts such as Evans (1992), Houck (1994), Walter and Proctor (1999), and Krantz and Walter (2009). For access to additional information on acarology organizations, see the websites listed below. For up-to-date information on research in acarology, see the journals listed below and be aware that every four years, since 1963, an International Congress of Acarology has been held for which proceedings volumes are produced that provide information on acarine biology, ecology, behavior, and evolution (see, for example, Halliday et al. 2001).

In this book, the focus primarily is on mites that are pests of production agriculture, especially agricultural plants (see Chapters 6 to 11). Mites attack honey bees, which are critical to the pollination of many crops (Needham et al. 1988) (see Chapters 20 and 21). Mites and ticks are pests of livestock (sheep, poultry, cattle, swine), either because they cause damage by their feeding or because they transmit diseases to them (see Chapters 22 and 23). Mites are important in the postharvest storage of crops, as they cause damage to stored grains, cheese, potatoes, ornamental bulbs, and vegetables (see Chapters 24). Not all mites are pests, and some beneficial mite species are effective natural enemies of plant-feeding mites and some plant-feeding insects (see Chapters

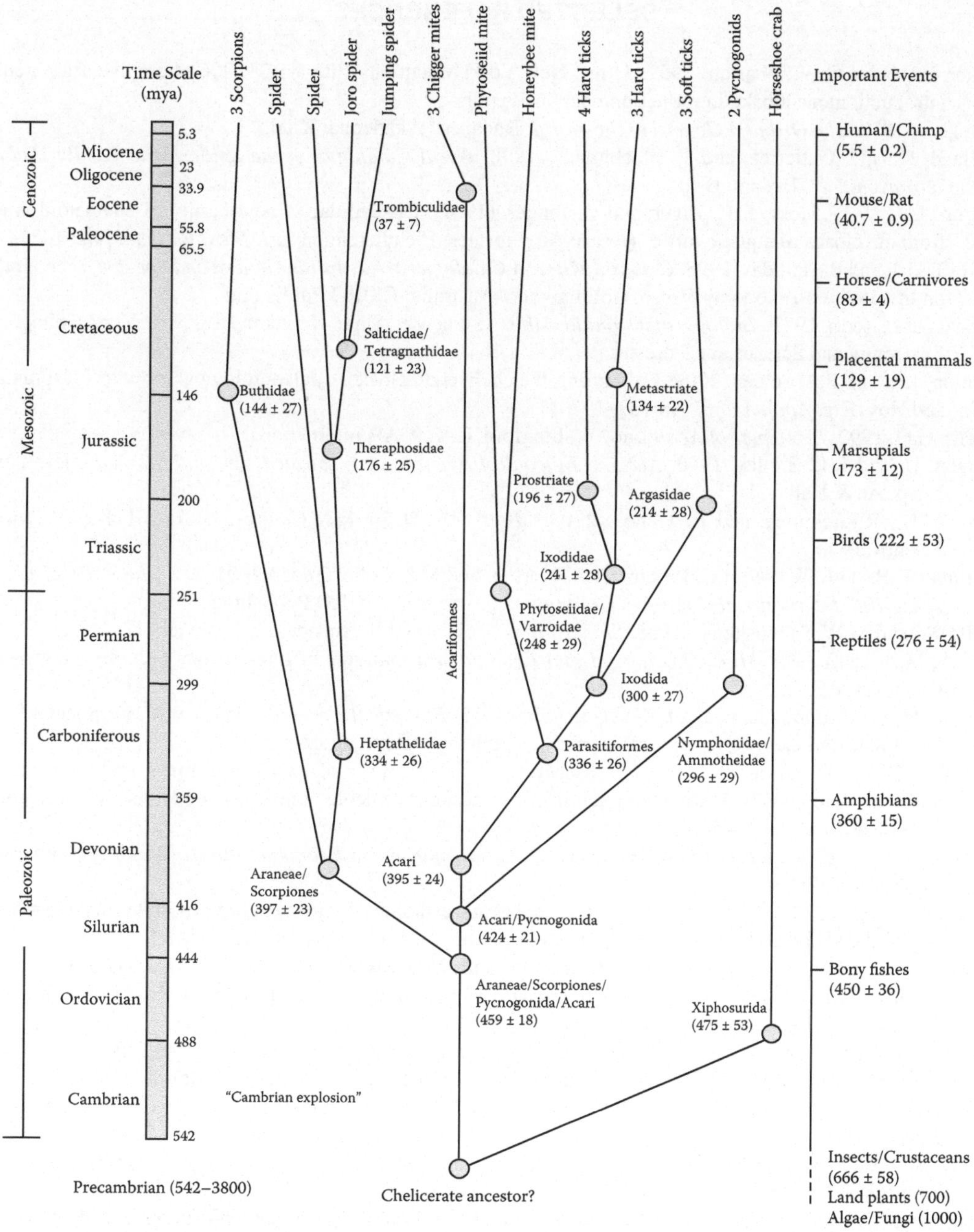

Figure 1.1 One estimate of the evolutionary age of mites places chelicerate ancestors in the Precambrian and the divergence of mites beginning at least 424 million years ago; thus, both insects and acarines are very ancient. (From Jeyaprakash, A. and Hoy, M.A., *Exp. Appl. Acarol.*, 47, 1–18, 2009. With permission.)

12 and 13). Soil mites (the Oribatida or Cryptostigmata) are important in breaking down organic material and in nutrient cycling in the soil (see Part V). For a brief overview of the diversity of form and function of mites, see Figures S1.1 through S1.12 on the accompanying CD. Whenever you see an S in front of the figure number, it means that the illustration is on the supplementary CD in a Chapter folder.

SELECTED REFERENCES

Baker, E.W. and G.W. Wharton. 1952. *Mites*. New York: Macmillan (http://ia311006.us.archive.org/2/items/introductiontoac00bake/introductiontoac00bake.pdf).

Balogh, J. 1972. *The Oribatid Genera of the World*. Budapest: Akademiai Kiado.

Bolland, H.R., J. Gutierrez, and C. Flechtmann. 1998. *World Catalogue of the Spider Mite Family (Acari: Tetranychidae)*. Boston: Brill.

Camin, J.H., W.W. Moss, J.H. Oliver, and G. Singer. (1967). Cloacaridae, a new family of cheyletoid mites from the cloaca of aquatic turtles (Acari: Acariformes: Eleutherengona). *J. Med. Entomol.* 4:261–272.

Colloff, M.J. and B. Halliday. 1998. *Oribatid Mites: A Catalogue of Australian Genera and Species*, Monograph on Invertebrate Taxonomy, Vol. 6. Collingwood, Australia: CSIRO Publishing.

Dindal, D.L. (ed.) 1977. *Biology of Oribatid Mites*. Syracuse: State University of New York College of Environmental Science and Forestry.

Dunlop, J.A. and P.A. Selden. 2009. Calibrating the chelicerate clock: A paleontological reply to Jeyaprakash and Hoy. *Exp. Appl. Acarol.* 48:183–197.

Evans, G.O. 1992. *Principles of Acarology.* Wallingford, U.K.: CAB International.

Gerson, U. and R.L. Smiley. 1990. *Acarine Biocontrol Agents: An Illustrated Key and Manual.* New York: Chapman & Hall.

Gerson, U., R.L. Smiley, and R. Ochoa. 2003. *Mites (Acari) for Pest Control.* Oxford, U.K.: Blackwell Scientific.

Halliday, R.B., D.E. Walter, H.C. Proctor, R.A. Norton, and M.J. Colloff (eds.) 2001. *Acarology: Proceedings of the 10th International Congress*. Collingwood, Australia: CSIRO Publishing.

Helle, W. and M.W. Sabelis (eds.) 1985. *Spider Mites*, Vols. 1 and II. Amsterdam: Elsevier.

Houck, M.A. (ed.) 1994. *Mites: Ecological and Evolutionary Analyses of Life-History Patterns.* New York: Chapman & Hall.

Hoy, M.A., G.L. Cunningham, and L. Knutson (eds.) 1983. *Biological Control of Pests by Mites*, Special Publ. No. 3304. Berkeley: University of California, Division of Agriculture.

Huffaker, C.B., M. van de Vrie, and J.A. McMurtry. 1970. Ecology of tetranychid mites and their natural enemies: A review. II. Tetranychid populations and their possible control by predators: An evaluation. *Hilgardia* 40(11):391–458.

Jeppson, L.R., H.H. Keifer, and E.W. Baker. 1975. *Mites Injurious to Economic Plants.* Berkeley: University of California Press.

Jeyaprakash, A. and M.A. Hoy. 2009. First divergence time estimate of spiders, scorpions, mites and ticks (subphylum: Chelicerata) inferred from mitochondrial phylogeny. *Exp. Appl. Acarol.* 47:1–18.

Keifer, H.H., E.W. Baker, T. Kono, M. Delfinado, and W.E. Styer. 1982. *An Illustrated Guide to Plant Abnormalities Caused by Eriophyid Mites in North America*, Agriculture Handbook No. 573. Washington, D.C.: U.S. Department of Agriculture.

Kostiainen, T. and M.A. Hoy. 1996. *The Phytoseiidae as Biological Control Agents of Pest Mites and Insects: A Bibliography [1960–1994]*, IFAS Monograph 17. Gainesville: University of Florida.

Krantz, G.W. and D.E. Walter (eds.) 2009. *A Manual of Acarology*, 3rd ed. Lubbock: Texas Tech University Press.

Lindquist, E.E., M.W. Sabelis, and J. Bruin (eds.) 1996. *Eriophyoid Mites: Their Biology, Natural Enemies and Control*, Vol. 6. Amsterdam: Elsevier.

McMurtry, J.A., C.B. Huffaker, and M. van de Vrie. 1970. Ecology of tetranychid mites and their natural enemies: A review. I. Tetranychid enemies: Their biological characters and the impact of spray practices. *Hilgardia* 40(11):331–390.

Meyer, M.K.P.S. (1996). *Mite Pests and Their Predators on Cultivated Plants in Southern Africa. Vegetables and Berries*, Plant Protection Research Institute Handbook No. 6. Pretoria, S. Africa: Agricultural Research Council.

Needham, G.R., R.E. Page, M. Delfinado-Baker, and C.E. Bowman (eds.) 1988. *Africanized Honey Bees and Bee Mites*. Chichester: Ellis Horwood.

Ochoa, R., H. Aguilar, and C. Vargas. 1994. *Phytophagous Mites of Central America: An Illustrated Guide.* Turrialba, Costa Rica: Centro Agronómico Tropical de Investigación y Enseñanza (CATIE).

Proctor, H.C. (ed.) 2006. *Aquatic Mites: From Genes to Communities*. Dordrecht: Kluwer Academic.

Treat, A.E. 1975. *Mites of Moths and Butterflies.* Ithaca, NY: Cornell University Press.

Walter, D.E. and H.C. Proctor. 1999. *Mites. Ecology, Evolution and Behaviour.* Wallingford, U.K.: CAB International.

Zhang, Z.-Q. 2003. *Mites of Greenhouses: Identification, Biology, and Control.* Wallingford, U.K.: CAB International.

Zhang, Z.-Q. and L.R. Liang. 1997. *An Illustrated Guide to Mites of Agricultural Importance.* Shanghai, China: Tongji University Press.

SELECTED WEBSITES RELEVANT TO ACAROLOGY

Acarology homepage: http://www.nhm.ac.uk/hosted_sites/acarology/ (This site includes links to the journals *Acarologia*, *Acarology Bulletin*, *International Journal of Acarology*, *Systematic and Applied Acarology*, and *Experimental and Applied Acarology* and other resources.)

Acarological Society of America: http://www.acariweb.com/ASA

Experimental and Applied Acarology journal articles: http://www.springerlink.com/content/100158/

European Association of Acarologists: http://euraac.boku.ac.at

Ohio State Acarology Laboratory (OSAL): http://www.biosci.ohio-state.edu/~acarolog/index.html

Ohio State UniversityAcarology Summer Program: http://www.biosci.ohio-state.edu/~acarolog/summerProgram/

CHAPTER 2

The Relationship of Mites to Other Arthropods

2.1 CHARACTERISTICS OF THE ARTHROPODA

Mites and ticks are chelicerate arthropods with ancient origins. This chapter reviews the characteristics of arthropods and their taxonomic relationships and provides an overview of their evolutionary history. The phylum Arthropoda consists of at least 700,000 described species, or about 80% of the known animals of the world. Some estimate that more than 2 million arthropod species will be described eventually. Both in number of species and number of individuals, arthropods far surpass all other multicellular phyla in both numbers and diversity, exceeded only by the unicellular microbiota.

Arthropod characters include a metamerically segmented body, chitinous exoskeleton, segments (somites) united by soft membranes, **tagmata** (distinct body regions), each **somite** (body segment) carrying a pair of appendages with at least one pair modified to act as mouthparts, a hemocoelic body (blood freely moves around the internal organs), a central nervous system consisting of a cerebral ganglion with a solid ventral ganglionated nerve cord, separate sexes with eggs usually rich in yolk, discontinuous growth associated with **ecdysis** (molting), and **metamorphosis** (a significant change in shape or structure) (Evans et al. 1961, Krantz 1979, Evans 1992, Krantz and Walter 2009).

The Arthropoda are considered by most to be monophyletic but, because there is disagreement about the evolutionary relationships of groups within the arthropods, there are a number of different ways to organize the groups, and different names have been given to them. The same is true of acarines, which makes the taxonomy of mites confusing. For example, ticks may be called Ixodida or Metastigmata. Arthropods often are divided into two subphyla, the Chelicerata and the Uniramia (or Mandibulata). Mites and ticks have chelicerate mouthparts (although often the chelicerae are highly modified) and insects (Hexapoda or Insecta) have mandibulate mouthparts (again often highly modified). The classification shown in Table 2.1 is used in this book to compare insects and mites, but other classifications have been used (Evans et al. 1961, Savory 1977, Krantz 1979, Krantz and Walter 2009).

The subclasses of arachnids are quite different from each other, and their evolutionary relationships are difficult to resolve (Savory 1977). The Acari are unusual among the arachnids because they include plant-feeding species (Walter and Proctor 1999). The Chelicerata, including mites and ticks, are ancient and have had a very long time to diversify. Based on fossil evidence and molecular analysis, the divergence time for the Atlantic horseshoe crab, *Limulus polyphemus* (Xiphosura), was estimated to be 475 ± 53 million years ago (mya) (Jeyaprakash and Hoy 2009) (see Figure 1.1 in Chapter 1). Dunlop and Selden (2009) summarized dates from the fossil record. The Eurypterida were present in the Ordovician about 460 mya (Table 2.2). Relatively modern-looking mites (Acari)

Table 2.1 One Classification of the Arthropoda

Phylum Arthropoda
- Subphylum Uniramia (Mandibulata)
  - Class Hexapoda (Insecta)
- **Subphylum Chelicerata**
  - Class Pycnogonida (sea spiders)
  - Class Eurypterida (extinct)
  - Class Xiphosura (horseshoe crabs)
  - **Class Arachnida** (arachnids have simple eyes, are mostly predaceous, and have chelicerae and pedipalps)
    - **Subclass Acari** (or **Acarina**) (unusual chelicerates because not all are predators; most diverse group of arachnids)
    - Subclass Amblypygi (whip spiders)
    - Subclass Araneae (spiders)
    - Subclass Opiliones (not all predatory)
    - Subclass Palpigradi (microwhip scorpions)
    - Subclass Pseudoscorpiones (pseudoscorpions)
    - Subclass Ricinulei (hooded tickspiders)
    - Subclass Schizomida (short-tailed whip scorpions)
    - Subclass Scorpiones (scorpions)
    - Subclass Solfugae (sun spiders)
    - Subclass Uropygi (whip scorpions, vinegaroons)

have been found in Devonian strata (410 mya). The oldest spiders (Aranae) have been found in Carboniferous fossils (312 mya), the oldest pycnogonids have been found in the Cambrian (501 mya), and scorpions have been found in the Silurian (428 mya).

The Chelicerata have an adult body *primitively* composed of 18 somites, which are divided into a **prosoma** of six segments and an **opisthosoma** of 12; however, segmentation may be obscured in both prosoma and opisthosoma by fusion or suppression of sclerites, and this is especially true in the Acarina. No Chelicerata (or mites) have compound eyes, and no chelicerates have more than 12 simple **ocelli**. The prosoma carries six pairs of appendages, including the **chelicerae**, **pedipalps**, and four pairs of legs. The chelicerae, which are in front of the mouth are of two or three parts and may be **chelate**, consisting of pincer-like parts, or unchelate. The pedipalps consist of six segments and may be chelate or leg-like. Legs have seven segments, with the tarsi ending in two or three claws. Sexes are separate with genital openings on the lower side of the opisthosoma. Courtship is common and often elaborate. Individuals may be terrestrial, carnivorous, nocturnal, and cryptic in their habits. The respiratory system consists of book lungs or **tracheae**, or both. Most chelicerates are predatory, but some mites (Acari) are phytophagous, saprophagous, or parasitic.

Table 2.2 Oldest Known Fossils of Some Chelicerate Orders

Taxon	Million Years Ago	Oldest Record
Acari	410	Devonian
Amblypygi	312	Carboniferous
Aranae	312	Carboniferous
Euryptidera	460	Ordovician
Opiliones	410	Devonian
Palpigradi	5	Pliocene
Pycnogonida	501	Cambrian
Pseudoscorpions	392	Devonian
Ricinulei	319	Carboniferous
Scorpiones	428	Silurian
Xiphosura	445	Ordovician

Source: Adapted from Dunlop, J.A. and Selden, P.A., *Exp. Appl. Acarol.*, 48, 183–197, 2009.

Table 2.3 A Classification of the Acari or Acarina

Phylum Arthropoda
  Subphylum Chelicerata
    Class Arachnida
      Subclass Acari or **Acarina** (unlike other arachnids, mites have a wide range of food habits; some are predators, fungivores, parasitic on vertebrates and invertebrates, and herbivores)
        Order Parasitiformes (or **Anactinochaeta**)
          Suborder Opilioacaridida (or Notostigmata) (not discussed further)
          Suborder Holothyrida (or Tetrastigmata) (not discussed further)
          Suborder Gamasida (or Mesostigmata) (includes predatory and parasitic mites)
          Suborder Ixodida (or Metastigmata) (soft and hard ticks)
        Order Acariformes (or **Actinochaeta**)
          Suborder Actinedida (or Prostigmata) (includes the plant-feeding Tetranychidae, Eriophyoidea, Tarsonemidae)
          Suborder Acaridida (or Astigmata) (includes many stored product and dust mites and parasitic mites)
          Suborder Oribatida (or Cryptostigmata) (moss or soil or beetle mites; rarely pests)

2.2 HIGHER CLASSIFICATION OF MITES

Taxonomic terminology for the Acari varies, is confusing, and is difficult to resolve into a single version because different taxonomists have different opinions as to the correct names, the correct taxonomic level, and the relationships of the groups within these levels. As a result, the literature is filled with conflicting versions of mite classifications. In this text, the Acari or Acarina are considered a subclass, with two orders, the Parasitiformes (or Anactinochaeta) and the Acariformes (or Actinochaeta) (Table 2.3) (Evans et al. 1961). Within these orders are suborders with at least two different names; for example, taxonomists use **Gamasida** or **Mesostigmata** to describe the suborder that contains the family Phytoseiidae, an important group of predatory mites, and the *Varroa* mite parasitic on honey bees. The literature contains both terms for the suborders, so you need to learn both. Without a complete fossil record and with many of the relationships obscured by convergence, there is considerable debate about acarine relationships. Mites are the most heterogeneous group of arachnids, with the body rarely showing traces of *true* segmentation; however, it appears that the Parasitiformes and the Acariformes are well-defined orders (Figure 2.1).

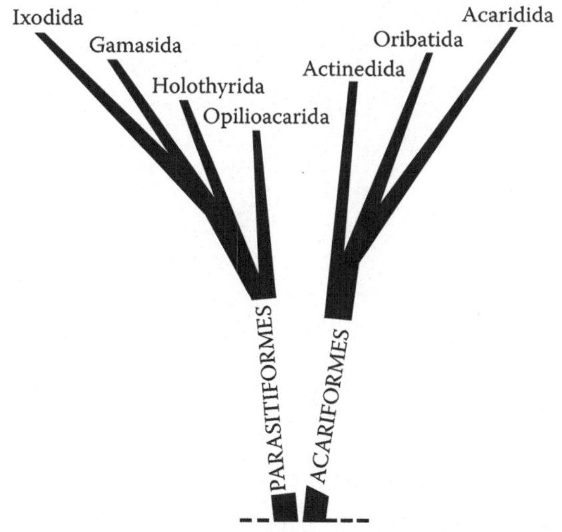

Figure 2.1 A proposed phylogenetic relationship among mite (acarine) groups. (Adapted from Krantz, G.W., *A Manual of Acarology*, 2nd ed., Oregon State University Press, Corvallis, 1979. With permission.)

The plant-feeding mites of interest to agriculture are in the **Prostigmata** (= **Actinedida**). The stored products and parasitic mites discussed in this text are in the **Acaridida** (= **Astigmata**). The predatory phytoseiids are in the **Gamasida** (= **Mesostigmata**), as are some parasitic pests of farm animals. The ticks (**Ixodida** or **Metastigmata**) will be discussed as pests of farm animals (and humans), and the soil mites (**Oribatida** or **Cryptostigmata**) will be discussed briefly for their role in soil fertility. The acarine suborders Tetrastigmata (Holothyrida) and Notostigmata (Opilioacarida) are rare mites that have no known economic importance in agriculture and will not be discussed further.

SELECTED REFERENCES

Dunlop, J.A. and P.A. Selden. 2009. Calibrating the chelicerate clock: A paleontological reply to Jeyaprakash and Hoy. *Exp. Appl. Acarol.* 48:183–197.

Evans, G.O. 1992. *Principles of Acarology*. Wallingford, U.K.: CAB International.

Evans, G.O., J.G. Sheals, and D. MacFarlane. 1961. *The Terrestrial Acari of the British Isles: An Introduction to their Morphology, Biology, and Classification.* London: British Museum of Natural History Press.

Jeyaprakash, A. and M.A. Hoy. 2009. First divergence time estimate of spiders, scorpions, mites and ticks (subphylum: Chelicerata) inferred from mitochondrial phylogeny. *Exp. Appl. Acarol.* 47:1–18.

Krantz, G.W. 1979. *A Manual of Acarology*, 2nd ed. Corvallis: Oregon State University Press.

Krantz, G.W. and D.E. Walter (eds.) (2009). *A Manual of Acarology*, 3rd ed. Lubbock: Texas Tech University Press.

Savory, T. 1977. *Arachnida*, 2nd ed. London: Academic Press.

Walter, D. and H. Proctor. 1999. *Mites: Ecology, Evolution, and Behaviour.* Wallingford, U.K.: CAB International.

CHAPTER 3

Basic Structure and Function of Mites

This chapter briefly introduces you to the basic morphology and physiology of mites and ticks, as well as their behavior and genetics. Their small size, typically approximately 0.5 to 1 mm in length, is associated with an apparent loss of true segmentation and a reduction in internal and external structures (Evans et al. 1961, Krantz 1971, Jeppson et al. 1975, Helle and Sabelis 1985, Evans 1992, Houck 1994, Walter and Proctor 1999, Krantz and Walter 2009).

3.1 MORPHOLOGY

Unlike insects, which have three major body regions or **tagmata** (head, thorax, and abdomen), the Acari have only two: the **gnathosoma**, which consists of a pair of chelicerae and a pair of palps, and the **idiosoma**, the rest of the body including the region with the legs (Figure 3.1) (Krantz and Walter 2009). The gnathosoma includes the **pedipalps** (or palps) and the **chelicerae** (mouthparts). *Mites have no head.* The idiosoma can be divided up further into a **podosoma**, the portion of the body containing the legs, and the **opisthosoma**, the region behind the legs. The **propodosoma** is the region containing the first two pairs of legs, and the **prosoma** consists of the gnathosoma and the podosoma. The **hysterosoma** consists of the podosoma and the opisthosoma.

The first pair of legs often is different in structure from legs II through IV, having numerous sensory setae (or hairs) or other receptors on them. To some extent, legs I function in a manner similar to the antennae and eyes of insects. Figure 3.1 shows a dorsal view of a gamasid mite, with the palps, chelicerae, four pairs of legs, and the dorsal shield (more heavily sclerotized area) with setae that are used as taxonomic characters. Note that, although there are two main tagmata (gnathosoma and idiosoma), there are no obvious segments or somites. Figure 3.2 provides a ventral view of a phytoseiid female, showing the ventral plates, setae, and position of legs I through IV.

Adult mites and ticks typically have four pairs of legs, and larvae typically have three pairs. Nymphs also have four pairs of legs. The **Eriophyoidea** (rust, gall, and bud mites) are exceptions, having only two pairs of legs, even as adults (Figure 3.3) (Keifer et al. 1982). The leg segments consist of the coxa, trochanter, femur, genu, tibia, tarsus, and pretarsus (Figure 3.4). The pretarsus often has setae and other structures that are of taxonomic importance (Figure 3.5). The exoskeleton of the typical mite is very similar to that of insects and is secreted by the **epidermis** (Krantz and Walter 2009). The epidermis, Schmidt layer, endocuticle, and the epicuticle (which consists of exocuticle, cuticulin, tetostracum, and cement layer) typically are present (Figure 3.6). Pore canals appear to arise from epidermal cells underlying the Schmidt layer and pass through the endo- and exocuticule. The pore canals may transport epidermal secretions to the cuticulin surface layer. The various layers of the cuticle are not evenly distributed over the body of mites. Some areas are more heavily sclerotized (and pigmented) than others and are named; for example, Figure 3.7 shows the

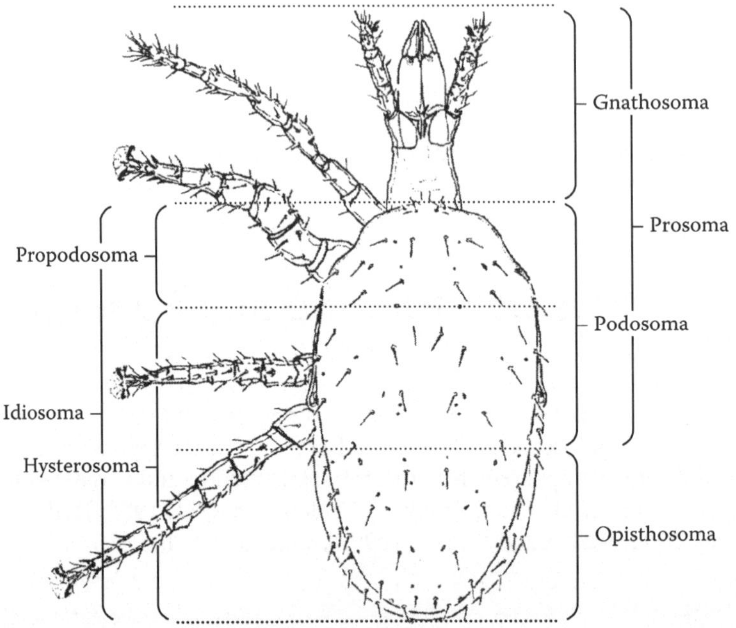

Figure 3.1 The dorsal aspect of a gamasid (mesostigmatid) mite, showing the major body divisions. The two main body regions are the gnathosoma (chelicerae + palps) and idiosoma. The chelicerae are located between the palps on the gnathosoma. Mites typically have sclerotized shields on the dorsal and ventral surfaces, but these do not bear any obvious relationship to the primary segmentation. This mite does not have eyes, and it perceives its environment using chemical, tactile, and other cues obtained through specialized receptors (usually setae) located on the palps, legs, and other regions of the body. (Adapted from Krantz, G.W. and Walter, D.E., Eds., *A Manual of Acarology*, 3rd ed., Texas Tech University Press, Lubbock, 2009. With permission.)

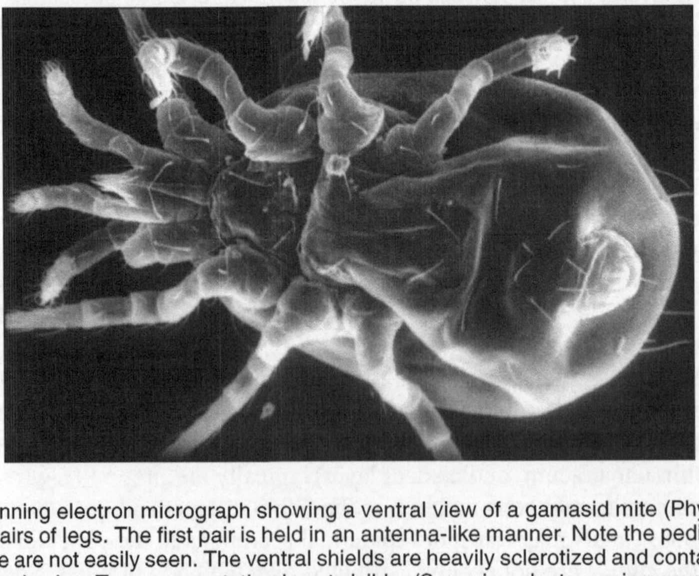

Figure 3.2 A scanning electron micrograph showing a ventral view of a gamasid mite (Phytoseiidae) and the four pairs of legs. The first pair is held in an antenna-like manner. Note the pedipalps, but the chelicerae are not easily seen. The ventral shields are heavily sclerotized and contain setae, as do the legs and palps. True segmentation is not visible. (Scanning electron micrograph by Ross P. Field, formerly at the University of California–Berkeley.)

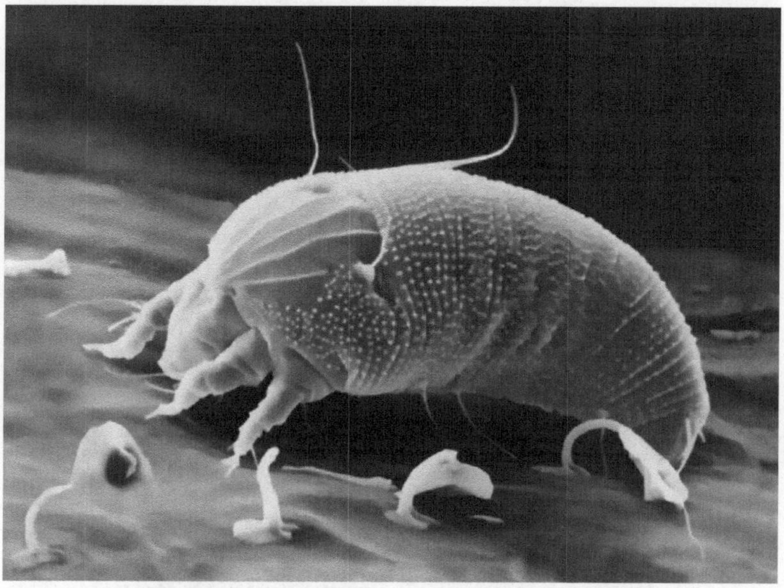

Figure 3.3 Eriophyoid mites (rust, bud, or gall mites) are unusual in having only two pairs of legs as adults. Eriophyoids walk on these legs. They disperse by walking, and aerially by standing on their posterior end (which has a sucker plate); they are then blown by the wind. Note that the exoskeleton has ring-like structures, but these are not true segments. (Scanning electron micrograph from the U.S. Department of Agriculture, Agricultural Research Service, Washington, D.C.)

ventral aspect of a gamasid mite, with the presternal shield, the sternal shield, the epigynial shield (covering the ovipositor), the ventral shield, and the anal shield. Parasitic forms often have greatly reduced sclerotinization, with very few plates or shields.

The epidermis produces **setae** (or hairs), and these show great diversity of form and function in mites. The location of and type of setae are commonly used to identify mites (see Figures 3.1, 3.2, 3.3, 3.5, and 3.7). Because mites lack complex eyes, setae are very important means by which they interact with their environment. Most mites have no eyes at all. Setae are tactile, and some are sensitive to CO_2, vibrations, heat, relative humidity, **pheromones** (chemical cues perceived within the species), and **kairomones** (chemical cues perceived between species). Some families of mites, such

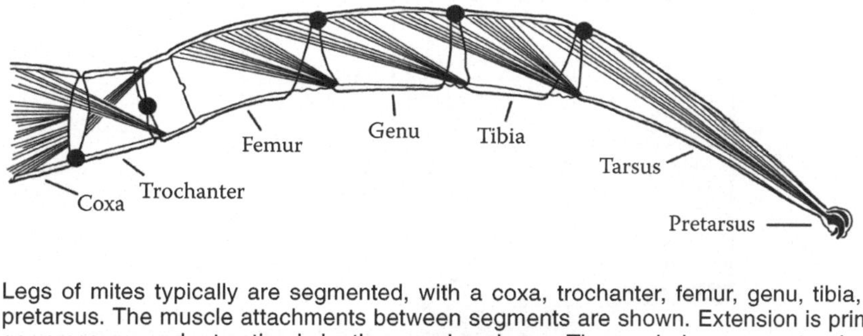

Figure 3.4 Legs of mites typically are segmented, with a coxa, trochanter, femur, genu, tibia, tarsus, and pretarsus. The muscle attachments between segments are shown. Extension is primarily by turgor pressure, and retraction is by the muscles shown. The symbol • represents where the leg is articulated. (Adapted from Krantz, G.W. and Walter, D.E., Eds., *A Manual of Acarology*, 3rd ed., Texas Tech University Press, Lubbock, 2009. With permission.)

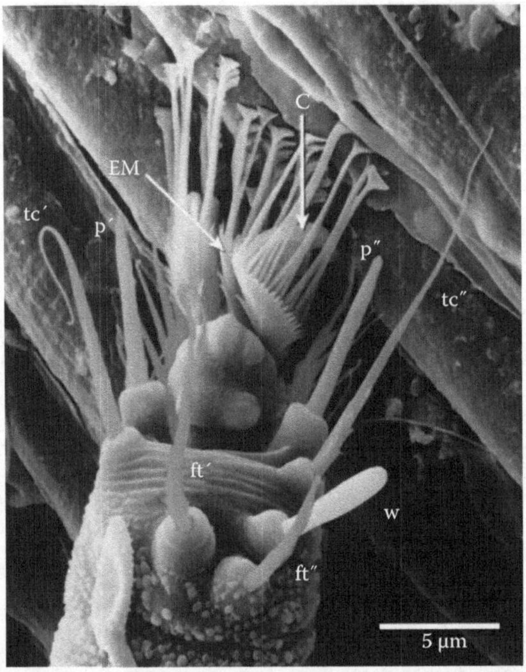

Figure 3.5 Scanning electron micrograph of the tarsus and pretarsus of a plant-feeding mite (*Brevipalpus phoenicis*) illustrating setae on the tarsus and other structures on the pretarsus. These structures are often used in taxonomic keys. EM = empodium, C = claw; the setae (tc', p', p", tc", ft', ft", w) have sensory functions. (Scanning electron micrograph by the U.S. Department of Agriculture, Agricultural Research Service, Washington, D.C.)

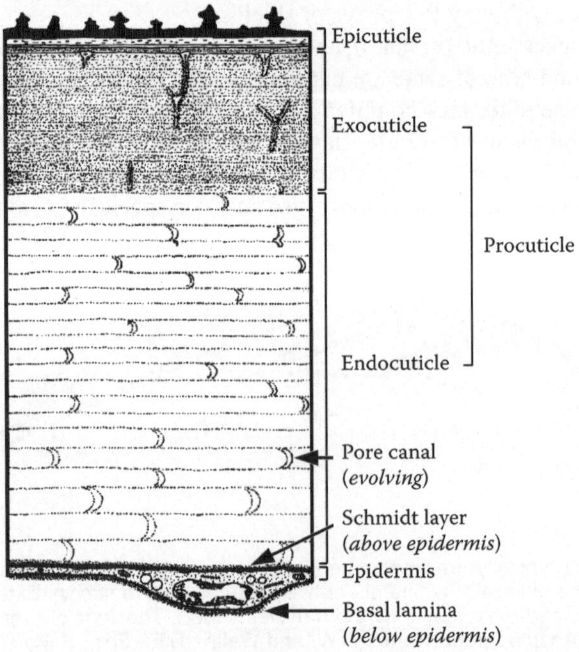

Figure 3.6 Cross-section of acarine cuticle. (Adapted from Krantz, G.W. and Walter, D.E., Eds., *A Manual of Acarology*, 3rd ed., Texas Tech University Press, Lubbock, 2009. With permission.)

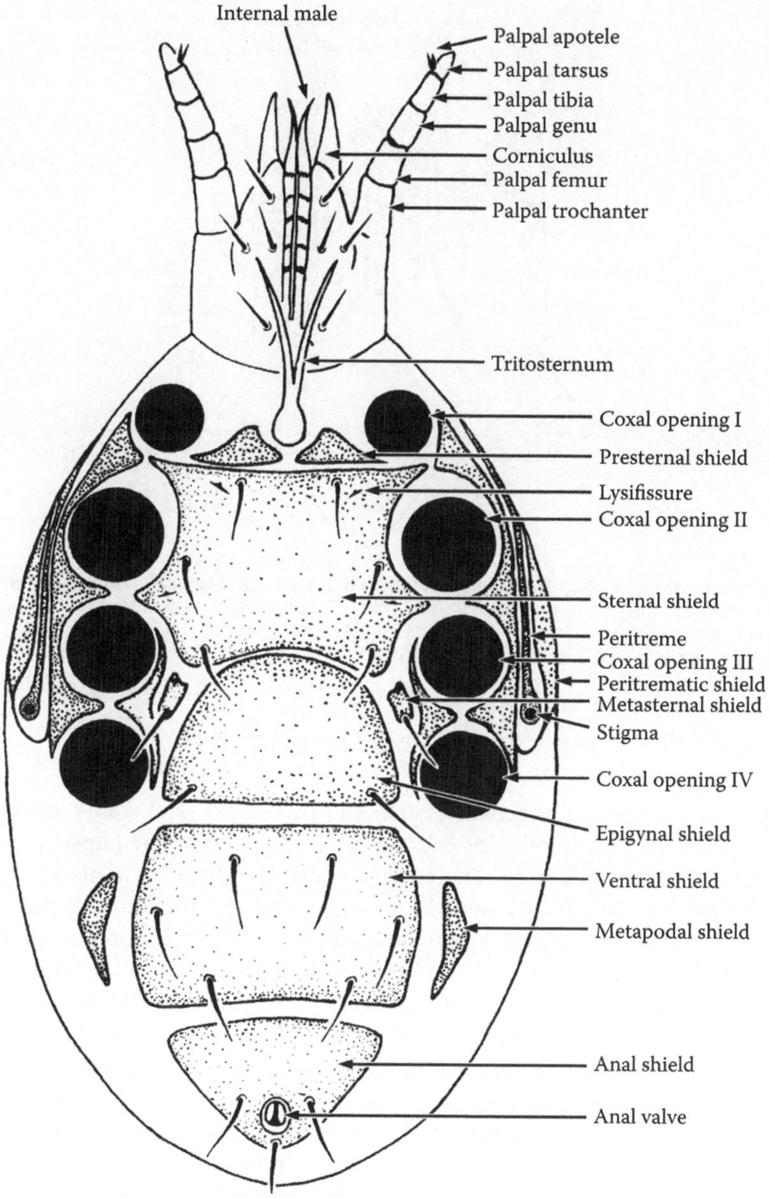

Figure 3.7 The venter of a typical gamasid (mesostigmatid) female. The gnathosoma includes the palps and the chelicerae. There are four coxal (leg) openings shown, with the opening to the peritreme (stigmata or respiratory opening) present between coxae III and IV. The genital opening is beneath the sclerotized epigynial shield, and the anus is present on the sclerotized anal shield. (Adapted from Krantz, G.W. and Walter, D.E., Eds., *A Manual of Acarology*, 3rd ed., Texas Tech University Press, Lubbock, 2009. With permission.)

as the Tetranychidae, have simple eyes (ocelli like). Despite the lack of eyes, mites are able to perceive daylength, which influences whether they enter **diapause** (a genetically determined suspension of development that occurs prior to the onset of unfavorable conditions) or develop continuously. Mites respond to light in both positive and negative ways. For species that are susceptible to desiccation, moving away from intense light (and its associated heat) is an important survival mechanism. It is likely that light is perceived through the exoskeleton directly by the brain for species lacking eyes.

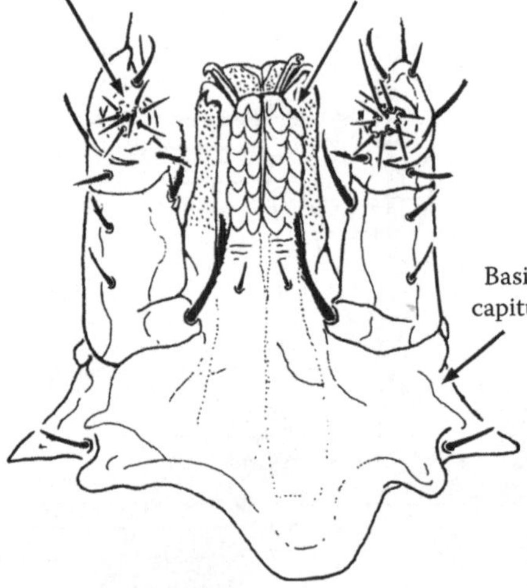

Figure 3.8 Tick mouthparts are highly modified compared to those of a typical mite, having a hypostome with retrorse teeth that make it difficult to dislodge ticks once they have attached to their host. (Adapted from Krantz, G.W. and Walter, D.E., Eds., *A Manual of Acarology*, 3rd ed., Texas Tech University Press, Lubbock, 2009. With permission.)

3.2 FEEDING AND FOOD TYPES

The gnathosoma (consisting of the chelicerae and pedipalps) is movably articulated with the idiosoma by a membrane. The mouthparts often lie within a cavity in the idiosoma called the **camerostome** when the mite is not feeding. There is considerable variability in morphology of the gnathosoma, with the palps and chelicerae often becoming highly modified for different feeding habits. Ticks, for example, have a specialized **hypostome** that is a barbed piercing organ; the hypostome has **retrorse teeth** that make it difficult to pull the mouthparts out of its host (Figure 3.8; also see Figure 22.2 in Chapter 22). Chelicerae of the plant-feeding spider mites (Acari: Tetranychidae) have become **stylet like** in appearance (Figure 3.9).

Most mites have chelicerae that are **chelate–dentate** (Figure 3.10). This structure allows mites to bite their food, and it is thought that the Chelicerata, including the mites or Acari, primitively were all predators. The chelicerae consist of a movable terminal digit opposing an anterodorsal prolongation of the subterminal segment. In some male mites (Gamasida or Mesostigmata), the chelicerae have become modified to include a **spermatodactyl**, a structure used for transferring spermatophores to the sperm receptacle of the female (see Chapter 12). The pedipalps are used as sensory appendages and sometimes become modified for grasping food. The pedipalps also are used for cleaning the chelicerae after feeding. Figures 3.1, 3.2, and 3.7 show a simple palpus with its segments.

The typical digestive tract of a mite includes the mouth, which opens into a muscular **pharynx**, which is connected to the stomach or **ventriculus** by a long **esophagus**. The esophagus passes through the "brain" (or **synganglion**) of the mite (Figure 3.11). A **hindgut**, divided into anterior and posterior regions, opens to the exterior either terminally or subterminally. The gut usually is provided with a number of paired **diverticula**. In some mites, a pair of **Malpighian tubules** may open into the hindgut. The stomach is small, and most of the midgut is formed by the large diverticulae or caecae in families in the Gamasida and in ticks (Ixodida). Spider mites and eriophyoid mites have no caecae. Paired salivary glands are associated with the alimentary canal (Figure 3.12). Salivary

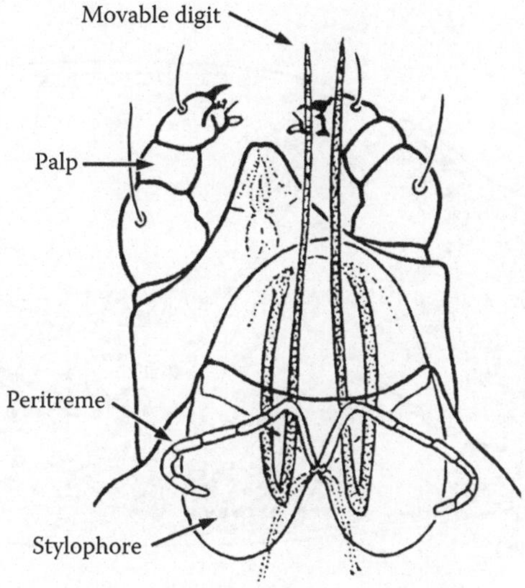

Figure 3.9 Spider mites have long, stylet-like chelicerae and shortened palps (Prostigmata or Actinedida: Tetranychidae). Note the peritreme is behind the gnathosoma. The stylets (movable digits) are inserted into plant cells and the contents are sucked out, leaving behind dead cells that lack chlorophyll, which causes stippling and yellowing of the leaf. (Adapted from Krantz, G.W. and Walter, D.E., Eds., *A Manual of Acarology*, 3rd ed., Texas Tech University Press, Lubbock, 2009. With permission.)

fluids are emptied into the mouth cavity or preorally. Salivary secretions of ticks have been studied extensively because of their economic importance and relatively large size, but little is known about salivary secretions of other acarines, even though salivary secretions are known to have interesting effects on plants or prey.

The stylet-like chelicerae of plant-feeding spider mites (Prostigmata or Actinedida: Tetranychidae) are able to penetrate plant tissues up to 100 microns deep (Figure 3.9). Feeding causes the epidermal cells of the plant to lose chloroplasts. The remaining contents tend to coagulate and turn an amber color. Spider mites can penetrate the parenchymous cells and damage

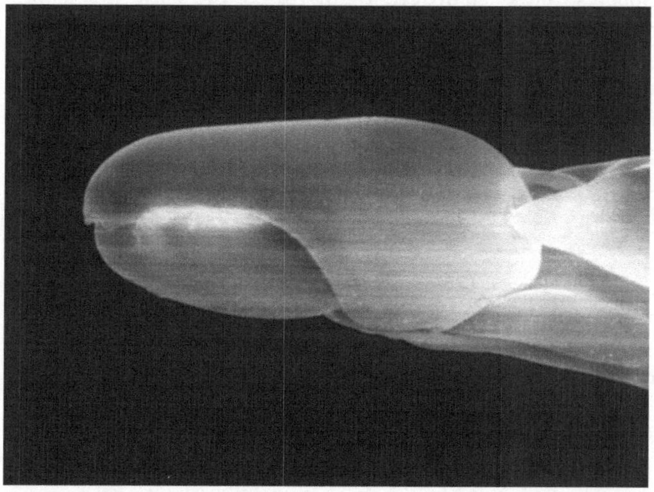

Figure 3.10 A typical chelate–dentate chelicera. (Scanning electron micrograph by Ross P. Field, formerly at the University of California–Berkeley.)

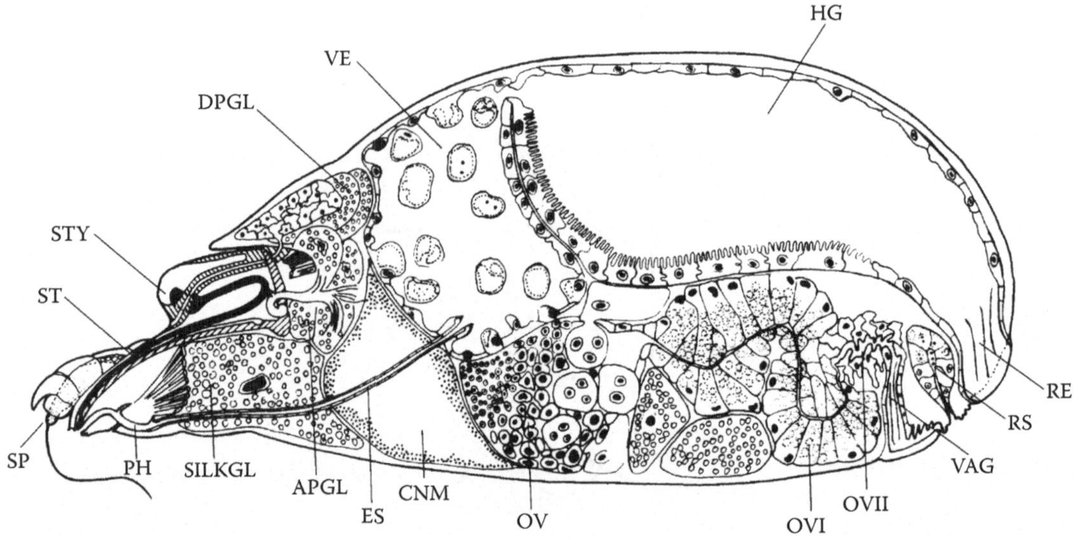

Figure 3.11 Diagram of a female two-spotted spider mite, *Tetranychus urticae*. Abbreviations: APGL, anterior podocephalic gland; CNM, central nervous mass; DPGL, dorsal podocephalic gland; ES, esophagus; HG, hindgut; OV, ovary with vitellogenic oocytes outside the ovary; OVI and OVII, anterior and posterior oviduct; PH, pharyngeal pump; RE, rectum; RS, receptaculum seminis where males insert the sperm packet; SILKGL, silk gland; SP, spinneret; ST, stylet; STY, stylophore; VAG, vagina; VE, ventriculus. Note that the esophagus goes through the central nerve mass, or brain. (Adapted from Blauvelt, W.E., *Mem. Cornell Univ. Agric. Exp. Stn.*, 270, 1–35, 1945. With permission.)

vascular bundles. Photosynthesis and transpiration are affected by spider mite feeding. The amount of damage to epidermal cells, palisade cells, or spongy mesophyll cells depends on the location of the mites and whether they are feeding on the upper or lower leaf surfaces. *Tetranychus urticae*, the two-spotted spider mite, can suck the contents of 18 to 22 cells per minute while feeding, resulting in dry, brittle foliage when populations are high. Their feeding affects photosynthesis and greatly enhances transpiration, causing additional water loss. Heavily damaged plants will defoliate. Different plant species (and even different cultivars of the same species) respond differently to being fed on by spider mites. These variations may be due to the chemicals in the saliva that affect

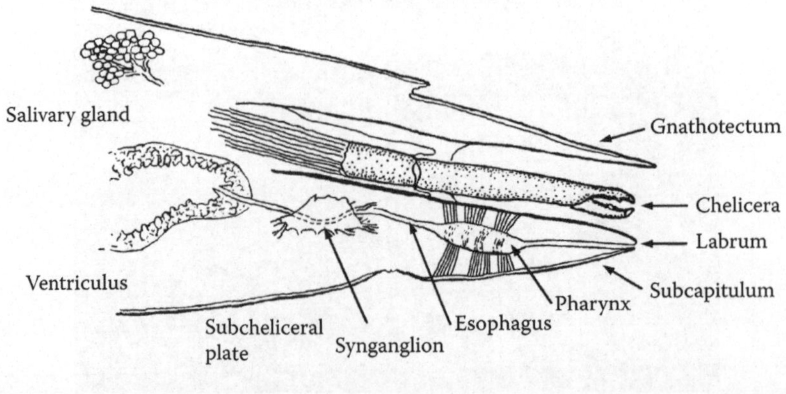

Figure 3.12 View of the anterior of a typical mite (with chelicerae) showing the location of the pharynx, esophagus, and synganglion (or brain). (Adapted from Krantz, G.W. and Walter, D.E., Eds., *A Manual of Acarology*, 3rd ed., Texas Tech University Press, Lubbock, 2009. With permission.)

the plant's immune response. For example, one variety of pears (Bartlett) is so sensitive to damage caused by the feeding of *T. urticae* that feeding by a single spider mite on that leaf can cause defoliation of that leaf (hypersensitive response). By contrast, many more *T. urticae* will be needed to cause defoliation on apple trees.

Mites feed on a variety of foods, ranging from decaying organic material to the living tissues of plants and animals. Ticks are exclusively ectoparasitic on vertebrates and feed on blood and tissue fluids, which is reflected in their highly modified mouthparts. The other groups (Prostigmata, Mesostigmata, and Astigmata) have species that are parasitic on vertebrates and invertebrates, while others are predators, **phytophages**, **mycetophages**, or **coprophages** (dung feeders).

Although some mites can feed on a broad variety of foods, others are very specialized; for example, some phytoseiids are obligatory predators, but other phytoseiids are able to feed on pollen, honeydew, or plant exudates as well as prey on spider mites, eriophyoid mites, tydeid mites, or small insects such as thrips. The quality of food affects the biology and ecology of the mite. For example, the phytoseiid *Metaseiulus occidentalis* is an obligatory predator and is unable to develop or reproduce on alternative foods such as pollen or honey. The prey it feeds on affects its survival and fecundity. When *M. occidentalis* was fed a diet consisting solely of *Tetranychus* eggs, it lived longer and had a higher rate of fecundity than when it was fed a diet consisting of active stages (Bruce-Oliver and Hoy 1990).

The details of how mites feed are not understood at present, and what we do know is usually based on morphological analyses of the mouthparts. The feeding behavior of ticks, because they are relatively large and are of major economic importance as vectors of human and animal diseases, has been studied extensively. Ticks pierce the skin of the host with their dentate chelicerae, which allows the hypostome to be pushed into the skin and the palps to splay out. Blood then is sucked up along the hypostomal groove by the action of the muscular pharynx. Tick saliva contains anticoagulants as well as an anesthetic and other chemicals that make transmission of disease-causing microorganisms more efficient and modulate the immune responses of their hosts.

It is thought that the Mesostigmata and Prostigmata do not ingest solid food. These mites have very large salivary glands, and this suggests that external preoral digestion occurs. The plant-feeding Tetranychidae have styliform chelicerae in which the stylets are probably formed by the elongation of the movable digits. The cheliceral bases in the Tetranychidae are fused to form an evertible stylophore (Figure 3.9).

3.3 EXCRETION

Excretion is the elimination of liquid and solid waste products of metabolism and is difficult to separate from the management of water concentration in the body. Excretion of waste products in mites primarily is accomplished by the Malpighian tubules, if present. In ticks, some waste products are excreted by the salivary glands. Some Prostigmata have excretory tubules that are separate from the digestive tract. The terminology used to describe excretory structures varies by mite group. Relatively little has been learned about excretion in mites. Many mites have paired Malpighian tubules connected to the digestive tract into which excretory products are discharged. The tubules typically are closed at their distal ends but may have loops along them that attach to the mid or hindgut. Excretory products are eliminated with the feces through the anus. The Malpighian tubules eliminate concretions of **guanine**, which is an insoluble final product of nitrogen metabolism. Guanine has a low solubility and often precipitates, even when present in low concentrations. Guanine particles are birefringent in polarized light and can be seen in the hindguts of many mites. In *T. urticae*, which has no Malpighian tubules, guanine is excreted in the hindgut and passed out with the black fecal pellets (McEnroe 1961a). In living mites, this white material may be seen to move back and forth in the lumen of the Malpighian tubules before it becomes mixed with feces in the rectum prior to expulsion.

Malpighian tubules are absent in many smaller mites (such as the Eriophyoidea). Because eriophyoid mites have very small mouthparts and feed on predigested liquid food, it is thought that the metabolic waste products remain in the hemolymph in the form of granules.

3.4 WATER BALANCE

Perhaps the most serious physiological problem that mites face is water balance due to their small size, which results in a high surface-to-volume ratio. Most terrestrial mite species continually lose water that must be replaced, and the aquatic species continually gain water that must be eliminated. Many terrestrial mites are cryptic, found in soil litter, under bark, within galls, or in other habitats with a high relative humidity. Water is lost by simple diffusion from the general body surface. Water is also lost by secretion of digestive fluids, defecation, excretion, production of pheromones or defensive fluids, and the production of reproductive products (oocytes, sperm).

Water is conserved by the **epicuticle**, which is thin, yet laminated into several distinct sublayers (Figure 3.6). Each sublayer is impregnated with wax, and a complex system of pore canals transports the wax from sites of production to the surface. If the waxy coverings are lost due to abrasive dusts, solvents, pesticides, or detergents, the ability of the epicuticle to protect the mite is seriously affected and death will occur due to desiccation.

Water is taken up by mites when drinking free water or when feeding on foods with high water content, in addition to the production of metabolic water and the active or passive uptake of water from the ambient air or aquatic habitat. **Coxal glands** are found in some mites and are involved in maintaining water balance and ionic balance. The coxal glands of ticks (Ixodida) are **osmoregulatory organs** as well as ionic regulators. They retain ions while they eliminate excess water. The blood meal thus is concentrated without causing an ionic imbalance. Less is known about the function of coxal glands in the Actinedida, Gamasida, Astigmata, and Oribatida. Some mites can absorb water from the atmosphere; for example, the grain mite *Acarus siro* (Astigmata or Acaridida) can take up water from the air as long as the relative humidity is not below the critical equilibrium humidity of 70%. As a result, a cultural method for controlling grain mites is to dry the grain and store it in a dry environment.

Spider mites (Tetranychidae) have an unusual problem in that they typically have *too much water* because they feed on plants and inhabit a microclimate on the leaf that has a high relative humidity (McEnroe 1961b). They ingest substantial quantities of plant fluids, and the nutrients therefore are diluted by water. Spider mites have a specialized digestive tract in which the excess fluids are shunted from the esophagus to the hindgut for elimination (Blauvelt 1945). The nutrients are concentrated in the midgut or ventriculus for digestion. Particulate matter, including chloroplasts, can be seen in the midgut. The two-spotted spider mite can eliminate the equivalent of 25% of its body weight in water in 30 minutes to maintain its proper water balance. The watery droplets eliminated may be seen on spider mite webbing when the relative humidity is high; if the relative humidity is low, the tiny droplets evaporate quickly. Spider mites also lose water from their tracheae when in low relative humidity, so they may close the stigmata by changing the position of the mandibular plate to reduce water loss (McEnroe 1961b). *Tetranychus urticae* can regulate its water loss even when the relative humidity is as low as 15 to 20%. At 30°C, but with relative humidities greater than 75%, *T. urticae* can maintain its water balance; however, if *T. urticae* is treated with a solvent that removes the lipids from the integument, water loss occurs rapidly. If mites are treated with abrasive silica gel, they also lose water rapidly through the integument.

Mites that are in diapause have an especially difficult water-balance problem, because they must survive for weeks and months without taking in any water; however, their reduced rate of metabolism requires less oxygen so adults in diapause can close off their stomata to avoid water loss. Diapausing eggs may have an extra-thick waxy covering to reduce water loss. The integument

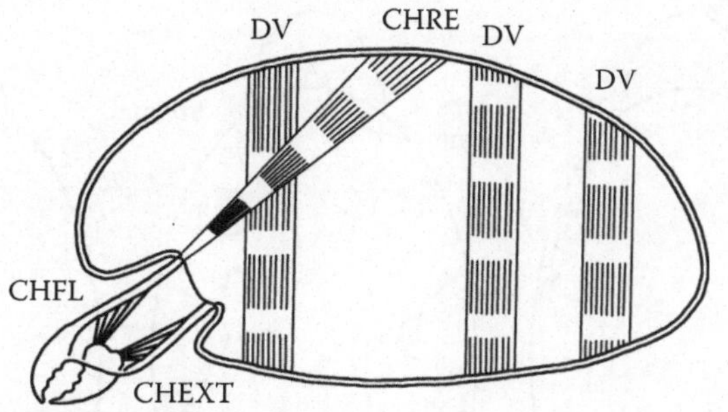

Figure 3.13 A diagram of extrinsic body muscles in a mite. Mites can contract their body with the dorsoventral muscles, which allows the legs and chelicerae to be extended by turgor pressure. Muscles retract the chelicerae. Abbreviations: CHFL, cheliceral flexors; CHEXT, cheliceral extensors; CHRE, cheliceral retractors; DV, dorsoventral muscle bands. (Adapted from Krantz, G.W. and Walter, D.E., Eds., *A Manual of Acarology*, 3rd ed., Texas Tech University Press, Lubbock, 2009. With permission.)

of diapausing tetranychid females is different from the integument of females that are not in diapause. Under substantial magnification, striations of the cuticle can be seen. These striations are solid in diapausing females but broken in active females (Pritchard and Baker 1952). The broken appearance of nondiapausing spider mite females is due to the presence of oval lobes at the upper edge of integumentary folds which presumably allow more water loss than can occur from the solid lobes.

3.5 MUSCLE SYSTEM

Mite muscles are striated and the number of muscles in mites is reduced as compared to insects (Krantz and Walter 2009). The two main types of muscles in mites are intrinsic and extrinsic. **Intrinsic muscles** extend over the joints of leg segments and are flexors (Figure 3.4). Intrinsic muscle actions result in flexion. **Extrinsic muscles** consist of dorsoventral, oblique, rotator, and elevator muscles (see Figure 3.4 and Figure 3.13). Extrinsic muscles originate and insert on the body wall. Most muscles in mites are dorsoventral and modify the shape of the body by modifying the turgor pressure of the body fluids (Figure 3.13). Hydrostatic pressure also causes the extension of the chelicerae, palps, and legs. Other muscles in mites include those associated with the Malpighian tubules, vaginal walls, and heart (if present), as well as the chelicerae (Figure 3.13).

3.6 RESPIRATION

Respiration involves the intake of oxygen, its transportation while dissolved in water or air into the body, and its subsequent use by cells and tissues in the body. The elimination of carbon dioxide produced from metabolic processes also must occur. There are diverse respiratory systems in mites in the sense that the location of **stigmata or stigma** (opening to the tracheal system) varies. The Astigmata (= Acaridida, which include stored products mites) typically lack stigmata. The Prostigmata (= Actinedida, which include plant-feeding mite families), have stigmata in the dorsal anterior region of the mite. The stigmatal openings of Cryptostigmata (= Oribatida, or soil or beetle mites) are hidden (or cryptic). The Metastigmata (= Ixodida) have the stigmatal openings on special plates just behind the fourth pair of legs. Finally, the Mesostigmata (= Gamasida) have their stigmata openings on the side of the body between legs III and IV on a peritrematal shield (Figure 3.14).

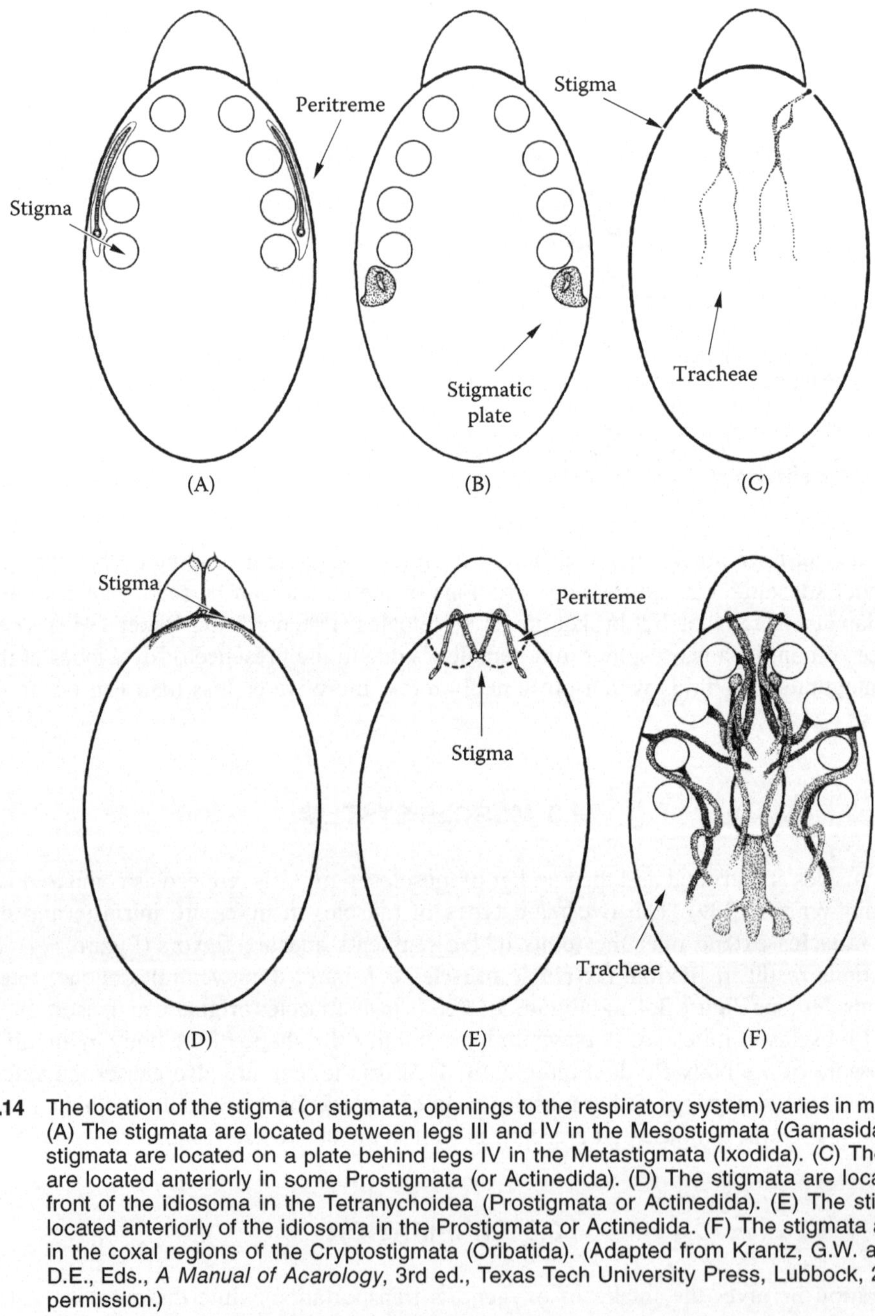

Figure 3.14 The location of the stigma (or stigmata, openings to the respiratory system) varies in mite groups. (A) The stigmata are located between legs III and IV in the Mesostigmata (Gamasida). (B) The stigmata are located on a plate behind legs IV in the Metastigmata (Ixodida). (C) The stigmata are located anteriorly in some Prostigmata (or Actinedida). (D) The stigmata are located at the front of the idiosoma in the Tetranychoidea (Prostigmata or Actinedida). (E) The stigmata are located anteriorly of the idiosoma in the Prostigmata or Actinedida. (F) The stigmata are hidden in the coxal regions of the Cryptostigmata (Oribatida). (Adapted from Krantz, G.W. and Walter, D.E., Eds., *A Manual of Acarology*, 3rd ed., Texas Tech University Press, Lubbock, 2009. With permission.)

3.7 THE NERVOUS AND SENSORY SYSTEMS

The typical arthropod nervous system consists of supra- and subesophageal ganglia connected by circumesophageal commissures. In larger arthropods, a chain of paired ventral ganglia is usually associated with each segment. The embryonic arachnid nervous system is similar to that of other arthropods and has ganglia in each segment; however, the nervous system of mites has become greatly consolidated.

The nervous system has become condensed into a compact, unsegmented mass (or **synganglion**) (Figure 3.12). No central nerve cord or ganglia are found in the opisthosoma, and nerves arise from this single concentrated mass. Most of the sensory structures on the body, palps, and legs of mites are seta like, although some pores have a sensory function (Figure 3.15). Many setae respond to tactile cues. Some are chemosensory, some may be involved in detecting relative humidity, and some may detect heat or CO_2, while others are of unknown function. Taxonomists use setal structure and location to identify mites at all taxonomic levels (species, genus, family, suborder). Identification of mites using setal patterns requires a phase-contrast microscope and well-mounted mites that show the appropriate characters clearly.

All eyes in the Arachnida (if eyes are present) are simple, with a cup-shaped retina (Krantz and Walter 2009). If mites have eyes, the eyes resemble the simple eyes of insects. Two pairs of eyes are found in some mites, including spider mites, but most lack eyes. Even if present, mite eyes do not form images but respond to light intensity. Light also probably is directly perceived by the brain through the cuticle.

The anterior pair of eyes of *Tetranychus urticae* can act as a scanning point detector but does not form an image (McEnroe 1969, McEnroe and Dronka 1969). Receptors for both green and ultraviolet light are present in the anterior eye. The posterior pair of eyes, with simple convex lenses, serves as a nondirectional receptor for near-ultraviolet light.

The biochemistry of the acarine nervous system has been studied relatively little, except in the larger ticks. Spider mites, specifically *Tetranychus urticae*, are known to have acetylcholine, choline acetylase, and cholinesterase, indicating that the complete cholinergic system necessary for nerve function and synaptic responses is present. A neurosecretory system is present in all mites, but it has been studied primarily in ticks because of their larger size and economic importance. The reasons why some pesticides, which often act on nerve function, are more toxic to mites (**acaricides** or **miticides**) than insects are unknown. Some pesticides (e.g., organophosphorus compounds, carbamates, pyrethroids) are toxic to both insects and mites, although the concentration that is lethal may differ between insects and mites or between different mite families, for unknown reasons.

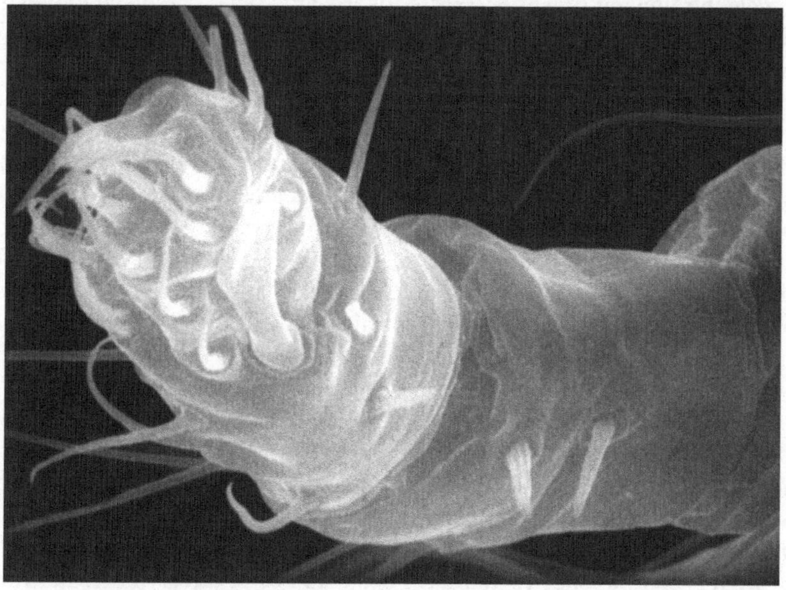

Figure 3.15 A scanning electron micrograph of the palp of the phytoseiid predator *Metaseiulus occidentalis* showing sensory setae. (Photograph by Ross P. Field, previously at the University of California–Berkeley.)

Despite the relative simplicity of their morphology, mites have surprisingly complex behaviors. Mites are able to respond to light, relative humidity, pheromones, kairomones, gravity, heat, and CO_2. Mites can respond to tactile cues and gradients of air temperature. Most mites are quite specific in the type of habitat they choose to frequent; for example, eriophyoid species inhabit specific parts of specific plants (buds, leaves, or twigs). Because eriophyoids cause specific types of feeding damage, they can often be identified to species level based on information about their host plant and their location on the host plant (Keifer et al. 1982).

3.8 THE CIRCULATORY SYSTEM

The mite circulatory system is an open system, in which the blood bathes the internal organs (Krantz and Walter 2009). The blood typically is colorless and contains numerous blood cells. The blood circulates due to contraction of the dorsoventral muscles or, in some larger mites, by a circulatory organ called a "heart." A heart can be found in ticks (Ixodida) and some Gamasida, but not in smaller mites. The heart is found in the anterior part of the opisthosoma; it is hexagonal or rhomboid in shape and is flattened dorsoventrally. It has a single chamber and one pair of ostia (slits) on the dorsal surface. The heart has a long anterior aorta. Many mites are so small that a functional heart is not necessary to maintain an adequate supply of blood throughout the body.

3.9 LIFE CYCLES

Most mites have two sexes, and many mite species are **dimorphic** (males and females readily distinguished from each other by size and shape, as well as by different setal patterns and genital structures). Most mites deposit eggs, but some retain their eggs and produce larvae or nymphs. There are changes in form from the larval to the adult stage, with changes in setal patterns, changes in size and shape of sclerotized plates, and the development of sexual characters such as an **aedeagus** (a sclerotized male intromittent organ for transfer of sperm to females) and an ovipositor in females of some species. Some male mites have modified chelicerae that can transfer spermatophores to the genital openings of females.

The life cycle of the Mesostigmata (Gamasida) typically includes an egg that produces a free-living larva (with three pairs of legs) that may feed before becoming quiescent prior to molting into the protonymph (with four pairs of legs), which then feeds and becomes quiescent before molting to the deutonymph (four pairs of legs). Male and female deutonymphs feed and become quiescent before molting to the adult stage. The male may develop more rapidly than the female, and the male may hover over a quiescent female deutonymph so that he may mate with her shortly after she molts. Ticks (Ixodida or Metastigmata) have a similar life cycle: egg → larva → several nymphal stages depending on the species of tick → adult male or female. The soil mites (Oribatida or Cryptostigmata) have a third nymphal stage (**tritonymph**) before becoming adults. The most complex life cycles are found in the Astigmata (or the Acaridida) and the Prostigmata (or the Actinedida).

Some species of Astigmata have a life cycle in which development proceeds from the protonymph to a specialized **hypopus** (a phoretic deutonymphal stage). A hypopus lacks a functional digestive system, usually has a thickened exoskeleton, and has specialized suckers or modified setae or legs for clasping onto its host. The hypopus often is able to disperse to another site by hitching a ride (= **phoresy**) on another organism such as an insect. The conditions that induce the hypopus probably vary from species to species and have been studied relatively little. For the two-spotted spider mite, however, the life cycle consists of egg → larva → protonymph → deutonymph → adult (male and female).

3.10 DIAPAUSE

Diapause is a genetically determined state of reduced metabolic activity, which is induced *prior to the onset of unfavorable conditions.* Diapause provides the insect or mite species with the ability to survive environmental extremes and results in altered or reduced activity (Veerman 1985). In some mites, diapause allows the mite to successfully overwinter (**hibernal diapause**). In other species, diapause allows the mites to survive hot, dry summer conditions (**aestival diapause**). **Quiescence**, which occurs in direct response to adverse physical conditions, is terminated as soon as favorable conditions return. In spider mites, diapause can occur in the egg or adult female stage (female reproductive diapause). Two-spotted spider mites (*Tetranychus urticae*) and Pacific spider mites (*T. pacificus*), for example, overwinter as mated adult females (males do not overwinter). Color changes may be noticeable in diapausing spider mite females. Diapausing *T. urticae* females are bright red, and diapausing *T. pacificus* females are bright orange. Females of both species typically lack the dark feeding spots on their sides when in diapause, do not feed, and lack the broken texture on their cuticular lobes, as described in Section 3.4 on water balance.

European red mite (*Panonychus ulmi*) females deposit summer eggs and winter (diapausing) eggs in different locations (Lees 1953, Cranham 1973). Diapausing eggs are laid on twigs of deciduous trees, and the summer eggs are laid on foliage. Individuals (eggs or adults) in a hibernal diapause typically have to experience a period of cold chill before the diapause is terminated.

Even in species that have a diapause, not all populations or all individuals in a population have the ability to enter diapause; for example, two-spotted spider mites in greenhouses often lose their genetic ability to enter diapause (through selection) and can reproduce all year around. The same is true for some populations of the two-spotted spider mite living in semitropical climates such as Florida and California. In the predatory mite family Phytoseiidae, a hibernal diapause has been found only in adult females in the species examined so far. In the phytoseiid *Metaseiulus occidentalis*, diapause induction occurs in the fall in California when temperatures decline and the daylengths shorten (Hoy 1975a,b, Hoy and Flaherty 1970, 1975). It is very easy, however, to select a nondiapausing strain of *M. occidentalis* because, even under the most inductive conditions tested, a portion (10 to 20%) of the females do not enter diapause (Figure 3.16). A nondiapause strain can be selected within two to three generations by rearing only those females that continue to reproduce under 8-hour daylengths at 19°C.

A hibernal diapause is found in eriophyoid females living on deciduous plants in temperate climates (Keifer et al. 1982, Lindquist 1996). Eriophyoid females in diapause have a different morphology than the summer females, which has caused taxonomic confusion. The **protogyne** is a nondiapausing female, and the **deutogyne** is the overwintering form. Because the morphology is different, the two forms were given two different species names in some cases (Figure 3.17). The environmental cues that induce diapause in the Eriophyoidea have not been studied, but diapause appears to be obligatory in some species. Deutogynes cannot lay eggs in the same growing season in which they appear but must experience winter chill before they can deposit eggs. Deutogynes have reduced microtuberculation of the idiosoma (probably to reduce water loss), narrow tergites, and fewer ridges, furrows, and projections than the protogynes of the same species.

3.11 DISPERSAL

Dispersal is important in the life cycles of mites, particularly those that live in temporary habitats, such as phytophagous species inhabiting deciduous plants (Jeppson et al. 1975, Keifer et al. 1982). Several dispersal mechanisms allow phytophagous mites to colonize distant plants and to escape natural enemies, at least temporarily. Spider mites, and other mites, can disperse by walking,

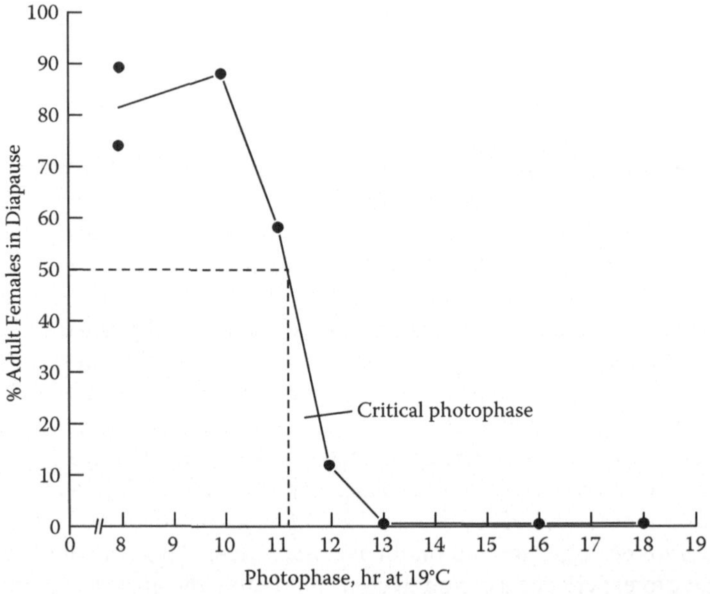

Figure 3.16 Diapause induction in a California population of the phytoseiid *Metaseiulus occidentalis* is influenced by the photoperiod when individuals are reared from egg to adult at 19°C. The critical photophase is the daylength at which approximately 50% of the females enter diapause under these temperature conditions (in this case, the critical photophase is approximately 11.2 hr). No diapause is induced under any photoperiod if these mites are reared at 25°C. (Adapted from Hoy, M.A. and Flaherty, D.L., *Ann. Entomol. Soc. Am.*, 63, 960–963, 1970. With permission.)

Figure 3.17 *Aceria* females (Eriophyoidea). The winter form (deutogyne) is in the center, with three summer forms (protogynes) surrounding her. (Scanning electron micrograph by the U.S. Department of Agriculture, Agricultural Research Service, Washington, D.C.)

although, due to their small size, the distances traveled are relatively small. Some mites, including spider mites and phytoseiid predatory mites, can disperse aerially (which involves silking or other specific behaviors). Some mites disperse by phoresy, which involves attaching themselves to insects or other hosts. Mites also may become aerial plankton if they are picked up by the wind and blown long distances. Such dispersal is hazardous, and we do not know how well mites survive this mode of dispersal. Survival likely depends on the temperature, the relative humidity, the length of time in the air stream, and whether the mites land on a suitable site where appropriate food can be found.

Some spider mite species (Tetranychidae) disperse aerially via silk strands (**silking**), but not all tetranychids do (Gerson 1985, Smitley and Kennedy 1988). Dispersing individuals of the two-spotted spider mite typically are newly mated females that have not yet begun to oviposit, so they are not yet weighed down by their relatively large eggs. Dispersing females climb to the top of a plant that has been heavily damaged by feeding from spider mites. Often large groups of newly emerged females may be seen. At certain wind speeds, they will be blown off to new sites. Such a dispersal mechanism is hazardous if the female does not find an appropriate host plant. Aerial dispersal of spider mites typically occurs when population densities are high and the host plant's quality has declined or changed. If densities are low, *Tetranychus urticae* females tend to disperse by walking to new leaves on the same plant, which is much less hazardous. Some immatures and males have been found on sticky panels used to detect aerial dispersal, but it is not clear that they undergo a specific dispersal behavior, and it appears that newly mated nongravid spider mite females are the primary dispersants.

Predatory mites (phytoseiids) also disperse aerially. Again, there is an explicit behavior involved in dispersal; for example, newly mated nongravid adult females of *Metaseiulus occidentalis* stand on their "hind legs" on leaves and orient into the wind (Hoy 1982, Field and Hoy 1985, Hoy et al. 1985). At certain wind speeds (not too high and not too low), they will release to be dispersed by the wind. By contrast, **gravid** (pregnant) females under the same conditions crouch down to avoid being blown off, and females in the dispersal mode may also crouch down if the wind speed is too great (Hoy et al. 1985).

Many mite species are phoretic on other organisms. These mites often have adaptations for phoresy, including the production of an anal stalk that is used to attach to an insect. Hypopi (deutonymphs of Acaridida or Astigmata) may have suckers and other structures that facilitate attachment to their insect (or other) host. Hypopi are not parasitic on the host because they lack mouthparts and functional digestive systems; however, if an insect has a very large number of hypopi on it, then that insect might be less able to fly due to the increased weight. Eriophyoid mites also may be phoretic on insects (Waite 1999).

3.12 REPRODUCTION

Mites have both direct and indirect sperm transfer (Walter and Proctor 1999, Krantz and Walter 2009). **Direct sperm transfer** is found in the Tetranychidae, Tenuipalpidae, Stigmaeidae, and Acaridae and at least one tydeid. Direct transfer involves males that transfer sperm directly into the sperm receptacle of the female. These males often have special structures for grasping the females during copulation, such as suckers on the posterior pair of legs or in the anal region. In some species, one or more pair of legs of males may be enlarged and have spurs. If a spermatophore is produced, the male may transfer it to the female genital opening in a variety of ways. In the Tetranychidae, males have an aedeagus, which is a sclerotized intromittent organ.

Mating behavior has been studied extensively in the two-spotted spider mite (*Tetranychus urticae*) (Cone et al. 1971a,b, Penman and Cone 1972, Potter et al. 1976). Males guard deutonymphal females prior to molting into the adult. Female deutonymphs produce a sex pheromone that elicits this guarding behavior. Males often fight over the right to guard a deutonymphal female, and they will attempt to mate with her even as she attempts to molt. The larger males typically win, but injuries or deaths from fighting are rare. The actual mating process involves the *T. urticae* male touching the female with his forelegs. Next, the male crawls under the female and holds on by clasping the female's third and fourth pairs of legs with his first two pair of legs. At the same time, the idiosoma of the male is arched upwards so the aedeagus is introduced into the genital opening of the female. At this stage, the female will permit a full copulation to occur or will terminate the mating. Copulation lasts about five minutes, and females may mate more than once. Studies using genetic markers indicate that the first mating is usually effective and the later ones are not (Helle 1967).

Unmated female tetranychids produce eggs that develop into haploid males (Helle 1985). This genetic system is called **arrhenotoky**. Mated arrhenotokous females produce both females (diploid) and males (haploid), whereas virgin females can deposit haploid (male) eggs only. The number of eggs deposited by mated and unmated tetranychids is approximately the same. Most spider mite species are arrhenotokous, although several species are **thelytokous** (producing females only). Thelytoky is common in the tetranychid subfamily Bryobiinae (Helle 1985). Some genera contain both arrhenotokous and thelytokous species (Jeppson et al. 1975).

Because males in the arrhenotokous Tetranychidae are haploid, any new mutations are immediately exposed to selection, whether the mutations are dominant or recessive. Selection on haploid males leads to a rapid loss of deleterious genes in the population. Arrhenotoky allows pesticide resistances to develop rapidly in spider mites because, if a haploid male is exposed to pesticides and he has a resistance allele, it will be selected for. A second factor that enhances the ability of tetranychids to become resistant to pesticides is the high rate of inbreeding they undergo. When brothers and sisters, or mothers and sons, mate with each other the likelihood that two copies (in females) of the resistance allele will be assembled is greater than if the species mate with unrelated individuals or are diplodiploid. Spider mite females typically deposit their eggs on a single leaf if the population density is relatively low. Under uncrowded conditions, very little migration of the progeny occurs; therefore, brother and sister (sib) matings probably occur frequently. The inbreeding that results increases the rate at which two rare resistance alleles can be assembled in the diploid females. The combination of sib mating and haplodiploidy leads to a rapid fixation of any new favorable (resistance) alleles in a population. In addition, haplodiploid spider mite species can respond rapidly to selection for the ability to survive on new host plants.

Spider mite species identifications are determined by examining the shape of the male aedeagus, which means that we need both males and females to identify to species. Males must be mounted on their sides on slides with the aedeagus extruded to get an accurate view of the aedeagus for identification. This can be difficult to achieve because the males are more or less triangular in shape.

Mating behavior in the Phytoseiidae has been studied in several species (Amano and Chant 1978, Hoy and Smilanick 1979, Hoy and Cave 1985, 1986, 1988, Tsunoda 1994). Deutonymphal females of *Metaseiulus occidentalis*, for example, produce a contact sex pheromone that causes males to guard them and mate with them shortly after they molt to adulthood (Hoy and Smilanick 1979). Males and females are dimorphic, with males being about half the size of adult females.

Indirect sperm transfer involves deposition of a sperm packet (spermatophore) on the substrate by males. In some cases, the spermatophores are deposited in groups. Females then pick up the spermatophores and place them in their sperm receptacle.

3.13 GENETICS AND SEX DETERMINATION

Mites have several different types of chromosomal systems and sex-determining mechanisms (Oliver 1971, 1977, Helle 1985, Hoy 1985). Some mites are diploid and have both sexes (diplodiploid). Others are haplodiploid (arrhenotokous), with males that are haploid and females that are diploid; for example, most spider mites (Tetranychidae) are arrhenotokous. Unmated females produce haploid (male) eggs. Mated females produce both male and female (diploid) progeny.

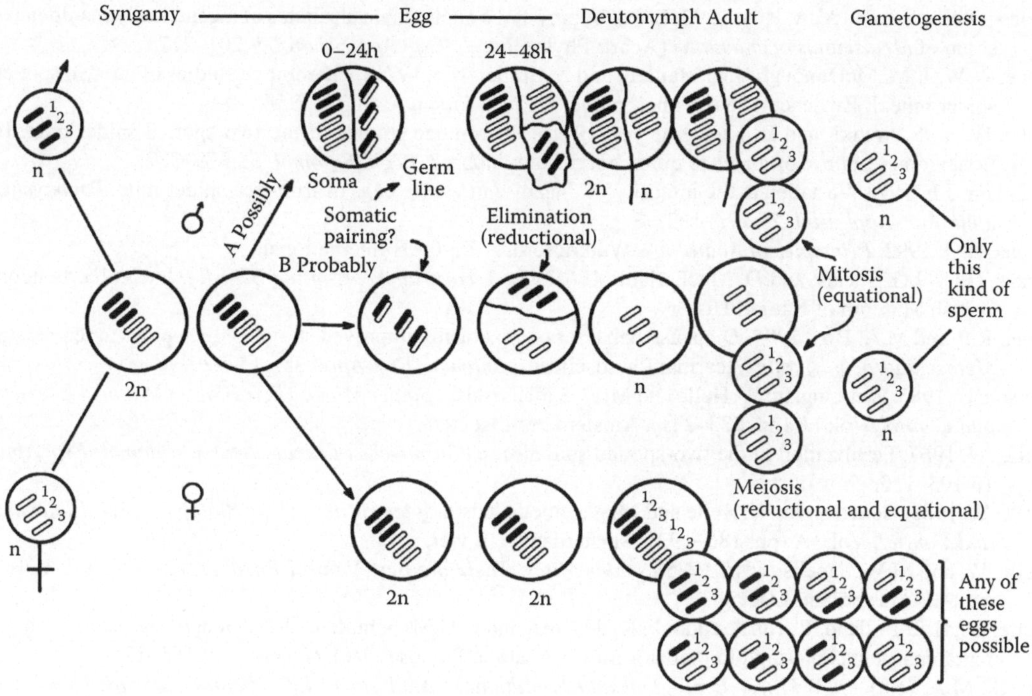

Figure 3.18 Outline of the parahaploid genetic system in the phytoseiid *Metaseiulus occidentalis*. Nearly all phytoseiids are parahaploid (or pseudoarrhenotokous). All females must mate, and all eggs are diploid. Females remain diploid, but male embryos lose half their chromosomes during embryogenesis; as a result, they are haploid and able to respond rapidly to selection. (Adapted from Nelson-Rees, W.A. et al., *Chromosoma*, 77, 263–276, 1980. With permission.)

Female phytoseiids that are unmated do not produce eggs, yet eggs appear, based on cytological evaluations, to have two different chromosome numbers (*n*, or haploid, and 2*n*, or diploid), which made people assume that they are haplodiploid, like the spider mites. The necessity of mating to produce any progeny, however, raised questions about the genetic system of phytoseiids, and experiments showed that they have an unusual genetic system called parahaploidy. In **parahaploidy** or **pseudoarrhenotoky**, all eggs must be fertilized (Helle et al. 1978, Hoy 1979, Nelson-Rees et al. 1980). Phytoseiid eggs that produce females remain diploid. In males, the diploid eggs undergo development, but about halfway through embryogenesis one half of the chromosomes are heterochromatinized (made inactive) and eliminated from the cells, leaving the male haploid for the rest of his life (he produces sperm by mitosis rather than meiosis) (Figure 3.18). Some evidence indicates that the chromosomes that are eliminated are inherited from his father. What does this mean for resistance development? It suggests that, like the Tetranychidae, phytoseiid males can undergo selection each generation as larvae, nymphs, and adults, and a new mutation can be selected for (or against) easily because the males are haploid after the eggs hatch.

SELECTED REFERENCES

Amano, H. and D.A. Chant. 1978. Mating behaviour and reproductive mechanisms of two species of predacious mites, *Phytoseiulus persimilis* Athias-Henriot and *Amblyseius andersoni* (Chant) (Acarina: Phytoseiidae). *Acarologia* 20:196–213.

Blauvelt, W.E. 1945. The internal morphology of the common red spider mite (*Tetranychus telarius* Linn.). *Mem. Cornell Univ. Agric. Exp. Stn.* 270:1–35.

Bruce–Oliver, S.J. and M.A. Hoy. 1990. Effect of prey stage on life table attributes of a genetically manipulated strain of *Metaseiulus occidentalis* (Acari: Phytoseiidae). *Exp. Appl. Acarol.* 9:201–217.

Cone, W.W.,L.M. McDonough, J.C. Maitlen, and Z. Burdajewic. 1971a. Pheromone studies of the two spotted spider mite. I. Evidence of a sex pheromone. *J. Econ. Entomol.* 64:355–358.

Cone, W.W., S. Predki, and E. Klostermeyer. 1971b. Pheromone studies of the two spotted spider mite. II. Behavioral response of males to quiescent deutonymphs. *J. Econ. Entomol.* 53:379–382.

Cranham, J.E. 1973. Variation in the intensity of diapause in winter eggs of fruit-tree spider mite, *Panonychus ulmi*. *Ann. Appl. Biol.* 75:173–182.

Evans, G.O. 1992. *Principles of Acarology.* Wallingford, U.K.: CAB International.

Evans, G.O., J.G. Sheals, and D. MacFarlane. 1961. *The Terrestrial Acari of the British Isles*, Vol. I. London: British Museum of Natural History.

Field, R.P. and M.A. Hoy. 1985. Diapause behavior of genetically improved strains of the spider mite predator *Metaseiulus occidentalis* (Acarina: Phytoseiidae). *Entomol. Exp. Appl.* 38:113–120.

Gerson, U. 1985. Webbing. In: W. Helle and M.W. Sabelis (eds.), *Spider Mites: Their Biology, Natural Enemies, and Control*, Vol. 1A (pp. 223–231). Amsterdam: Elsevier.

Helle, W. 1967. Fertilization in the two-spotted spider mite (*Tetranychus urticae*: Acari). *Entomol. Exp. Appl.* 10:103–110.

Helle, W. 1985. Genetics. In: W. Helle and M.W. Sabelis (eds.), *Spider Mites: Their Biology, Natural Enemies, and Control*, Vol. 1A (pp. 185–192). Amsterdam: Elsevier.

Helle, W. and M.W. Sabelis (eds.) 1985. *Spider Mites: Their Biology, Natural Enemies, and Control*, Vols. 1 and 1B. Amsterdam: Elsevier.

Helle, W., H.R. Bolland, R. van Arendonk, R. de Boer, and G.G.M. Schulten. 1978. Genetic evidence for biparental males in haplo-diploid predator mites (Acarina: Phytoseiidae). *Genetica* 49:165–171.

Houck, M.A. (ed.) 1994. *Mites: Ecological and Evolutionary Analyses of Life-History Patterns.* New York: Chapman & Hall.

Hoy, M.A. 1975a. Diapause in the mite *Metaseiulus occidentalis*: Stages sensitive to photoperiodic induction. *J. Insect Physiol.* 21:745–751.

Hoy, M.A. 1975b. Effect of temperature and photoperiod upon the induction of diapause in the mite *Metaseiulus occidentalis*. *J. Insect Physiol.* 21:605–611.

Hoy, M.A. 1979. Parahaploidy of the "arrhenotokous" predator, *Metaseiulus occidentalis* (Acarina: Phytoseiidae) demonstrated by x-irradiation of males. *Entomol. Exp. Appl.* 26:97–104.

Hoy, M.A. 1982. Aerial dispersal and field efficacy of a genetically improved strain of the spider mite predator *Metaseiulus occidentalis*. *Entomol. Exp. Appl.* 32:205–212.

Hoy, M.A. 1985. Recent advances in genetics and genetic improvement of the Phytoseiidae. *Annu. Rev. Entomol.* 30:345–370.

Hoy, M.A. and F.E. Cave. 1985. Mating behavior in four strains of *Metaseiulus occidentalis* (Acari: Phytoseiidae). *Ann. Entomol. Soc. Am.* 78:588–593.

Hoy, M.A. and F.E. Cave. 1986. Screening for thelytoky in the parahaploid phytoseiid, *Metaseiulus occidentalis* (Nesbitt). *Exp. Appl. Acarol.* 2:273–276.

Hoy, M.A. and F.E. Cave. 1988. Premating and postmating isolation among populations of *Metaseiulus occidentalis* (Nesbitt) (Acarina: Phytoseiidae). *Hilgardia* 56(6):1–20.

Hoy, M.A. and D.L. Flaherty. 1970. Photoperiodic induction of diapause in a predaceous mite, *Metaseiulus occidentalis*. *Ann. Entomol. Soc. Am.* 63:960–963.

Hoy, M.A. and D.L. Flaherty. 1975. Diapause induction and duration in vineyard-collected *Metaseiulus occidentalis*. *Environ. Entomol.* 4:262–264.

Hoy, M.A. and J.M. Smilanick. 1979. A sex pheromone produced by immature and adult females of the predatory mite, *Metaseiulus occidentalis*, Acarina: Phytoseiidae. *Entomol. Exp. Appl.* 26:291–300.

Hoy, M.A., J.J.R. Groot, and H.E. van de Baan. 1985. Influence of aerial dispersal on persistence and spread of pesticide–resistant *Metaseiulus occidentalis* in California almond orchards. *Entomol. Exp. Appl.* 37:17–31.

Jeppson, L.R., H.H. Keifer, and E.W. Baker. 1975. *Mites Injurious to Economic Plants.* Berkeley: University of California Press.

Keifer, H.H., E.W. Baker, T. Kono, M. Delfinado, and W.E. Styer. 1982. *An Illustrated Guide to Plant Abnormalities Caused by Eriophyid Mites in North America*, USDA Handbook 573. Washington, D.C.: U.S. Department of Agriculture.

Krantz, G.W. 1971. *A Manual of Acarology*. Corvallis: Oregon State University Bookstores.

Krantz, G.W. and D.E. Walter (eds.) 2009. *A Manual of Acarology*, 3rd ed. Lubbock: Texas Tech University Press.

Lees, A.D. 1953. Environmental factors controlling the evocation and termination of diapause in the fruit tree red spider mite *Metatetranychus ulmi* Koch (Acarina: Tetranychidae). *Ann. Appl. Biol.* 40:449–486.

Lindquist, E.E., M.W. Sabelis, and J. Bruin (eds.) 1996. *Eriophyoid Mites: Their Biology, Natural Enemies and Control*, Vol. 6. Amsterdam: Elsevier.

McEnroe, W.D. 1961a. Guanine excretion by the two-spotted spider mite. *Ann. Entomol. Soc. Am.* 54:925–926.

McEnroe, W.D. 1961b. The control of water loss by the two-spotted spider mite (*Tetranychus telarius*). *Ann. Entomol. Soc. Am.* 54:883–887.

McEnroe, W.D. 1969. Eyes of the female two-spotted spider mite, *Tetranychus urticae*. I. Morphology. *Ann. Entomol. Soc. Am.* 62:461–466.

McEnroe, W.D. and K. Dronka. 1969. Eyes of the two-spotted spider mite, *Tetranychus urticae*. II. Behavioral analysis of the photoreceptors. *Ann. Entomol. Soc. Am.* 62:466–469.

Nelson-Rees, W.A., M.A. Hoy, and R.T. Roush. 1980. Heterochromatinization, chromatin elimination and haploidization in the parahaploid mite *Metaseiulus occidentalis* (Nesbitt) (Acarina: Phytoseiidae). *Chromosoma* 77:263–276.

Oliver, J.H. 1971. Parthenogenesis in mites and ticks (Arachnida: Acari). *Am. Zool.* 11:283–299.

Oliver, J.H. 1977. Cytogenetics of mites and ticks. *Annu. Rev. Entomol.* 22:407–429.

Penman D.R. and W.W.Cone. 1972. Behavior of male two-spotted spider mites in response to quiescent female deutonymphs and to web. *Ann. Entomol. Soc. Am.* 65:1289–1293.

Potter, D.A., D.L. Wrensch, and D.E. Johnston. 1976. Guarding, aggressive behavior and mating success in male two-spotted spider mites. *Ann. Entomol. Soc. Am.* 68:707–711.

Pritchard, A.E. and E.W. Baker. 1952. A guide to the spider mites of deciduous fruit trees. *Hilgardia* 21:253–287.

Smitley, D.R. and G.G. Kennedy. 1988. Aerial dispersal of the two-spotted spider mite (*Tetranychus urticae*) from field corn. *Exp. Appl. Acarol.* 5:33–46.

Tsunoda, T. 1994. Mating behavior of the predacious mite, *Amblyseius womersleyi* Schicha (Acari: Phytoseiidae). *Appl. Entomol. Zool.* 29:141–147.

Veerman, A. 1985. Diapause. In: W. Helle and M.W. Sabelis (eds.), *Spider Mites: Their Biology, Natural Enemies, and Control*, Vol. 1A (pp. 279–316). Amsterdam: Elsevier.

Waite, G.K. 1999. New evidence further incriminates honey-bees as vectors of lychee erinose mite *Aceria litchii* (Acari: Eriophyiidae). *Exp. Appl. Acarol.* 23:145–147.

Walter, D. and H. Proctor. 1999. *Mites: Ecology, Evolution and Behaviour.* Wallingford, U.K.: CAB International.

CHAPTER 4

Collection, Identification, and Culturing of Mites

4.1 COLLECTION GOALS

As indicated in Chapters 1 and 2, mites can be found in nearly every habitat. Each type of habitat may require different collecting methods (Evans et al. 1961, Evans 1992, Krantz and Walter 2009). However, the focus here is on collecting mites of agricultural importance, and the collection method chosen will depend on your goals. Are you collecting mites from foliage so they can be provided to a taxonomic specialist for identification? Are the mites to be reared in the laboratory to monitor their biology? Are the mites crop pests and you want to quantify their abundance to evaluate population densities and economic injury levels? Are you collecting mites from honey bees, stored foods, or livestock?

4.2 COLLECTING PLANT-FEEDING OR PREDATORY MITES

Plant-feeding or predatory mites may be found on the foliage, bark, or twigs of orchard, vineyard, or field crops, and they can be collected using a fine sable-hair brush moistened with alcohol, if the goal is to preserve the mites for identification, or with a dry sable-hair brush, if they are to be kept alive (Jeppson et al. 1975). Fine sable-hair brushes are recommended because the hairs have tapered tips that make it easier to pick up the mites (Figure 4.1). Brushes suitable for handling individual mites can be purchased in art-supply stores; sizes 00 to 00000 are recommended. Larger brushes, perhaps 2 to 4 cm in width and soft in texture, can be used to brush large numbers of mites from foliage onto a sheet of paper to subsequently be placed into vials containing alcohol.

Be sure to label your samples; record the date, host plant or animal, and details of location (bark, bud, foliage, feather), location (town, county, state, and GPS location, if available), and name of collector. If the label will be placed into 70% or 95% ethanol, use a dark pencil to make the label because most ink will dissolve in alcohol. Because colors of spider mites are removed in alcohol, note the color of the living mites on the label, as well.

The use of mouth aspirators to collect mites always is hazardous because it is possible to inhale tiny mites, even if mesh screening is placed over the aspirator tube. Aspirators powered by a squeeze bulb or by a vacuum pump are much safer and easier to use, especially if large numbers of mites are to be collected. Some people develop severe allergic responses to mites, so it is important to avoid inhaling their proteins.

If you are collecting mites from foliage, fruits, or stems in order to identify them, it usually is easier to bring the foliage into the laboratory in a sealed plastic bag that has been chilled in an ice chest in the field. Chilling foliage reduces the likelihood that the predatory and plant-feeding mites will run off the plant material and become lost within the bag. Examining the foliage under a

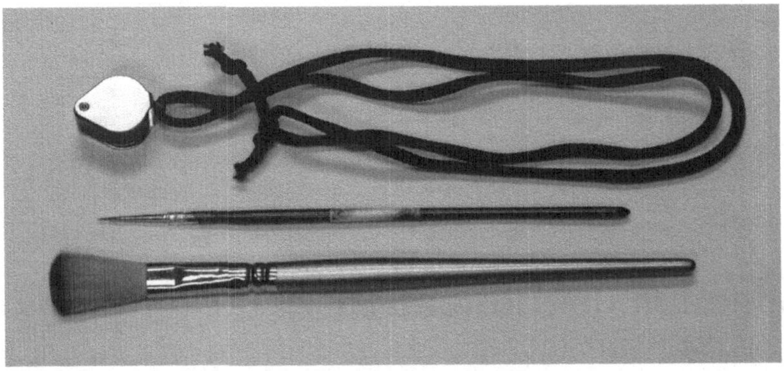

Figure 4.1 A 10× hand lens is useful for locating mites on foliage. A fine sable-hair brush (sizes 00 to 00000) is desirable when handling individual mites because the sable hairs are tapered. A larger, soft brush is useful for brushing large quantities of mites from foliage or from glass plates containing mites brushed from foliage with a mite-brushing machine. (Photograph by Lyle Buss, Department of Entomology and Nematology, University of Florida, Gainesville.)

dissecting microscope allows removal of the specimens with a fine sable-hair brush for further study. Most plant-feeding mites are found on the undersurface of foliage, unless densities are extremely high, in which case the mites may also be on the top of the leaf (Sabelis 1985).

A mite-brushing machine can be used to monitor the densities of predatory and plant-feeding mites (Henderson and McBurnie 1943, Putman 1966, Sabelis 1985; also see Chapter 5 for additional discussion of mite-brushing machines) (Figure 4.2). Wear a dust mask, gloves, and laboratory coat; if possible, place the mite-brushing machine into a hood that can remove the small particles produced when brushing mites to reduce the likelihood of developing allergies to spider mites (Astarita et al. 1994). It is common for workers to develop minor to severe allergic reactions to mites if such precautions are not taken. Leaf-washing methods also have been developed (Jedicskova 1997).

To monitor mites that are dispersing aerially, place greased or oiled glass microscope slides or plastic panels on vertical poles or hang them on ropes (Figure 4.3). Using a light oil, such as sewing machine oil, to grease the slides or panels will prevent larger insects from being trapped on the panels, making it easier to find and remove mites. A thick oil or grease will make it difficult to remove mites from the panels. Alternatively, potted "trap plants" can be placed in the field to monitor both plant-feeding and predatory mites that are attracted to the plants, although the plants may have to be watered periodically in some climates. Placing a potted plant with a specific plant-feeding mite on it provides a way to monitor the natural enemies that may be attracted to that mite. Plant foliage may be sampled using sweep nets or beating sheets. Mite-washing machines allow large amounts of foliage to be monitored (Leigh et al. 1984).

To monitor eriophyoid mites that inhabit galls or buds, the mites must be dissected out (Lindquist et al. 1996). Eriophyoid mites in erinea (hairy patches on the foliage) may be gently heated so the eriophyoid mites walk out of the erinea to be sampled. A single human eyelash attached to a wooden applicator stick can be used to pick up these tiny mites.

4.3 MONITORING VERTEBRATES FOR PARASITIC MITES AND TICKS

When sampling dead vertebrates for parasitic mites or ticks, combing or brushing the body over a shallow white pan can produce samples that can be picked up more readily than if having to part fur or feathers. If the vertebrate host is alive, it may be useful to anesthetize the mites by holding only the body of the host in a box containing chloroform (Evans et al. 1961).

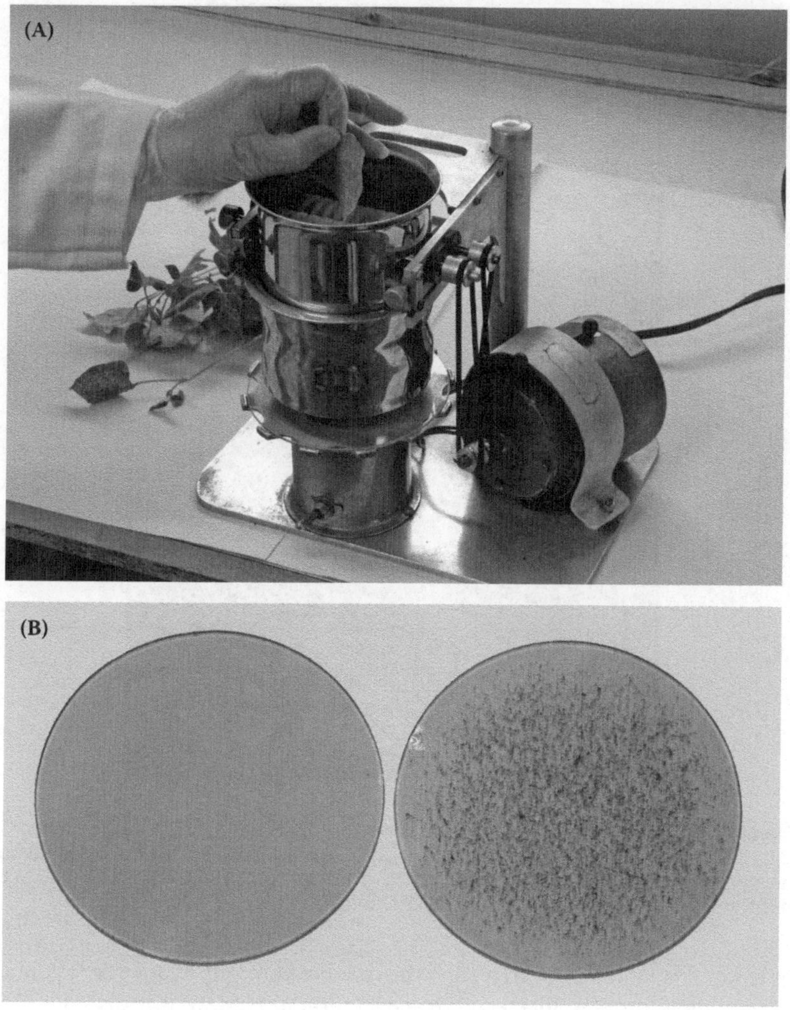

Figure 4.2 (A) A mite-brushing machine. While brushing spider mites from foliage, wear gloves and a lab coat, as well as a dust mask (not shown), to reduce the likelihood of developing allergies to mites. The two brushes at the top of the machine rotate and the leaf containing the mites is inserted between them. Mites brushed off the leaf are collected on the rotating glass plate at the bottom. (B) The mites on the glass plate can be counted when assaying population densities in a crop. Alternatively, the mites can be brushed off the glass plate to feed phytoseiid predator colonies. A clean plate is on the left, and the one on the right contains spider mites. (Photograph by Lyle Buss, Department of Entomology and Nematology, University of Florida, Gainesville.)

If you are examining bird or mammal nesting materials (or soil litter, mosses, or tree bark), a Berlese funnel is useful (Walter and Krantz 2009). A Berlese funnel (or modifications thereof) consists of a funnel with an overhead light bulb; the sampled material is placed on a mesh screen at the wide-mouthed portion of the funnel, and the heat and desiccation of the light bulb will drive the mites out of the sample (Figure 4.4). The sloping sides of the funnel direct the mites down into a collecting container. If the mites are to be preserved, the collecting container can contain alcohol and a small amount (1 to 3%) of glycerin, but if the mites are to be kept alive a moistened pad or filter paper should be placed in the collecting container. Berlese funnels allow determination of what species are present, but they may not be suitable for a quantitative sampling program (Krantz and Walter 2009). The amount of time the Berlese funnels are allowed to operate is determined by the size of the light

Figure 4.3 Towers placed within an almond orchard contain greased plastic panels to monitor aerial dispersal of spider mites and phytoseiid predators. The grease (sewing machine oil) is sufficiently fluid that larger arthropods do not stick to the panels. (Photograph by M.A. Hoy, formerly at the Department of Entomology, University of California–Berkeley.)

bulb, the sample size, its moisture content, and the relative humidity in the room containing the funnel. Plant-feeding mites can be collected from vegetation in Berlese funnels, although they do not provide the quantitative data that mite-brushing machines can. For additional details on collecting and preparing mites from other habitats, see Evans et al. (1961)and Krantz and Walter (2009).

4.4 IDENTIFICATION OF MITES

Specialized training is required for a novice to identify mites to species using taxonomic keys. Identification of mites, even to family level, requires learning to slide-mount mite specimens in the correct orientation so relevant morphological characters are visible. You will need to focus on the tarsi, mouthparts, dorsal and ventral plates, setal patterns, and other setal structures (Figure 4.5). You need a slide-mounting medium, such as Hoyer's mounting medium (25 g distilled water; 15 g gum arabic; 100 g of chloral hydrate, which is difficult to obtain; and 10 g of glycerol). Often, both adult males and females are required to key mites to the species level. You should have access to a phase-contrast microscope or other high-quality compound microscope to see the structures on both the dorsum and venter of the mite (Figure 4.5). Taxonomically oriented courses, such as those offered by the Ohio State Acarology Laboratory during the summer, provide training in how to identify mites and are highly recommended. Information about the courses is available at http://www.biosci.ohio-state.edu/~acarolog/index.html.

Figure 4.4 One example of a Berlese funnel system for extracting mites and other arthropods from soil, leaf litter, moss, or foliage. Funnels can be of different sizes, and the intensity of light can be varied, depending on the sample size and sampling goals. (Photograph by Lyle Buss, Department of Entomology and Nematology, University of Florida, Gainesville.)

If you have training in slide-mounting mites and wish to identify mites, see the recent taxonomic keys to all mite families in Krantz and Walter (2009). Gerson and Smiley (1990) provide a key limited to the families of mites that are natural enemies. Zhang (2003) provides keys to families (and species) of mites found in greenhouses, including both pest and beneficial mites. Efforts are made in these publications to define and describe the terminology used in the keys. Bolland et al. (1998) provide keys to the genera of spider mites (Tetranychidae) of the world and a catalog of the world's species. Recall that most keys to species require both males and females. Amrine et al. (2003) provide keys to the world genera of the Eriophyoidea, and Keifer et al. (1982) have provided a key based on the host plants inhabited by eriophyoid mites of North America (see the Chapter 8 folder on the supplementary CD). A key to the genera of the world Tenuipalpidae was published by Mesa et al. (2009), and Lin and Zhang (2002) published a key to the tarsonemid genera of the world. Colloff (2009) provides a key to the mites found in house dust. A website devoted to the identification of the many species of mites associated with honey bees also contains a key (http://insects.ummz.lsa.umich.edu/beemites/), although two mite species are of primary importance (see Chapters 20 and 21). A key to tick species can be found at http://webpages.lincoln.ac.uk/fruedisueli/FR-webpages/parasitology/Ticks/TIK/tick-key/index.htm. Information on mite pests of crops in South Africa is available in Meyer (1981).

The photographs provided in the text in Part III of this volume, or the supplementary illustrations on the CD, will help you to recognize families of plant-feeding mites (Tetranychidae, Eriophyoidea, Tarsonemidae, Tenuipalpidae) as well as some beneficial species (Phytoseiidae, Stigmaeidae) using a hand lens or a dissecting microscope. You will also learn to recognize two mite species that attack honey bees (see Chapters 20 and 21) (Needham et al. 1988).

Many crops in specific geographic areas have a limited number of pest mite species (often only two or three), and substantial literature resources exist in the form of pest-management manuals or websites with photographs (e.g., Meyer 1981, University of California 2010). Such resources allow the identification of many plant-feeding mites and their natural enemies on crop plants, sometimes

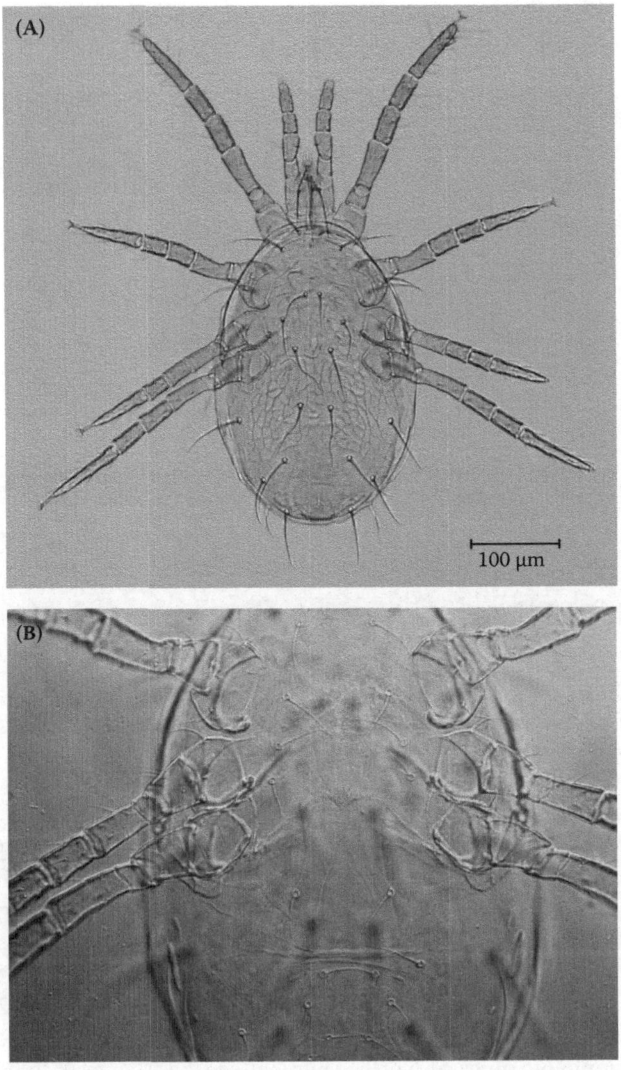

Figure 4.5 (A) View of a slide-mounted phytoseiid mite. (B) View of the venter of a female showing setal patterns. (Photographs by Lyle Buss, Department of Entomology and Nematology, University of Florida, Gainesville.)

to the species or genus level, unless it is a new invasive pest. Specialist taxonomists always should be consulted if it is critical to obtain a species identification. Unfortunately, such specialists are becoming a rare and endangered species themselves due to a lack of funding for such positions in academia and in governmental agencies. Sources of information and names of relevant taxonomists can be found on several of the websites cited below.

4.5 CULTURING MITES

The most common method for rearing plant-feeding mites, such as tetranychids, tenuipalpids, and tarsonemids, to observe their biology in the laboratory is the leaf-disc or leaf-arena method (Helle and Overmeer 1985, Abou-Setta and Childers 1987). Leaves are cut with a razor blade or

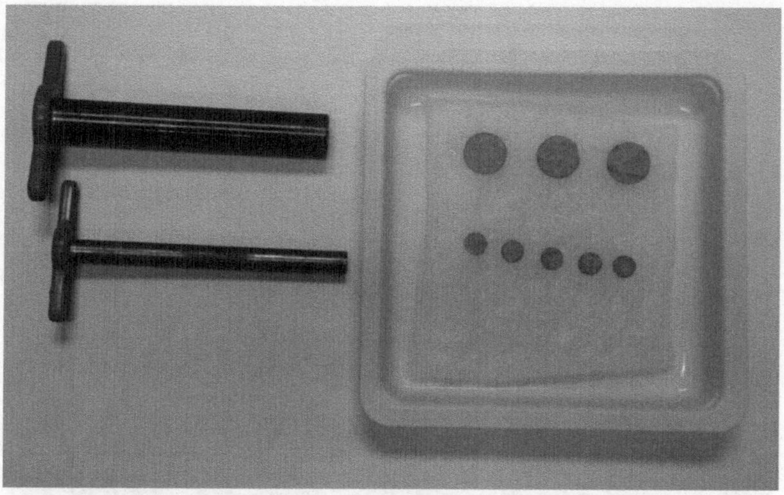

Figure 4.6 Plant-feeding mites can be isolated and reared on leaf discs of their host plant. Leaf discs can be cut with a single-edge razor blade or with cork borers. This method also allows study of some species of predatory mites that feed on the plant-feeding mites. (Photograph by Lyle Buss, Department of Entomology and Nematology, University of Florida, Gainesville.)

cork borer and placed bottom-surface up on water-soaked cotton or a moist sponge (Figure 4.6). Most plant-feeding mites will be contained on the leaf disc until it begins to degrade. At that time, you can move individual mites to a fresh disc with a sable-hair brush or you can move multiple mites by placing the old disc on top of the new and allowing the mites to move off on their own as the old leaf disc dries up. Different sizes of containers and leaf discs can be used. Alternatively, potted plants can be infested with plant-feeding mites and held in cages to prevent contamination by predators. Maintaining the leaf discs or potted plants under controlled temperatures, daylengths, and relative humidity allows biological studies to be conducted and replicated.

As noted by McMurtry and Croft (1997), members of the family Phytoseiidae have varying lifestyles; some species are obligatory predators on small arthropods (especially tetranychids, tydeids, eriophyoids, tenuipalpids, tarsonemids, and immature thrips), while others are able to feed and develop on alternative foods such as pollen and nectar. Many methods have been developed for rearing phytoseiids, due largely to the varying behavior of the different species or the goals of particular studies (e.g., mass rearing for releases of phytoseiids in classical biological control vs. rearing for biological studies in the laboratory). Overmeer (1985), Argov et al. (2002), and Aratchige et al. (2010) offer additional examples of rearing options for phytoseiids.

Phytoseiid mites that are obligatory predators often can be reared on two-spotted spider mites (*Tetranychus urticae*), which are reared on a suitable host plant. In my laboratory, we use an inexpensive host plant (pinto beans, which is a cultivar of *Phaseolus vulgaris*); the beans are available in bulk from grocery stores throughout North America. The bean seeds can be planted near the soil surface in flats in a greenhouse (Figure 4.7). Within 3 to 5 days at 27 to 30°C, the plants will emerge and the dicotyledon leaves will expand. When the leaves have expanded, spider mites can be placed on them and allowed to multiply (Figure 4.7). After about another week at 30 to 35°C, the spider mites have multiplied and can be used as prey for phytoseiid colonies. The spider mites can be brushed off the leaves with a mite-brushing machine onto glass plates, and the spider mites on the glass plates can be brushed onto the predator colonies (see Figure 4.2).

Colonies of several phytoseiids, such as *Metaseiulus occidentalis*, can be reared on dark-colored paraffin-coated paper discs resting on water-soaked cotton (Figure 4.8). The paraffin-coated paper discs are made by dipping dark-colored paper into melted paraffin wax and then allowing the discs

Figure 4.7 Rearing of the two-spotted spider mite *Tetranychus urticae* on pinto bean plants in a greenhouse to obtain prey for phytoseiid colonies. Seeds are sowed shallowly and thickly in a mixture of soil and vermiculite. Within a week the plants have emerged and the dicotyledon leaves have expanded. Subsamples of foliage from older spider mite cultures are examined to confirm that no contamination with predators has occurred. Cut foliage containing spider mites only is then laid on the top of the clean bean plants. As the cut foliage dries, the active stages of spider mites walk onto the young plants and multiply. Typically, these flats are ready for harvesting within a week in the greenhouse. (Photograph by Lyle Buss, Department of Entomology and Nematology, University of Florida, Gainesville.)

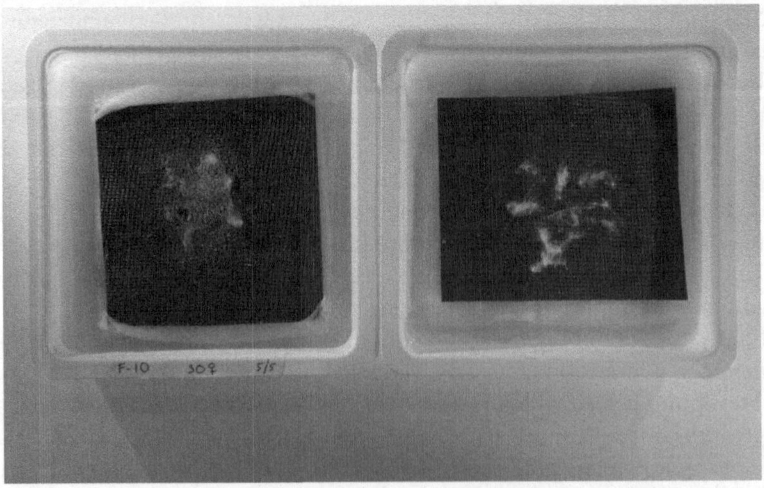

Figure 4.8 One method to maintain phytoseiid colonies in the laboratory involves rearing them on an artificial substrate that rests on water-soaked cotton. In this case, black artist's paper is dipped in melted paraffin wax and allowed to dry on a mesh so small grooves are present (pseudo leaf veins), because phytoseiids often are found along leaf veins. Small tufts of cotton are added to the paraffin disk, as well, because many phytoseiids like to deposit eggs on or in leaf hairs (right). The dish on the left contains phytoseiids that have been fed spider mites of all stages obtained by brushing mites off bean leaves with a mite-brushing machine. Such colonies can be maintained for several weeks before subculturing must be done. The dish on the right lacks mites. (Photograph by Lyle Buss, Department of Entomology and Nematology, University of Florida, Gainesville.)

to cool and dry on a mesh-covered tray. The mesh creates grooves in the paraffin that mimic leaf veins, along which the predators like to rest. Small tufts of cotton can be added to the paraffin disc to mimic leaf hairs, on which many phytoseiids like to deposit eggs. Some phytoseiids, such as *Phytoseiulus persimilis*, are difficult to contain on paraffin-coated discs and may be reared on other substrates (Galazzi and Nicoli 1996). Phytoseiids that disperse readily can be reared on plastic discs with grooves around the edges that contain castor or other oils that are repellent (Overmeer 1985).

Rearing methods for various species of eriophyoid mites are reviewed by Oldfield and Perring (1996). Because of the mite's specialized host-plant requirements and locations (buds, in galls, or as vagrants), rearing methods for eriophyoid mites must be developed on a species-by-species basis.

4.6 ADDITIONAL ACAROLOGICAL INFORMATION

For additional information about mites, Acarology.org (http://www.acarology.org/) has links to national and international acarology societies. A directory of acarologists of the world can be found at the Acarology home page (http://www.nhm.ac.uk/hosted_sites/acarology). This site provides links to some journals that specialize in publications on mites and ticks, including *Experimental and Applied Acarology*, *International Journal of Acarology*, and *Systematic and Applied Acarology*. In addition, many papers on mites and ticks are published in entomological journals, and mites are featured on the Tree of Life web project (http://tolweb.org/tree?group=Acari&contgroup=Arachnida).

SELECTED REFERENCES

Abou-Setta, M.M. and C.C. Childers. 1987. A modified leaf arena technique for rearing phytoseiid or tetranychid mites for biological studies. *Florida Entomol.* 70:245–248.

Amrine, J.W., T.A.H. Stasny, and C.H.W. Flechtmann. 2003. *Revised Keys to the World Genera of the Eriophyoidea (Acari: Prostigmata)*. West Bloomfield, MI: Indira Publishing House.

Aratchige, N.S., L.C.P. Fernando, P.H.P.R.de Silva, K.F.G. Perera, C.S. Hettiarachchi, K.P. Waidyarathne, and S.M.V. Jayawardena. 2010. A new tray-type arena to mass rear *Neoseiulus baraki*, a predatory mite of coconut mite, *Aceria guerreronis*, in the laboratory. *Crop Prot.* 29:556–560.

Argov, Y., S. Amitai, G.A.C. Beattie, and U. Gerson. 2002. Rearing, release and establishment of imported predatory mites to control citrus rust mite in Israel. *BioControl* 47:399–409.

Astarita, C., A. Franzese, G. Scala, S. Sproviero, and G. Raucci. 1994. Farm workers' occupational allergy to *Tetranychus urticae*: Clinical and immunologic aspects. *Allergy* 49:466–471.

Bolland, H.R., J. Gutierrez, and C.H.W. Flechtmann. 1998. *World Catalogue of the Spider Mite Family (Acari: Tetranychidae)*. Leiden: Brill Academic.

Colloff, M.J. 2009. *Dust Mites*. Dordrecht: CSIRO Publishing and Springer Science.

Evans, G.O. 1992. *Principles of Acarology*. Wallingford, U.K.: CAB International.

Evans, G.O., J.G. Sheals, and D. MacFarlane. 1961. *The Terrestrial Acari of the British Isles: An Introduction to Their Morphology, Biology, and Classification*. London: British Museum of Natural History.

Galazzi, D. and G. Nicoli. 1996. Comparative study of strains of *Phytoseiulus persimilis* Athias-Henriot (Acarina Phytoseiidae). I. Development and adult life. *Boll. 1st. Entomol. "G. Grandi" Univ. Bologna* 50:215–231.

Gerson, U. and R.L. Smiley. 1990. *Acarine Biocontrol Agents: An Illustrated Key and Manual*. London: Chapman & Hall.

Helle, W. and W.P.J. Overmeer. 1985. Rearing techniques. In: W. Helle and M.W. Sabelis (eds.), *Spider Mites: Their Biology, Natural Enemies and Control*, Vol. 1A (pp. 331–335). Amsterdam: Elsevier.

Henderson, C.F. and H.V. McBurnie. 1943. *Sampling Techniques for Determining Populations of the Red Citrus Mite and Its Predators*. USDA Circular No. 671. Washington, D.C.: U.S. Department of Agriculture.

Jedicskova, J. 1997. Modification of a leaf-washing apparatus for the recovery of mites. *Exp. Appl. Acarol.* 21:273–277.

Jeppson, L.R., H.H. Keifer, and E.W. Baker. 1975. *Mites Injurious to Economic Plants.* Berkeley: University of California Press.

Keifer, H.H., E.W. Baker, T. Kono, M. Delfinado, and W.E. Styer. 1982. *An Illustrated Guide to Plant Abnormalities Caused by Eriophyid Mites in North America.* Agriculture Handbook No. 573. Beltsville, MD: U.S. Department of Agriculture.

Krantz, G.W. and D.E. Walter (eds.) 2009. *A Manual of Acarology*, 3rd ed. Lubbock: Texas Tech University Press.

Leigh, T.F., V.L. Maggi, and L.T. Wilson. 1984. Development and use of a machine for recovery of arthropods from plant leaves. *J. Econ. Entomol.* 77:271–276.

Lin, J. and Z.-Q. Zhang. 2002. *Tarsonemidae of the World (Acari: Prostigmata): Key to Genera, Geographical Distribution, Systematic Catalogue and Annotated Bibliography*. London: Systematic & Applied Acarology Society.

Lindquist, E.E., M.W. Sabelis, and J. Bruin. 1996. *Eriophyoid Mites: Their Biology, Natural Enemies, and Control*, Vol. 6. Amsterdam: Elsevier.

McMurtry, J.A. and B.A. Croft. 1997. Life-styles of phytoseiid mites and their roles in biological control. *Annu. Rev. Entomol.* 42:291–321.

Mesa, N.C., R. Ochoa, W.C. Welbourn, G.A. Evans, and G.J. De Moraes. 2009. A catalog of the Tenuipalpidae (Acari) of the world with a key to genera. *Zootaxa* 2098:1–185.

Meyer, M.K.P.S. 1981. Mite pests of crops in Southern Africa. *Sci. Bull. Dept. Agric. Fish.* 397:65–67.

Needham, G.R., R.E. Page, M. Delfinado-Baker, and C.E. Bowman (eds.) 1988. *Africanized Honey Bees and Bee Mites*. New York: Ellis Horwood.

Oldfield, G.N. and T.M. Perring. 1996. Rearing techniques. In: E.E. Lindquist, M.W. Sabelis, and J. Bruin (eds.), *Eriophyoid Mites: Their Biology, Natural Enemies, and Control*, Vol. 6 (pp. 377–382). Amsterdam: Elsevier.

Overmeer, W.P.J. 1985. Rearing and handling. In: W. Helle and M.W. Sabelis (eds.), *Spider Mites: Their Biology, Natural Enemies, and Control*, Vol. 1B (pp. 161–170). Amsterdam: Elsevier.

Putman, W.L. 1966. Sampling mites on peach leaves with the Henderson–McBurnie machine. *J. Econ. Entomol.* 59:224–225.

Sabelis, M.W. 1985. Sampling techniques. In: W. Helle and M.W. Sabelis (eds.), *Spider Mites: Their Biology, Natural Enemies, and Control*, Vol. 1A (pp. 337–350). Amsterdam: Elsevier.

University of California. 2010. IPM manuals are available for apples, pears, walnuts, almonds, cotton, tomatoes, alfalfa, cole crops, strawberries, rice, potatoes, floriculture and nurseries, and other crops. The manuals contain many useful descriptions and color photographs of mite pests and their natural enemies. Much of the information is relevant throughout the United States and the world; the UC-IPM Statewide Integrated Pest Management Program website (www.ipm.ucdavis.edu) also contains very useful information. Hard copies can be purchased online or from ANR Communication Services, 1301 S. 46th Street, Bldg. 478, MC3580, Richmond CA 94804.

Walter, D.E. and G.W. Krantz. 2009. Collecting, rearing and preparing specimens. In: G.W. Krantz and D.E. Walter (eds.), *A Manual of Acarology*, 3rd ed. (pp. 83–96). Lubbock: Texas Tech University Press.

Zhang, Z.-Q. 2003. *Mites of Greenhouses: Identification, Biology, and Control.* Wallingford, U.K.: CAB International.

PART II

Integrated Mite Management Strategy and Tactics

For many years, mites primarily have been controlled with pesticides; however, mites can be managed using integrated pest management (IPM) tactics that are compatible with each other. Sometimes the use of chemical pesticides will be required; if they are chosen and applied judiciously, chemical control can be compatible with other management tools. The following chapter introduces you to the concepts of integrated pest management and integrated mite management (IMM). The tactics and tools used in IMM are described, and some relevant references are provided.

A seminal paper by Stern et al. (1959) illustrated how IPM tactics were combined in an early program developed for alfalfa in California. This milestone paper was published in *Hilgardia*, which is no longer in print, but a PDF of the paper is provided on the CD that accompanies this book, with the permission of the Regents of the University of California.

In Part IV, the IMM tactics used in several crop systems (cassava in Africa, almonds in California, apples in Washington State, citrus in Florida and California, and ornamentals grown in greenhouses) are described to illustrate how IMM tools can be combined for specific programs to manage plant-feeding mites.

CHAPTER 5

The Strategy of Integrated Mite Management

5.1 HISTORICAL OVERVIEW

Prior to World War II, spider mites and other plant-feeding mites generally were sporadic or minor pests of agricultural crops. This changed rapidly after the war, with the extensive use of synthetic organic pesticides, such as DDT and organophosphate (OP) insecticides, and fertilizers (Stern et al. 1959, Huffaker et al. 1970). Spider mites became especially serious pests in greenhouses during the 1950s and 1960s (van de Vrie et al. 1972). Furthermore, spider mites rapidly developed resistances to multiple pesticides, especially in greenhouse ornamental and vegetable plant production systems, where pesticides were used intensively (Georghiou and Saito 1983). It was common for a new pesticide to lose its usefulness within a year after its introduction (Cranham and Helle 1985).

Three hypotheses were developed to explain the sudden increase in problems in managing spider mites (Huffaker et al. 1969, 1970): (1) *stimulation* of mite populations by pesticides and fertilizers, (2) *destruction of natural enemies* by pesticides, and (3) *escape of prey from predators in time and space* due to disruption by pesticides and fertilizers. The debate as to which hypothesis was correct was vigorous and, at times, acrimonious. As is often the case in such scientific debates, all three hypotheses proved to be correct, at least in certain situations.

It eventually was discovered that many of the new synthetic organic pesticides destroyed natural enemy populations (both predatory insects and mites), leading to the subsequent escape of spider mite populations from their control (Huffaker et al. 1970). It eventually was resolved that the overuse of synthetic organic fertilizers led to plants that were more suitable for high rates of reproduction by spider mites. In addition, some pesticides (such as carbaryl and DDT) were shown to have a stimulatory effect on spider mite reproduction when present in low concentrations, leading to a *higher* reproductive rate!

This stimulatory effect on mite reproduction is called **hormoligosis** and is a general physiological phenomenon that occurs when living organisms are exposed to very low concentrations of otherwise toxic materials (Luckey 1968). Hormoligosis is an ongoing problem, although it may not be recognized. More recently, James (1997) and James and Price (2002) found that the insecticide imidacloprid increases egg production in the phytoseiid *Amblyseius victoriensis* and in the two-spotted spider mite, suggesting that this product has a hormoligotic effect. Hormoligosis occurs in many organisms as a response to low concentrations of stressors such as chemicals or low doses of radiation (Luckey 1968). The basis for the effect is thought to be the activity of detoxifying enzymes (or DNA repair mechanisms for the radiation). The induced physiological activity allows the organism to respond to higher concentrations of toxins or other stressors later. For example, hormoligotic effects may occur when spider mites are exposed to 1/100 of the LC_{50} (concentration that typically kills 50% of the population) of a pesticide.

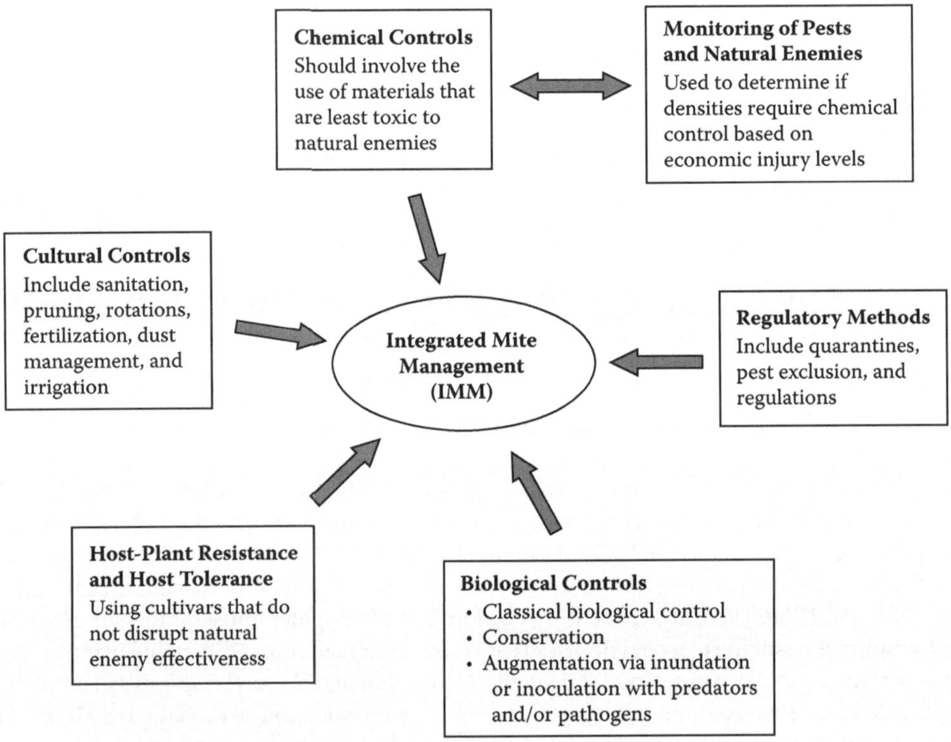

Figure 5.1 Components of an integrated mite management (IMM) program. Although chemical control may be needed, it should only be used when monitoring indicates that natural enemies and all other tactics will not adequately suppress pests. If chemical control is required, the products ideally will be selective, so secondary pest outbreaks do not occur due to destruction of natural enemies.

The third hypothesis was correct in many situations. Pesticide applications applied to control insect pests can kill the predators of the spider mites without killing the spider mites (or the product is sufficiently less toxic so that a greater proportion of spider mites survives than their predators). This outbreak of spider mites is called a **secondary pest outbreak** and occurs because the few surviving predators cannot suppress the spider mites.

The rapid growth of spider mite problems, especially the rapid development of resistance to pesticides, contributed to the pressure to develop and implement a different philosophy of pest control. This philosophy led to the strategy of **integrated pest management (IPM**, or **IMM** when the pests are mites), which has the goal of manipulating pest populations to keep them from reaching damaging levels without attempting to eliminate them (Stern et al. 1959, Dent 1995, Flint and Gouveia 2001). IPM is different from pest *control*, in which we attempt to control or eliminate the pest. IPM is concerned with *managing* beneficial, as well as pest, species (Figure 5.1). IPM is also concerned with *preventing* pest population increases by managing the components of a cropping system so pests never become a serious problem. IPM reflects a strategy of relying on multiple tactics: host-plant resistance, cultural controls, biological controls, genetic controls, quarantines, and biorational controls, including mating disruption and mass trapping.

The strategy of relying solely on chemical pesticides to control agricultural mites is unsustainable. Many pesticides are harmful to natural enemies, the soil, the water supply, and agricultural workers. Consumers are concerned about pesticide residues on foods (Pimentel and Lehman 1993). Pesticides are expensive, although they are more simple to deploy when compared to IPM, which requires detailed information about the crop, the pests, and their natural enemies. Fewer new

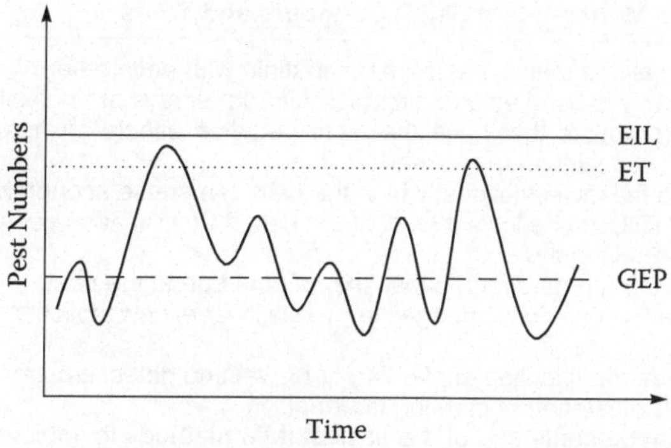

Figure 5.2 Pest populations typically fluctuate around a general equilibrium position (GEP) and may only reach pest status in some years. Some years they reach the economic threshold (ET) (or action threshold), indicating that the economic injury level (EIL) may be reached. (Adapted from Stern, V.M. et al., *Hilgardia*, 29(2), 81–101, 1959.)

products are becoming available due to the very high cost and length of time to develop and register a new product (typically more than $180 million and 8 to 10 years). For small specialty crops, registration of new pesticides is especially problematic.

An IPM program should use tactics that are compatible with each other and have the fewest negative effects on nontarget organisms and the environment, yet allow the crop to be produced economically. Significant social and ecological issues are imbedded in IPM (Pimentel and Lehman 1993). The strategy of pest control primarily relies on one tactic—chemical control.

Another one-tactic approach involves a total reliance on **biological control** and the elimination of chemical control. Just as a total reliance on chemical control is not usually viable, relying solely on biological control may not be an option for some cropping systems. There usually are pests in the crop that cannot be controlled effectively by natural enemies alone. This is especially true if a pest has just invaded and is without its specialist natural enemies. Generalist natural enemies in the new environment may not be fully effective, or their populations may be too low to adequately reduce populations of the new pest.

Both IPM and IMM rely on knowledge of the economic injury level of pests and effective monitoring methods for both pest and natural enemy populations (Figure 5.2). Pest populations typically fluctuate from year to year, but over time they display an average general equilibrium position. The **economic threshold** is the pest level at which control measures should be applied to prevent pest levels from reaching the economic injury level. The **economic injury level** (EIL) is the lowest pest level that will cause economic damage or loss (Higley and Pedigo 1996). One equation for calculating the EIL is EIL = cost of control ÷ (market value of the crop × yield loss per pest). Economic injury levels rarely are known for tree crops and vines because measuring crop losses in perennial crops requires 2 or 3 years. Perennial plants have extensive stored reserves that can make up for losses in photosynthetic rate, or other damage, for a year or two so that growth deficits or yield losses are delayed. The **economic damage** is the amount of injury that will justify the cost of control measures; one must include the cost of applying the control and, ideally, would include the costs of negative environmental and human health effects (although these can be difficult to quantify). IPM or IMM is information intensive, requiring considerable knowledge about the biology, ecology, economics, and cultural practices associated with crop production in a specific geographic region. Typically, IPM or IMM programs require multiple tactics; however, problems for each crop usually are site specific, so the exact combination of appropriate components will vary. An IMM program for mites in citrus in Florida will not be identical to an IMM program for mites in California citrus (see Chapter 18). Likewise, an IMM program will have to be modified as conditions change, such as

Table 5.1 Integrated Mite Management (IMM) Concepts and Tools

1. *Integrated* means that all the tactics used are compatible with each other.
2. Chemical control typically is used as a last resort; calendar sprays are unlikely to be appropriate.
3. Selective pesticides (products that have the least negative effects on natural enemies in the crop) should be employed if pesticides are required.
4. *Economic threshold* is the lowest level at which the pest can cause economic damage.
5. *Economic injury level* (EIL) is the lowest level of the pest that will cause economic loss or damage that justifies the cost of controlling the pest.
6. *Economic damage* is the amount of crop loss that occurs due to the pest.
7. IMM programs change as conditions change; new pests invade, new tools become available, resistances to pesticides develop.
8. Cultural controls include modification of planting or harvesting dates, crop rotations, dust management, pest exclusion, and disinfestation of planting material.
9. Host-plant resistance represents one of the easiest IMM methods to implement because it typically is implemented by providing the crop seed.
10. Biological control can be used in three ways: classical, augmentative, and conservation. Each method is appropriate under different conditions, but conservation of natural enemies is always appropriate.
11. Growing a healthy crop is integral to IMM; grow crops in climates for which they are suited and manage water, soil, and fertilizers effectively.
12. Quarantines can exclude pests and are important at the national, state, and local levels; excluding pests from greenhouses and other enclosed environments can be very effective if rigorously practiced.

when new cultivars are introduced, new pests are added to the cropping system, or new pesticides are introduced (or eliminated due to resistance or regulatory problems). Agricultural entomologists and acarologists are unlikely to run out of work.

Ideally, chemical control is used as a last resort in an IPM program and is used only when economic damage has occurred (Table 5.1). Ideally, the chemicals used will have the least negative effects on humans, the environment, and the natural enemies in the cropping system. Host-plant resistance, cultural practices, and biological controls usually will be very important tactics. Understanding the role of abiotic factors, such as climate, and effective management of soil, water, and fertilizers is important. Biorational controls may be effective in certain programs and may include genetic control of pest species, mating disruption, or mass trapping. Cultural practices deployed may include modifying the timing of planting and harvesting, dust management, irrigation, crop rotations, quarantines, and other forms of pest exclusion (see Figure 5.1). Cultivars of crops that are resistant to pests or diseases are often crucial to eliminating or reducing pesticide use (McNab and Jerie 1991). Section 5.2 through Section 5.9 describe the tactics that can be used in IPM programs in agricultural crops.

5.2 CLASSICAL, AUGMENTATIVE, AND CONSERVATION BIOLOGICAL CONTROL

Biological control, from an entomological point of view, is the use of parasitoids, predators, and pathogens to manage pests (DeBach 1964, Van Driesche and Bellows 1995, Hajek 2004). Biological control of mites is achieved primarily by predators (insects, spiders, or mites) and, less often, by pathogens. There are no parasitoids of mites (although a parasitoid in the family Encyrtidae attacks ticks). **Predators** are free-living organisms, each of which will consume a number of pests (prey) in its lifetime. Many predators attack a variety of prey species and thus are generalists. Predators of pest mites include spiders, insects (lacewings, lady beetles, predatory thrips, syrphids), and other mites, including the Phytoseiidae and the Stigmaeidae (see Chapters 11, 12, and 13). Pathogens include viruses, bacteria, fungi, and protozoa (Boucias and Pendland 1998, Bruin and van der Geest 2009). Pathogens of pest mites are primarily viruses and fungi (see Chapter 14). Relatively few pathogens of mites have been studied sufficiently to enable them to be used by pest managers but many are important in suppressing pest populations if the environmental conditions are conducive.

Table 5.2 Three Types of Biological Control Can Be Used in Integrated Mite Management Programs

Type	Definition
Classical	*Import natural enemies and evaluate risks prior to obtaining permission to release.* Permanent establishment is the goal so the natural enemy can provide long-term suppression of the target pest (and have few or no non-target effects). Costs and benefits typically are high (i.e., a $250 benefit per year for each $1 spent). Projects may take up to 10 or 15 years to complete.
Augmentation	
Inoculative	*Release commercially produced natural enemies.* The goal is for the released natural enemies to multiply and suppress pest populations within the growing season. Long-term establishment is not expected but could occur. Because release rates are lower than for inundative releases, it is less expensive.
Inundative	*Release large numbers of commercially produced natural enemies.* The goal is to achieve rapid suppression of pests. Costs are higher than for inoculative releases. Long-term establishment is possible but not the goal.
Conservation	*Preserve (conserve) natural enemies in the crop.* Conservation should always be a goal; without providing environmental conditions that allow the natural enemies to survive, classical and augmentative biological controls cannot succeed. Conservation most often involves modification of spray practices. Providing refuges, alternative foods, and shelters may be helpful.

Three major tactics are used in biological control: classical, augmentative, and conservation (Table 5.2). **Classical biological control** involves importation, evaluation, release, and permanent establishment in the environment of natural enemies from the area of origin of a foreign pest. It assumes that natural enemies from the area of the pest's origin will be more effective and specific than natural enemies in the pest's new environment (DeBach 1964, McMurtry 1983, Kauffman and Nechols 1992, Van Driesche and Bellows 1993, Bellows and Fisher 1999, Gurr and Wratten 2000, Hajek 2004). Compared to insect pests, relatively little work has been done with mites as targets of classical biological control, with the exception of phytoseiid predators in orchard systems (McMurtry 1983). Another major exception is the Cassava Green Mite Classical Biological Control Project in Africa (see Chapter 15).

Augmentative biological control involves the mass rearing and release of natural enemies to control a target pest (Ridgway et al. 1998, van Lenteren 2003). The natural enemies must be capable of being mass reared and must be released at an appropriate time and in sufficient numbers so they can achieve effective control before economic damage has been caused. They are not expected to establish permanently in the environment (although that may occur). Augmentation can be thought of as the release of living pesticides. Several insect and mite predators are commercially available and can be used successfully in augmentative biological control, especially in crops with a high economic value such as greenhouse vegetables and field-grown strawberries (Table 5.3). Predators are expensive, however, and information about appropriate release rates and timing may not available for every crop and specific geographic location, so the reliability of augmentative releases can be questionable until confirmed by onsite experiments. Two approaches are taken in augmentation.

Inundation involves releasing large numbers of natural enemies for *immediate* reduction of a damaging or near-damaging pest population. These natural enemies are used in a manner analogous to an insecticide, and inundation is the more expensive option of augmentative biological control.

Inoculation involves releasing smaller numbers of natural enemies, starting when the pest population is *very low.* The natural enemies are expected to reproduce to provide control before economic injury is caused, but they are not expected to establish permanently. Because they reproduce after release, this option usually is less expensive than inundation.

Substantial information (Table 5.4) is necessary when ordering natural enemies for an augmentative release, including accurate taxonomic identification of the pest to order the appropriate natural enemy. Sometimes knowing the biotype of the pest or natural enemy is important. Certainly, it is necessary to know whether the environmental conditions are conducive to the survival of the

Table 5.3 Some Natural Enemies Produced Commercially to Control Greenhouse Pests

Natural Enemy	Pest
Phytoseiulus persimilis	***Tetranychus urticae***
Encarsia formosa	*Trialeurodes vaporariorum*, *Bemisia tabaci*
Opius pallipes	*Liriomyza bryoniae*
Amblyseius barkeri	*Thrips tabaci*, *Frankliniella occidentalis*
Dacnusa sibirica	*Liriomyza bryoniae*, *L. trifolii*, *L. huidobrensis*
Diglyphus isaea	*L. bryoniae*, *L. trifolii*, *L. huidobrensis*
Bacillus thuringiensis	Lepidoptera
Heterorhabditis spp.	*Otiorrhynchus sulcatus*
Steinernema spp.	Sciaridae
Amblyseius cucumeris	*Thrips tabaci*, *Frankliniella occidentalis*
Chrysoperla carnea	Aphids
Aphidoletes aphidimyza	Aphids
Verticillium lecanii	Aphids
Aphelinus abdominalis	*Macrosiphum euphorbiae*
Aphidius colemani	*Aphis gossypii*
Trichogramma spp.	Lepidoptera
Leptomastix dactylopii, *Cryptolaemus montrouzieri*	*Planococcus citri*
Anthocorus nemorum	Thrips
Metaphycus helvolus	Scales
Trichoderma harzianum	*Fusarium* spp.
Amblyseius cucumeris or ***A. degenerans***	Thrips
Eretmocerus californicus	*Bemisia tabaci*
Metaseiulus occidentalis	***Tetranychus urticae***
Hippodamia convergens	Aphids
Nuclear polyhedrosis virus (NPV)	*Spodoptera exigua*
Orius sp.	*Frankliniella occidentalis*

Note: Species in bold are mites.
Source: Adapted from van Lenteren, J.C. and Woets, J., *Annu. Rev. Entomol.*, 33, 239–269, 1988.

released natural enemies (e.g., temperature, relative humidity, lack of pesticide residues). Release rates are crucial; accurately estimating the pest population density by sampling is necessary for ordering and releasing sufficient natural enemies to achieve control in a timely manner. These natural enemies must be properly handled when they are received and released so they survive and reproduce. Multiple releases may be necessary, and monitoring the results of the releases will determine whether the releases were effective.

Table 5.4 What You Need to Know When Ordering Natural Enemies for Augmentative Releases

1. Identify the pest so you can order the appropriate natural enemy species. Different natural enemy species or biotypes may be needed for various environmental conditions.
2. Know how to monitor the pest population so you know how many natural enemies to purchase.
3. Know the release rates for each natural enemy in the crop. Are you releasing inoculatively or inundatively?
4. Calculate the number of natural enemies required based on release rate, area to be covered, and density of pest.
5. Handle the natural enemies on receipt appropriately and release them so their quality remains high.
6. Release at the appropriate time; *earlier is better* if the releases are inoculative.
7. Monitor the results of the release.
8. Know the frequency of releases necessary if multiple releases are required.

The costs of augmentation can be difficult to assess because adequate numbers often are not released. In certain high-value crops, such as strawberries, predatory mite releases can cost as much as $600/acre/season but, if the two-spotted spider mite is resistant to all of the registered pesticides, then this may be the only management option and is justified because the value of the crop is high. For low-value crops, the direct costs of augmentative releases often are higher than the costs of applying pesticides, but the indirect costs of applying pesticides should be included for comparison. Indirect costs include the cost of losing an effective product to resistance, the potential negative effects of pesticides on worker health, the negative environmental effects, and the costs of added pesticide applications required after secondary pest outbreaks.

Augmentative releases may not always be successful, for reasons outlined in Table 5.4 and Table 5.5. Some problems may be attributed to the quality and quantity of natural enemies chosen and their source. Some problems may be due to the fact that the shipping, handling, and release methods (including rates and timing of releases) were not appropriate. It may be necessary to experiment with augmentative releases in each new crop or environment to achieve reliable pest suppression. When making experimental releases, consider carefully how to design the experiments. Both release and no-release plots, which may be treated with pesticides, will be necessary. No drift of pesticides into the release plots should be allowed.

To determine the pest density and release ratio, the pest populations in both release and no-release plots must be sampled prior to making releases. After the releases, perhaps at more than one rate, monitor the results. Monitoring involves taking multiple samples over time and evaluating predator and prey densities, which should be compared to those in the no-release plots over time. If possible, estimate the degree of damage to the crop in each plot at the end of the experiment and evaluate the crop yield.

Unfortunately, augmentative biological controls raise certain ethical issues (Hoy et al. 1991). In some cases, natural enemies have been sold to growers not to control pests but to provide an economic incentive so the grower does not apply disruptive pesticides to the endemic natural enemies. Such releases are termed placebo releases and are costly (see Table 5.5). It is more ethical to train growers to monitor their own natural enemy populations so they reduce or eliminate disruptive pesticide applications and do not spend money on making unnecessary releases.

Table 5.5 Logistical and Ethical Issues Related to Augmentative Biological Control

1. Placebo effects may be occurring when releases are made to convince the grower to avoid pesticide applications, even though the endemic natural enemies will actually provide the control. This causes growers to spend money for unneeded natural enemies. Proper education of growers can eliminate this unnecessary cost.
2. Quality control problems may occur in commercial rearing facilities; the natural enemies shipped may be the wrong species, diseased, malnourished, too old to be effective, or mishandled so they arrive in poor condition.
3. The effective species or biotypes may not have been used or may not be available, perhaps because the colonies became contaminated with similar-looking species or the specific biotype was not used.
4. Timely availability of natural enemies is essential if releases are to be made when prey densities are low, thereby reducing costs. It is difficult for producers to maintain rearing facilities all year around, yet huge numbers of a natural enemy may be needed within a brief period when the pests increase.
5. Effective release rates and timing information are required to obtain optimal results yet may not be available for all pests and climatic conditions.
6. Natural enemies may take time to build up in sufficient numbers to control the pests, so accurate and rapid monitoring methods must be provided to growers or pest control advisers so they can monitor the effectiveness of the releases.
7. The quality of a crop may not be maintained when effective control is achieved too late.

Source: Based on information from Hoy et al. (1991), van Lenteren (2000), and van Lenteren and Woets (1988).

Research is needed to improve the reliability and cost-effectiveness of augmentative biological control. Good quality-control programs are necessary for all natural enemy species, as well as the development of reliable production methods sufficient to meet peak demand (van Lenteren and Woets 1988, Hoy et al. 1991, van Lenteren 2003). The development of high-quality, inexpensive artificial diets for natural enemies could greatly reduce costs. Improved shipping methods would ensure that the natural enemies arrive in good condition. Better storage methods might allow natural enemies to be stockpiled so they can be shipped at peak demand times. Finally, information on release rates and timing for each crop and environment would be valuable. Quality control is important. Studies have shown that it is common for natural enemies shipped by commercial producers to be the wrong species (not surprising, because natural enemies such as phytoseiid predators are tiny and difficult to identify). It is also possible that too few have been shipped or that the packaging and shipping conditions were inadequate so many arrive dead or stressed. Diets may be so inadequate that longevity and fecundity are reduced, and the natural enemies may arrive diseased due to crowding and lack of adequate sanitation. Many companies indicate they will ship new natural enemies if there are problems, but that creates a delay in making effective releases, which can result in poor control.

A key rule in both inundative and inoculative releases is to *make releases early* when pest populations are relatively low; for example, you may need a ratio of approximately 1 predator female (healthy) to 5 or 10 spider mites (all stages) to get rapid and effective biological control. If your crop consists of 1000 plants with 30 leaves each, and each leaf averages 10 spider mites, you have approximately 300,000 spider mites and will need to release approximately 30,000 healthy predators if you use a 1:10 release ratio. In contrast, if you make releases when there are only 0.1 spider mites per leaf, you will need only 3000 predators, thereby reducing costs and damage to the crop.

Successful augmentation programs are most commonly conducted in protected agriculture (van Lenteren and Woets 1988, van Lenteren 2000). The control of various pests in greenhouse-grown vegetables in Europe is highly effective due to a relatively condensed industry, scientists who develop information on release rates and timing, and producers of natural enemies that have close ties to end users. (See Table 5.3 for a partial list of pests and their natural enemies for which considerable information is available.) Greenhouse-grown crops are grown on about 200,000 hectares around the world. This cropping system offers several advantages and disadvantages. Disadvantages include the fact that many crops are high value, with a very low economic injury threshold. For ornamentals, the threshold is nearly zero. As a result, biological control must be very effective or growers will be unwilling to adopt this tactic. Growers have strong incentives to adopt a biological control-based IPM program because greenhouse pests have developed resistance to pesticides rapidly. If proper sanitation and quarantines are employed, the number of pests able to invade greenhouses is relatively limited. Because the area contained in greenhouses is small and concentrated, augmentative biological control efforts are feasible and, potentially, cost effective. Other incentives are the health benefits to workers of a pesticide-free environment and the benefits of pesticide-residue-free foods and other products—which increasingly are being demanded by consumers. Effective IMM or IPM programs in greenhouses require the use of a variety of tactics (Table 5.6).

Table 5.6 Integrated Mite Management Programs in Greenhouses Require Multiple Tactics

1. Quarantines to prevent pest problems by inspections and disinfections
2. Crop rotations to reduce pest populations
3. Host-plant resistance (cultivars vary in sensitivity to pests and suitability for controlling them) to reduce pest populations
4. Cultural controls (heat, modification of relative humidity) to make conditions less favorable for pests
5. Biological controls (inoculative or inundative) to suppress a variety of pests
6. Chemical control with selective pesticides that are not disruptive to natural enemies when other tactics are inadequate
7. Treatments based on monitoring rather than the calendar
8. Treatments based on economic injury levels

Greenhouse-grown vegetables have the most pest species that can be controlled by augmentation (van Lenteren and Woets 1988, van Lenteren 2000). The European greenhouse industry has adopted augmentation as a primary tactic for control of pests of vegetables and some ornamentals; however, in countries where greenhouse industries are relatively small and scattered, as they are in the United States, it is unattractive to commercial producers of natural enemies to develop all of the information necessary to implement augmentation. If more producers begin to use augmentation, company sizes may increase and result in an economy of scale in the production of natural enemies. Another option involves the organization of grower cooperatives to produce specific natural enemies for that crop.

Augmentation requires that the pest manager have considerable information on pest and natural enemy biology, an understanding of pest-natural enemy dynamics, and the availability of adequate numbers of high-quality natural enemies for release at the right time. Thus, it is an information-intensive and logistically challenging management approach. Augmentation has the potential to become a more dominant component of pest management programs if sufficient resources are devoted to obtaining the necessary information and to improvements in rearing and deployment. These resources will have to be funded by private corporations as well as universities, departments of agriculture, and cooperative extension.

Conservation of natural enemies involves changing agricultural production practices to enhance the effectiveness of natural enemies in the cropping system (McMurtry et al. 1970, Barbosa 1998, Pickett and Bugg 1998, Landis et al. 2000). Conservation is most often achieved by modifying pesticide applications, including the rates applied, the product type, and the timing of application. Conservation also may require altering irrigation practices or using ground-cover plants between rows to reduce dust and cool the crop. Conservation may involve adding an essential component to the environment, such as plants with pollen or nectar sources or overwintering sites for natural enemies. Conservation of natural enemies is critical to the development of effective IPM programs.

5.3 QUARANTINES

Quarantines against foreign pests may keep them from entering the state or country (see Figure 5.1). Quarantines that keep pests out of greenhouses or other contained areas also are important IMM tools. Quarantines may be effective if sufficient diligence is employed, but it can be difficult and costly to keep pests out forever. Too many people carry fruit and other plant materials from country to country or from state to state illegally, and border inspections and quarantines may miss these very small arthropods if they are present in low densities. The commercial transport of huge numbers of plants and agricultural products from around the world makes it difficult to keep all exotic pests out. It has been estimated that only about 2% of all agricultural products coming into the United States, for example, are inspected. Growers producing crops in greenhouses, or other contained environments, may be successful in excluding pests if they maintain sanitation, have appropriate exclusion methods in place (e.g., screen vents to prevent invasions by larger pests, eliminate weeds that harbor pests around the greenhouse, use pest-free plants to initiate crop production) and have properly trained personnel that understand the importance of maintaining pest-free crop production areas (see Table 5.6).

5.4 CULTURAL CONTROLS

Cultural controls involve all modifications to agronomic practices that are intended to reduce pest damage (Flint and Gouveia 2001, Vincent et al. 2003, Gurr et al. 2004). Cultural practices include changing the timing of planting and harvest to avoid or minimize pest damage and cultivation to control weeds and other pests in the soil. Proper management of temperature and relative humidity can be useful to suppress pests in greenhouse crop production (see Table 5.6). Another example of cultural control that is important in managing spider mite populations in irrigated agricultural regions is **dust management**. Dust management is important for managing spider mites, especially in climates where crop irrigation is necessary. Dust on foliage makes it easier for spider mites to become serious pests. There is controversy as to whether the dust makes the foliage more suitable for spider mites or the dust interferes with the effectiveness of spider mite predators. In Mediterranean-type climates, such as California, where there is little or no rain between April and October, managing dust on roads is an accepted cultural practice, and growers are urged to drive slowly or to resurface them to reduce dust (see Figure S5.1 on the CD). Spider mite outbreaks are commonly observed first on the edges of orchards, vineyards, or fields or along roadsides in California where dust can be a problem due to a lack of adequate irrigation and a lack of rain during the summer (see Figure S5.2 on the CD). **Cover crops**, such as legumes, can be planted between rows of trees or vines to reduce the amount of dust. Cover crops may add nitrogen to the soil and provide shelter, alternative prey or hosts, and nectar for adult parasitoids. Negative aspects of cover crops include the extra costs of fertilizer or water to sustain them and the possibility that a vector of a plant disease will be fostered in them.

Effective irrigation is crucial in many climates and may affect spider mite populations. It appears that water-stressed plants allow spider mite populations to increase more rapidly than they might on well-watered plants, perhaps because their food is more concentrated and they can reproduce at a higher rate (Youngman and Barnes 1986, English-Loeb 1990). A lower relative humidity typically favors most spider mite species, although some species do well under humid conditions.

Managing fertilizer applications is another important cultural practice; spider mite populations can be stimulated if the plant is receiving a higher rate of nitrogen fertilizer than is needed (van de Vrie et al. 1972). Succulent, tender growth is favorable for most spider mite species. The elimination of crop residues can destroy pests and prevent them from damaging subsequent crops. **Crop rotation** can be used to reduce pest population buildup by planting crops that do not harbor the same pests continuously. **Polycropping** (planting crop mixtures) is traditionally found in many tropical and subtropical areas. Polycropping, in theory, reduces pests because the presence of the host plant is not as apparent if the crops that are planted do not have the same pests (Pickett and Bugg 1998). It is not clear that polycropping is useful in managing plant-feeding mites, but if natural enemies are retained in the crops, it could be useful.

5.5 GENETIC CONTROL

Genetic control methods include the use of sterilized males in the sterile insect release (**SIR**) or sterile insect technique (**SIT**), as well as the release of genetically incompatible strains that could result in reproductive incompatibility. **Autocidal (genetic) control** (i.e., SIT or SIR) typically involves the use of irradiation or chemosterilants to sterilize pest insects or mites that, if released in adequate numbers, can result in pest population reductions or elimination (Geier 1966, Gould and Schliekelman 2004). Unfortunately, we have no examples of mites that have been successfully managed or eradicated through autocidal control methods. Tests with arrhenotokous (haplodiploid) spider mites showed that wild females that mated with sterile (haploid) males simply remated with their haploid (fertile) sons. Unmated spider mite females can deposit haploid eggs that develop into males and the sons can mate with their mothers.

Dieleman and Overmeer (2009) evaluated the possibility of releasing a population of two-spotted spider mite females that was genetically incompatible with the existing population in the greenhouse in order to reduce the population. In controlled matings, males and females of the two populations produced sterile hybrids; however, they discovered that males of the two populations were more effective in mating with their own females, so this control method would not be effective in greenhouse conditions where mate-choice could occur.

de la Fuente et al. (2006) proposed to control ticks by silencing a single gene that affects reproductive development in both males and females so they are unable to reproduce. Control by release of mass-reared female and male ticks could potentially reduce pest populations. So far, no genetic control programs have proven to be effective when directed against mites; however, only a few studies have been conducted and it could be possible that genetic control projects could be effective in the future.

5.6 CHEMICAL CONTROL

Chemical control should be considered as the tactic of last resort in an IMM program. Pesticides should be used only when necessary, and then only the least toxic and disruptive product available so natural enemies are not disrupted (Stern et al. 1959). So-called calendar applications of acaricides (made whether or not the pest is present at an economic level) are not appropriate in an IMM program. Chemical control should be applied only when the pest population exceeds the economic injury threshold and other control tactics have not worked or are not appropriate (see Figure 5.1). Although this principle is designed to minimize pesticide use, sometimes the number of chemical applications can increase in an IMM program; however, under such conditions the pesticides applied ideally are **selective pesticides**, meaning they suppress pest populations without disrupting natural enemies.

Pesticide applications can be designed to be selective in several ways. The objective is to kill the pest but not the beneficial organisms. Pesticides can be **intrinsically selective**, in that their chemistry makes them more toxic to pests than to beneficial species. Unfortunately, intrinsically selective pesticides often have a very narrow toxicity spectrum, so it may not be economical for a commercial producer to develop and register such a product, and these products are less often available. Examples of selective pesticides include some microbial pesticides, such as *Bacillus thuringiensis*, and some insect growth regulators.

Pesticide selectivity can be achieved in creative ways; for example, it is possible to use **systemic pesticides**, which are applied to the plant roots and taken up throughout the plant. The pesticide kills pests feeding on the plant but may not kill beneficial species because there are no toxic residues on the outside of the leaves. Partial treatment of only a portion of the field (**spot treatment**) can be done to preserve natural enemies and reduce costs. Spider mite outbreaks commonly are found in the same sites in a crop year after year, often along roadsides or in regions where water penetration or management is poor. A wise pest manager monitors those sites early in the season and makes applications to suppress the spider mite populations in that localized area. By making timely applications to trouble spots, subsequent treatment of the entire orchard or field may be avoided.

Bait sprays can allow pest managers to use pesticides selectively to control insect pests. Several species of fruit flies are suppressed by bait sprays, which may have less toxic effects on nontarget species (Urbaneja et al. 2009) than pesticides that are applied to the entire plant. I am unaware, however, of any bait sprays for mite control, although some bait sprays applied to control insects can suppress predatory mite (Phytoseiidae) populations.

Sometimes portions of fields are left unsprayed to provide a refuge for natural enemies. In apple orchards in Pennsylvania, only alternate rows of trees are sprayed at one time to control the codling moth; in this way, ladybird beetles (*Stethorus punctum*) are preserved and can control the European red mite *Panonychus ulmi* (Hull and Beers 1985).

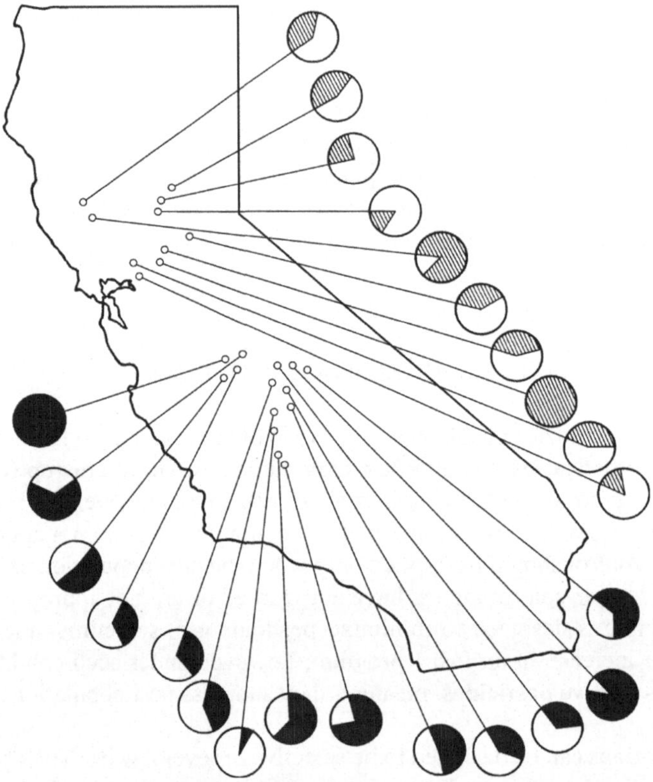

Figure 5.3 Variability in resistance of the phytoseiid *Metaseiulus occidentalis* to organophosphate insecticides used in pears and grapes in California was achieved through selection in the field. The degree of shading in each circle represents the relative level of resistance in the populations collected from that crop. The fully shaded circles indicate the highest levels of resistance found. Note the substantial variability in resistance levels within the species and within the crops. This indicates that dispersal is slow (or rare) and that local selection in orchards or vineyards yields populations that vary. Pest managers need to know the past treatment history of the field or must test the population themselves when planning on using specific pesticides in a selective manner. (Adapted from Hoy, M.A., *Annu. Rev. Entomol.*, 30, 345–370, 1985. With permission.)

In some cropping systems, pesticide-resistant natural enemies have developed through field selection, and several species of phytoseiids have been selected in the laboratory for resistance to pesticides (Hoy 1985) (Figure 5.3). Pesticide-resistant natural enemies, including especially predatory mites (Phytoseiidae) that are able to survive pesticides such as azinphosmethyl, were able to survive and control spider mites in apples and almonds (see Chapters 16 and 17). This reduced the need to apply pesticides to control the spider mites. Additional species of predatory mites have been selected in the laboratory for resistances to a variety of pesticides and have been used in IPM programs for apples, grapes, almonds, and strawberries (Chapter 12).

Acaricides (or **miticides**) are pesticides that provide economic control of pest mites and ticks. Some acaricides can act as insecticides or fungicides. The toxicity of an acaricide is evaluated by a **dose–response curve** or a **concentration–response curve**. Examples of concentration–response curves are shown in Figure 5.4. Such curves are obtained by exposing test mites or insects to increasing concentrations of the pesticide and recording the resulting mortality after a given time interval (Yu 2008). In a concentration–mortality bioassay, the mites are exposed to a particular *concentration* of pesticide, but the amount taken up by the mite is unknown. For example, if the pesticide is applied to foliage and the mite walks about on the foliage, the actual amount of toxicant

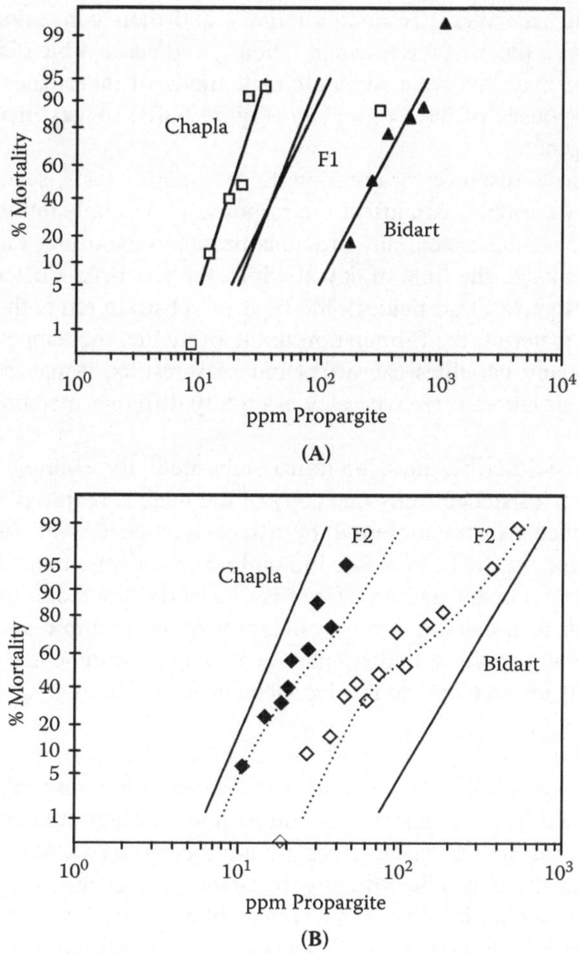

Figure 5.4 (A) Concentration–response curve showing the responses of colonies of *Tetranychus pacificus* that are resistant (Bidart) and susceptible (Chapla) to the acaricide propargite. The two types of F_1 progeny are intermediate in their responses, indicating that the resistance is not fully dominant or fully recessive. (B) The concentration–mortality curves of two types of F_2 progeny suggest that a major gene is involved in the resistance. (Adapted from Hoy, M.A. and Conley, J., *J. Econ. Entomol.*, 82, 11–16, 1989. With permission.)

to which the mite is exposed depends on the activity of the mite, the amount taken up through the integument, or by feeding. By contrast, a dose–response bioassay applies a specific *dose* of the pesticide to the mite.

Useful estimates of toxicity are **LD_{50}**, the *dose* required to kill 50% of the test population, and **LC_{50}**, the *concentration* required to kill 50% of the test population. If the dose is introduced through the insect's mouth it is an **oral LD_{50}**, if it is introduced through the skin or integument it is a **dermal LD_{50}**, and if it is introduced through the respiratory system it is the inhalation LD_{50}. A measured dose can be applied to an arthropod by inserting a measured amount of toxicant into the gut or by applying a measured amount to the integument. Concentrations are typically applied by spraying the substrate. The lower the LD_{50} or LC_{50} values, the more toxic the poison.

Figure 5.4A shows a concentration–response curve in parts per million (ppm) for the acaricide propargite exhibited by adult females from colonies of the Pacific spider mite *Tetranychus pacificus* (Hoy and Conley 1989). The two types of F_1 females (produced by crossing Chapla males and Bidart females, and *vice versa*) respond similarly and their concentration–response curves are about midway between those of the resistant (Bidart) and susceptible (Chapla) colonies, which indicates that resistance may involve a semidominant mode of inheritance, or is determined by multiple genes. The responses of the F_2 progeny (Figure 5.4B) suggest that the resistance could involve a major single gene.

When conducting dose–response or concentration–response tests, several experimental controls are required. Many variables can affect the response of a population to pesticides, including the age of the individual being tested, any previous pesticide exposure, the size of the individuals, the sex of the individuals, the time of day at which the test is conducted (mites can have diel periodicities in their response to pesticides), the type of substrate (smooth or rough), the type of application of the toxic material, the formulation of the pesticide, the temperature, and the relative humidity. Because so many variables can affect bioassay results, it may be difficult to compare data obtained by different laboratories where even slightly different methods could yield different results.

Mode of inheritance describes how the trait is inherited; for example, a single major dominant gene can determine resistance (only one copy of the gene is required for the mite to express the resistance). Resistance can be conferred by a recessive gene, with two copies of the gene required for the resistance trait to be expressed in diploid organisms. In haplodiploid spider mites, all resistance genes are dominant and expressed in males, because only one gene copy is present. Resistance also can be a quantitative trait determined by multiple genes of equal and additive effect. In Figure 5.4, a single semidominant gene may determine the propargite resistance, although additional tests are required to resolve whether more than one gene actually contributes to this resistance.

The mortality of adult females at different concentrations has been transformed into a straight line because the data are graphed on a log scale. The concentration–response lines from the reciprocal F_1 females, obtained by crossing the susceptible and resistant populations, are intermediate and similar, indicating that the resistance genes are not located on sex-determining chromosomes and that it is likely that the genes determining resistance are semidominant. If the genes were recessive, the F_1 females should have given a response like that of the susceptible Chapla females. If the genes were completely dominant, the F_1 females should give a response similar to the resistant Bidart females (Preisler et al. 1990).

5.6.1 Acaricide Classifications

Pesticides are classified in several ways, including: (1) their mode of entry into the target pest, (2) their chemical structure, or (3) their source (Table 5.7).

5.6.1.1 Mode of Entry

Pesticides can enter and kill mites as *stomach* poisons, *contact* poisons, or *fumigants.* A *systemic* acaricide is absorbed into a plant or animal and protects that plant or animal from pests after the pesticide is translocated throughout the plant or animal.

5.6.1.2 Chemical Structure

Pesticides are classified as *organic* or *inorganic.* Inorganic pesticides do not contain the element carbon (but include arsenic, mercury, zinc, sulfur, boron, or fluorine). Few inorganic pesticides are used, except for sulfur, an exceptionally effective acaricide for eriophyoids and most tetranychids.

Table 5.7 Acaricide Types and Their Classifications

Classification	Type (Examples)
Mode of entry	Stomach Contact Fumigant (rarely relevant in mites) Systemic in plants
Chemical structure	
Organic	Botanicals (pyrethrin, neem as azadirachtin, nicotine, rotenone) Essential oils from plants (various) Products from microorganisms (avermectins, milbemectin)
Synthetic organic	Organosulfurs (tetradifon, chlorfenson, propargite) Organotins (cyhexatin, azocyclotin, fenbutatin oxide) Pyrethroids (bioresmethrin, fenpropathrin, bifenthrin) Pyrazoles (tebufenpyrad, fenpyroximate) Quinazolines (fenazaquin) Methoxyacrylates (fluacrypyrim) Naphthoquinones (acequinocyl) Tetronic acids (spirodiclofen) Tetrazines (clofentezine) Oxazoles (etoxazole) Carbazates (bifenazate) Benzoylacetonitriles (cyflumetofen) Trifluoromethanesulfonanilides (amidoflumet) Bridged diphenyls (benzoximate)
Inorganic	Sulfur
Source	
Plant-derived	Essential oils from plants
Microorganisms	Avermectins, milbemectins

Source: Adapted from Yu, S.J., *The Toxicology and Biochemistry of Insecticides*, CRC Press, Boca Raton, FL, 2008.

5.6.1.3 Source

Organic pesticides include *botanicals* (natural organic pesticides) produced by plants (natural pyrethrums, nicotine, rotenone, essential oils such as those from the neem tree, soybean oil). Essential oils are any volatile oil that gives distinctive odor or flavor to a plant, flower, or fruit, such as lavender, rosemary, or citrus oil. Essential oils have been registered as pesticides since 1947, and at least 24 different ones are available in registered products in the United States. These are used as repellants, feeding depressants, insecticides, and miticides. Botanicals have relatively high LD_{50} values to mammals, so they usually are considered safe to humans. Some newer pesticides are derived from microbes, such as avermectin or spinosad.

5.6.2 Insecticides as Acaricides

Many insecticides have acaricidal properties. Sometimes an insecticide is more effective as an insecticide than as an acaricide (lower concentrations are required to kill the insect than are required to kill the mite species). Some products are more toxic (often for unknown reasons) to mites than to insects. It is likely that mites have the same fundamental physiological responses to toxic chemicals as insects, although mite physiology and responses to pesticides have been studied less. Different mite species respond differently to different products, which could be due to behavioral differences (feeding behavior, location on plant, activity levels), differences in cuticle thickness, differences in detoxification rates, or other biochemical, morphological, or behavioral factors. Different formulations can influence toxicity to different species of insects and mites.

Many insecticides were effective acaricides before resistance to them developed; for example, many organophosphates (OPs) were toxic to spider mites until resistances developed. Likewise, carbamates, formamides, and many pyrethroids have both insecticidal and acaricidal properties. Other products have both fungicidal and acaricidal properties.

5.6.3 Acaricide Types

Pesticide registrations change frequently, so some of the materials discussed here may be obsolete or illegal. Always check with your government authorities to be sure products containing these active ingredients are registered for use. Always read labels carefully, and follow the directions completely. In the United States, the Environmental Protection Agency (http://epa.gov/pesticides/) and EXTOXNET (http://extoxnet.orst.edu/) websites provide information about pesticides and their toxicology.

5.6.3.1 Synthetic Organic Pesticides

Synthetic organic pesticides can be separated into groups based on their chemistry (Yu 2008). The main groups are *chlorinated hydrocarbons* (such as DDT and chlordane, which are banned from use in many countries); *organophosphates* (such as malathion, parathion, and azinphosmethyl), most of which are going off the market due to concerns about food safety; *carbamates* (e.g., carbaryl, propoxur); *pyrethroids* (e.g., permethrin, bifenthrin); and a variety of newer products with very different chemistries, including *neonicotinoids*, *pyrroles*, *carbazates*, and *pyridazinones*. Table 5.7 lists products used primarily for control of plant-feeding mites.

5.6.3.1.1 Chlorinated Hydrocarbons

Dienochlor is an acaricide with long residual activity that has been used in greenhouses and on outdoor ornamentals. Dienochlor cannot be used on food crops and has short residual activity when used outdoors. It has a rapid effect on mites, stopping their feeding within hours. Endosulfan and DDT historically were used as acaricides (as well as insecticides), but many mites have developed resistance to these products. These products no longer are registered for use in most areas.

5.6.3.1.2 Organophosphates

The organophosphates (pesticides that include phosphorus) are derived from phosphoric acid and are the most toxic of all pesticides to vertebrates. They are related to nerve gases by structure and mode of action, but OPs are less persistent in the environment than the organochlorines such as DDT. The use of OPs has been nearly eliminated in the United States, and elsewhere, due to concerns about human health and other environmental effects. OPs inhibit cholinesterases in the nervous system. Acetylcholine is the chemical signal that is carried across synapses, where the electrical signal is transmitted across a gap to a muscle or another neuron. After the electrical signal (nerve impulse) has been conducted across the gap by acetylcholine, the cholinesterase enzyme removes the acetylcholine so the circuit will not be kept on. When OPs poison an organism, the OP attaches to the cholinesterase and it cannot remove the acetylcholine. The circuits then remain on because acetylcholine accumulates. This gives rise to rapid twitching of the voluntary muscles and to paralysis, which can be lethal if it persists in the vertebrate respiratory system. Not all OPs are highly toxic to vertebrates; if the phosphorus is modified by esterification (adding oxygen, carbon, sulfur, and nitrogen), various classes of OPs can be produced. Some of these are relatively safe to vertebrates, such as malathion.

5.6.3.1.3 Carbamates

Introduced in the 1950s, carbamates (aldicarb, carbofuran, methomyl, propoxur) are derivatives of carbamic acid. The mode of action of carbamates is to inhibit cholinesterase. Carbaryl has been one of the most popular products available to home gardeners for controlling a variety of insect pests, and it has low mammalian oral and dermal toxicity. Carbaryl is known to induce outbreaks of spider mites after applications are made to control other pests. The outbreaks are due to two factors: (1) carbaryl kills many phytoseiid species and other natural enemies of spider mites, and (2) it stimulates reproduction of spider mites through a process called *hormoligosis*. Very low doses of carbaryl have been found to act like a hormone to stimulate reproduction of the two-spotted spider mite *Tetranychus urticae*. It is likely that the use of carbamates will be eliminated or greatly reduced in the United States due to the Food Quality Protection Act.

Hormoligotic effects may involve newer pesticides, as well. Imidacloprid, a chloronicotinyl insecticide introduced in the early 1990s, is widely used. It is a broad-spectrum systemic pesticide that acts on sucking insects (aphids, whiteflies, leafhoppers), coleopterans, dipterans, and lepidopterans. At normal field rates, it is not toxic to phytophagous mites. The impact of imidacloprid on various natural enemies is controversial; some reports indicate it is highly toxic but other species are reported to be tolerant. Sublethal effects of imidacloprid on natural enemies include reduction in prey consumption and alteration of walking behavior. The predator *Amblyseius victoriensis* increased egg production by up to 54% when exposed to imidacloprid, a potentially beneficial effect (James 1997), but James and Price (2002) reported that the fecundity of *Tetranychus urticae* was increased by 10 to 26% during the first 12 days of adult life and by 19 to 23% during adulthood. These data suggest that imidacloprid may affect mite management in a variety of crops. The possibility that imidacloprid, and other pesticides not yet evaluated for hormoligotic effects, will stimulate reproduction in mites needs to be thoroughly assessed due to its potential impact on IMM programs.

Benomyl is a carbamate that has been used primarily as a fungicide, but it has acaricidal properties. Benomyl is interesting because it acts to sterilize at least some phytoseiid predators. Phytoseiid females treated with benomyl survive, but they do not deposit eggs, apparently because benomyl disrupts spindle fiber formation in cells and interferes in the synthesis of DNA.

5.6.3.2 Organosulfurs

Tetradifon, chlorfenson, and propargite are organosulfurs. These products contain sulfur as a central atom with two phenyl rings. Tetradifon is particularly toxic to mites but has very low toxicity to insects. Organosulfurs are often ovicidal. Propargite was used for more than 20 years in the United States and appeared to be immune to the development of resistance in spider mite populations; however, propargite resistance has now developed in many populations of spider mites around the world, which is unfortunate because propargite is less toxic to phytoseiids than to spider mites and thus could be used in IMM programs.

5.6.3.3 Organotins

Cyhexatin and fenbutatin-oxide are examples of tin compounds that are primarily acaricides and fungicides. Cyhexatin was introduced in 1967 and was widely used for many years before resistance developed in spider mites, although some people had concluded that the organotins were immune to resistance problems. The organotins were useful in IMM programs because they were more toxic to spider mites than to phytoseiids. These products have been taken off the market in the United States.

5.6.3.4 Formamides

Formamides include chlorodimeform, amitraz, and formetanate. These products are effective against the eggs of Lepidoptera and against most stages of mites and ticks. The mode of action is unclear but is thought to be due to the inhibition of monoamine oxidase, which results in the accumulation of compounds called biogenic amines.

5.6.3.5 Pyrethroids

Many synthetic pyrethroids have acaricidal activity. Some (such as bioresmethrin, fenpropathrin, and bifenthrin) are considered effective acaricides. Unfortunately, pyrethroids usually are very toxic to beneficial arthropods, including phytoseiids. These detrimental effects can be very long lasting because the residues persist a long time. Few have been found useful for IMM programs for this reason (Croft 1990). Laboratory selection of phytoseiids (*Amblyseius fallacis*, *Metaseiulus occidentalis*, *Typhlodromus pyri*) for resistance to some pyrethroids has been successful using both laboratory and field-selection methods (Hoy 1985).

5.6.3.6 Pyrroles

Pyridaben is a novel pesticide that works as a mitochondrial electron transport inhibitor to block cellular respiration, causing pests to become uncoordinated and die. It can be used on both insects and mites.

5.6.3.7 Azadirachtin

Azadirachtin is a triterpenoid extracted from the seeds of the neem tree (*Azadirachta indica*). Extracts include a combination of compounds, the proportion of which varies from tree to tree. Such variability makes it difficult to predict the precise effect of the product when extracted by local farmers. Commercial products may be more consistent in their effects because they have been tested to confirm their quality and are blended to achieve a consistent product. Azadirachtin blocks the action of the molting hormone ecdysone.

5.6.3.8 Avermectin

Avermectin is a natural product containing a macrocyclic lactone glycoside that is a fermentation product of *Streptomyces avermitilus*, a microorganism isolated from soil. Avermectin is actually a mixture of two homologs, both of which have biological activity. Avermectin has insecticidal and acaricidal properties and is closely related to ivermectin, which kills nematodes. At appropriate rates, avermectin is less toxic to beneficial phytoseiids than to spider mites. It paralyzes active spider mite stages but is not toxic to eggs. Avermectin has **translaminar** activity (meaning it is taken up by the plant tissue and subsequently by spider mites feeding on the plant tissues), but it has a short residual toxicity to phytoseiids. Resistance to this product has been reported in spider mites. A resistant strain of *Metaseiulus occidentalis* was obtained after laboratory selection, suggesting that resistance mechanisms may be present in field populations. The mode of action of avermectin involves blocking the neurotransmitter gamma-aminobutyric acid (GABA) at the neuromuscular junction. Mites that are exposed to abamectin become paralyzed and, although they do not die immediately, the paralyzed mites stop feeding.

5.6.3.9 Clofentezine and Hexythiazox

Clofentezine and hexythiazox are very interesting growth regulators of mites; they kill eggs (ovicides)of spider mites but not the active stages. The products have different chemistries, but both are nontoxic to phytoseiid eggs or active stages. In fact, when the phytoseiid *Metaseiulus occidentalis* is fed a diet consisting solely of spider mite eggs that have been killed with these products, the predator females reproduce and their progeny develop normally (Hoy and Ouyang 1986). This selectivity makes the products useful for IMM programs because predators can be maintained while suppressing spider mite populations. Unfortunately, resistance to these products has developed in spider mite populations in several locations (Thwaite 1991, Herron et al. 1993, Grosscurt et al. 1994, Yamamoto et al. 1995, Pree et al. 2002).

5.6.3.10 Tebufenpyrad

Tebufenpyrad is a phenoxypyrazole that has been evaluated in Australia, where it was shown to be useful in IMM programs in apples because it is selective (relatively nontoxic) to phytoseiid predators.

5.6.3.11 Essential Oils

Soybean oil was first registered in 1959 for use as an insecticide and a miticide, but other plant-based oils can be used as pesticides. Three products currently are registered to control mites on fruit trees, vegetables, and a variety of ornamentals. Soybean oil is not **phytotoxic** (causing plant injury) under most conditions, but care should be taken to ensure that organic oils are not phytotoxic. Many essential oils are approved for organic farming.

5.6.3.12 Inorganics

Sulfur is a good acaricide and fungicide, although it can be phytotoxic, especially if plants are not well watered during hot weather. Sulfur is probably the oldest known acaricide. Sulfur (dusts, wettable powders, and flowable formulations) usually is highly effective against spider mites and rust mites (eriophyoids), with one known exception. Spider mites (*Tetranychus pacificus* and *Eotetranychus willamettei*) in California vineyards developed resistance to sulfur, probably because sulfur was applied up to 20 times a season over many years to control powdery mildew (see Figure S5.3 on the CD). After a number of years, these spider mites became pests because they were no longer controlled by sulfur that was applied to control powdery mildew. A number of years later, the predatory mite *Metaseiulus occidentalis* was shown to have developed resistance to sulfur (Hoy and Standow 1981, 1982). The resistance to sulfur in this natural enemy of spider mites is based on a single major dominant gene. Once the predator became resistant to sulfur, it became an effective predator of the sulfur-resistant spider mites in San Joaquin Valley vineyards in California. The resistance to sulfur in *M. occidentalis* is unusual; even very high application rates of sulfur are nontoxic to the resistant populations so we do not get a linear response when we do a concentration–mortality test. Interestingly, populations of this predator collected from nearby almond orchards in California are susceptible to sulfur, indicating that populations are subjected to local selection and evolution and dispersal must be relatively low (Hoy 1985). No genetic analyses have been conducted on the resistance to sulfur in the spider mites, so the mode of inheritance of sulfur resistance remains unknown. The biochemical mechanisms of resistance to sulfur are unknown for both spider mites and the phytoseiid. Sulfur is toxic to most phytoseiid species.

5.6.3.13 Petroleum Oils

Petroleum oils are excellent insecticides, acaricides, and fungicides for IMM programs and have been used in pest management programs for over 100 years (Johnson 1985, Davidson et al. 1991). One of the virtues of petroleum oils is that there is no documented case of resistance in mites to them (Table 5.8). Also, only minimal protective clothing, such as overalls, goggles, and masks, must be worn when handling petroleum oils. Their toxicity to vertebrates is low, and petroleum oils have fewer detrimental effects on natural enemies (both insect and mite predators) than do most synthetic pesticides. Petroleum oils do not stimulate secondary pest outbreaks, and they break down and are no longer toxic after the spray dries.

Most oils used are distillations of petroleum, although some oils derived from plants (sesame, almond, citrus) are used. Crude petroleum oil is a complex mixture of hydrocarbons with both straight-chain and ring molecules. Crude oil is separated into a range of products by distillation and refining. The lightest fractions include gasoline, kerosene, diesel, and jet fuel. As these lighter fractions distill or boil, they are separated into different fractions. Spray oils are derived from the lighter lubricating oil fraction and distill at a temperature range of 600 to 900°C.

Currently used petroleum oils in the United States are narrow-range oils that have had the waxes, sulfur, and nitrogen compounds removed. The sulfur compounds are removed because they are likely to cause phytotoxic effects. The degree of removal of these compounds, referred to as the *unsulfonated residue* (UR) rating, is an important piece of information on the label and commonly is greater than 92%. Labels on sprays usually describe the degree to which the sulfur compounds have been removed and the percentage of active oil, which should be greater than 60%. In some countries, the purity of petroleum oils can vary and cause phytotoxicity; read the labels carefully and test the products in small areas to confirm the product does not damage the crop.

Since the mid-1960s, narrow-range horticultural oils have been used both as dormant or summer oil sprays. These highly refined and narrow-range petroleum oils rarely cause phytotoxicity and increasingly are used for controlling both insect and mite pests on deciduous and citrus trees, as well as on ornamental trees and shrubs. They have a wide range of activity against scales, mites, psyllids, mealybugs, whiteflies, leafhoppers, and eggs of mites, aphids, and some Lepidoptera. **Dormant sprays** are used to control overwintering pests in deciduous trees and vines, and **summer oils** are used to control pests during the growing season.

Petroleum oil kills mites and their eggs by contact. The toxicity appears to be due to suffocation of the pest, although it also may be due to chemical effects. Oils block spiracles, reducing the availability of oxygen, and suffocation occurs within 24 hours. Penetration and corrosion of tracheae and damage to muscles and nerves may contribute to the toxicity of oils. Oils are sometimes a repellent

Table 5.8 Petroleum Oils as Acaricides: Pros and Cons

Pros	Cons
Highly refined, narrow-range oils can be used as dormant or summer sprays.	There is no residual activity; multiple applications may be necessary to achieve control.
Petroleum oils have a wide range of activity against insects and mites.	It is essential to obtain good coverage or efficacy is reduced.
Petroleum oils kill by contact; unsprayed natural enemies will survive.	Phytotoxicity can occur when applied to heat- or drought-stressed plants. The susceptibility of cultivars to phytotoxicity can vary.
No residue remains to kill natural enemies; when the oil dries it is no longer toxic.	Petroleum oils are not compatible with sulfur or some other pesticides, causing phytotoxicity.
Adjuvants can improve coverage and efficacy of oils.	
Oils have few negative effects on vertebrates and are safe to handle.	
Oils dissipate after spraying, and leave little or no residue on crops.	
No resistance to oils is known in arthropods.	

to pests. Once the oil dries it is no longer toxic to most natural enemies. The very short residual activity of oil makes it a useful material for IMM programs, although it also means that there is no residual toxicity to the pests.

The efficacy of oil for mite control can be enhanced by the addition of an organosilicone adjuvant such as Silwet® L-77 (Cating et al. 2009). The adjuvant itself has acaricidal properties, and the combination of 0.05% Silwet and 2% oil increased the mortality of mites (and scales and other insects) compared to the application of oil alone. It appears that the enhanced efficacy promoted by the adjuvant is due to the increased coverage achieved.

No resistance to petroleum oils has been reported in mites, perhaps because oils have a relatively short residual activity. Oils are easy to apply, inexpensive, and safe to handle. They are relatively harmless to vertebrates, dissipate quickly after spraying, and leave little or no residue on crops. Organic (plant) oils may be used by organic farmers.

Disadvantages to the use of petroleum oils include their lack of residual activity and their ability to *kill only upon contact*, so thorough and precise coverage is necessary to achieve effective control (Table 5.8). It is likely that oil applications will have to be applied more than once to achieve complete control, and they may have to be applied at higher volumes than other types of pesticides because they have limited residual activity. Phytotoxicity can occur, even with these narrow-range oils, if plants are weakened or under moisture stress. Applications should not be made during droughts or periods of very high temperatures. Some varieties of plants are more susceptible to phytotoxicity, so caution should be taken when using oils for the first time on a particular crop or cultivar. Petroleum oils are not compatible with sulfur or some other pesticides and can cause serious phytotoxicity problems. Petroleum oils are typically applied at rates of 0.25 to 2.00% (250 mL to 2 L of oil per 100 L of water).

5.6.3.14 Control of Ticks and Chiggers

Ticks and chiggers can be repelled from attacking humans by the use of DEET (*N,N*-diethyl-3-methylbenzamide). Other repellents that are less effective include citronella and other plant oils and IR3535 (ethyl butylacetylaminopropionate) (Stafford 2004). The application of pyrethroids (permethrin, bifenthrin, cyfluthrin) to clothing (not skin) can kill ticks before they are able to attach to their human host. Many of these products are used on livestock, as well. Other products can be used in the landscape to suppress ticks if applied at the correct time, including carbaryl, chlorpyrifos, diazinon, and carbaryl (Stafford 2004).

5.7 PESTICIDE RESISTANCE

Resistance to pesticides is an increasingly serious problem around the world (Georghiou 1972, Watson and Brown 1977, Georghiou and Saito 1983, Cranham and Helle 1985, Roush and Tabashnik 1990, Tabashnik 1990, Yu 2008). Resistance to one or more pesticides has been documented in more than 440 species of insects and mites. Spider mite and tick species have developed resistance to all classes of pesticides (Thwaite 1991, Stumpf and Nauen 2001); for example, high levels (>2000-fold) of resistance to clofentezine were identified in the European red mite in apple orchards in Ontario, Canada, after about 5 years of use (Pree et al. 2002). The same populations were resistant to hexythiazox and to cyhexatin and fenbutatin-oxide. Resistance to hexythiazox and clofentezine has been found in European red mite in Australia, as well (Thwaite 1991).

Resistance is the decreased response of a population to a pesticide or control agent after its application. It is an evolutionary or genetic response to selection, and most scientists assume that resistance alleles are present in the population and that the pesticide applications select for those individuals that carry the rare resistance alleles. It is possible, however, that some pesticides are

Table 5.9 Genetic Assumptions that Underlie Resistance Management Models

Model	Assumptions are essential to the success of each model; if violated, the program may not delay resistance.
Mixtures	Resistance to each product is monogenic (determined by one gene). No cross-resistance occurs between the products in the mixture. Resistant mites are rare in the population. Products have equal persistence in the environment. Some mites remain untreated (in refuges). Resistance is functionally recessive (only homozygotes survive exposure).
Rotations	The frequency of individuals resistant to one product will decline during use of the alternative product, which will be true if there is negative cross-resistance (which is rare), a substantial fitness cost is associated with resistance, or immigration of susceptible individuals occurs.
Mosaics	Susceptible individuals are maintained and able to move into surrounding patches. Negative cross-resistance may be required or fitness costs may be associated with the resistance.
High-dose approach	Complete coverage is assumed, which may be very difficult to obtain, as well as the effective kill of all individuals.[a]

[a] High-dose approach ignores negative effects on secondary pests, natural enemies, or the environment.

Source: Adapted from Tabashnik, B.E., in *Pesticide Resistance in Arthropods*, Roush, R.T. and Tabashnik, B.E., Eds., Chapman & Hall, New York, pp. 153–182; Hoy, M.A., *Phil. Trans. R. Soc. Lond. B*, 353(1376), 1787–1795, 1998.

mutagenic and can actually induce genetic changes that result in resistance to the pesticides. **Cross-resistance** is a genetic response to selection after exposure to one compound that generates resistance to both that compound and to other compounds. **Multiple resistances** involve resistance to different compounds due to the coexistence of different resistance mechanisms in the same individuals. Multiple resistances usually are generated by sequential or simultaneous selection by more than one type of pesticide. The development of multiple resistances can occur even if rotations or mixtures are used in an attempt to manage resistance (Table 5.9).

Tolerance is an innate ability to survive a given toxicant dose without prior exposure and evolutionary change. Different species have naturally occurring differences in their responses to pesticides. Recent data indicate that different host plants can affect the susceptibility of *Tetranychus urticae* to different pesticides. Presumably, the differential responses are related to different levels of metabolizing enzymes in the mites induced by the plants. Thus, physiological responses of herbivores to host plants may lead to enhanced metabolism of pesticides because mechanisms that function in the detoxification of plant allelochemicals may be effective in detoxifying pesticides (Yang et al. 2001).

One of the stars in developing resistance is *Tetranychus urticae*, an important pest of fruits, vegetables, grapes, and ornamentals. Failures in chemical control of spider mites have been reported for organophosphates, dicofol, organotins, hexythiazox, clofentezine, and abamectin only a few years after their introduction. Resistance mechanisms in mites are similar to those found in insects, including reduced penetration of acaricides through the mite cuticle; enhanced degradation through esterases, glutathione *S*-transferases, or cytochrome-P450-dependent monooxygenases; and a change in the target site so the target is no longer sensitive to the pesticide (Knowles 1997).

5.7.1 Managing Resistance May Be a Myth

Resistance is especially serious in spider mites, ticks, and other mites that are treated frequently with pesticides. Resistance in spider mites is especially likely because tetranychids are haplodiploid, and any resistance alleles in the males are exposed to selection every generation (see Figure 6.2 in Chapter 6). Recall that diplodiploid organisms (which include most animals) have two sets of

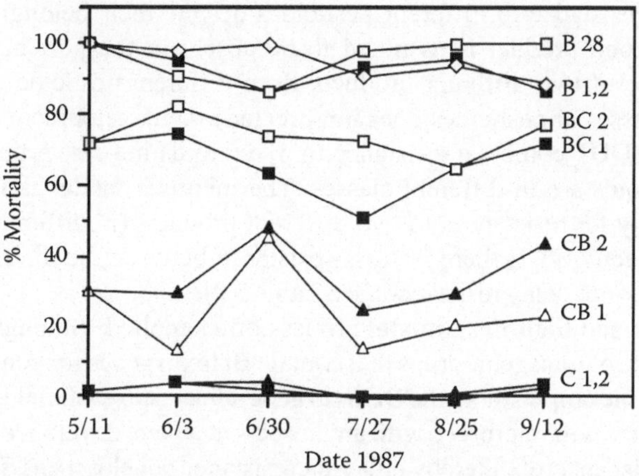

Figure 5.5 The dogma that resistance genes *always* confer lowered fitness is not true. It is assumed that a product can be reused after several years of no use because the former resistance will be lost if there is no selection. These data show that propargite resistance was very stable in this population of *Tetranychus pacificus*, even without selection on the heterogeneous colonies over 20 generations in greenhouse trials. Lines C1 and C2 are the susceptible Chapla colony; lines CB1 and CB2 are a colony derived from F_1 females from the cross Chapla female × Bidart male. Lines BC1 and BC2 are from the reciprocal F_1 cross (Bidart female × Chapla male). B1 and B2 are of the resistant (Bidart) colony without additional selection, and B28 was selected 3 times more with propargite. (Adapted from Hoy, M.A. and Conley, J., *J. Econ. Entomol.*, 82, 11–16, 1989. With permission.)

chromosomes, with one set being inherited from their fathers and one set from their mothers. Each chromosome set has the same genes, but differences in the genes may occur due to mutations so the mite may inherit a different **allele** (or version) of a specific gene from its mother or father. However, haploid male spider mites have only one set of chromosomes that they inherit from their mother. If a male inherits a resistance allele (a version of the normal gene that confers resistance), then it will be expressed in that haploid male and can be selected for rapidly. Haplodiploidy, in combination with the ability of tetranychids (and other mites) to inbreed, can lead to rapid fixation of rare recessive resistance alleles.

A number of genetic models have been developed that attempt to provide advice regarding how to manage resistance, although the better term might be to *delay* resistance (Hoy 1998). Two tactics, rotations and mixtures, are most often recommended to growers, but they may not actually delay resistance, especially if assumptions regarding a lack of cross-resistance and fitness costs do not prove to be true (Table 5.9).

The **rotation of pesticide classes model** assumes that the resistance allele frequency for product A will decline during the period when product B is being used. During the interval when product B is being used, no selection for resistance to product A is being conducted, so resistance allele A will not increase (assuming no cross-resistance). If there is a fitness cost to the resistance allele for product A, then the frequency should even decline. Likewise, the rotations model assumes that there is a **fitness cost** (a reduction in an organism's ability to survive and reproduce associated with a mutation in a gene) associated with the resistance allele. Although fitness costs can occur, not all resistance alleles cause reduced fitness (Figure 5.5). In addition, fitness costs can occur early on, but over time selection can operate to eliminate the fitness costs by selecting on other genes that improve survival and reproduction. The rotations model also assumes that susceptible mites can migrate into the crop to dilute the resistance alleles, but if crops are treated with the same products over a wide geographic area this may not be possible or sufficient. If the assumptions of a lack of cross-resistance and the presence of fitness costs are valid, then the rotations model may work to delay resistance.

Cross-resistance (resistance to different pesticides despite their belonging to different chemical classes) to the second product is assumed to be absent in the pest in the **mixtures model**, which involves the use of two different products from different pesticide classes. Unfortunately, the use of different classes of pesticides does not preclude the development of cross-resistance; for example, resistance to DDT confers a resistance to pyrethroids in some arthropods despite the fact that DDT and pyrethroids are in different classes. The mixtures model assumes that it is difficult to select simultaneously for resistances to two different products (in different classes and with different mechanisms of activity). If there *is* cross-resistance between the different products, then the mixtures approach may not delay resistance (see Table 5.9).

The **mosaic model** and **high-dose model** are less often applied, although the mosaics model is being applied in the case of transgenic crops that contain Bt toxin genes for controlling insect pests. By maintaining refuges of the crop without the Bt toxin gene adjacent to each field, it is hoped that susceptible insects feeding there will interbreed with any insects that have developed resistance (which must be recessive) in the adjacent crop, thereby delaying resistance development. The high-dose approach is very difficult to achieve because it is difficult to obtain *total* coverage of the crop at a high dose (especially on tall fruit trees). The high-dose approach is likely to disrupt natural enemies, leading to secondary pest outbreaks, and to create environmental and human health concerns (see Table 5.9).

It is difficult to anticipate whether a particular resistance management approach will work in one locality based on the results of a resistance management program in another (Hoy 1998). The reason is that the genetic basis of resistance to pesticides can be different in different populations and species (the resistance to the same product could be dominant in one location, recessive in another, and based on different modes of action in the two locations). If resistances are different in mechanism or mode of inheritance, a particular resistance management model that was developed in one region likely will not work in another.

A common question growers ask is whether it is possible to reuse a pesticide if a resistant population has not been exposed to the product for some years. In general, the resistance allele probably is present in low frequencies (perhaps 5%) in the population even after several years without selection with the pesticide. It is possible to reselect a fully resistant population with the application of a few sprays (Edge and James 1986), although exceptions may occur (Flexner et al. 1988, 1989).

Only a few field trials have been conducted to compare the results of various resistance management approaches. Flexner et al. (1995) monitored resistances in the two-spotted spider mite over 7 years in Oregon pear orchards when mites were treated with one of five treatments: (1) rotation of organotins (cyhexatin, fenbutatin oxide) with hexythiazox, (2) continuous hexythiazox use, (3) alternation of both hexythiazox and organotin within a season, (4) between-year rotations of both organotins and hexythiazox, and (5) half rates of both types of compound in replicated field plots where acaricides were applied twice a season. The authors concluded that, "Overall, use in the field was not extended by rotations or by half rate combinations compared with consecutive uses, but benefits from these programs may occur because of slow registration of new acaricides. Alternate, consecutive uses may give 8 years of control (if hexythiazox is used first) compared with 5 to 6 years of control for other programs. Resistance to organotins conferred cross-resistance to hexythiazox." Clearly, additional replicated field trials on resistance-management models for spider mites are needed, but these data do not suggest that rotations are especially useful.

5.7.2 Methods for Evaluating Resistance

A variety of methods are used to assess resistance to pesticides in mites. The test method chosen will depend upon the goals of the researcher, and each method has strengths and weaknesses (Robertson et al. 2007). Bioassays must be carefully designed to obtain statistically valid data (Abbott 1925). Because resistance is a genetically determined change in the population's ability to tolerate a pesticide, one must have at least two different populations to test—one that is putatively resistant and one that exhibits the normal, wild-type response. Unless these two populations can be compared under identical laboratory conditions, it can be difficult to document resistance. Historical data are of questionable value in assessing whether a population is resistant, because it is very difficult to conduct identical bioassays in two different laboratories, even when attempts are made to use the same methods.

Very small differences in techniques may result in very large differences in toxicity responses. Spider mite populations tested on smooth leaves may respond differently than the same populations tested on the same plant species, but on a cultivar with hairy leaves. Small differences in formulations and temperature influence the responses of mites to pesticides. Small differences in age or feeding status also influence toxicity responses. The time of day at which the tests are conducted can affect results; mites exhibit **diel periodicity**, like most organisms, and their detoxification rate varies at different times of the day. Ideally, conclusions about resistance should be based on comparative data obtained by the same researcher under identical conditions.

The apparent failure of a product to control a mite population under field conditions is not adequate evidence for resistance. **Field failure** is a reason to investigate further, but field failures can occur for reasons that have nothing to do with resistance. Failures could occur because the applicator may have mixed the product improperly, coverage may have been inadequate, the pH of the water used to mix the pesticide could have altered the toxicity of the product, or the product could have been old or degraded due to improper storage. Spray-rig nozzles may have been calibrated incorrectly, or the grower may have driven the rig too fast so too little product was applied.

5.7.2.1 *Slide-Dip Bioassays*

Slide-dip bioassays of adult females of spider mites and phytoseiids have been proposed as a standard method for assessing resistance or tolerance. This method involves placing healthy young adult females on their backs on double-sided sticky tape applied to glass microscope slides and dipping the slides into a specific pesticide concentration (often a serial dilution series), then holding the slides under controlled temperature and relative humidity before scoring. This method has the virtue of being relatively rapid and easy to conduct and allows comparisons of different populations under consistent conditions; however, measuring toxicity to adult females after 24 or 48 hours is not an appropriate assay for some pesticide types (e.g., ovicides, growth regulators). The results probably bear little relation to the field toxicity of the product because it is rare for coverage to be complete in the field, and rain or ultraviolet radiation that can degrade pesticide residues is not incorporated into most laboratory trials. Slide dips, because they stress the mite and involve complete coverage, often cause high rates of mortality at relatively low concentrations. As a result, many products appear more toxic to mites using this assay than they would under field conditions, where mites can feed and move around and coverage is rarely complete. The slide-dip method is a good *comparative* method when testing a susceptible and a resistant strain, but this method may not be useful in predicting whether the resistance level induced is relevant to field application rates.

5.7.2.2 *Leaf-Dip or Leaf-Spray Bioassays*

Leaf-dip or leaf-spray bioassays involve placing mites on leaf disks, which are then sprayed or dipped into a specific concentration of pesticide. This type of bioassay provides an exposure that is more similar to what the mites would experience under field conditions, and it is possible to measure survival, fecundity, and ability of progeny to successfully develop on pesticide residues. The results, however, do not give a reliable way to predict whether the putatively resistant population will survive at field rates, because the toxicants are not exposed to rain, sun, and other hazards. As a result, resistant mites may not survive the field rate in the tests but could do so in the field.

5.7.2.3 *Whole-Plant Bioassays*

Whole-plant bioassays involve spraying entire plants and can be fairly realistic, unless the plants (and pesticide residues) are not exposed to sunlight or rain. This test has limitations similar to the leaf-dip or leaf-spray assays.

5.7.2.4 *Field Trials*

Field trials are the most realistic method for assessing resistance, but it can be difficult to determine why the predators or spider mites died (e.g., did tolerant predators fly in and eliminate the pest?). If adequately replicated over time and space, field trials provide very relevant information. The relevance of application method (high or low volume), coverage, formulation, and droplet size can be assessed. Unfortunately, field trials are expensive to carry out, so the methods described above are used to save time and funds.

5.7.3 Pesticides as Last Resort

Although pesticides will not be eliminated in agricultural crop production, they "should not automatically be given the highest priority. Whether they should be considered tools of last resort depends on features of the particular system in which pest management is being used" (National Research Council Commission on Life Sciences 2000).

5.8 HOST-PLANT RESISTANCE

Plants have evolved under multiple selection pressures including predation by arthropods. Plants escape by emigration to new habitats via seeds or by evolving mechanisms to tolerate, reduce, or repel the attacks of arthropods. The heritable traits of a plant that reduce the likelihood that the plant will be eaten by an arthropod are known as **host-plant resistance**. Plants can become resistant by a variety of mechanisms, and host-plant resistance can be an important method of managing pests in agriculture (Painter 1951, Stern et al. 1959, Maxwell and Jenkins 1980, dePonti 1985). Extensive research has been conducted to determine how plants resist attack by insects and mites, although relatively few crops have been selected specifically for resistance to plant-feeding mites. More often, cultivars are developed for other reasons and subsequently are shown to have differing degrees of tolerance or resistance to mites.

Painter (1951) categorized three broad types of plant resistance: tolerance, nonpreference, and antibiosis. **Tolerance** describes the ability of a plant to repair or endure damage by the arthropod. Tolerance is accomplished by replacing damaged tissue or lost nutrients. The plant actively compensates for the damage but does not affect the arthropod directly. **Nonpreference** affects the behavior or biology of the arthropod. Nonpreference is due to physical or chemical characteristics of the plant that repel or cause an arthropod to prefer another plant for feeding, shelter, or oviposition. Heavy pubescence or spines may render the plant repellent. Chemicals in the plant that have a repellent effect are known as secondary plant compounds. These compounds include essential oils, alkaloids, quinones, and glycosides. **Antibiosis** occurs when the physical and chemical traits of a plant result in decreased survivorship, fecundity, or developmental rate of the arthropod. Antibiosis may be caused physically by heavy pubescence, increased thickness of the cuticle, or sticky exudates that slow feeding and subsequent growth of the arthropod. Antibiosis may also be caused by the presence of secondary plant compounds or by the lack of a nutrient needed by the arthropod, such as an essential amino acid, that reduces the growth or reproduction of the insect or mite.

Different varieties of a crop could utilize more than one resistance mechanism. A crop may have varieties that are completely avoided by the arthropod (nonpreference). There may be varieties of the same crop that are partly avoided by the arthropod (lower level of nonpreference), but antibiosis then affects those individuals that do feed. Such a combined type of resistance may be common, with nonpreference, antibiosis, and tolerance interacting to provide useful levels of resistance. Various plant parts, such as roots, stems, and leaves, may have different modes of resistance. Resistance mechanisms also may change as a plant matures.

When breeding for host-plant resistance, crop breeders should consider the importance of plant attributes, such as domatia, glandular trichomes, pubescence, and extrafloral nectaries, with regard to enhancing the roles of biological control agents (Agrawal and Karban 1997, Cortesero et al. 2000). Plants benefit from both chemical and morphological defenses against plant pests; however, natural enemies use chemical cues produced by plants after attack by their pests. Plants may provide shelter (domatia) and supplemental food sources for natural enemies. Thus, crop breeders and acarologists need to share information to enhance the management of plant pests.

Induced host-plant resistance (immunization or vaccination) can result from the inoculation of bacteria, fungi, and viruses into the plant, as well as by feeding by arthropods on the plant (Karban and Carey 1984, Karban 1986, Karban and Meyers 1989, Karban and English-Loeb 1990). Attacks by herbivores, such as insects or mites, can induce chemical and physical changes in host plants. The induced changes are associated with lower rates of feeding by arthropods; for example, phytoalexins deter feeding by herbivorous arthropods, although their role is not well understood. In some cases, plants that have been artificially defoliated and allowed to refoliate will support smaller insect populations; the mortality, fecundity, and development rates may be lower on these refoliated plants. Caution is necessary when interpreting these experiments because the results from simulated herbivory and real herbivory may be different because herbivores may themselves induce plant changes that reduce feeding damage by arthropods later.

Induced resistance to mite feeding has been reported in several crops (e.g., cotton, grapes). In experiments, mite populations grew more rapidly on new growth of cotton seedlings never exposed to mites than on new growth of plants whose cotyledons were exposed earlier to mites (Karban and Carey 1984). Experiments in which a second mite introduction on the exposed plants involved a different species produced the same result, indicating that the unknown substance responsible for the response is transported systemically among cotton seedling leaves. Vineyards in the San Joaquin Valley of California that contain Willamette spider mites (*Eotetranychus willamettei*) tend not to have outbreaks of Pacific spider mites (*Tetranychus pacificus*), which are the more serious pests.

Karban and English-Loeb (1990) purposefully introduced Willamette mites into a commercial vineyard that had a history of chronic problems with Pacific mites. The Pacific mite populations were reduced significantly on rows on which the releases were made compared to the control rows. The following year, however, there were no differences in mite populations between the two treatments.

If induced resistance is to become a practical IMM tool, the induced resistance must be repeatable in the field and must not require a high level of damage to induce the desired response. Damage must not reduce growth or yield of the crop. Induced resistance has not yet been used in IMM programs, but the phenomenon indicates that plants have an ability to respond to a variety of stresses (e.g., mechanical wounding, application of plant growth regulators, herbicides) that affect performance of plant-feeding insects and mites. In the future, cultural manipulations of the host plant may become a tool for managing economically important pests.

5.9 SAMPLING AND MONITORING METHODS

An IPM program should be based on knowing what pests (and natural enemies) are present in the crop and how many are present. Calendar applications of pesticides generally are not appropriate in an IPM program. Knowing what pests are present and how many there are must be combined with the knowledge of how many mites the crop can tolerate without causing economic damage. Thus, monitoring is a key component of IMM and can be done to answer a number of questions, ranging from very basic to applied. Researchers may monitor crops using time-consuming counts of individual mites on a per-leaf basis, but growers or pest control advisors may want to use more rapid, yet reasonably accurate methods such as presence–absence (or binomial) sampling methods.

Researchers often collect foliage from the crop and place it into plastic bags that are sealed, immediately refrigerated, and held in ice chests or refrigerators until the plant-feeding mites (and their natural enemies) can be counted. When counting, individual leaves (top and bottom) may be scanned under a dissecting microscope, and the number of eggs, larvae, nymphs, and adults of both pest mite and predator are usually recorded on a leaf-by-leaf basis. Monitoring of predator and prey populations on individual leaves over time allows researchers to determine the distribution patterns of both predators and their prey as densities change, and whether the predators are suppressing the pest mites. Once this detailed information is obtained, indices can be developed that will allow growers or pest control advisors to use a presence–absence system for that crop.

Presence–absence or binomial sampling of mites on crops is a method for making pest-management decisions that relies on the relationship between the pest population density and the number of samples with *any* individuals. Using a hand lens in the field, researchers note either "leaves infested" or "not infested." In some cases, the sampling scheme involves monitoring only the pest, but in other crops it includes a decision tree for the presence of natural enemies. Presence–absence monitoring methods are considerably more rapid than the method of counting mites on foliage using a dissecting microscope (Wilson et al. 1981, 1983, 1984, Jones and Parrella 1984, Zalom et al. 1984, 1986, Hergstrom and Niall 1990, Hepworth and MacFarlane 1992, Harris et al. 2000, Opit et al. 2003, Hall et al. 2007). In crops where a presence–absence sampling method has not been developed, inspecting a portion of the leaf can reduce counting time in the laboratory, although this may introduce some error.

A machine that brushes mites from leaves may provide a good estimate of the population densities (Henderson and McBurnie 1943, Putnam 1966). Mite-brushing machines have two brushes with soft bristles that rotate (Figure 5.6). Plant leaves are passed between the rollers at the top of the machine, and the rollers brush the surface of the leaf, removing the mites onto a glass plate positioned below on a turntable that rotates so the mites are evenly distributed. The glass plates can be coated with an adhesive to fix the mites, but if the plates are placed into a refrigerator immediately an adhesive is not necessary, and subsequent counting can be done under a dissecting microscope while the mites are still inactive. A grid can be placed under the glass plate and only a portion of the plate can be counted to reduce time. For example, assuming that the rotating glass plate has an even distribution of mites on it, counting every fourth section would allow you to estimate total mites per leaf, per plant, or per plot when multiplied by four. The actual sample size can be based on random samples from individual plants or field sections. This count is used to estimate the mite population density based on the sampling universe, sample size, and sampling technique.

Other techniques for finding, counting, and estimating populations of mites on crops include pressing crop leaves between blotter paper and counting the blotches left behind by crushed mites and eggs (Venables and Dennys 1941), washing mites from leaves with solutions and aliquoting (Leigh et al. 1984, Zacharda et al. 1988, Jedicskova 1997), and photographing the leaves (Sircom 2000). All sampling methods for arthropods must be compared to reliable estimates of populations in the environment to establish reliability (Southwood and Henderson 2000). See Southwood and Henderson (2000) for a detailed discussion of monitoring methods.

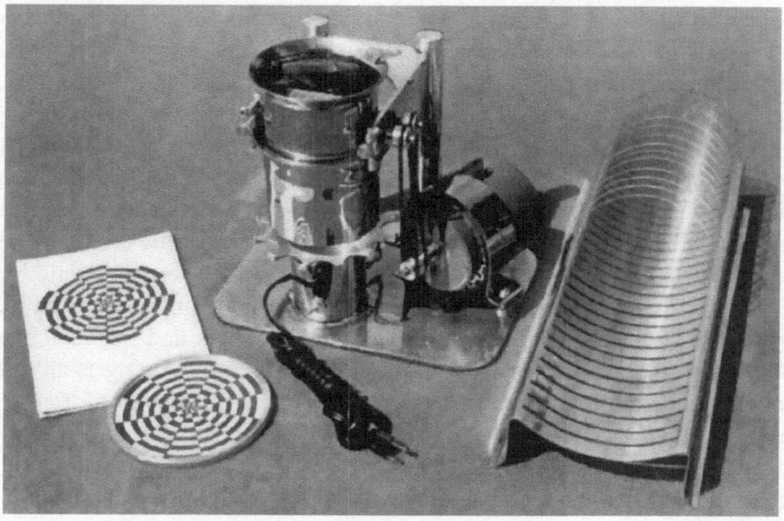

Figure 5.6 A mite-brushing machine and its components. On the left are grids that allow keeping track of the counts in the various regions of the glass plate. Alternatively, it is possible to count some fraction of the total number of sections and multiply to estimate the total number of mites per unit of sample. The machine has two removable metal cylinders that slide into the stand to surround the brushes and to ensure that the mites are directed evenly onto the glass plate below. At the top are two soft brushes that rotate and remove mites from leaves that are placed between the brushes. The glass plate at the bottom is rotated so the mites are distributed evenly. The metal cylinders and brushes can be removed for cleaning. The glass plates (right in the metal tray) can be lightly greased or placed into a refrigerator to immobilize the mites for counting.

5.10 CROP PEST CONTROL CONSULTANTS AND INTEGRATED PEST MANAGEMENT

Crop pest control consultants are specialists in pest management who usually have earned a master's or doctoral degree in agronomy, horticulture, entomology, plant pathology, weed science, or nematology. Some are employed by large agricultural companies to provide pest-management advice for the company's agricultural production systems. Others may go into business for themselves and provide consulting services to agricultural producers. In some cases, companies hire a number of consultants to provide services to multiple farm managers or farmers. Crop pest control consultants provide a service to farm managers, who must deal with diverse and complex pest and disease problems, in addition to addressing labor, economics, agronomy, and marketing issues. The benefits derived from crop pest control consulting services are expected to include improved crop production and reduced pest-management costs in order to justify the cost of the services.

Many crop pest control consultants specialize in insect and mite problems, although they also must be knowledgeable about all aspects of the relevant crop production practices, including fertilization, irrigation, and weed, nematode, and disease control. Crop pest control consultants should be familiar with potential insect and mite pests in the crop so they can accurately identify them and their natural enemies; they should monitor their population fluctuations, estimate when the economic injury level (if known) will be exceeded, and be able to evaluate all pest management options available to them. Consultants need to be familiar with the most recent research and improvements in pest management tools.

The crops must be monitored in a timely and efficient manner so the consultant can inform the farm manager of the need for some type of intervention to reduce the potential for negative economic effects on crop production. Monitoring may involve sampling plant parts, soil, or traps for the pests at the appropriate season. Adequate sample sizes must be taken so reliable conclusions can be made. Most consultants will describe the options available to remedy a pest problem, although many pest control consultants consider it unethical to provide the actual pest control service (such as pesticide applications) because this could be construed as a conflict of interest.

Pest control consultants should know what management options are available and suitable for each pest-management situation. This includes knowing which pesticide products are registered for use on the crop, the proper application methods, and the possible unintended side effects of the pesticide, such as toxicity to natural enemies, phytotoxic effects on specific crop varieties, incompatibility with other pest control practices, and unintended effects on the environment. If augmentative biological control is an option, then the consultant should provide information on the appropriate natural enemy to release for the target pest, effective release rates and methods, and monitoring methods for evaluating the efficacy of the natural enemy. If mass trapping is an option, then the consultant should provide information on effective trap placement, number of traps necessary, and sampling schedules.

In some locations, pest control consultants are licensed. As a result, growers may find it easier to evaluate the qualifications of consultants with whom they have no prior experience. Typically, pest control consultants are required to obtain continuing training to maintain their licenses. Approved training may be available from seminars and courses offered by universities, cooperative extensions, or state or federal departments of agriculture. Pest control consultants are members of professional organizations offering a platform for the discussion of a variety of professional issues, such as obtaining professional insurance that would provide assistance if legal actions are taken against a consultant should crop losses occur. Ethical issues are also discussed, and most organizations have a code of ethics providing guidelines regarding the consultant's relationship to the public, to the employer or client, and to each other. Many pest control consultants have degrees in specific disciplines (such as entomology), but most must still obtain a considerable amount of on-the-job training because agricultural pest problems, and solutions, can be quite site specific. Thus, many new consultants work for an established consultant to gain experience with local pest problems for a season or two before striking out on their own.

Outstanding pest control consultants will be highly knowledgeable about all pest management options, including cultural controls, host-plant resistance, biological control, and least-toxic methods of chemical control. Consultants or their assistants monitor fields sufficiently often that solutions can be provided in a timely manner. Ideally, pest problems will be anticipated, and, in consultation with the farm manager, cultural controls or other controls will be introduced before the problem occurs. Spider mites, for example, are often more serious pests in the margins of crops grown along dusty roads. Reducing the dust can significantly reduce the pest spider mite populations, so the pest control consultant may recommend paving a road or other practices that will reduce dust levels.

Pest control consultants need to have excellent communication skills because most farm managers want the best and most complete information transmitted to them in a clear and timely manner. This can take the form of newsletters, personal reports, written reports, or web-based newsletters. Sometimes the consultant must simply indicate that the information is unavailable or inappropriate for the particular situation facing the farm manager. Such areas of uncertainty must be described in an open and honest manner so that the farm manager can trust the advice of the consultant.

Often, farm managers hire pest control consultants because they want to change their pest management practices, perhaps from a pesticide-based management strategy to an integrated pest management strategy, in which pesticides are used less often and only when other management tactics are unable to maintain low pest populations. Such a change in strategy must be approached carefully, with both the farm manager and consultant listening carefully and learning to trust each other. Unfortunately, IPM is sometimes as much art as science because pest problems differ slightly in every field every year. Cropping systems are dynamic, with ever-changing weather, economic, environmental, and legal issues. Most consultants find that a hands-on, one-on-one, field-by-field approach is necessary when a grower adopts an IPM-based pest management strategy. A change in strategy may require several years to achieve.

Several professional organizations exist in the United States, including the National Alliance of Independent Crop Consultants (NAICC), which was founded in 1978 and consists of more than 450 members in 40 states and several foreign countries. The Certified Professional Crop Consultant (CPCC) and the Independent Certified Professional Crop Consultant (CPCC-I) programs are administered by the NAICC.

Crop consulting is a challenging and demanding profession, requiring considerable knowledge, interpersonal skills, and site-specific experience. The goal of the consultant is to help the farm manager grow the best crop possible under the conditions in that specific field.

SELECTED REFERENCES

Abbott, W.S. 1925. A method of computing the effectiveness of an insecticide. *J. Econ. Entomol.* 18:265–267.

Agrawal, A.A. and R. Karban. 1997. Domatia mediate plant–arthropod mutualism. *Nature* 387:852–563.

Barbosa, P. (ed.) 1998. *Conservation Biological Control*. San Diego, CA: Academic Press.

Bellows, T.S. and T.W. Fisher (eds.) 1999. *Handbook of Biological Control: Principles and Applications of Biological Control.* San Diego, CA: Academic Press.

Boucias, D.G. and J.C. Pendland. 1998. *Principles of Insect Pathology*. Boston, MA: Kluwer Academic.

Bruin, J. and L.P.S. van der Geest (eds.) 2009. *Diseases of Mites and Ticks*. Heidelberg: Springer.

Cating, R.L., M.A. Hoy, and A.J. Palmateer. 2009. Silwet L-77 improves the efficacy of horticultural oils for control of Boisduval scale *Diaspis boisduvalii* (Hemiptera: Diaspididae) and the flat mite *Tenuipalpus pacificus* (Arachnida: Acari: Tenuipalpidae) on orchids. *Florida Entomol.* 93:100–106.

Cortesero, A.M., J.O. Stapel, and W.J. Lewis. 2000. Understanding and manipulating plant attributes to enhance biological control. *Biol. Control* 17:35–49.

Cranham, J.E. and W. Helle. 1985. Pesticide resistance in Tetranychidae. In: W. Helle and M.W. Sabelis (eds.), *Spider Mites: Their Biology, Natural Enemies, and Control*, Vol. 1B (pp. 405–421). Amsterdam: Elsevier.

Croft, B.A. 1990. *Arthropod Biological Control Agents and Pesticides*. New York: John Wiley & Sons.

Davidson, N.A., J.E. Dibble, M.L. Flint, P.J. Marer, and A. Guye. 1991. *Managing Insects and Mites with Spray Oils*, UC ANR Publication 3347. Oakland: Agriculture and Natural Resources, University of California.

de la Fuente, J., C. Almazan, V. Naranjo, E.F. Blouin, J.M. Meyer, and K.M. Kocan. 2006. Autocidal control of ticks by silencing a single gene by RNA interference. *Biochem. Biophys. Res. Commun.* 344:332–338.

DeBach, P. (ed.) 1964. *Biological Control of Insect Pests and Weeds*. London: Chapman & Hall.

Dent, D. 1995. *Integrated Pest Management*. London: Chapman & Hall.

dePonti, O.M.B. 1985. Host-plant resistance and its manipulation through plant breeding. In: W. Helle and M.W. Sabelis (eds.), *Spider Mites: Their Biology, Natural Enemies, and Control* (pp. 395–403). Amsterdam: Elsevier.

Dieleman, J. and W.P.J. Overmeer. 2009. Preferential mating hampering the possibility to apply a genetic control method against a population of *Tetranychus urticae* Koch. Zeitsch. *Angew. Entomol.* 71:156–161.

Edge, V.E. and D.G. James. 1986. Organotin resistance in two-spotted mite, *Tetranychus urticae* Koch (Acarina: Tetranychidae) in Australia. *J. Econ. Entomol.* 79:1477–1483.

English-Loeb, G.M. 1990. Plant drought stress and outbreaks of spider mites: A field test. *Ecology* 71:1401–1411.

EXTOXNET. 2010. This site (extoxnet@ace.orst.edu) provides information about pesticides, including concepts in toxicology and environmental chemistry. It is a cooperative effort of the University of California–Davis, Oregon State, Michigan State, Cornell University, and the University of Idaho.

Flexner, J.L., P.H. Westigard, and B.A. Croft. 1988. Field reversion of organotin resistance in the two-spotted spider mite (Acari: Tetranychidae) following relaxation of selection pressure. *J. Econ. Entomol.* 81:1516–1520.

Flexner, J.L., K.M. Theiling, B.A. Croft, and P.H. Westigard. 1989. Fitness and immigration: Factors affecting reversion of organotin resistance in the two-spotted spider mite (Acari: Tetranychidae). *J. Econ. Entomol.* 82:996–1002.

Flexner, J.L., P.H. Westigard, R. Hilton, and B.A. Croft. 1995. Experimental evaluation of resistance management for two-spotted spider mite (Acari: Tetranychidae) on southern Oregon pear: 1987–1993. *J. Econ. Entomol.* 86:1517–1524.

Flint, M.L. and P. Gouveia. 2001. *IPM in Practice: Principles and Methods of Integrated Pest Management*, UC ANR Publication 3418. Oakland: Agriculture and Natural Resources, University of California.

Geier, P.W. 1966. Management of insect pests. *Annu. Rev. Entomol.* 11:471–490.

Georghiou, G.P. 1972. The evolution of resistance to pesticides. *Annu. Rev. Ecol. Syst.* 3:133–168.

Georghiou, G.P. and T. Saito (eds.) 1983. *Pest Resistance to Pesticides*. New York: Plenum Press.

Gould, F. and P. Schliekelman. 2004. Population genetics of autocidal control and strain replacement. *Annu. Rev. Entomol.* 49:193–217.

Grosscurt, A.C., R.A.J. Wixley, and M. ter Haar. 1994. Cross-resistance between flucycloxuron, clofentezine and hexythiazox in *Panonychus ulmi* (fruit tree red spider mite). *Exp. Appl. Acarol.* 18:445–458.

Gurr, G. and S. Wratten (eds.) 2000. *Biological Control: Measures of Success*. Dordrecht: Kluwer Academic.

Gurr, G.M, S.D. Wratten, and M.A. Altieri (eds.) 2004. *Ecological Engineering for Pest Management: Advances in Habitat Manipulation for Arthropods*. Melbourne, Australia: CSIRO.

Hajek, A.E. 2004. *Natural Enemies: An Introduction to Biological Control*. Cambridge, U.K.: Cambridge University Press.

Hall, D.G., C.C. Childers, and J.E. Eger. 2007. Binomial sampling to estimate rust mite (Acari: Eriophyidae) densities on orange fruit. *J. Econ. Entomol.* 100:233–240.

Harris, M.A., M.J. Brewer, and J.A. Meyer. 2000. Presence–absence sequential sampling plan for northern fowl mite, *Ornithonyssus sylviarum* (Acari: Macronyssidae) on caged-layer hens. *J. Econ. Entomol.* 93:544–549.

Henderson, C.F. and H.V. McBurnie. 1943. *Sampling Techniques for Determining Populations of the Red Citrus Mite and Its Predators*, USDA Circular 671. Washington, D.C.: U.S. Department of Agriculture.

Hepworth, G. and R. MacFarlane. 1992. Systematic presence–absence sampling method applied to two-spotted spider mite (Acari: Tetranychidae)on strawberries in Victoria, Australia. *J. Econ. Entomol.* 85:2234–2239.

Hergstrom, K. and R. Niall. 1990. Presence–absence sampling of two-spotted spider mite (Acari: Tetranychidae) in pear orchards. *J. Econ. Entomol.* 83:2032–2035.

Herron, G.A., V. Edge, and J. Rophail. 1993. Clofentezine and hexythiazox resistance in *Tetranychus urticae* Koch in Australia. *Exp. Appl. Acarol.* 17:433–440.

Higley, L.G. and L.P. Pedigo (eds.) 1996. *Economic Thresholds for Integrated Pest Management*. Lincoln: University of Nebraska Press.

Hoy, M.A. 1985. Recent advances in genetics and genetic improvement of the Phytoseiidae. *Annu. Rev. Entomol.* 30:345–370.

Hoy, M.A. 1998. Myths, models and mitigation of resistance to pesticides. *Phil. Trans. R. Soc. Lond. B*, 353(1376):1787–1795.

Hoy, M.A. and J. Conley. 1989. Propargite resistance in Pacific spider mite (Acari: Tetranychidae): Stability and mode of inheritance. *J. Econ. Entomol.* 82:11–16.

Hoy, M.A. and Y.L. Ouyang. 1986. Selectivity of the acaricides clofentezine and hexythiazox to the predator *Metaseiulus occidentalis* (Acari: Phytoseiidae). *J. Econ. Entomol.* 79:1377–1380.

Hoy, M.A. and K.A. Standow. 1981. Resistance to sulfur in a vineyard spider mite predator. *Calif. Agric.* 35(5/6):8–10.

Hoy, M.A. and K.A. Standow. 1982. Inheritance of resistance to sulfur in the spider mite predator *Metaseiulus occidentalis*. *Entomol. Exp. Appl.* 31:316–323.

Hoy, M.A., R.M. Nowierski, M.W. Johnson, and J.L. Flexner. 1991. Issues and ethics in commercial releases of arthropod natural enemies. *Am. Entomol.* 37:74–75.

Huffaker, C.B., M. van de Vrie, and J.A. McMurtry. 1969. The ecology of tetranychid mites and their natural control. *Annu. Rev. Entomol.* 14:125–174.

Huffaker, C.B., M. van de Vrie, and J.A. McMurtry. 1970. Ecology of tetranychid mites and their natural enemies: A review. II. Tetranychid populations and their possible control by predators: An evaluation. *Hilgardia* 40(11):391–428.

Hull, L.A. and E.H. Beers. 1985. Ecological selectivity: Modifying chemical control practices to preserve natural enemies. In: M.A. Hoy and D.C. Herzog (eds.), *Biological Control in Agricultural IPM Systems* (pp. 103–122). Orlando, FL: Academic Press.

James, D.G. 1997. Imidacloprid increases egg production in *Amblyseius victoriensis* (Acari: Phytoseiidae). *Exp. Appl. Acarol.* 21:75–82.

James, D.G. and T.S. Price. 2002. Fecundity in two-spotted spider mite (Acari: Tetranychidae) is increased by direct and systemic exposure to imidacloprid. *J. Econ. Entomol.* 95:29–732.

Johnson, W.T. 1985. Horticultural oils. *J. Environ. Hort.* 3:188–191.

Jones, V.P. and M.P. Parrella. 1984. Dispersion indices and sequential sampling plans for the citrus red mite (Acari: Tetranychidae). *J. Econ. Entomol.* 77:75–79.

Jedicskova, J. 1997. Modification of a leaf-washing apparatus for the recovery of mites. *Exp. Appl. Acarol.* 21:273–277.

Karban, R. 1986. Induced immunelike resistance to spider mites in cotton. *Calif. Agric.* 40(11/12):13–15.

Karban, R. and J.R. Carey. 1984. Induced resistance of cotton seedlings to mites. *Science* 225:53–54.

Karban, R. and G.M. English-Loeb. 1990. A "vaccination" of Willamette spider mites (Acari: Tetranychidae) to prevent large populations of Pacific spider mites on grapevines. *J. Econ. Entomol.* 83:2252–2257.

Karban, R. and J.H. Meyers. 1989. Induced plant responses to herbivory. *Annu. Rev. Ecol. Syst.* 20:331–348.

Kauffman, W.C. and J.E. Nechols (eds.) 1992. *Selection Criteria and Ecological Consequences of Importing Natural Enemies*. Lanham, MD: Entomological Society of America.

Knowles, C.O. 1997. Mechanisms of resistance to acaricides. In: V. Sjut (ed.), *Molecular Mechanisms of Resistance to Agrochemicals*, Vol. 13 (pp. 57–77). Heidelberg: Springer.

Landis, D.A., S.D. Wratten, and G.M. Gurr. 2000. Habitat management to conserve natural enemies of arthropod pests in agriculture. *Annu. Rev. Entomol.* 45:175–201.

Leigh, T.F., V.L. Maggi, and L.T. Wilson 1984. Development and use of a machine for recovery of arthropods from plant leaves. *J. Econ. Entomol.* 77:271–276.

Luckey, T.D. 1968. Insecticide hormoligosis. *J. Econ. Entomol.* 61:7–12.

Maxwell, F.G. and J.N. Jenkins (eds.) 1980. *Breeding Plants Resistant to Insects.* New York: John Wiley & Sons.

McMurtry, J.A. 1983. Phytoseiid predators in orchard systems: A classical biological control success story. In: M.A. Hoy, G.L. Cunningham, and L. Knutson (eds.), *Biological Control of Pests by Mites* (pp. 21–26), Special Publ. 3304. Berkeley: University of California Press.

McMurtry, J.A., C.B. Huffaker, and M. van de Vrie. 1970. Ecology of tetranychid mites and their natural enemies: A review. I. Tetranychid enemies: Their biological characters and the impact of spray practices. *Hilgardia* 40(11):331–390.

McNab, S.C. and P.H. Jerie. 1991. Leaf scorch responses of "Sensation" and "Bartlett" pear to two-spotted spider mite (Acari: Tetranychidae). *J. Econ. Entomol.* 84:1334–1338.

National Research Council Commission on Life Sciences. 2000. *The Future Role of Pesticides in U.S. Agriculture*. Washington, D.C.: National Academy Press.

Opit, G.P., D.C. Margolies, and J.R. Nechols. 2003. Within-plant distribution of spider mite, *Tetranychus urticae* Koch (Acari: Tetranychidae), on ivy geranium: Development of a presence–absence sampling plan. *J. Econ. Entomol.* 96:482–488.

Painter, R.H. 1951. *Insect Resistance in Crop Plants.* New York: Macmillan.

Pickett, C.H. and R.L. Bugg (eds.) 1998. *Enhancing Biological Control: Habitat Management to Promote Natural Enemies*. Berkeley: University of California Press.

Pimentel, D. and H. Lehman (eds.) 1993. *The Pesticide Question: Environment, Economics, and Ethics*. New York: Chapman & Hall.

Pree, D.J., L.A. Bittner, and K.J. Whitty. 2002. Characterization of resistance to clofentezine in populations of European red mite from orchards in Ontario. *Exp. Appl. Acarol.* 27:181–193.

Preisler, H.K., M.A. Hoy, and J.L. Robertson. 1990. Statistical analysis of modes of inheritance for pesticide resistance. *J. Econ. Entomol.* 83:1649–1655.

Putman, W.L. 1966. Sampling mites on peach leaves with the Henderson–McBurnie machine. *J. Econ. Entomol.* 59:224–225.

Ridgway, R.L., M.P. Hoffmann, M.N. Inscoe, and C.S. Glenister (eds.) 1998. *Mass-Reared Natural Enemies: Application, Regulation and Needs*. Lanham, MD: Entomological Society of America.

Robertson, J.L., N.E. Savin, H.K. Preisler, and R.M. Russell. 2007. *Pesticide Bioassays with Arthropods*, 2nd ed. Boca Raton, FL: CRC Press.

Roush, R.T. and B.E. Tabashnik (eds.) 1990. *Pesticide Resistance in Arthropods.* New York: Chapman & Hall.

Sircom, J. 2000. Photographic sampling: A photographic sampling method for mites on plants. *Exp. Appl. Acarol.* 254:55–61.

Southwood, T.R.E. and P.A. Henderson. 2000. *Ecological Methods*, 3rd ed. Oxford: Blackwell Scientific.

Stafford III, K.C. 2004. *Tick Management Handbook.* New Haven: Connecticut Agricultural Experiment Station.

Stern, V.M., R.F. Smith, R. van den Bosch, and K.S. Hagen. 1959. The integrated control concept. *Hilgardia* 29(2):81–101. (A PDF of this classic paper is on the CD and is provided with permission of the Regents of the University of California.)

Stumpf, N. and R. Nauen. 2001. Cross-resistance, inheritance and biochemistry of mitochondrial electron transport inhibitor–acaricide resistance in *Tetranychus urticae* (Acari: Tetranychidae). *J. Econ. Entomol.* 94:1577–1583.

Southwood, T.R.E. and P.A. Henderson. 2000. *Ecological Methods*, 3rd ed. Oxford, U.K.: Blackwell Scientific.

Tabashnik, B.E. 1990. Modeling and evaluation of resistance management tactics. In: R.T. Roush and B.E. Tabashnik (eds.), *Pesticide Resistance in Arthropods* (pp. 153–182). New York: Chapman & Hall.

Thwaite, W.G. 1991. Resistance to clofentezine and hexythiaxoz in *Panonychus ulmi* from apples in Australia. *Exp. Appl. Acarol.* 11:73–80.

Urbaneja, A., P. Chueca, H. Monton, S. Pascual-Ruiz, O. Dembilio, P. Vanaclocha, R. Abad-Moyano, T. Pina, and P. Castanera. 2009. Chemical alternatives to malathion for controlling *Ceratitis capitata* (Diptera: Tephritidae), and their side effects on natural enemies in Spanish citrus orchards. *J. Econ. Entomol.* 102:144–151.

van de Vrie, M., J.A. McMurtry, and C.B. Huffaker. 1972. Ecology of tetranychid mites and their natural enemies: A review. III. Biology, ecology and pest status and host-plant relations of tetranychids. *Hilgardia* 41:343–432.

Van Driesche, R.G. and T.S. Bellows, Jr. (eds.) 1993. *Steps in Classical Biological Control.* Lanham, MD: Entomological Society of America.

Van Driesche, R.G. and T.S. Bellows, Jr. 1995. *Biological Control.* New York: Chapman & Hall.

van Lenteren, J.C. 2000. A greenhouse without pesticides: Fact or fantasy? *Crop Prot.* 19:375–384.

van Lenteren, J.C. (ed.) 2003. *Quality Control and Production of Biological Control Agents: Theory and Testing Procedures*. Wallingford, UK: CAB International.

van Lenteren, J.C. and J. Woets. 1988. Biological and integrated pest control in greenhouses. *Annu. Rev. Entomol.* 33:239–269.

Venables, E.P. and A.A. Dennys. 1941. A new method of counting orchard mites. *J. Econ. Entomol.* 34:324.

Vincent, C., G. Hallman, B. Panneton, and F. Fleurat-Lessard. 2003. Management of agricultural insects with physical control methods. *Annu. Rev. Entomol.* 48:261–281.

Watson, D.L. and A.W.A. Brown (eds.) 1977. *Pesticide Management and Insecticide Resistance*. New York: Academic Press.

Wilson, L.T., T.F. Leigh, and V. Maggi. 1981. Presence–absence sampling of spider mite densities on cotton. *Calif. Agric.* 35(7):10.

Wilson, L.T., D. Gonzalez, T.F. Leigh, V. Maggi, C. Foristiere, and P. Goodell. 1983. Within-plant distribution of spider mites (Acari: Tetranychidae) on cotton: A developing implementable monitoring program. *Environ. Entomol.* 12:128–134.

Wilson, L.T., M.A. Hoy, F.G. Zalom, and J.M. Smilanick. 1984. Sampling mites in almonds. I. Within-tree distribution and clumping patterns of mites with comments on predator–prey interactions. *Hilgardia* 52(7):1–13.

Yamamoto, A., H. Yoneda, R. Hatano, and M. Asada. 1995. Genetic analysis of hexythiazox resistance in the citrus red mite, *Panonychus citri* (McGregor). *J. Pestic. Sci.* 20:513–519.

Yang, X., D. Margolies, K.Y. Zhu, and L.L. Buschman. 2001. Host plant-induced changes in detoxification enzymes and susceptibility to pesticides in the two-spotted spider mite (Acari: Tetranychidae). *J. Econ. Entomol.* 94:381–387.

Youngman, R.R. and M.M. Barnes. 1986. Interaction of spider mites (Acari: Tetranychidae) and water stress on gas-exchange rates and water potential of almond leaves. *Environ. Entomol.* 15:594–560.

Yu, S.J. 2008. *The Toxicology and Biochemistry of Insecticides*. Boca Raton, FL: CRC Press.

Zacharda, M., O. Pultar, and J. Muska. 1988. Washing technique for monitoring mites in apple orchards. *Exp. Appl. Acarol.* 5:181–183.

Zalom, F.G., M.A. Hoy, L.T. Wilson, and W.W. Barnett. 1984. Sampling mites in almonds. II. Presence–absence sequential sampling for *Tetranychus* mite species. *Hilgardia* 54(7):14–24.

Zalom, F.G., L.T. Wilson, C.E. Kennett, N.V. O'Connell, D.L. Flaherty, and J.G. Morse. 1986. Presence–absence sampling of citrus red mite. *Calif. Agric.* 40(3):15–16.

PART III

Pest Mites and Their Natural Enemies on Plants

Part III introduces the most important families of plant-feeding mites and their natural enemies. Discussion of the pest mites, including the Tetranychidae (Chapter 6), Tarsonemidae (Chapter 7), Eriophyoidea (Chapter 8), Tenuipalpidae (Chapter 9), and Penthaleidae (Chapter 10), is followed by an overview of families often discovered on plants but having different roles (Chapter 11). Some are pests, some serve as alternative prey for predators, and others are predators. The last three chapters of Part III cover beneficial species found on plants, including the Phytoseiidae, the most important acarine predators of plant-feeding mites (Chapter 12). Chapter 13 discusses the insects that are predators of plant-feeding mites, and Chapter 14 deals with pathogens of plant-feeding mites.

[illegible]

[illegible]

CHAPTER 6

Tetranychidae: Premier Plant Pests

6.1 SYSTEMATICS

The Tetranychidae (Actinedida or Prostigmata) include species that are the most important plant-feeding mite pests of agriculture around the world, attacking food crops, trees, and ornamentals (Jeppson et al. 1975, Meyer 1981, Hussey and Scopes 1985, Zhang and Liang 1997, Bolland et al. 1998, Zhang 2003). Outbreaks of spider mites can cause serious yield losses and even death of the crop plant. About 1250 species are known to feed on 3877 different plants, although only about 100 species are considered economically important. The number of species is unclear because the taxonomy of this group is not fully resolved, and additional species may be discovered in new habitats. Some species probably include cryptic species. A relative few species (e.g., *Tetranychus urticae, T. cinnabarinus, T. pacificus, T. kanzawai, Panonychus ulmi, P. citri, Oligonychus punicae, O. coffeae, Eutetranychus orientalis*) are considered major pests in multiple locations and on multiple crops. The most notorious of all is the two-spotted spider mite (*Tetranychus urticae*), which has a worldwide distribution (Figure 6.1).

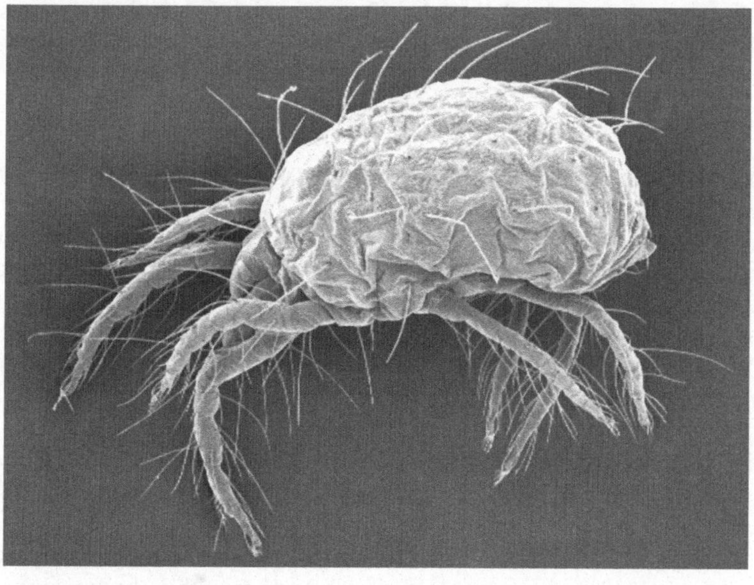

Figure 6.1 Scanning electron micrograph of a female two-spotted spider mite (*Tetranychus urticae*). Spider mites are relatively soft bodied and covered with setae, many of which have sensory functions. (Photograph by Drion Boucias, Department of Entomology and Nematology, University of Florida, Gainesville.)

All tetranychids are plant feeders, ranging in size from about 0.2 to 0.9 mm in length. The chelicerate mouthparts of the Tetranychidae are highly modified for feeding, with stylet-like chelicerae that can penetrate cells and suck out the contents (see Figure S6.1 on the CD and Figure 3.9 in Chapter 3). The Tetranychidae are called spider mites because species in the subfamily Tetranychinae produce silk from glands opening on the tarsus of the pedipalps. Spider mites in the subfamily Bryobinae do not produce silk, however.

The Tetranychidae consists of approximately 70 genera divided into two subfamilies, the Bryobinae and Tetranychinae; most agricultural pests are in the Tetranychinae. Tetranychids are soft bodied, medium in size (for mites), and often brightly colored (red, green, yellow, orange) (see Figure S6.2A–D on the CD). The most abundant and widely distributed species, *Tetranychus urticae*, may be a species complex because it contains distinct populations with different biological attributes (see Figure S6.3 on the CD). This species is found on over 250 plants around the world and is highly variable. In fact, originally *T. urticae* was described under 50 different names. This variability can be important to pest managers because biological attributes of *T. urticae* (often called *T. telarius* in older literature) can affect management decisions. Resistances to pesticides, host-plant range, diapause traits, and other important attributes relevant to pest management may vary in *T. urticae* (and other spider mites) (Helle 1968, Gould 1979, Cranham and Helle 1985, Hill and O'Donnell 1991a,b, Gotoh et al. 1993, 2003, 2007, Kostiainen and Hoy 1996, Tsagkarakou et al. 1996, Kawakami et al. 2009, Li et al. 2009). It is likely that many tetranychid species are genetically diverse; for example, hawthorn spider mite (*Amphitetranychus viennensis*) populations from Japan and France exhibit substantial genetic differences (Navajas et al. 1999). Whether these populations are different biotypes, races, strains, or cryptic species is unresolved. To add to the confusion, some species may be synonyms (Gotoh et al. 2009).

The division of tetranychids into subfamilies, tribes, and genera is based mainly on the empodium (the structure located between the tarsi) of females. To identify spider mites to species, males are required (if they exist). A few species (especially in the Bryobini) are thelytokous, having only females. The shape of the male aedeagus (sclerotized male intromittent organ), the chaetotaxy of legs, the shape of the peritreme, and the shape of the structure on the palpal tarsus used for spinning silk are important characters. For a specialist to identify spider mites, adults of both sexes are required.

6.2 BIOLOGY

The tetranychid life cycle is egg → larva → protonymph → deutonymph → adult male and female. Eggs typically are round (see Figures S6.4 and S6.5 on the CD). There is usually a quiescent stage between each instar before molting occurs. The life cycle is dependent on the temperature, relative humidity, and host plant and can be as brief as 3 to 5 days at high temperatures (25 to 30°C). **Diapause**, a genetically determined state of arrested development, may occur in populations in temperate climates to allow them to survive extreme summer or winter conditions. In greenhouses and in tropical or subtropical environments, reproduction may occur all year. Selection for nondiapausing *Tetranychus urticae* can occur rapidly in greenhouses and can result in populations that are genetically unable to enter diapause.

Unlike many mites, tetranychids have two pairs of eyes closely arranged on the propodosoma (see Figures S6.2D, S6.5, and S6.6B–D on the CD). The eyes of the two-spotted spider mite are simple and consist of a lens and a retina (McEnroe 1969). The anterior eye appears to act as a scanning-point detector; it has a movable lens that is able to shift and contains receptors for both green and ultraviolet light. The dense mass of red oil droplets within the first eye shields the retinal cells (McEnroe 1969). The posterior eye lacks the red oil, suggesting that light need not enter through the lens to stimulate the retinal cells. McEnroe (1969) indicated that the "eyes obviously do not resolve an image but serve as point detectors. The few sensory cells present would restrict the dynamic range of the eyes." Subsequently, McEnroe and Dronka (1969) found that the anterior eyes have receptors for both near-ultraviolet and green spectra, and the posterior eyes have a receptor for the near-ultraviolet spectrum. All live stages have red eyes, and they even can be seen in larvae just prior to hatching from eggs (see Figure S6.5 on the CD).

Most tetranychids are arrhenotokous. Unmated females produce haploid eggs that develop into males. If a female is unmated, she may survive to mate with her son, so inbreeding can be common. Also, because females tend to deposit their eggs on a single leaf, brothers and sisters often mate, providing another opportunity for inbreeding. Despite the frequency of inbreeding in tetranychids, which in many organisms results in the expression of deleterious mutations and death of populations, tetranychids have sufficient genetic variability to adapt rapidly to new environmental conditions (Helle 1968, Helle and Sabelis 1985).

The sexes are dimorphic, with the male being rather triangular in shape and the female broadly oval. Males typically are about half the size of females (see Figure S6.6A–D on the CD). When spider mites mate, the male inserts his aedeagus into the female to deposit sperm by bending his idiosoma up (see Figure S6.6C on the CD). Mated females deposit both fertilized and unfertilized eggs. The sex ratio of progeny produced by mated females is usually three diploid females to one haploid male. Two-spotted spider mite females deposit up to 30 eggs per day for approximately a week, depending on the species and temperature. Between each molt, the mites become inactive (quiescent) and attach themselves to the substrate. Males develop more rapidly than females, and a male may then hover over and guard his deutonymphal sister in order to mate with her as soon as she molts.

Spider mite males guard female deutonymphs in response to the production of a contact sex pheromone (see Figure S6.6A–D on the CD). Adult females also produce the sex pheromone and may mate more than once to have sufficient sperm for fertilizing their eggs (Cone et al. 1971a,b, Sonenshine 1985, Royalty et al. 1992, 1993, Rasmy and Hussein 1994). Several males will attempt to guard the same deutonymph and will fight (see Figure S6.6D on the CD), although males rarely are injured severely. Usually the larger and more aggressive male will win and mate with the female (Potter et al. 1976). Krainacker and Carey (1989) showed that, under laboratory conditions in which males were provided unlimited numbers of virgin females over the 8 days of their adult life, the average male could inseminate 70 females and contributed to the production of 1145 daughters. Spider mite females typically have a preoviposition period of approximately 24 hours. During this time they may disperse by walking or by aerial movements (see Figure S6.7A,B on the CD). On average, *Tetranychus urticae* females live for 7 to 10 days.

The European red mite (*Panonychus ulmi*) is found on deciduous trees (see Figure S6.8 on the CD). European red mite eggs have a layer that is resistant to penetration and attack by chemicals (see Figure S6.9 on the CD). The developing mite embryo waterproofs the summer egg by secreting a wax layer into the inside of the shell about 6 hours before the egg is laid. The shell of the winter egg is similar, but winter eggs are held in the female until a later stage of embryonic development so they also are waterproofed at the base of the egg when it is deposited on bark. Eggs of *P. ulmi* have a small **stipe** (stalk), and females use silk strands to help attach the egg stipe to the host plant (see Figure S6.9 on the CD). Chemical control is difficult because the winter egg is so well protected. Pest managers typically wait to apply pesticides to control the European red mite after larvae have hatched but before damage has been done to the crop.

Life cycles vary with the species, population, climate, and host plant. Some populations, especially in tropical and subtropical areas, remain on their host plants throughout the year, only stopping oviposition when temperatures fall below their developmental threshold. All stages may be present on the plants throughout the year. In temperate climates, spider mites typically enter a diapause during the winter.

Diapause can be induced by environmental cues (temperature, daylength, and food quality) prior to the actual onset of adverse conditions (Lees 1953, Veerman 1977, 1985). Saito et al. (2005) discovered that in some tetranychid species adult females enter diapause even when reared under nondiapause-inducing conditions. Termination of **hibernal diapause** (occurring in the winter) often requires a period of low temperatures and short daylengths before termination can occur. Diapause termination does not happen immediately when diapausing females or eggs are transferred to conditions favorable for development; they must go through a period of diapause development, the length of which is genetically determined and based on local climatic conditions. Populations in different climatic conditions have different intensities of diapause that are appropriate for the local environment (Helle 1968, Kawakami et al. 2009), so the timing of spring emergences will vary from site to site. This variability affects pest management programs directed toward controlling the newly hatched larvae and nymphs of species, such as *Panonychus ulmi*, that overwinter as eggs in diapause. The optimal timing of sprays to control *P. ulmi* larvae varies due to the local climate and the genetic constitution of the population.

Hibernal diapause occurs either in the adult female or in the egg in tetranychids. *Eotetranychus* and *Tetranychus* species overwinter as adult mated females; such females move off deciduous trees or other plants into sheltered sites. *Panonychus*, *Bryobia*, and *Oligonychus* species overwinter in the eggs, which typically are deposited on the trunk of the trees. *Tetranychus urticae* females in diapause are bright red and lack their normal food spots. They are inactive, nonfeeding, and typically found in crevices or other hidden areas. Diapausing *Tetranychus pacificus* females are orange in color and, in California vineyards, are found in cracks in the bark on the trunk of the vine. *Eotetranychus willamettei* females in diapause are typically lemon yellow in color and also are found on the trunks of vines in California vineyards. **Aestivation** is a form of diapause that occurs during the hot dry weather in the summer. *Petrobia latens* and *P. apicalis* are active during the cooler seasons, and their diapausing eggs hibernate during hot, dry summers.

6.3 ROLE OF SILK

Silk protects spider mites from many predator species and from desiccation under low relative humidity (see Figure S6.7A,B on the CD). It may protect them from rainfall and pesticide applications, as well (Mothes and Seitz 1981,Gerson 1985). Applications of acaricides to dense populations of spider mites protected by a silk mat are likely to fail to control the mites due to a lack of penetration. Silk may protect *Tetranychus urticae* when they encounter some pesticide residues (such as synthetic pyrethroids) because the pesticides cause them to spin down to areas lacking pesticide residues. This behavior makes it possible for them to survive some sprays and allows their populations to rebound (Margolies and Kennedy 1988).

Silk is used by some species in dispersal. Some spider mites 'balloon' in a manner similar to that of spiders, but not all spider mite species that produce silk balloon. Webbing also is used in courtship; for example, female deutonymphs of *Tetranychus* species spin a silk web that attaches them to the leaf surface while they are quiescent before undergoing a molt. The females also secrete a sex pheromone, and the web may serve as a pheromone substrate or carrier (Penman and Cone 1972). The behavior of males is modified by the presence of the female's web. After the male finds a quiescent female deutonymph, he spins a thin silk mat over her and guards her until she emerges as an adult. Webbing may help species to displace other mite species that do not web. Species of *Bryobia* do not produce webbing and cannot walk on leaf surfaces webbed by *T. urticae*. Webbing may protect *Tetranychus* species from generalist predators or, in some cases, may be used as cues by predators in locating their prey. Webbing may entangle some predatory phytoseiids, and these predators may avoid foliage with heavy webbing. Other phytoseiids seem to be well adapted to walking through webbing and appear to use the webbing as a cue to search intensively for prey. Ladybird beetles (Coccinellidae) such as *Stethorus*, which are specialist predators of spider mites, can feed under webs.

6.4 DISPERSAL

Spider mites lack wings but they can disperse in several ways, and their dispersal has implications for pest management programs. Spider mite species can invade countries on plants or their products; for example, apples must be disinfected of spider mite eggs before they can be shipped overseas. Pesticide-resistant populations of spider mites can move into new crops or regions, leading to changes in resistance-allele frequencies that can affect acaricide use (Grafton-Cardwell et al. 1991, Dunley and Croft 1992). Field hot spots can be the sources of spider mite infestations over the entire crop if not controlled before they disperse. In the case of invasive species, quarantines and disinfestation methods for crops may prevent some pests from invading new areas, whereas monitoring crops to locate spider mite hot spots can allow spot treatments using predator releases or selective acaricides.

Mites move shorter distances by walking from leaf to leaf on a plant, from foliage to overwintering sites in the soil, or on the trunks of trees or vines. Mites also walk from overwintering sites, even over the soil surface, to new host plants. Contamination of farm equipment, insects, and birds also serves to disperse spider mites to new host plants. In the case of contaminated farm equipment, proper sanitation can reduce movements of spider mites from field to field. Barriers against crawling spider mites are necessary to prevent contamination of greenhouses; having weed-free areas around a greenhouse can help reduce infestations. Longer distance movements of tetranychids occur on plant materials and by aerial means. Spider mite species have moved around the world on plant material, and a number of species (e.g., *Tetranychus urticae*, *T. cinnabarinus*, *Panonychus citri*, *P. ulmi*, *Bryobia rubrioculus*) are so widely distributed that taxonomists cannot determine their areas of origin.

Several species, including *Panonychus citri* and *Eotetranychus sexmaculatus*, lower themselves from the plant on silk threads (ballooning) (see Figure S6.7B on the CD). A slight breeze releases the mites from the host, and the mites can be carried substantial distances. For mites that do not balloon, a very explicit behavior has been observed (Smitley and Kennedy 1985, 1988, Osakabe et al. 2008). Newly mated *Tetranychus urticae* females that have not begun to deposit eggs move to the top of a plant and may form clumps of mites. Slight breezes will disperse the females. Males and immatures are rare in these dispersal groups. *Tetranychus urticae* does not appear to truly balloon because they do not have a trailing strand of silk (Li and Margolies 1994).

Aerial dispersal can be very hazardous. It is unlikely that a dispersing mite can choose to land on a specific host plant. If the mite lands on an unsuitable plant and is unable to relaunch itself into the air for further dispersal or walk to a suitable host plant, that mite may fail to survive and reproduce. Of course, a single mated female mite that disperses from a hot spot in a crop and lands on a surrounding suitable host can initiate a new colony. At present, we do not know if spider mites are able to relaunch themselves should they land on unsuitable plants. Fans that blow mites from hot spots where high-density populations occur can aid the movement of spider mites in greenhouses. Frequent monitoring and control of the hot spots by predator releases or use of selective acaricides can reduce damage to the crop.

Unfavorable host-plant quality due to drought or high population densities or to changes in barometric pressure can result in mass dispersal of spider mites in the field (Smitley and Kennedy 1985, Margolies 1987, Margolies and Kennedy 1988, Li and Margolies 1994). Relatively calm winds seem to stimulate mass dispersal of spider mites in the field, and crop physiological status may trigger aerial dispersal, as well. Aerial dispersal from fields into adjacent crops can result in an explosion of spider mite populations within a short period of time, especially if there are few natural enemies in the adjacent crop. As a result, pest managers should monitor crops often (at least once a week or more frequently) to be sure that such aerial dispersal does not cause economic losses.

6.5 POPULATION DYNAMICS

Reproductive rates in spider mite species that are economic pests are influenced by crop cultivar, temperature, relative humidity, water, fertilizers, and some pesticides (van de Vrie et al. 1972). Some pesticides stimulate fecundity of spider mites or favor them indirectly by killing their natural enemies (Huffaker et al. 1970). The intrinsic rate of increase (r_m) of these pests varies between species and host plants (and even cultivars). *Tetranychus urticae* females have been documented, in the laboratory, to produce more than 312 eggs. On average, *T. urticae* females produce about 200 eggs, resulting in an average of 176 adult progeny within a week. For a review of reproductive strategies in tetranychids that are crop pests, see Sabelis (1985). The biology of the many tetranychid species that are not crop pests has been studied only rarely.

6.6 TETRANYCHID ANATOMY

Internal and external tetranychid anatomy has been reviewed by Blauvelt (1945), Lindquist (1985), and Alberti and Crooker (1985) (also see Figure 3.11 in Chapter 3). Females have a single ovary, a single oviduct, a vagina, and a single seminal receptacle or spermatheca. These fill almost the entire ventral half of the hysterosoma. The reproductive system of males consists of a pair of testes, a pair of vasa deferens, a seminal vesicle, and an ejaculatory duct entering the aedeagus. Sensory receptors of spider mites include two pairs of nonfaceted (simple) ocelli on the propodosoma (see Figures S6.2D and S6.6A–D on the CD). The setae of spider mites are varied and include tactile and chemoreceptors. The palps and legs contain contact chemoreceptors, the palps contain mechanoreceptors, and most of the body and leg setae are mechano- and olfactory receptors. Males use contact chemoreceptors to respond to pheromones produced by quiescent female deutonymphs and adults. The intense colors exhibited by spider mites are caused by carotenoid pigments in the viscera and hemolymph, but the cuticle is usually colorless. In addition, chlorophyll and chlorophyll-derived pigments can be seen in the gut or in food residues. Due to the lack of color in the cuticle, spider mites preserved in alcohol or cleared for slide mounting usually lack color. The brain of *T. urticae* is a single fused ganglion consisting of the supraesophageal and subesophageal ganglia. It also includes optic, cheliceral, pedipalpal, rostral, and leg ganglia. As can be seen in Figure 3.11 in Chapter 3, the esophagus passes through the brain.

6.7 SELECTED SPECIES OF PLANT PESTS BY GENUS

The following sections describe some of the most important plant-feeding tetranychids, organized by genus.

6.7.1 *Bryobia*

Bryobia rubrioculus, the brown mite, is an important pest of pome and stone fruit trees (peaches, almonds) around the world; it is sometimes called the brown almond mite (see Figure S6.10 on the CD). This species does not spin webs. It overwinters as eggs on twigs, and the larvae migrate to buds, where they cause white feeding spots on young leaves. Damaged leaves may bronze. Mites migrate back to the twigs to molt, oviposit, or rest, so pest managers who sample only the foliage may miss a large proportion of the pest population. The brown mite has two generations per year in some parts of the world and up to six in South Africa. Typically, the last generation deposits diapausing eggs on twigs, often as early as June or July in the Northern Hemisphere (early summer in temperate climates), and the eggs hatch the following spring. ***Bryobia praetiosa***, the clover mite, may consist of a complex of races or species. It is often confused with *B. rubrioculus*. *Bryobia praetiosa* is common around the world and feeds on a variety of herbaceous species, including clovers, lawn grasses, ornamental flowers, alfalfa, wheat, rye, barley, and other grains. Damage includes leaf stippling and fine, meandering streaks on leaves. The leaves become yellow or brown and may wilt. *Bryobia praetiosa* can be either **univoltine** or **multivoltine** (one or many generations per year). The adult female is broadly oval, flattened, and dark green to brownish green with pale red legs. The first pair is longer than the body. Newly hatched larvae are bright red but become green after feeding. This species may invade houses in large numbers in the fall and become a nuisance because they leave unsightly spots on the walls if a large number are crushed. They are, however, harmless to humans. This species can overwinter in both the egg and active stages. During the summer, the eggs may enter an aestival diapause. Aestivating eggs hatch in late summer or early fall.

6.7.2 *Eotetranychus*

This genus represents approximately 180 species. ***Eotetranychus willamettei*** is found in vineyards in California, especially early in the spring (Flaherty 1969, Hanna et al. 1997). This species causes minimal damage to most grape varieties in California, but it serves as important prey for the phytoseiid *Metaseiulus occidentalis*, especially early in the growing season. *Metaseiulus occidentalis* subsequently is present in adequate numbers to suppress populations of the more serious vineyard pest, *Tetranychus pacificus*. ***Eotetranychus hicoriae*** is a pest of pecan, hickory, oak, and chestnut trees throughout the eastern United States. Trees heavily infested appear to have scorched foliage in large areas. ***Eotetranychus lewisi*** feeds on the leaves and fruits of a variety of hosts around the world, including poinsettias (see Figure S6.11 on the CD). It also is found on citrus (especially the fruit). Superficially, it can be mistaken for the two-spotted spider mite in appearance, but it is smaller. ***Eotetranychus sexmaculatus***, the six-spotted mite, is a pest of citrus, avocados, maples, pyracanthas, azaleas, and camphor. ***Eotetranychus yumensis*** is a pest of citrus, grain sorghum, grapes, primrose, puncture vine, and *Atriplex* (salt bush) in desert areas of California, Nevada, and Mexico. It produces copious webbing to which dust adheres. These mites increase in the fall, and their numbers remain high during the winter but then decline in spring or early summer. The mites aestivate during the summer under the bark of the trunk.

6.7.3 *Eutetranychus*

Eutetranychus banksi is the Texas citrus mite, which is found in the Americas on citrus, almond, croton, fig, and castor beans (see Figure S6.12 on the CD). ***Eutetranychus orientalis*** is the Oriental red mite, which is a pest of citrus in Asia and the Middle East and South Africa.

6.7.4 *Mononychellus*

The ***Mononychellus*** genus includes the cassava green mite, which will be discussed in Chapter 15 as an example of a successful classical biological control program in Africa (see Figure S6.13 on the CD).

6.7.5 *Oligonychus*

This genus includes ***Oligonychus punicae***, the avocado brown mite, which is a pest of avocados in California and of grapes and pomegranate in tropical Asia and the Americas (see Figure S6.14 on the CD). ***Oligonychus pratensis*** is the Banks grass mite, a serious pest of many grass species and of dates in southern California. Hosts include wheat, corn, sugarcane, sorghum, dates, bluegrass, and Bermuda grass. The Banks grass mite is a serious pest in the central plains states of the United States and in Washington and Oregon on corn. These mites spin copious webbing so the lower portions of the plant may become matted. This webbing protects eggs and makes it difficult for predators or pesticides to control this mite.

6.7.6 *Panonychus*

The fruit tree red spider mite, ***Panonychus ulmi*** (or *Metatetranychus ulmi*), is found around the world on a variety of trees (especially Rosacae), including apples, pears, plum, almonds, peaches, and sometimes grapes. *Panonychus ulmi* overwinters as a diapausing egg, which is deposited on the smaller twigs and branches (Lees 1953). Summer eggs, which develop without interruption, are usually deposited on leaves. Both eggs are similar in appearance, but the winter eggs are larger and more heavily pigmented, and they have a thick wax layer and an intermediate cement layer. The inner layer is very resistant to chemicals (see Figure S6.9 on the CD). The egg is glued to the substrate, and the female provides additional anchorage by arranging threads of silk as guy lines running from the "spike" (stipe) on the top of the egg to the substrate. Winter eggs hatch in spring; the duration of diapause and hatch is synchronized with the local climatic conditions (Tsugawa et al. 1966, Cranham 1973, Herbert and McRae 1982). Models have been developed to predict when the eggs will hatch so pesticide applications can be made at an appropriate time (Broufas and Koveos 2000). ***Panonychus citri***, the citrus red mite, is the most serious mite pest of citrus in California, South Africa, and Japan (see Figure S6.2B on the CD). It is found in Florida, China, South America, and India. In addition to citrus, *P. citri* can be found on almonds, pears, castor beans, and broadleaf ornamentals. This mite is red, and the bases of the setae are red rather than white, as is the case with *P. ulmi* (see Figure S6.8 on the CD).

6.7.7 *Petrobia*

Petrobia latens is the brown wheat mite or onion mite. It is a pest of small grains around the world. It also can cause damage to onions, carrots, cotton, endive, lettuce, iris, gladiolus, sorghum, alfalfa, clover, and strawberry. The mite is a dry-weather pest, and damage is similar to that caused by drought. Heavily infested fields appear dried out even when there is sufficient soil moisture. Females are oval, brown to greenish brown, with pale yellow legs. The first pair of legs is markedly longer than the body. No males are known (thelytoky).

6.7.8 *Tetranychus*

The most widespread and abundant species in this genus is ***Tetranychus urticae*** (also known as *T. telarius*). This may be a complex of species and is highly variable in its biology, behavior, and ecology. The taxonomy is confused in part due to partial reproductive incompatibilities found between some of the populations. In some cases, the incompatibilities between different populations are caused by *Wolbachia* microbial symbionts (Breeuwer 1997, Gotoh et al. 2003, 2007). *Tetranychus urticae* females are straw colored or green, with a dark blotch on each side (sometimes referred to as *feeding spots*). The legs are pale to yellowish (see Figures S6.3 and S6.6B,D on the CD). Adult males are smaller than females and triangular in shape. Initially, these mites feed on lower leaf surfaces but can cover an entire plant as populations increase. Damage initially appears as chlorotic stipples on the leaves, but large areas will turn yellow and curl as feeding damage accumulates. Leaves also may become bronzed, and the plant can defoliate. Webbing is often very visible, giving a shiny appearance to the plant. Other pests in this genus include ***Tetranychus ludeni*** (dark-red spider mite), which is found in warmer parts of the world as a pest of vegetables as well as wild plants. ***Tetranychus evansi*** (the tomato spider mite), found in North and South America, attacks tomato plus other solanceous plants such as potatoes, eggplant, peppers, and sweet potatoes. The females are orange-red, with an indistinct dark blotch on each side and reddish legs (see Figure S6.2C on the CD). Males are straw to orange colored. *Tetranychus evansi* prefers the lower surface of leaves but can occur on both the top and bottom. Damage first appears as stipples that later result in a silvery or yellowish appearance to the leaves. These mites can kill plants rapidly. ***Tetranychus pacificus*** is a pest of grapes, cotton, and other crops in the western United States (see Figure S6.2D on the CD).

6.8 TETRANYCHIDAE AND PLANT DISEASES

Plant disease, in its broadest sense, refers to all injuries or abnormalities, including those caused by tetranychids. Spider mites remove chlorophyll and other cell contents, causing economic injury when high populations inflict sufficient damage over a period of days. The population density and feeding interval required to produce visible or measurable injury are influenced by mite species, variety of plant, temperature, relative humidity, plant vigor, and transpiration rate. The loss of chlorophyll results in chlorotic patches of leaf tissue (stipples) (see Figure S6.1 on the CD). Most damage is caused by a reduced photosynthetic rate. Very high levels of damage can lead to complete defoliation and death of the plant. Anecdotal evidence suggests that spider mites inject toxins or growth regulators while they feed. Little is known, however, about the nature of these chemicals or the precise mechanism by which they are introduced into plants.

Different plants respond very differently to feeding by the same species; for example, Bartlett pear leaves take on a scorched appearance due to the feeding of one or a few *Tetranychus urticae* per leaf, and leaves can drop off the tree with as few as a single mite per leaf, especially if the weather is hot and dry. Leaf scorch is the development of brown necrotic areas that expand from feeding damage and lead to leaf abscission or drying. Leaf scorch can lead to premature flowering in autumn, reduced fruit production, and reduced fruit set. A pear cultivar called Sensation, thought to be a bud mutation of Bartlett, is tolerant of feeding by *T. urticae*. Feeding levels appear to be the same, fecundity of the mites appears to be the same, and mite numbers developing on the leaves appear to be the same. The scorch response thus appears to be a difference in the plant response. Clearly, Sensation would be more suitable for use in an integrated mite management (IMM) program. Managing *T. urticae* and predatory mite populations in Bartlett pear orchards is very difficult because the economic injury threshold is extremely low. Predatory phytoseiids cannot be maintained effectively in the Bartlett pear orchards at the very low prey densities of *T. urticae* that can be tolerated. The rapid development of leaf scorch is most likely a **hypersensitive response** by the plant. Hypersensitivity is associated with a rapid plant reaction to pathogens that result in less favorable conditions for the attacking pathogen. Hypersensitivity to *T. urticae* also has been shown to occur in soybean cultivars, where the rate of increase of the mite population on the sensitive cultivar was reduced by rapid tissue death, resulting in less long-term damage.

By contrast, apple leaves are relatively tolerant of feeding damage by a single *Tetranychus urticae* and show only slight leaf stippling. Different grape varieties in California vary in their response to *Eotetranychus willamettei*. On vigorous grape varieties such as Thompson seedless, *E. willamettei* causes little damage and is considered a beneficial species because it is important prey for the phytoseiid *Metaseiulus occidentalis* (Flaherty 1969, Kinn and Doutt 1972, Hanna et al. 1997). By contrast, *E. willamettei* can cause economic injury to wine grape cultivars that have fewer leaves (Hanna et al. 1997).

Environmental conditions influence the effects of spider mite feeding. Citrus trees can tolerate considerable mite feeding during periods of high relative humidity, but low populations can cause serious water loss during dry conditions, as well as leaf and fruit drop and twig dieback. It is always appropriate to reduce damage by making sure the crop is properly watered. For perennial crops, two

Table 6.1 Methods for Managing Tetranychid Mite Pests

1. Quarantines and disinfestations of plants or plant products can reduce invasions by new species into countries, into greenhouses, and into fields.
2. Proper sanitation (e.g., maintaining plant-free areas around greenhouses, regular maintenance of farm equipment, and eliminating weeds around field margins that are suitable hosts for spider mites) can reduce crop infestations by spider mites.
3. Cultivars or crops should be tolerant of or resistant to spider mites.
4. Irrigation of agricultural crops should be managed to ensure that the crop is properly watered to reduce damage from spider mites.
5. Minimizing dust in irrigated cropping systems will reduce spider mite populations.
6. Monitoring for spider mite outbreaks (hot spots) and treating with natural enemies or with selective acaricides can reduce dispersal into the rest of the greenhouse or field crop.
7. Augmentative releases of natural enemies of spider mites in greenhouses can be very effective if release rates and timing are appropriate.
8. Classical biological control of spider mites can be effective when a new species invades a country.
9. Conservation of natural enemies (e.g., predatory insects, phytoseiids and other acarine predators, pathogens) can maintain effective control in some crops. Conservation most often involves using selective pesticides or cultural practices that preserve natural enemies.
10. Monitoring predator and prey populations provides information as to whether the economic injury level has been exceeded and whether a pesticide application is warranted to suppress spider mites.
11. Chemical control of spider mites is risky because they can develop resistance to pesticides rapidly. Pesticide products used should ideally be less toxic to spider mite predators (and pathogens) than to the target spider mite. Monitoring should be conducted and chemical treatments applied only in hot spots to suppress spider mite outbreaks before they can spread to adjacent crop plants.

or more years may be required to show reduced tree or vine growth or yield because so much energy is stored in the roots. In annual vegetable and field crops, feeding can result in complete destruction of the crop or affect crop quality very quickly, and it is easier to estimate economic injury levels. Mites that feed as colonies on the underside of leaves, such as the six-spotted mite *Eotetranychus sexmaculatus* on citrus, can retard local tissue growth underneath the leaf while the upper leaf tissue continues to grow. This results in deformed leaves.

The two-spotted spider mite (as well as the aphid *Myzus persicae*) can transmit potato virus Y, at least in greenhouses (Jeppson et al. 1975). This disease is found in potatoes, tomatoes, and tobacco. Tobacco ring-spot virus, tobacco mosaic virus, southern bean mosaic virus, and cotton curliness are apparently transmitted by *Tetranychus urticae*. When healthy mites were placed on leaves sprayed with diseased plant sap, the plants developed virus symptoms. The spider mite can carry and excrete viruses, but they do not inject the virus during feeding; however, the mites cause sufficient injury while feeding on the leaf that virus particles present on leaf surfaces can be carried into the plant. In 1982, a disease of barley was discovered in Montana. The disease is associated with a virus-like particle and appears to be transmitted by the spider mite *Petrobia latens* (Robertson and Carroll 1988).

Spider mites can cause allergic responses in humans in both the laboratory and the field. Individuals who are sampling mites and using mite-brushing machines should take precautions to reduce the likelihood of allergies. When the mites are brushed off the leaves, they are sometimes damaged so small particles are released. Working in a fume hood, where the allergens can be sucked away, and wearing laboratory coats, gloves, and dust masks will reduce exposure to allergenic materials. Allergies to *Panonychus ulmi*, *Tetranychus mcdanieli*, and *T. urticae* have been documented in workers in both fields and greenhouses (Astarita et al. 1994). The main symptoms of allergic responses include rhinitis, bronchial asthma, and contact urticaria.

The methods used to manage tetranychid mite pests should include the multiple tactics outlined in Chapter 5 and in Table 6.1. The use of pesticides as the primary control strategy is not sustainable due to the ability of spider mites to develop resistance, as well as to the loss of registration or lack of registration of new products for small specialty crops.

6.9 TETRANYCHIDAE AS WEED CONTROL AGENTS

Tetranychids have been used in at least two biological control projects to suppress weeds, but uncertainties about the taxonomy of spider mite species and their host range have limited their use so far. The taxonomic difficulties arise from the recognition that some species may, in fact, be species complexes. The degree of reproductive isolation between apparently different species has resulted in considerable uncertainty about the validity of some taxonomic entities.

6.9.1 Biological Control of Gorse

The gorse spider mite (***Tetranychus lintearius***) is native to Europe and causes serious damage to gorse (*Ulex europaeus*, Leguminosae), which is a pest of pastures in New Zealand (Ireson et al. 2003). The mite has an unusual colonial behavior. Mites of all stages gather closely together in groups of many thousands, inside a heavy communal web (see Figure S6.15 on the CD). This species was imported into New Zealand from the United Kingdom in 1988 following 5 years of debate over whether biological control of gorse was desirable. This is the first tetranychid mite purposefully introduced into any country for biological control of weeds. *Tetranychus lintearius* was evaluated before the releases were conducted to confirm its host specificity and stability as a species (Hill and O'Donnell 1991a,b). *Tetranychus lintearius* had never been recorded from any plant other than gorse and thus appeared to be very host specific. Experimental studies indicated that this species could develop on beans (*Phaseolus vulgaris*) and soybeans (*Glycine max*) but could not complete a second generation on beans and never remained on bean plants in the field. Cross-breeding tests were conducted between *T. lintearius*, *T. urticae*, and *T. turkestani* to determine if *T. lintearius* was reproductively isolated from these closely related species. As expected, no viable female progeny were produced in the between-species crosses. Only haploid male progeny were produced from unfertilized eggs, which is normal for haplodiploid (arrhenotokous) tetranychids. *Tetranychus lintearius* was released in Australia after testing on an additional 32 plant species (Ireson et al. 2003); however, predation by the introduced phytoseiid *Phytoseiulus persimilis* and a native coccinellid, *Stethorus histrio*, could disrupt the usefulness of *T. lintearius* as a biological control agent of gorse (Pratt et al. 2003).

6.9.2 Biological Control of *Opuntia*

Tetranychus desertorum was introduced into Australia accidentally and colonized a portion of Queensland. Its establishment and spread resulted in reductions in fruiting of the introduced prickly pear cactus (weed). The spider mite reduced cactus density substantially in a large region within a 2-year period; however, the subsequent release of host-specific insects resulted in a decline in *T. desertorum* populations.

6.10 HOST-PLANT RESISTANCE TO TETRANYCHIDAE

Resistance against spider mites is known to occur in many crops, including almonds, avocado, barley, bean, cassava, castor bean, chrysanthemum, citrus, corn, cotton, cucumber, eggplant, geranium, grapes, hops, peanuts, potatoes, roses, soybean, strawberry, sugar beet, sunflower, tea, tobacco, and tomatoes (De Ponti 1985, Harman et al. 1996, Bynum et al. 2004, Lopez et al. 2005). This diversity of resistant crops suggests that host-plant resistance is likely to be found if looked for. Initially, resistance in existing cultivars or breeding lines can be discovered by comparing counts of mite populations on different crop varieties grown under the same conditions and with equivalent initial mite populations.

Spider mites can be tested on leaf discs of different plant varieties to determine whether the resistance is due to **antibiosis**, which results in reduced fecundity or developmental rate or survival. Antibiosis to *Tetranychus urticae* has been found in cassava, cotton, hops, tobacco, geraniums, and apple varieties. Crop varieties also have been shown to exhibit a **nonpreference** mode of resistance to spider mites. Nonpreference can be demonstrated by providing mites with a choice of different types of leaves and counting the number that end up on each. Nonpreference by *T. urticae* was found in strawberry and tobacco cultivars. If neither nonpreference nor antibiosis is found, or if their effects are too low to account for the observed resistance, then it is often assumed that the plant variety possesses a high level of **tolerance** (the ability of the host to suffer feeding but thrive).

Several sorghum varieties are resistant to the Banks grass mite *Oligonychus pratensis*; however, when 19 lines were evaluated, no antibiosis was found (Foster et al. 1977). Some varieties were more readily colonized than others and nonpreference was concluded to be the mode of resistance. The mechanism causing nonpreference was the presence of upright leaves; these mites prefer the undersides of bent leaves, which receive indirect sunlight and maintain a higher relative humidity. Additional research indicated that tolerance also plays a role; resistant varieties have a higher sugar content and maintain healthy green stalks and leaves longer despite feeding by the mites.

Different tomato cultivars resistant to *Tetranychus urticae* have different mechanisms. Some cultivars result in decreased fecundity of mites, resulting in lower plant damage (antibiosis). The resistance is due, in part, to the glandular hairs on leaves; the exudate of the hairs decreases fecundity but is not the sole mechanism. The density of the glandular hairs contributes to the resistance. Repellency is highest on leaves with greater pubescence and sticky exudate. The hairs thus have both antibiotic and nonpreference effects. Furthermore, ethanol extracts from tissue, hairs, and fruit repelled spider mites, indicating that the essential oils of tomato leaves had both an antibiotic and a repellent effect. As a result, tomato resistance is due to pubescence, exudates, and essential oils that repel and reduce spider mite attack. Some varieties may have low pubescence (which should lead to susceptibility) but have a very high tolerance and are resistant (Rodriguez et al. 1972, Van Haren et al. 1987, Guo et al. 1993, Fernandez-Munoz et al. 2000).

Cucumbers contain secondary plant compounds called cucurbitacins. Breeding has removed these chemicals from the fruit of cultivated cucumbers but left them in the stems and leaves. Varieties with higher cucurbitacin content are highly repellent to honey bees and yellow jackets, but cucumber beetles are highly attracted to the bitter varieties with high levels of cucurbitacins. Immatures of *Tetranychus urticae* do not develop well (antibiosis) on such leaves, and *T. urticae* mites exhibit a nonpreference for the bitter varieties. The curcurbitacins thus appear to be a strong repellent and toxin for many arthropods, including mites (Gould 1979).

Multiple mechanisms for resistance to arthropods (and diseases) in plants are apparently common, probably due to the multiple selection pressures on plants. The plant must evolve protective mechanisms to deal with multiple stresses, even though it is costly for the plant to develop these structures and chemicals. As a result, the plant must limit the number of these resistance mechanisms. This argues for a limited repertoire of resistance mechanisms in any plant; however, it is disadvantageous for a plant to produce only one defense. If an arthropod or other organism overcomes that single defense, then the plant has no protection. It probably is safer to be moderately efficient at maintaining several defenses than to be highly efficient in a single defense mechanism that may eventually be overcome. In other words, the evolution of defense mechanisms in plants must remain flexible in response to changing selection pressures.

Host-plant resistance potentially has the ability to solve some difficult pest management problems in crops for which other management options are limited; for example, crops such as wheat or corn are difficult and expensive to treat with pesticides. Resistance to pesticides in the pest mites is always a threat. Releases of natural enemies are too expensive in wheat and corn, and few other management options are available. If plant resistance to spider mites develops, acarologists should be involved in the evaluation of new cultivars to ensure that the resistance does not disrupt the effectiveness of natural enemies. Glandular hairs, for example, can be a useful resistance mechanism, but these glandular hairs could also disrupt the effectiveness of natural enemies, such as phytoseiids.

Another attribute breeders need to consider is the importance of **acarodomatia** in crop plants (Walter 1996, English-Loeb et al. 2002). Acarodomatia are tufts of hairs or invaginations in the leaf surface of perennial plants (such as fruit trees, grapevines) that serve as residences of predatory and mycophagous mites (see Figure S6.16 on the CD). These domatia may protect predatory mites and result in reduced densities of pest mites. Crop breeding programs often eliminate or reduce domatia, but breeders should attempt to maintain them in new cultivars as a refuge for beneficial mite species.

According to De Ponti (1985), deliberate breeding for resistance to spider mites is limited to a few crops; however, this IMM tool should be considered more often in the future and used in combination with other integrated pest management tactics to delay the development of resistance to host-plant resistance.

6.11 RESISTANCE TO HOST-PLANT RESISTANCE

Mites have developed resistance to many acaricides and insecticides (Helle 1965, Edge and James 1982, 1986, Goodwin et. al. 1995, Herron et al. 1998, Nauen et al. 2001, Stumpf and Nauen 2002). It has been argued that insecticide resistance in phytophagous arthropods is derived from their long association with secondary plant compounds. Tolerance to secondary plant compounds may confer a cross-resistance to some pesticides.

Gould (1979) tested the rate of adaptation of *Tetranychus urticae* to host-plant compounds by placing mites on a preferred host (beans) and a poor host (a bitter cucumber cultivar). Within 21 months (approximately 80 generations), the mites reared on cucumber had adapted to cucumber and had a higher fecundity when reared on cucumber than when reared subsequently on beans. Furthermore, the *T. urticae* colony reared on the bitter cucumber cultivars was more tolerant to several insecticides than lines reared on the standard susceptible cultivar; thus, some plant-defense factors could trigger the development of resistance to pesticides. These results illustrate that extensive genetic variation exists in *T. urticae* populations, and the development of resistance to host-plant resistance is possible.

Resistance to host-plant resistance is more likely to develop when crops are planted over a large geographical area over a number of years. This continuity in time and space will give mites the opportunity to develop resistance to host-plant resistance, especially if the resistance is based on a single mechanism. If host-plant resistance to mites is to be used effectively in IMM programs, a diversity of crops or resistant varieties should be deployed. Ideally, these varieties should not have serious negative effects on arthropod natural enemies. Relying on a single IMM tactic (host-plant resistance or chemical control) can be dangerous. Concerns about resistance to pesticides and host-plant resistance suggest that diversity in management tactics is essential in IMM programs. Biological control and host-plant resistance must be maintained as compatible and complementary tactics within an IMM strategy.

6.12 PESTICIDE RESISTANCE IN TETRANYCHIDS

As discussed in Chapter 5, resistance develops rapidly in tetranychid mites to many classes of pesticides (Helle 1965, Edge and James 1982, 1986, Hussey and Scopes 1985, Goodwin et. al. 1995, Herron et al. 1998, Nauen et al. 2001, Stumpf and Nauen 2002, Van Leeuwen et al. 2010). Reasons include the fact that spider mite populations have multiple generations a year, so they can undergo

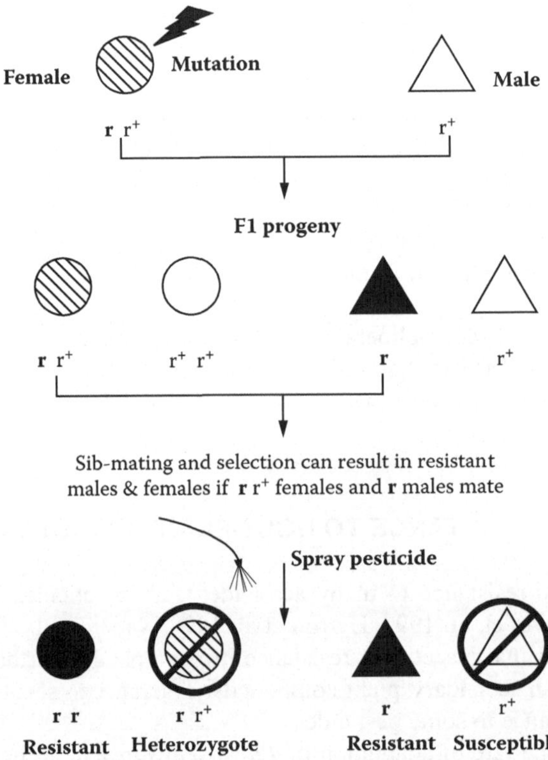

Figure 6.2 If a spider mite female (diploid) has a recessive mutation conferring resistance *r*, she can produce two types of haploid eggs (sons): *r* (resistant) and *r+* (wild-type). If the *r* (resistant) sons mate with their diploid sister (inbreeding) who is heterozygous (*r/r+*) for the resistance allele, then some of their progeny will be resistant to the pesticide. Resistant progeny are the *r/r* daughters and *r* son. The *r/r+* daughter and the *r+* son are susceptible and will die.

selection multiple times a year. Tetranychids produce large populations and, despite the fact that they commonly inbreed, which should reduce genetic variability, they exhibit great genetic variability as a result of their large population size. Because spider mites inbreed, if a recessive resistant mutation occurs, it will rapidly become homozygous in females (Figure 6.2). Finally, males of tetranychids are haploid; as a result, any new mutations that confer resistance to pesticides can be exposed to selection each generation (Figure 6.2). The combination of inbreeding, male haploidy, high reproduction rate, and large population sizes contributes to the propensity of spider mites to develop resistances to pesticides.

SELECTED REFERENCES

Alberti, G. and A.R. Crooker. 1985. Internal anatomy. In: W. Helle and M.W. Sabelis (eds.), *Spider Mites: Their Biology, Natural Enemies, and Control*, Vol. 1A (pp. 29–62). Amsterdam: Elsevier.

Astarita, C., A. Franzese, G. Scala, S. Sproviero, and G. Raucci. 1994. Farm workers' occupational allergy to *Tetranychus urticae:* Clinical and immunologic aspects. *Allergy* 49:466–471.

Blauvelt, W.E. 1945. The internal morphology of the common red spider mite (*Tetranychus telarius* Linn.). *Mem. Cornell University of Agric. Exp. Stn.* 270:1–35.

Bolland, H.R., J. Gutierrez, and C.H.W. Flechtmann. 1998. *World Catalogue of the Spider Mite Family Tetranychidae*. Leiden: Brill Academic.

Breeuwer, J.A.J. 1997. *Wolbachia* and cytoplasmic incompatibility in the spider mites *Tetranychus urticae* and *T. turkestani*. *Heredity* 79:41–47.

Broufas, G.D. and D.S. Koveos. 2000. Threshold temperature for post-diapause development and degree-days to hatching of winter eggs of the European red mite (Acari: Tetranychidae) in Northern Greece. *Environ. Entomol.* 29:710–713.

Bynum, E.D., W. Xu, and T.L. Archer. 2004. Diallel analysis of spider mite resistant maize inbred lines and F_1 crosses. *Crop Sci.* 44:1535–1541.

Cone, W.W., L.M. McDonough, J.C. Maitlen, and Z. Burdajewic. 1971a. Pheromone studies of the two spotted spider mite. I. Evidence of a sex pheromone. *J. Econ. Entomol.* 64:355–358.

Cone, W.W., S. Predki, and E. Klostermeyer. 1971b. Pheromone studies of the two spotted spider mite. II. Behavioral response of males to quiescent deutonymphs. *J. Econ. Entomol.* 53:379–382.

Cranham, J.E. 1973. Variation in the intensity of diapause in winter eggs of fruit-tree spider mite, *Panonychus ulmi*. *Ann. Appl. Biol.* 75:173–182.

Cranham, J.E. and W. Helle. 1985. Pesticide resistance in Tetranychidae. In: W. Helle and M.W. Sabelis (eds.), *Spider Mites: Their Biology, Natural Enemies, and Control*, Vol. 1B (pp. 405–422). Amsterdam: Elsevier.

De Ponti, O.M.B. 1985. Host-plant resistance and its manipulation through plant breeding. In: W. Helle and M.W. Sabelis (eds.), *Spider Mites: Their Biology, Natural Enemies, and Control*, Vol. 1B (pp. 395–404). Amsterdam: Elsevier.

Dunley, J.E. and B.A. Croft. 1992. Dispersal and gene flow of pesticide resistance traits in phytoseiid and tetranychid mites. *Exp. Appl. Acarol.* 14:313–325.

Edge, V.E. and D.G. James. 1982. Detection of cyhexatin resistance in two-spotted mite, *Tetranychus urticae* Koch (Acarina: Tetranychidae) in Australia. *J. Aust. Entomol. Soc.* 21:198.

Edge, V.E. and D.G. James. 1986. Organotin resistance in *Tetranychus urticae* (Acari: Tetranychidae) in Australia. *J. Econ. Entomol.* 79:1477–1483.

English-Loeb, G., A.P. Norton, and M.A. Walker. 2002. Behavioral and population consequences of acarodomatia in grapes on phytoseiid mites (Mesostigmata) and implications for plant breeding. *Entomol. Exp. Appl.* 104:307–319.

Fernandez-Munoz, R., E. Dominguez, and J. Cuartero. 2000. A novel source of resistance to the two-spotted spider mite in *Lycopersicon pimpinellifolium* (Jusl.) Mill.: Its genetics as affected by interplot interference. *Euphytica* 111:169–173.

Flaherty, D.L. 1969. Ecosystem trophic complexity and densities of the Willamette mite, *Eotetranychus willamettei* Ewing (Acarina: Tetranychidae). *Ecology* 50:911–916.

Foster, D.G., G.L. Teetes, J.W. Johnson, D.T. Rosenow, and C.R. Ward. 1977. Field evaluation of resistance in sorghums to banks grass mite. *Crop Sci.* 17:821–823.

Gerson, U. 1985. Webbing. In: W. Helle and M.W. Sabelis (eds.), *Spider Mites: Their Biology, Natural Enemies, and Control*, Vol. 1A (pp. 223–231). Amsterdam: Elsevier.

Goodwin, S., G. Herron, N. Gough, and T. Wellham. 1995. Relationship between insecticide–acaricide resistance and field control of *Tetranychus urticae* (Acari: Tetranychidae) infesting roses. *J. Econ. Entomol.* 88:1106–1112.

Gotoh, T., J. Bruin, M.W. Sabelis, and S.B.J. Menken. 1993. Host race formation in *Tetranychus urticae*: Genetic differentiation, host plant preference, and mate choice in a tomato and a cucumber strain. *Entomol. Exp. Appl.* 68:171–178.

Gotoh, T., H. Noda, and X.Y. Hong. 2003. *Wolbachia* distribution and cytoplasmic incompatibility based on a survey of 42 spider mite species (Acari: Tetranychidae) in Japan. *Heredity* 91:208–216.

Gotoh, T., J. Sugasawa, H. Noda, and Y. Kitashima. 2007. *Wolbachia*-induced cytoplasmic incompatibility in Japanese populations of *Tetranychus urticae* (Acari: Tetranychidae). *Exp. Appl. Acarol.* 42:1–16.

Gotoh, T., R. Araki, A. Boubou, A. Migeon, F. Ferragut, and M. Navajas. 2009. Evidence of co-specificity between *Tetranychus evansi* and *Tetranychus takafujii* (Acari: Prostigmata, Tetranychidae): Comments on taxonomic and agricultural aspects. *Int. J. Acarol.* 35:485–501.

Gould, F. 1979. Rapid host range evolution in a population of the phytophagous mite *Tetranychus urticae* Koch. *Evolution* 33:791–802.

Grafton-Cardwell, E.E., J. Granett, and S.M. Normington. 1991. Influence of dispersal from almonds on population dynamics and acaricide resistance frequencies of spider mites infesting neighboring cotton. *Exp. Appl. Acarol.* 10:187–212.

Guo, Z., P.A. Weston, and J.C. Snyder. 1993. Repellency to two-spotted spider mite, *Tetranychus urticae* Koch, as related to leaf surface chemistry of *Lycopersicon hirsutum* accessions. *J. Chem. Ecol.* 19:2965–2979.

Hanna, R., L.T. Wilson, F.G. Zalom, and D.L. Flaherty. 1997. Effects of predation and competition on the population dynamics of *Tetranychus pacificus* on grapevines. *J. Appl. Ecol.* 34:878–888.

Harman, J., P. Paul, R. Craig, D. Cox-Foster, J. Medford, and R.O. Mumma. 1996. Development of a mite bioassay to evaluate plant resistance and its use in determining regeneration of spider mite resistance. *Entomol. Exp. Appl.* 81:301–305.

Helle, W. 1965. Resistance in the Acarina: Mites. *Adv. Acarol.* 2:71–93.

Helle, W. 1968. Genetic variability of photoperiodic response in an arrhenotokous mite (*Tetranychus urticae*). *Entomol. Exp. Appl.* 11:101–113.

Helle, W. 1985. Genetics. In: W. Helle and M.W. Sabelis (eds.), *Spider Mites: Their Biology, Natural Enemies, and Control*, Vol. 1A (pp. 185–192). Amsterdam: Elsevier.

Helle, W. and M.W. Sabelis (eds.) 1985. *Spider Mites: Their Biology, Natural Enemies, and Control*, Vols. 1A and 1B. Amsterdam: Elsevier.

Herbert, H.J. and K.B. McRae. 1982. Predicting eclosion of overwintering eggs of European red mite *Panonychus ulmi* (Acarina: Tetranychidae), in Nova Scotia. *Can. Entomol.* 114:703–712.

Herron, G.A., V.E. Edge, L.J. Wilson, and J. Rophail. 1998. Organophosphate resistance in spider mites (Acari: Tetranychidae) from cotton in Australia. *Exp. Appl. Acarol.* 22:17–30.

Hill, R.L. and D.J. O'Donnell. 1991a. The host range of *Tetranychus lintearius* (Acarina: Tetranychidae). *Exp. Appl. Acarol.* 11:253–269.

Hill, R.L. and D.J. O'Donnell. 1991b. Reproductive isolation between *Tetranychus lintearius* and two related mites, *T. urticae* and *T. turkestani* (Acarina: Tetranychidae). *Exp. Appl. Acarol.* 11:241–251.

Huffaker, C.B., M. van de Vrie, and J.A. McMurtry. 1970. Ecology of tetranychid mites and their natural enemies: A review. II. Tetranychid populations and their possible control by predators: An evaluation. *Hilgardia* 40(11):391–458.

Hussey, N.W. and N. Scopes (eds.) 1985. *Biological Pest Control: The Glasshouse Experience.* Ithaca, NY: Cornell University Press.

Ireson, J.E., A.H. Gourlay, R.M. Kwong, R.J. Holloway, and W.S. Chatterton. 2003. Host specificity, release, and establishment of the gorse spider mite, *Tetranychus lintearius* Dufour (Acarina: Tetranychidae), for the biological control of gorse, *Ulex europaeus* L. (Fabaceae), in Australia. *Biol. Control* 26:117–127.

Jeppson, L.R., H.H. Keifer, and E.W. Baker. 1975. *Mites Injurious to Economic Plants.* Berkeley: University of California Press.

Kawakami, Y., H. Numata, K. Ito, and S.G. Goto. 2009. Dominant and recessive inheritance patterns of diapause in the two-spotted spider mite *Tetranychus urticae*. *J. Hered.* 101:20–25.

Kinn, D.N. and R.L. Doutt. 1972. Natural control of spider mites on wine grape varieties in northern California. *Environ. Entomol.* 1:513–518.

Kostiainen, T. and M.A. Hoy. 1996. *The Phytoseiidae as Biological Control Agents of Pest Mites and Insects: A Bibliography (1960–1994)*, Monograph 17. Gainesville: Institute of Food and Agricultural Sciences, University of Florida. (Bibliography includes references to published articles involving the control of spider mites; indices separate the publications by crop and by species. Monograph is available on the CD in the Chapter 12 folder.)

Krainacker, D.A. and J.R. Carey. 1989. Reproductive limits and heterogeneity of male two-spotted spider mites. *Entomol. Exp. Appl.* 50:209–214.

Lees, A.D. 1953. Environmental factors controlling the evocation and termination of diapause in the fruit tree red spider mite *Metatetranychus ulmi* Koch (Acarina: Tetranychidae). *Ann. Appl. Biol.* 40:449–486.

Li, J. and D.C. Margolies. 1994. Barometric pressure influences initiation of aerial dispersal in the two-spotted spider mite. *J. Kansas Entomol. Soc.* 67:386–393.

Li, T., X.L. Chen, and X.Y. Hong. 2009. Population genetic structure of *Tetranychus urticae* and its sibling species *Tetranychus cinnabarinus* (Acari: Tetranychidae) in China as inferred from microsatellite data. *Ann. Entomol. Soc. Am.* 103:674–683.

Lindquist, E.E. 1985. External anatomy. In: W. Helle and M.W. Sabelis (eds.), *Spider Mites: Their Biology, Natural Enemies, and Control*, Vol. 1A (pp. 3–28). Amsterdam: Elsevier.

Lopez, R., A. Levi, B.M. Shepard, A.M. Simmons, and D.M. Jackson. 2005. Sources of resistance to two-spotted spider mite (Acari: Tetranychidae) in *Citrullus* spp. *HortScience* 40:1661–1663.

Margolies, D.C. 1987. Conditions eliciting aerial dispersal behavior in Banks grass mite, *Oligonychus pratensis* (Acari: Tetranychidae). *Environ. Entomol.* 16:928–932.

Margolies, D.C. and G.G. Kennedy. 1988. Fenvalerate-induced aerial dispersal by two-spotted spider mite. *Entomol. Exp. Appl.* 46:233–240.

McEnroe, W.D. 1969. Eyes of the female two-spotted spider mite, *Tetranychus urticae.* I. Morphology. *Ann. Entomol. Soc. Am.* 62:461–466.

McEnroe, W.D. and K. Dronka. 1969. Eyes of the two-spotted spider mite, *Tetranychus urticae*. II. Behavioral analysis of the photoreceptors. *Ann. Entomol. Soc. Am.* 62:466–469.

Meyer, M.K.P.S. 1981. *Mite Pests of Crops in Southern Africa*, Science Bulletin 397. Pretoria: Department of Agriculture and Fisheries, Republic of South Africa.

Mothes, U. and K.A. Seitz. 1981. Fine structure and function of the prosomal glands of the two-spotted spider mite, *Tetranychus urticae* (Acari: Tetranychidae). *Cell Tissue Res.* 221:339–349.

Nauen, R., N. Stumpf, A. Elbert, C.P.W.Zebitz, and W. Kraus. 2001. Acaricide toxicity and resistance in larvae of different strains of *Tetranychus urticae* and *Panonychus ulmi* (Acari: Tetranychidae). *Pest Manag. Sci.* 57:253–261.

Navajas, M., J. Gutierrez, J. Lagnel, G. Fauvel, and T. Gotoh. 1999. DNA sequences and cross-breeding experiments in the hawthorn spider mite *Amphitetranychus viennensis* reveal high genetic differentiation between Japanese and French populations. *Entomol. Exp. Appl.* 90:113–122.

Osakabe, M., H. Isobe, A. Kasai, R. Masuda, S. Kubota, and M. Umeda. 2008. Aerodynamic advantages of upside down take-off for aerial dispersal in *Tetranychus* spider mites. *Exp. Appl. Acarol.* 44:165–183.

Penman D.R. and W.W.Cone. 1972. Behavior of male two-spotted spider mites in response to quiescent female deutonymphs and to web. *Ann. Entomol. Soc. Am.* 65:1289–1293.

Potter, D.A., D.L. Wrensch, and D.E. Johnston. 1976. Guarding, aggressive behavior and mating success in male two-spotted spider mites. *Ann. Entomol. Soc. Am.* 68:707–711.

Pratt, P.D., E.M. Coombs, and B.A. Croft. 2003. Predation by phytoseiid mites on *Tetranychus lintearius* (Acari: Tetranychidae), an established weed biological control agent of gorse (*Ulex europaeus*). *Biol. Control* 26:40–47.

Rasmy, A.H. and H.E. Hussain. 1994. Effect of age and mating on release of female sex pheromones and male response in the two-spotted spider mite. *J. Appl. Entomol.* 117:109–111.

Robertson, N.L. and T.W.Carroll. 1988. Virus-like particles and a spider mite intimately associated with a new disease of barley. *Science* 240:1188–1190.

Rodriguez, J.G., D. Knavel, and O.J. Aina. 1972. Studies in the resistance of tomatoes to mites. *J. Econ. Entomol.* 65:50–53.

Royalty, R.N., P.L. Phelan, and F.R. Hall. 1992. Arrestment of male two-spotted spider mite caused by female sex pheromone. *J. Chem. Ecol.* 18:137–153.

Royalty, R.N., P.L. Phelan, and F.R Hall. 1993. Quantitative and temporal analysis of effects of two-spotted spider mite (Acari: Tetranychidae) female sex pheromone on male guarding behavior. *J. Chem. Ecol.* 19:211–223.

Sabelis, M.W. 1985. Reproductive strategies. In: W. Helle and M.W. Sabelis (eds.), *Spider Mites: Their Biology, Natural Enemies, and Control*, Vol. 1A (pp. 265–275). Amsterdam: Elsevier.

Saito, Y., K. Ito, and T. Sakagami. 2005. Imaginal induction of diapause in several "adult-female diapausing" spider mites. *Physiol. Entomol.* 30:96–101.

Smitley, D.R. and G.G. Kennedy. 1985. Photo-oriented aerial dispersal behavior of *Tetranychus urticae* (Acari: Tetranychidae) enhances escape from the leaf surface. *Ann. Entomol. Soc. Am.* 78:609–614.

Smitley, D.R. and G.G. Kennedy. 1988. Aerial dispersal of the two-spotted spider mite (*Tetranychus urticae*) from field corn. *Exp. Appl. Acarol.* 5:33–46.

Sonenshine, D.E. 1985. Pheromones and other semiochemicals of the Acari. *Annu. Rev. Entomol.* 30:1–28.

Stumpf, N. and R. Nauen. 2002. Biochemical markers linked to abamectin resistance in *Tetranychus urticae* (Acari: Tetranychidae). *Pest. Biochem. Physiol.* 72:111–121.

Tsagkarakou, A., M. Navajas, L. Jacques, J. Gutierrez, and N. Pasteur. 1996. Genetic variability in *Tetranychus urticae* (Acari: Tetranychidae) from Greece: Insecticide resistance and isozymes. *J. Econ. Entomol.* 89:1354–1358.

Tsugawa, C., M. Yamada, S. Shirasaki, and N. Oyama. 1966. Forecasting the outbreak of destructive insects in apple orchards. VII. Termination of diapause in hibernating eggs of *Panonychus ulmi* (Koch) in relation to temperature. *Jpn. J. Appl. Entomol. Zool.* 10:174–180.

van de Vrie, M., J.A. McMurtry, and C.B. Huffaker. 1972. Ecology of tetranychid mites and their natural enemies: A review. III. Biology, ecology and pest status and host–plant relations of tetranychids. *Hilgardia* 41(13):343–432.

Van Haren, R.J.F., M.M. Steenhuis, M.W. Sabelis, and O.M.B. De Ponti. 1987. Tomato stem trichomes and dispersal success of *Phytoseiulus persimilis* relative to its prey *Tetranychus urticae*. *Exp. Appl. Acarol.* 3:115–121.

Van Leeuwen, T., J. Vontas, A. Tsagkarakou, W. Dermauw, and L. Tirry. 2010. Acaricide resistance mechanisms in the two-spotted spider mite *Tetranychus urticae* and other important Acari: A review. *Insect Biochem. Molec. Biol.* 40:563–572.

Veerman, A. 1977. Photoperiodic termination of diapause in spider mites. *Nature* 266:526–527.

Veerman, A. 1985. Diapause. In: W. Helle and M.W. Sabelis (eds.), *Spider Mites: Their Biology, Natural Enemies, and Control*, Vol. 1A (pp. 279–316). Amsterdam: Elsevier.

Walter, D.E. 1996. Living on leaves: Mites, tomenta, and leaf domatia. *Annu. Rev. Entomol.* 41:101–114.

Zhang, Z.-Q. 2003. *Mites of Greenhouses: Identification, Biology, and Control.* Wallingford, U.K.: CAB International.

Zhang, Z.-Q. and L. Liang. 1997. *An Illustrated Guide to Mites of Agricultural Importance.* Shanghai, China: Tongji University Press.

Spider Mite Website

A spider mite website that offers information by species and host plant, as well as a bibliography, is at http://www1.montpellier.inra.fr/CBGP/spmweb/index.php.

CHAPTER 7

The Tarsonemidae

7.1 BIOLOGY OF THE PLANT-FEEDING TARSONEMIDAE

The family Tarsonemidae (Prostigmata or Actinedida) is very diverse in terms of its feeding habits, with some species feeding on algae, fungi, and plants (Meyer 1981, Lindquist 1986, Lin and Zhang 1999, 2002, Krantz and Walter 2009). Some tarsonemid mites are parasites. Other species are predators found in house dust and may be potential sources of human allergy. Tarsonemidae consists of three subfamilies and contains about 530 species in 40 genera. The family is primarily tropical and subtropical in its distribution, but some species are found in the Nearctic and Palearctic regions. The tarsonemid mites found as pests on plants primarily are in the Tarsoneminae and the Pseudotarsonemoidinae (Zhang and Liang 1997).

Tarsonemid mites are very small, ranging from 0.1 to 0.3 mm in length. Mature mites appear hard and shiny, and most are semitranslucent and pale or white in appearance, although the food eaten can affect their color (see Figures S7.1 and S7.2 on the CD). The body and posterior legs have few setae. Sexual dimorphism is pronounced, with males being smaller than females. The general body contour for females is oval, with the dorsum convex and the anterior two pairs of legs separated from the posterior two pair by a distinct space. The mouthparts consist of stout palps and slender stylet-like chelicerae. Adult males have a unique structure known as the genital papilla or genital plate, which is located on the opisthosoma and contains the aedeagus, accessory genital organs, and appendages.

The life cycle of plant-feeding tarsonemid species is modified compared to that of the Tetranychidae. The life cycle usually requires about a week (Jeppson et al. 1975). Females deposit eggs, laid singly, which are white, oval, opaque, and large in comparison to the size of the female (see Figures S.7.1 and S7.2 on the CD). Females deposit 1 to 5 eggs per day, with most species producing 10 to 20 eggs in total. The opisthosoma of the six-legged larva has a peculiar enlargement into a triangular platelike structure, which is accentuated in males. After the active larval stage, the mites enter a quiescent pupal stage in which they are transformed into the adult. The pupal stage is inactive, and the integument appears inflated. During this stage, the fourth pair of legs and the genital structures develop.

Females walk on all four pairs of legs, but the males appear to use their fourth pair for walking only rarely; rather, the male uses the fourth pair to transport female pupae and adult females on his back. Males hold female pupae with the fourth pair of legs, and the pupae are fastened to the male by his genital papillae. Males do not carry larvae, and male pupae are rarely carried.

Most species appear to be arrhenotokous, with unfertilized eggs producing males and fertilized eggs producing females, although a few are thelytokous. Tarsonemids appear to develop best under warm temperatures, high relative humidity, and low light intensity. None is known to diapause. Three tarsonemid genera are known to feed on tender succulent parts of higher plants. Two species, *Steneotarsonemus* (= *Phytonemus*) *pallidus* (cyclamen mite or strawberry mite) and

Polyphagotarsonemus (=*Hemitarsonemus*) *latus* (broad mite, citrus silver mite, or yellow tea mite), may inject a toxin during feeding, causing abnormal plant growth. Plant-feeding pests of particular importance to agriculture include the cyclamen mite and the broad mite; however, one of the major pests of honey bees is a parasitic tarsonemid (*Acarapis woodi*) that lives within the tracheae of honey bees (see Chapter 21).

7.2 *STENEOTARSONEMUS* (OR *PHYTONEMUS*) *PALLIDUS*

The cyclamen mite or strawberry mite causes the greatest amount of damage to agriculture, followed closely by *Polyphagotarsonemus latus*. Both species have large host-plant ranges, including many commercially important crops. Cyclamen mite is especially important as a pest of strawberries, cyclamen, *Gerbera*, begonia, African violets, and ivy. It is widespread around the world (Denmark 2004). The mite is pinkish orange as an adult, translucent in immature stages, and very small (see Figure S7.1 on the CD). Females (250 microns in length) deposit their eggs on young, unfolding leaves at the crown of the strawberry plant, and when the leaves emerge they are stunted, crinkled, and malformed. Leaf stems do not elongate (see Figure S7.3 on the CD). When infestations are large, the whole plant can become dwarfed, leaves may turn brownish green, and fruit are small, dry, and withered. Mite populations on *Gerbera* foliage produce bronzed patches along the midribs and cause the foliage to curl. Flowers attacked in the bud stage become distorted and unfit for sale.

Cyclamen mites avoid light, require high relative humidity, and most often are found in unopened leaflets in the crown of their host plant. Adults overwinter in temperate climates, but in mild climates development may continue all year. Mites disperse in strawberry fields by crawling along runners, by hitchhiking on transplanted daughter plants, or by being blown by the wind. Tarsonemids are very susceptible to drying, so dispersal by walking or by the wind is hazardous.

Eggs are laid in clusters; females deposit 1 to 3 eggs per day for a total of 12 to 16. The elliptical eggs are relatively large, almost half the size of the adult female (see Figure S7.1 on the CD). The life cycle requires approximately 2 to 3 weeks. Larvae are white, and the hind end is triangular. Females have reduced legs IV with a distal threadlike seta. The male is smaller than the female and has well-developed legs IV that terminate in a claw.

Cyclamen mites are difficult to control because they are so well protected in the small leaf buds in the plant crown (Herron et al. 1996). Managing cyclamen mites in strawberries involves removing and destroying infested plants as soon as they are spotted. New plantings should be made from mite-free stock, and new plants should not be planted near infested ones. Care should be taken to avoid transporting mites on tools. Fumigation of planting stock with methyl bromide was formerly practiced, but immersing plants in water at 43.5°C for 30 minutes can kill most mites on planting stock. Some phytoseiids are effective predators of cyclamen mites in California strawberries if pesticide applications are modified so the predators are not eliminated (Huffaker and Kennett 1953, Croft et al. 1968).

7.3 *POLYPHAGOTARSONEMUS* (OR *HEMITARSONEMUS*) *LATUS*

Polyphagotarsonemus latus, also known as the broad mite, citrus silver mite, yellow tea mite, or tropical mite, is found throughout the subtropics and tropics and in greenhouses in temperate regions on many agricultural crops, as well as ornamentals and wild plants (Gerson 1992, Zhang 2003). It is especially important as a pest of cotton, tea, rubber, citrus, tobacco, potatoes, beans, peppers, gerberas, dahlias, zinnias, and chrysanthemums (Alagarmalai et al. 2009).

This mite is small and cannot be seen easily with the naked eye (see Figure S7.4 on the CD). Adult females are broadly oval and pale yellow or yellow-green with an indistinct white stripe in the middle. The males are the same color but lack the stripe and are only about half the size of females; however, males have with relatively longer legs. Eggs are oval, flattened, and translucent and have multiple rows of round white knobs (see Figures S7.4 and S7.5 on the CD). The six-legged larva is white immediately after hatching but later becomes translucent.

Tarsonemids usually are found in damp, shady places, and these mites feed almost entirely on the lower leaf surface, causing leaves to become rigid and rolled under at the edges. Feeding damage is restricted to young foliage or flowers and may result in curling and crinkling of leaves followed by patches of blisters (Peña and Bullock 1994). The mites are thought to inject toxins during feeding, and the damage may persist for weeks after the removal of the mites (Zhang 2003). Severe attacks stunt plants and may kill them.

Polyphagotarsonemus latus multiplies rapidly, and only 4 to 5 days are required to complete a generation in the summer. The sex ratio is usually one male to four females, and the mites are arrhenotokous (Gerson 1992). Males emerge first and locate the pharate females. Each male places its female at a right angle on the posterior part of its dorsum, held there by the male's genital papillae. The males carry the pharate females about, usually moving upward toward the plant's apical parts. Mating occurs immediately after females emerge.

Dispersal is by walking, wind, insects (phoresy), and humans. Adult females of *Polyphagotarsonemus latus* can disperse by attaching themselves to the tibiae and tarsi of several insects, including green peach aphid (*Myzus persicae*), sweet potato whitefly (*Bemisia tabaci*), greenhouse whitefly (*Trialeurodes vaporariorum*), and the silver leaf whitefly (*Bemisia argentifolii*) (Flechtmann et al. 1990, Parker and Gerson 1994, Fan and Petitt 1998) (Figure 7.1). In one

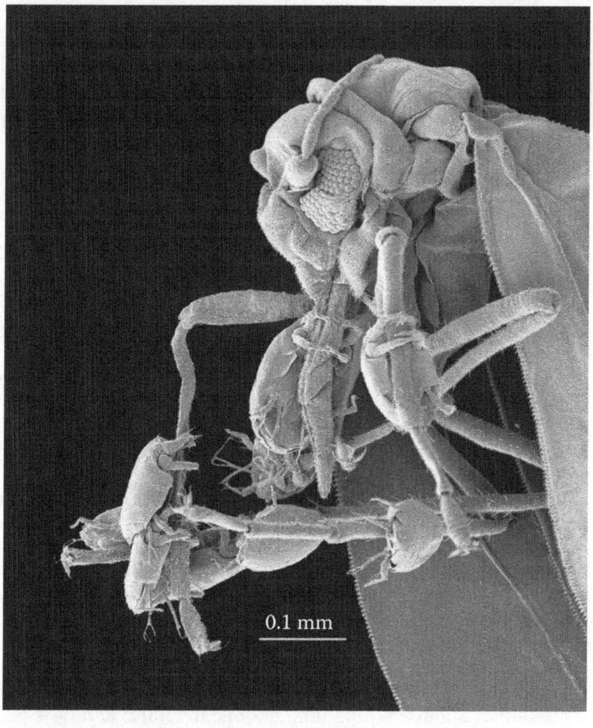

Figure 7.1 Broad mites (*Polyphagotarsonemus latus*) are attached to the whitefly (*Bemisia tabaci*) for dispersal by phoresy. (The mites were collected from Toowoomba, Queensland, Australia, and photographed by David E. Walter.)

experiment, 48% of *B. argentifolii* had broad mites on them within one hour after landing on infested plants. These insect pests and *P. latus* feed and oviposit on the young leaves of many common host plants, suggesting that broad mites may be vectored by insects more often than previously recognized.

7.4 CONTROL

Tarsonemid mites can often be controlled using a combination of cultural practices, releases of predators, and modification of spray practices to conserve endemic predators (Table 7.1). Sulfur dusts, oils, or heat treatments will control broad mites and cyclamen mites. The egg stage is not very susceptible to sulfur, so treatments need to be repeated two or three times after 4- to 5-day intervals. Oils must be applied several times because oils do not kill the eggs. Other chemical controls will have to be evaluated for efficacy, registration, and availability.

A variety of phytoseiid species are effective against tarsonemids; for example, Badii and McMurtry (1984) found that *Typhlodromus annectens*, *T. porresi*, and *T. rickeri* preferred broad mite larvae, while *Euseius stipulatus* fed on all but the nymphal stages in California. Gerson (1992) listed a number of phytoseiid predators of the broad mite, including *Amblyseius agrestis*, *A. delhiensis*, *A. largoensis*, *A. nicholski*, *A. ovalis*, *Euseius victoriensis*, *Typhlodromus annectens*, *T. porresi*, *T. rickeri*, and *T. stipulatus*. Peña et al. (1989) surveyed lime trees in southern Florida for predators of the broad mite and found that the phytoseiid *Typhlodromalus peregrinus* accounted for 72% of the predatory mites, outnumbering *Typhlodromips dentilis*, *A. aeralis*, and *Galendromus helveolus*. *Typhlodromalus peregrinus* also preyed on the citrus red mite (*Panonychus citri*) and on scale insects (*Chrysomphalus aonidum* and *Lepidosaphes beckii*) and could survive on fungi and other organic materials. *Typhlodromalus peregrinus* appeared to prefer broad mite to the citrus rust mite (Peña 1992).

The fungal pathogen *Hirsutella nodulosa* infects the broad mite on citrus in Cuba (Gerson 1992). Peña et al. (1996) found that the fungal pathogen *Beauveria bassiana* killed the broad mite more rapidly than *Hirsutella thompsonii* or *Paecilomyces fumosoroseus* in the laboratory under controlled temperature and relative humidity conditions.

Table 7.1 Integrated Approaches for Controlling Tarsonemid Mites

1. Control of tarsonemid pests of crop plants is based on biological control in many parts of Europe. Releases of *Neoseiulus cucumeris*, *N. reductus*, *N. californicus*, *N. fallacis*, and *Metaseiulus occidentalis* have been successful, using a predator-to-prey ratio of 1 predator to 10 tarsonemids as soon as mite damage is noticed in strawberries.
2. Naturally occurring fungal pathogens can suppress tarsonemid populations in humid environments, but high relative humidities also favor tarsonemid populations.
3. Spray practices can be modified to conserve endemic predators.
4. Crop residues that can serve as reservoirs of infestation should be cleaned up; weeds that are hosts to tarsonemids should be removed.
5. Farm machinery and other equipment should be cleaned regularly to avoid moving mites from an infested area to an uninfested one.
6. An effective fertilizer rate should be maintained, as overfertilizing can enhance mite buildups by making the crop more lush and suitable for feeding.
7. Plant material should be free of mites; plants such as strawberry can be disinfected by immersing them into warm water at 45°C for 15 to 30 minutes.
8. Cultivars tolerant of or resistant to tarsonemid mites should be planted.
9. Pest populations can be monitored by examining young unfolded leaves, under the calyx of the flower buds, or within leaf sheaths. Look for stunted and crinkled foliage, especially in the center of the plant.
10. If predators are not present in adequate numbers, chemical control can be effective. Use products that are selective (such as sulfur or oil), if possible. Excellent coverage is essential for products that are not systemic because tarsonemids usually are present in the unfolded, crinkled leaves and are difficult to reach. Resistance to pesticides has been documented in some populations.

7.5 OTHER PEST SPECIES OF TARSONEMIDS

7.5.1 *Steneotarsonemus ananas*

The pineapple leathery pocket mite is apparently specific to pineapples (Jeppson et al. 1975). It may initiate fungal infections as well as injure fruit by feeding. The result is that some segments of the fruit remain green and become rotten. The mites also can injure young plants in Hawaii and Australia.

7.5.2 *Steneotarsonemus bancrofti*

The West Indian sugar cane mite is host specific and occurs where sugar cane is grown. The mites live under the young green leaf sheaths and produce transparent brown craterlike depressions on the young cane stalk, giving the plant a scabby or scarred appearance. This mite also can introduce spores of red rot in sugar cane.

7.5.3 *Steneotarsonemus laticeps*

The bulb scale mite is a pest of narcissus bulbs and other plants in the Amaryllidaceae. Females deposit 5 to 28 eggs, and the life cycle from egg to adult in the field is about 7 weeks. The mites feed on the epidermal surfaces of the scales of the bulbs, especially around the bulb neck. Feeding by *S. laticeps* causes distortion, stunting, and, sometimes, mortality of leaves and flowers. Feeding by *S. laticeps* may cause bronze streaks and cracks on foliage and flower stems as a result of scar tissue formed by the plant. The mites are unable to penetrate vigorously growing bulbs in the spring, but in the fall the bulbs lose moisture and shrink, which allows the mites to penetrate into the soft tissues. When the bulbs begin to grow in the spring, the bulbs swell so much that the pressure crushes many mites. In vigorous bulbs, mite mortality may be nearly complete. This pest is spread to new bulbs while they are in storage. Control can be achieved by treating the bulbs with hot water at 43°C for one hour.

7.5.4 *Steneotarsonemus spinki*

The panicle rice mite (or the rice tarsonemid mite, rice white mite, rice mite, or spinki mite) is an important mite attacking rice throughout Asia (Hummel et al. 2009). It is also found in the Caribbean and can cause significant crop losses. It is a direct pest, but it also transmits a number of plant pathogens. It has been collected from a delphacid insect in Louisiana, so it could be phoretic on this insect. Recently, this pest has been rediscovered in Louisiana rice fields after a long interval in which it did not appear to be present (see Figure S7.6 on the CD).

SELECTED REFERENCES

Alagarmalai, J., M. Grinberg, R. Perl-Treves, and V. Soroker. 2009. Host selection by the herbivorous mite *Polyphagotarsonemus latus* (Acari: Tarsonemidae). *J. Insect Behav.* 22:375–387.

Badii, M.H. and J.A. McMurtry. 1984. Feeding behavior of some phytoseiid predators on the broad mite, *Polyphagotarsonemus latus* (Acari: Phytoseiidae, Tarsonemidae). *Entomophaga* 29:49–53.

Croft, B.A., P.D. Pratt, C. Koskela, and D. Kaufman. 1968. Predation, reproduction, and impact of phytoseiid mites (Acari: Phytoseiidae) on cyclamen mite (Acari: Tarsonemidae) on strawberry. *J. Econ. Entomol.* 91:1307–1314.

Denmark, H.A. 2009. *Featured Creatures: Cyclamen Mite*. Gainesville: Department of Entomology and Nematology, University of Florida Institute of Food and Agricultural Sciences (http://entnemdept.ifas.ufl.edu/creatures/orn/cyclamen_mite.htm).

Fan, Y.Q. and F.L. Pettit. 1998. Dispersal of the broad mite, *Polyphagotarsonemus latus* (Acari: Tarsonemidae), on *Bemisia argentifolii* (Homoptera: Aleyrodidae). *Exp. Appl. Acarol.* 22:411–415.

Flechtmann, C.H.W., J.M. Guerrero, J.A. Arroyave, and L.M. Constantino. 1990. A little known mode of dispersal of *Polyphagotarsonemus latus* (Banks). *Int. J. Acarol.* 16:181–190.

Gerson, U. 1992. Biology and control of the broad mite, *Polyphagotarsonemus latus* (Banks) (Acari: Tarsonemidae). *Exp. Appl. Acarol.* 13:163–178.

Herron, G., L. Jiang, and R. Spooner-Hart. 1996. A laboratory-based method to measure relative pesticide and spray oil efficacy against broad mite, *Polyphagotarsonemus latus* (Banks) (Acari: Tarsonemidae). *Exp. Appl. Acarol.* 20:495–502.

Huffaker, C.B. and C.E. Kennett. 1953. Developments toward biological control of cyclamen mite on strawberries in California. *J. Econ. Entomol.* 46:802–812.

Hummel, N.A., B.A. Castro, E.M. McDonald, M.A. Pellerano, and R. Ochoa. 2009. The panicle rice mite, *Steneotarsonemus spinki* Smiley, a re-discovered pest of rice in the United States. *Crop Prot.* 28:547–560.

Jeppson, L.R., H.H. Keifer, and E.W. Baker. 1975. *Mites Injurious to Economic Plants.* Berkeley: University of California Press.

Krantz, G.W. and D.E. Walter (eds.) 2009. *A Manual of Acarology*, 3rd ed. Lubbock: Texas Tech University Press.

Lin, J.Z and Z.-Q. Zhang. 1999. *Tarsonemidae of China (Acari: Prostigmata): An Annotated and Illustrated Catalogue and Bibliography*, Special Publ. 3. London: Systematic & Applied Acarology Society.

Lin, J.Z. and Z.-Q. Zhang. 2002. *Tarsonemidae of the World: Key to Genera, Geographical Distribution, Systematic Catalogue and Annotated Bibliography.* London: Systematic & Applied Acarology Society.

Lindquist, E.E. 1986. The world genera of Tarsonemidae (Acari: Heterostigmata): Morphological, phylogenetic, and systematic revision, with a reclassification of family-group taxa in the Heterostigmata. *Mem. Entomol. Soc. Canada* 126:1–517.

Meyer, M.K.P.S. 1981. *Mite Pests of Crops in Southern Africa*, Science Bulletin 397. Pretoria: Department of Agriculture and Fisheries, Republic of South Africa.

Parker, R. and U. Gerson. 1994. Dispersal of the broad mite, *Polyphagotarsonemus latus* (Banks) (Heterostigmata: Tarsonemidae), by the greenhouse whitefly, *Trialeurodes vaporariorum* (Westwood) (Homoptera: Aleyrodidae). *Exp. Appl. Acarol.* 18:581–585.

Peña, J.E. 1992. Predator–prey interactions between *Typhlodromalus peregrinus* and *Polyphagotarsonemus latus*: Effects of alternative prey and other food resources. *Florida Entomol.* 75:241–248.

Peña, J.E. and R.C. Bullock. 1994. Effects of feeding of broad mite (Acari: Tarsonemidae) on vegetative plant growth. *Florida Entomol.* 77:180–184.

Peña, J.E., R.M. Baranowski, and H.A. Denmark. 1989. Survey of predators of the broad mite in southern Florida. *Florida Entomol.* 72:373–377.

Peña, J.E., L.S. Osborne, and R.E. Duncan. 1996. Potential of fungi as biocontrol agents of *Polyphagotarsonemus latus* (Acari: Tarsonemidae). *Entomophaga* 41:27–36.

Zhang, Z.-Q. 2003. *Mites of Greenhouses: Identification, Biology, and Control.* Wallingford, U.K.: CAB International.

Zhang, Z.-Q. and L. Liang. 1997. *An Illustrated Guide to Mites of Agricultural Importance.* Shanghai, China: Tongji University Press.

CHAPTER 8

The Eriophyoidea: The Good, the Bad, and the Unknown

8.1 BASIC BIOLOGY

The Eriophyoidea is a superfamily consisting of three families of mites that contain species of agricultural importance (Baker et al. 1996). Excellent review articles on the Eriophyoidea, published in 2010 in the journal *Experimental and Applied Acarology* (see References), updated information provided earlier in the book *Eriophyoid Mites: Their Biology, Natural Enemies, and Control*, edited by Lindquist et al. (1996). A great deal of detailed information is available in these references on the morphology, taxonomy, biology, ecology, behavior, rearing methods, sampling methods, and control of these mites. Eriophyoid mites are considered to be second in importance as agricultural pests after the Tetranychidae.

According to de Lillo and Skoracka (2010), about 4000 species of eriophyoid mites are known to attack a diverse array of plants, but "a huge number of species remains undiscovered," especially species in the tropics and subtropics, areas that have been studied very little (Amrine and Stasny 1994, Amrine 1996, 2003). de Lillo and Skoracka (2010) noted that 70 new eriophyoid species have been described each year between 1996 and 2007. Although eriophyoid mites can be crop pests, the relationship between many of these species and their host plants is more benign. It appears that, in some lineages, the relationship has evolved to allow both plant and mite to coexist without significant harm.

Eriophyoids are tiny plant-feeding mites placed in the families Eriophyidae, Phytoptidae, and Diptilomiopidae. Some species can be significant crop pests because they cause direct damage, while others can transmit plant viruses as well. Gamliel-Atinsky et al. (2010) implicate some eriophyoids in the mechanical transmission of spores of fungal plant pathogens. Because eriophyoid mites are so small, they are difficult to detect in shipments of crop commodities. As a result, they are easily transported around the world on their host plants, creating significant economic damage (Navia et al. 2010); for example, the pear leaf blister mite *Eriophyes pyri* is a pest of fruit trees around the world, as is the tomato russet mite *Aculops lycopersici.*

Some eriophyoids are beneficial to humans for controlling weeds or serving as prey for phytoseiid predators (Smith et al. 2009b, Skoracka et al. 2010). Most pest eriophyoids attack perennial plants important in agriculture and forestry, but some also are found on annual plants such as tomatoes (Lindquist et al. 1996, de Lillo and Skoracka 2010).

The Eriophyoidea are unusual because they have only two pairs of legs as adults (Figure 8.1 and Figure 8.2) (also see Figures S8.1 and S8.2 on the CD). The life cycle consists of egg → larva → nymph → adult males and females. Actually, though, the life cycle of eriophyoids is more complex than this simple statement. In a few species, the production of only males by virgin females has been

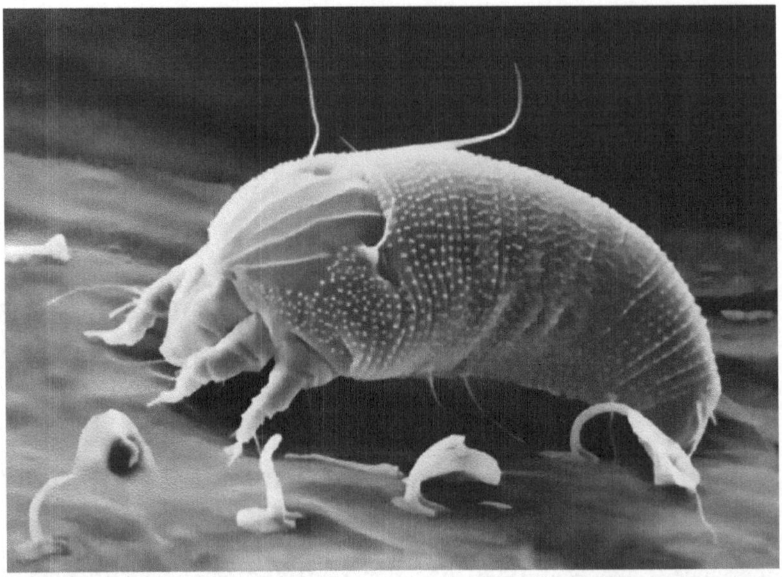

Figure 8.1 Scanning electron micrograph of the eriophyid rust mite (*Aceria anthocoptes*) on Canada thistle. Note the two pairs of legs on this eriophyoid. (Photograph by U.S. Department of Agriculture, Agricultural Research Service.)

observed and has led to the conclusion that eriophyoids are arrhenotokous, with a female-biased sex ratio. Females pick up spermatophores deposited by males on the host plant, and the males of some species deposit spermatophores around quiescent female deutonymphs to enhance the rate of discovery of their spermatophores (indirect sperm transfer). In some species, **ovoviviparity** (retention of the egg within the mother's body until the egg is ready to hatch) or **viviparity** (in which the egg is retained inside until the egg hatches) has been reported. Females typically deposit about 50 eggs. Because eriophyoids have only two pairs of legs, it is difficult to determine if the immature stages are larvae or nymphs except for size differences. In some species, two morphologically distinct forms of adult females, the **protogyne** and **deutogyne**, are known. These different females have, in the past, led to considerable taxonomic confusion, and they were given two different species names. In a few species, two forms of males are produced, one resembling the protogyne and the other the deutogyne female. Protogyne females occur in the summer on deciduous plants, and deutogyne females overwinter in diapause in regions with well-defined winters. Diapause appears to occur primarily in adult females, but diapause in eggs may occur.

Eriophyoids are the smallest arthropods that feed on plants, averaging 100 to 500 μm (0.1 to 0.5 mm) in length, and they cannot be seen without substantial magnification. Eriophyoid mouthparts consist of a pair of small stylet-like chelicerae and a pair of accessory stylets, which are modified palps. The small size of the stylets allows eriophyoids to penetrate only about 5 μm (0.005 mm) into the leaf, so feeding occurs in the plant epidermis. Eriophyoids disperse by walking, by farm machinery or farm workers, and by the wind. One species, *Aceria litchii*, has been found in the hairs of honey bees after the bees foraged in infested flower panicles of lychees in Australia. Subsequently, they were found in new flower panicles (Waite 1999). It is thought that it is only the adult females that disperse.

Life histories of some eriophyoid mites are known in detail, but much remains to be learned about key pests. The common names of eriophyoids are blister mites, rust mites, bud mites, or gall mites, because their feeding on tender new growth causes distinctive types of damage. The Eriophyoidea are found most often on perennial plants, and most are generally quite host specific.

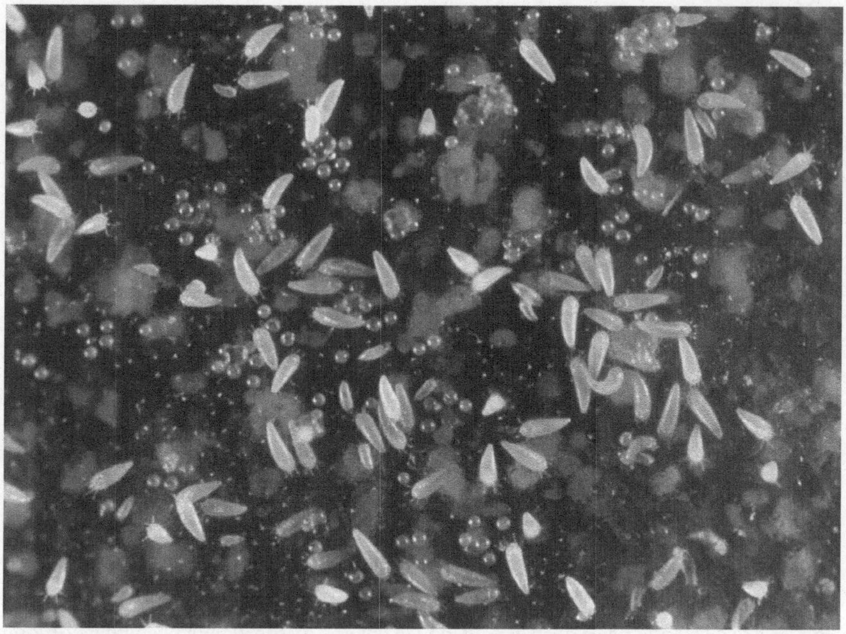

Figure 8.2 The tomato rust mite (*Aculops lycopersici*) is a leaf vagrant on tomato leaves, stems, and fruit. (Photograph by Lyle Buss, Department of Entomology and Nematology, University of Florida, Gainesville.)

In fact, eriophyoids are sufficiently host specific that a key is available that shows photographs of the damage on each host plant in North America. If you know the name of the host plant and the type of damage the mite causes, it is possible to identify the mite species (Keifer et al., 1982), unless they are species that have invaded North America since 1982. Even where two eriophyoid species occur on a plant, their feeding damage and location on the plant may be so distinctive that they can be discriminated.

Fusiform eriophyoids that are found wandering on the leaf or bud surfaces of their hosts are classified as **rust mites** or **leaf vagrants** (e.g., citrus rust mite, tomato russet mite, or pear rust mite) (Figure 8.1 and Figure 8.2) (also see Figures S8.1 and S8.2 on the CD). These mites cause a bleaching and drying of leaf tissues from their feeding (see Figure S8.3 on the CD). Eriophyoids are found within buds, in blisters, or in galls (Figure 8.3 and Figure 8.4) and are called **gall** or **bud mites** (see Figures S8.4A–D and S8.5 on the CD). Leaf injury may involve only the surface (epidermis) of the leaf because the feeding stylets are short; for example, the peach rust mite (*Vasates cornutus*) causes browning or silvering of the peach leaf surfaces. Walnut blister mites (*Eriophyes erinea*) cause blister-like swellings on walnut leaves (Figure 8.3). *Eriophyes pyri*, the pear blister mite, invades the mesophyll of pear leaves and causes serious injury. The formation of hairy patches or **erinea** on the underside of leaves (see Figure S8.6 on the CD) produces sites favorable for survival of the eriophyoids. Buds are injured when eriophyoids feed on their surface or when they cause gall formation. The citrus bud mite (*Aceria sheldoni*) causes fruit and leaf malformation in lemons. When eriophyoids feed in buds, they can cause a **witches' broom** of twigs, flower galls, shortening of internodes, or secondary development of leaf hairs (see Figure S8.5 on the CD).

All eriophyoids appear to prefer young, soft tissues and meristems of host plants (Petanovic and Kielkiewicz 2010a,b). During feeding, saliva is injected into plant cells. It is believed that the saliva contains enzymes that are used to digest the cuticle and cellulose of the cell wall. The mite then sucks out the cell contents during a 10- to 20-minute feeding interval; however, damage varies by

Figure 8.3 Blister-like swellings on English walnut leaves caused by the walnut blister mite (*Eriophyes erinea*). (Photograph by Jack Kelly Clark, University of California Statewide Integrated Pest Management Program.)

mite species and host plant. In some cases, damage is limited to the cells that have been punctured by the mites. After feeding, rust mites, for example, move away to feed on new cells, and this can result in large-scale damage to the epidermis, resulting in bronzing, silvering, or **russeting**. In some cases, feeding by these mites can result in degeneration of the cells of the spongy parenchyma, thickening of cell walls, and deposition of lignin-like compounds. In addition, plants may produce defensive proteins in response to feeding by these vagrant mites.

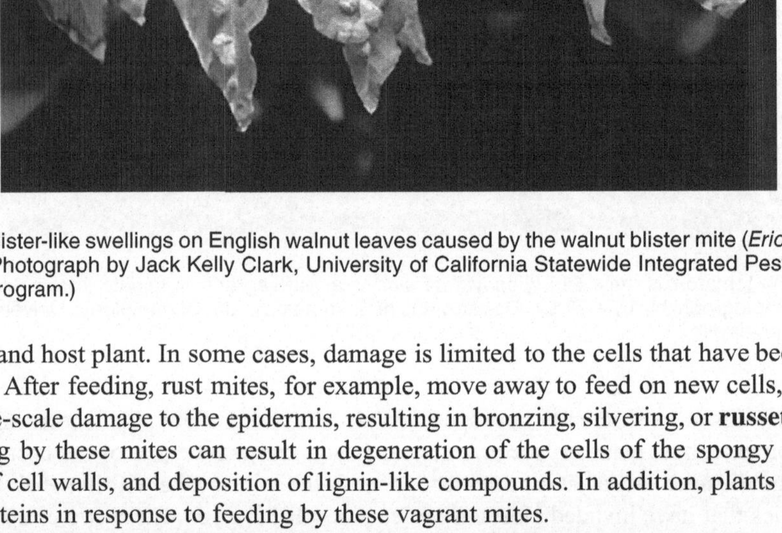

Figure 8.4 Damage to lemon caused by the citrus bud mite, *Aceria* (= *Eriophyes*) *sheldoni*. (Photograph by Jack Kelly Clark, University of California Statewide Integrated Pest Management Program.)

The induction of galls in host plants is much more complex, and the biochemistry involved is not yet fully resolved. What is known is discussed by Petanovic and Kielkiewicz (2010a). Galling first involves puncture of the cell wall, followed by a change in cellular pH, changes in cell-membrane permeability, modification of nuclear DNA, and other cytological modifications. The plant cell produces callose tissue induced by the wounding, followed by the formation of layers of nutritive tissues from which the gall mites obtain their food. Often, many hairs grow on the surface of the gall due to abnormal mitoses. Other complex biochemical reactions occur, depending on the mite species and their host plant.

Presumably, eriophyoids living in galls, erinea, or in leaf sheaths under bud scales are protected, to some degree, from predation by phytoseiids or other predators. These modifications also provide microhabitats that increase the relative humidity. Eriophyoids may modify their behavior to increase their survival; for example, *Rhinophytoptus concinnus* were shown to climb to the top of elm leaf trichomes shortly before becoming quiescent, remaining there until they molted. This perching behavior on the tips of leaf hairs by quiescent stages has been observed in other eriophyoids. The anal sucker appears to be sticky, which seems to help the mites stay on the tip of the leaf hair. Michalska (2003) demonstrated that this behavior allows the mites to hide from predators, although once the predators detect them they are eaten readily. This climbing behavior could be a method to avoid wetting of the leaf in regions where there is high relative humidity, which can foster infection with entomopathogenic fungi (Michalska 2003). The taxonomy of the Eriophyoidea continues to change. Table 8.1 provides a partial list of the genera of eriophyoids by family (Lindquist et al. 1996).

Table 8.1 Some Genera of Eriophyoid Mites in the Three Families

Family	Genera	
Phytoptidae	*Acathrix*	*Retrarcus*
	Mackella	*Septoptus*
	Nalepella	*Trisetacus*
	Phytoptus	
Eriophyidae	*Abacarus*	*Colopodacus*
	Acalitus	*Coptophylla*
	Acaricalus	*Cosella*
	Acaphylla	*Cosetacus*
	Acaphyllisa	*Eriophyes*
	Acera	*Epitrimerus*
	Acerimina	*Ioracarus*
	Acitonotus	*Keiferophyes*
	Aculus	*Neocalacarus*
	Aculops	*Paracolomerus*
	Anthocoptes	*Paraphytoptus*
	Calacarus	*Phyllocoptruta*
	Calepitrimerus	*Phyllocoptes*
	Cecidophyes	*Tegolophuys*
	Cecidophyopsis	*Tetra*
	Ciolomerus	*Vasates*
	Cisaberoptus	
Diptilomiopidae	*Apodiptacus*	
	Diptacus	
	Trimoptes	

Source: Adapted from Lindquist, E.E. et al., Eds., *Eriophyoid Mites: Their Biology, Natural Enemies, and Control*, Vol. 6, Elsevier, Amsterdam, 1996.

8.2 VECTORS OF DISEASE: DEFINITELY UNDESIRABLE

At least nine eriophyoids are known to be vectors of 12 plant viruses: potyviruses from the genera *Rymovirus* and *Tritimovirus*, alexiviruses, and nepoviruses (Comoviridae) (Oldfield 1970, Nault 1997, de Lillo and Skoracka 2010). Each virus has a limited host-plant range and, generally, a single mite vector. Recent studies suggest that several eriophyoids can transmit more than one virus, but it is possible that there are cryptic species in these mite "species," with each one transmitting a different virus.

One of the best-studied eriophyoid-vectored diseases is wheat streak mosaic virus (WSMV), which is transmitted by *Aceria tosichella* (Eriophyidae). Carew et al. (2009) indicated recently that *A. tosichella* might be a species complex in Australia, with two separate, closely related species present. Both types were found on wheat, but one was found on other host plants, as well. The mite known now as *A. tosichella* originally was identified as *A. tulipae*, a species that fed on wheat, onions, garlic, and tulips. Later, the species on tulips was found to be distinct from the wheat species, and the wheat species was named *A. tritici* (but *A. tritici* is considered to be a synonym of *A. tosichella*). It is unclear whether the two lineages in Australia both transmit virus, whether they both transmit the virus efficiently, and whether they overwinter on the same host plants. Adequate control of alternative host plants during winter could reduce the ability of these mite species to transmit WSMV.

Eriophyoid mites can acquire viruses after a few minutes of feeding, but longer access to plants results in higher transmission rates. Viruses are retained by the mites following a molt, but they are not transovarially transmitted. Viruses persist for about a week at room temperature or for several weeks at lower temperatures in the mite vectors. The virus must be acquired by one of the two nymphal stages if subsequent transmission by adults is to occur. Because adults cannot acquire and transmit viruses, spread of the viruses must originate from the original host of the individual mite.

Dispersal of eriophyoids, and thus spread of a virus, depends on temperature and wind. Although eriophyoids are wingless, during daylight hours they may stand up on their anal papillae and allow themselves to be distributed by wind. In addition, they can attach themselves to farm machinery or to farm worker's clothing and be moved shorter distances.

High plains disease is consistently associated with a virus and the wheat curl mite (*Aceria tosichella*) (Brown and Skoglund 1999). Prior to 1993, no corn virus diseases were found in Colorado, but a disease, initially diagnosed as wheat streak mosaic, was found that year. Tests for known corn viruses in the United States were negative, indicating it was a new virus, and the disease has now been found in other parts of the United States and in other countries. Infected corn plants show chlorotic flecking, chlorotic and reddened stripes running parallel to the veins, severe stunting, and eventually death. In wheat, symptoms resemble wheat streak mosaic virus, with linear flecks and yellow streaks, necrosis, stunting and death. Unfortunately, mixed infections of high plains disease and wheat streak mosaic virus can be found, which makes field diagnosis on the basis of symptoms difficult.

Plants infected with the virus can be confirmed by enzyme-linked immunosorbent assay (ELISA) and western blots (another method of identifying specific proteins). The host range of the virus appears to include corn, wheat, barley, rye, and oats, as well as cheat grass (*Bromus secalinus*) and yellow foxtail (*Setaria glauca*). The virus can be seed transmitted. Resistant cultivars have been identified in wheat, and integrated management is possible using host-plant resistance, elimination of volunteer plants during summer and winter, isolation from small grains (sources of the virus and vector), and modification of planting date (planting after weeds and volunteer wheat are eliminated).

8.3 SELECTED ERIOPHYOID PESTS

8.3.1 *Phyllocoptruta oleivora*

Citrus rust mite is one of the most important and widespread citrus pests in the world (Allen 1978, 1979, Hall et al. 1997). These mites reproduce rapidly; the life cycle requires about 7 to 10 days in summer, and females live about 20 days, depositing groups of 20 to 30 eggs in indentations on fruits and on the ventral surfaces of leaves. Severely infested fruits have a powdery appearance as a result of the exuviae left by the molting immatures. Mites do not tolerate bright sunlight well, so they inhabit the undersurfaces of leaves and shaded areas of the fruit; they do well in warm, humid conditions. Feeding damage affects the exterior of the fruit, which causes fresh market fruit to be downgraded; damaged fruits become silver, reddish brown, or purplish black as time goes by. The interior quality of the fruit is not affected, so when the fruit is grown for juice higher mite densities can be tolerated. The rust mite is able to colonize very young fruit, so control measures, if required, should be applied early. For additional information on citrus rust mite, see Chapter 18. In South Africa, a predatory mite, *Amblyseius citri*, feeds on this rust mite, although it is not fully effective. In Florida, the rust mite is controlled in part by the fungus *Hirsutella thompsonii*. Unfortunately, the fungus often is eliminated if sprays are applied to control fungal diseases of citrus. The rust mite is thought to have originated in Southeast Asia, but little information is available on the natural enemies of this pest there, and classical biological control of this pest should be explored to determine if effective natural enemies in Asia could be identified.

8.3.2 *Aculus cornutus*

The peach silver mite attacks peaches and almonds. These mites are found on leaves, and their feeding causes the leaf surface to appear silver when high populations are present. High densities can reduce tree growth and reduce fruit size. Peach silver mites can serve as prey for the phytoseiid *Metaseiulus occidentalis* in California almond orchards and thus rarely become a pest in almond orchards under integrated mite management (IMM) there (see Chapter 17).

8.3.3 *Aculops lycopersici*

The tomato rust mite is a widespread pest of tomatoes that attacks a variety of plants in the Solanaceae, making it an unusual eriophyoid (Duso et al. 2010). It is very common on tomatoes and can easily kill tomato plants. It causes damage to chilies, potatoes, eggplants, tobacco, and petunias. Tomato rust mite is often found on *Ipomoea purpurea* (morning glory) in field margins, as well as on a variety of other weeds, which can serve as a reservoir. Tomato rust mites have multiple generations during the growing season, tolerate fluctuations in temperature and relative humidity, and have a short generation time of approximately 6 to 7 days. Females can deposit 10 to 53 eggs, and populations increase best at approximately 27°C and 30% relative humidity. Mites disperse by wind, by farm workers, and on agricultural tools. Early damage symptoms on tomato are silvering of the undersurface of lower leaves, browning of leaves, followed by cracking. Tomato rust mites prefer warm weather and are able to feed on the upper surfaces of leaves in direct sunlight (an unusual feature for eriophyoids) (see Figure 8.2) (also see Figure S8.2 on the CD). Symptoms include bronzing of leaves, withering, and change of stem color from green to brown (see Figure S8.2B on the CD). Defoliation takes place, so eventually only young growth remains. The fruits become sunburned, plant growth is slowed, and fruit production is reduced. If the fruits are attacked, the skin becomes brown. On potatoes, the tomato rust mite does not cause browning, but the leaves may become dry and the whole plant can succumb.

Distributions within fields can be patchy, making it difficult to monitor crops. In greenhouses, the pest can increase rapidly at high temperatures and low relative humidity. When the tomato plant begins to die, the rust mite females migrate to the top of the plant, holding their bodies perpendicular to the leaf surface and adhering to the surface by their anal papillae. This posture allows them to be blown off the leaf surface. On occasion, they form chains by crawling up each other to disperse.

Management plans vary with the geographic region. Growing the crop with sufficient water is important, because often the most damaged plants are water stressed. In northern Europe, IMM can be achieved, but in southern Europe the mild climate allows the mites to move to alternative host plants, and chemical control is more often used to suppress the mite. Sulfur or oils can control the tomato rust mite. Research is under way to determine how to use selective pesticides and phytoseiids on a commercial scale. Unfortunately, glandular trichomes on tomatoes can limit the effectiveness of some phytoseiids, as can the webbing produced by spider mites on the crops.

8.3.4 *Aculus schlechtendali*

The apple rust mite feeds on flowers, fruits, and leaves of apple in Europe and North America (Hoyt 1969, Walde et al. 1997, Spieser et al. 1998). It causes apple leaves to roll up longitudinally and become rusty brown on the lower side, in addition to causing russeting of the fruit. This mite may increase transpiration rates and decrease photosynthesis when densities are greater than 50 mites per square centimeter, but it is unclear how this affects tree growth or yield (Duso et al. 2010). Different apple cultivars appear to respond differently to feeding by this mite, and the mite seems to be a more serious pest in Europe where susceptible cultivars are grown. By contrast, this mite does not appear to be a serious pest of apples in North America, perhaps because different cultivars are grown or because predatory mites that are maintained in these orchards through the use of selective pesticides use these mites as prey. A variety of predators, including phytoseiids, anthocorids, mirids, and lacewings, feed on these mites. In Europe, the phytoseiids *Typhlodromus pyri*, *Amblyseius andersoni*, and *Euseius finlandicus* are important predators. In North America, *T. pyri*, *Metaseiulus occidentalis*, *T. caudiglans*, and *M. citri* prey on these mites. Other acarine predators known to feed on these mites include *Zetzellia mali* (Stigmaeidae), the anystid *Anystis baccarum*, and the tarsonemid *Dendroptus* near *suskii* (Duso et al. 2010).

8.3.5 *Aceria sheldoni*

The citrus bud mite can be a serious pest of citrus. It occurs where citrus is grown and attacks all types of citrus, but it is most injurious to lemons (see Figure 8.4). These mites occur in sheltered places such as under bud bracts, within buds at the base of the leaf petioles close to the buds, and under the fruit calyx. They avoid living in exposed sites and can invade buds of all ages, even dormant buds on old wood. One to three mites per bud can retard growth. Leaves developing from infested buds may have strange shapes, and flowers may be misshapen and underdeveloped. Lemons that develop from damaged fruit may drop off prematurely or take on strange shapes. The symptoms on navel and Valencia oranges are similar, but not so severe. Natural enemies of *Aceria sheldoni* include a fungus disease (*Hirsutella* sp.) and predatory mites in the family Stigmaeidae (*Agistemus africanus* and *A. transatalensis* are predators of this mite in Africa). Where broad-spectrum pesticides are applied in citrus, the importance of the bud mite probably has been underestimated.

8.3.6 *Aceria guerreronis*

The coconut mite attacks the very young fruits (nuts) of the coconut palm, which is thought to be its primary host (Howard and Moore 2006). Very large populations can build up rapidly, causing scarring and distortion of the nuts and premature nut drop. The mite is a serious pest of coconuts in many tropical countries, although its original distribution is thought to be tropical America; however, the coconut mite has been found in West Africa, so it is unclear whether the mite was originally from the Eastern or Western Hemisphere. It might have made its way to the Americas several hundred years ago along with the introduction of coconuts, but this is controversial. Navia et al. (2005) evaluated 29 populations of the mite from the Americas, Africa, and the Indo-Pacific regions using molecular methods and found that the highest genetic diversity occurred in Brazil, suggesting that the coconut mite originated in South America, perhaps on a different palm species. Because the coconut is thought to have originated in the Indo-Pacific region, it appears that the coconut palm is a new host of this mite, which could also explain its virulence to the coconut palm. New-host associations often result in significant damage to the host. What is not clear is why the mite has become invasive within the last 50 years (Navia et al. 2005).

The mites develop from egg to adult in about 10 days. Populations build up on the lower surfaces of the perianth (outer envelope of a flower) during the first 6 months of development of the coconut but decline thereafter while the coconut matures, in about 12 months. The mites may disperse by air currents or by crawling. Varieties or cultivars differ in their susceptibility to the coconut mite, but all are damaged to some degree. The coconut may become distorted, stunted, and discolored, and up to 30% of the copra (coconut meat within the nut) can be lost. The clear liquid in immature coconuts is often harvested and sold as coconut milk, but malformed coconuts are not readily purchased.

Control of *Aceria guerreronis* varies from country to country. This mite typically is not a serious pest in the Western Hemisphere. A number of predatory mites have been found to prey on coconut mites in Florida, including the phytoseiids *Amblyseius largoensis*, *Neoseiulus mumai*, and *N. paspalivorus* (Howard and Moore 2006). Other predators (e.g., *Bdella distincta*, *Steneotarsonemus furcatus*) have been found associated with the coconut mite in Puerto Rico. *Hirsutella thompsonii* has been found attacking coconut mites, but its efficacy is dependent upon high relative humidity. In Brazil, the predatory mites *Neoseiulus paspalivorus* and *Proctolaelaps bickleyi* are associated with the coconut mite (Lawson-Balagbo et al. 2007, 2008). In Benin, *N. paspalivorus* is an important predator of *A. guerreronis* (Negloh et al. 2010). In some cases, farmers prune off the coconuts in all stages of development to eliminate the coconut mites. This control method reduces yield but is appropriate where coconuts are pruned to prevent the falling nuts from hurting tourists. Chemical control of the mites is difficult in these tall trees, especially because sprays have to be repeated frequently. Systemic miticides may persist longer but could result in residues within the coconuts.

Efforts to conduct classical biological control programs are under way, and methods for rearing the mite and phytoseiids are being investigated (De Silva and Fernando 2008). Also, life history studies of candidate species are being conducted (Lawson-Balagbo et al. 2007, 2008, Domingos et al. 2010), and methods to monitor populations (Siriwardena et al. 2005) are being developed.

8.4 COLLECTING AND SAMPLING ERIOPHYOIDS

Extracting eriophyoid mites from leaf hairs, galls, or buds and counting them is very tedious. These tiny mites are very susceptible to damage and desiccation. It can require considerable ingenuity to collect eriophyoids and estimate their abundance. A variety of sampling methods have been developed for these very tiny mites (Perez-Moreno and Moraza-Zorrilla 1998, Monfreda et al. 2010). Their small size, tendency to occur in cryptic habitats on their host (buds, erinea, galls), and propensity for using tender new growth make monitoring eriophyoids difficult. So far, no molecular methods have been used to detect eriophyoid mites; rather, foliage or plant samples are collected and, most frequently, examined under a dissecting microscope. Other collection methods include sticky tape or adhesive traps (plates thinly coated with grease), brushing machines, or washing. Aerial dispersal can be studied by monitoring glass slides or plates coated with grease or with detergent-water pan traps. See Monfreda et al. (2010) and de Lillo and Skoracka (2010) for reviews of these methods.

8.5 ERIOPHYOIDS AS ALTERNATIVE PREY: POTENTIALLY GOOD

Aculus schlechtendali (Eriophyidae), the apple rust mite, is considered a beneficial species in Washington apple orchards under an IMM program because these mites serve as prey for the phytoseiid *Metaseiulus occidentalis* early in the growing season (Hoyt 1969). Apple rust mites allow predator populations to increase so they can suppress the spider mite pests (*Panonychus ulmi* and *Tetranychus* spp.) on apples later in the growing season (Hoyt 1969). Apple rust mite populations must become relatively large before they cause significant damage to apple foliage in the northwestern United States. Likewise, in the eastern United States, *A. schlechtendali* is an alternative food source for *Typhlodromus pyri* early in the growing season. *Typhlodromus pyri* is an important predator of the European red mite (*Panonychus ulmi*) in apples there.

8.6 INVASIVE ERIOPHYOID SPECIES: CLEARLY BAD

The small size of eriophyoids makes them easy to transport on fruits and foliage without detection; for example, *Tegolophus perseaflorae* has been found in flowers and fruits of avocado in south Florida (Peña and Denmark 1996). The mites feed on buds, causing necrotic spots on apical leaves and subcircular, irregular openings on mature leaves. Mites have also been found on petioles, the underside of leaves, and fruitlets. Their feeding causes fruit deformation and discoloration. Peña and Denmark (1996) speculated that avocado grafting material is often transported and that this new pest was introduced unintentionally into Florida. Another invasive eriophyoid, *Aceria guerreronis*, recently invaded coconuts grown in India and Sri Lanka, where it is causing substantial economic problems (Howard and Moore 2006).

Navia et al. (2010) evaluated the impact of eriophyoid invasions on agriculture, the mechanisms by which the mites invaded, and the need to raise awareness of these pests so that additional invasions are prevented through adequate quarantine and detection methods. They include a table listing invasive species, their origins, their host plants, and the countries into which the mites have been introduced. Because eriophyoids are so tiny, relatively little effort has been given to preventing invasions, and rapid methods for identifying these mites by quarantine officers are needed. Improved methods also are needed to disinfest plants of eriophyoids before their shipment to new countries.

8.7 BIOLOGICAL CONTROL OF WEEDS BY ERIOPHYOID MITES: POTENTIALLY GOOD

Eriophyoid mites have several useful attributes for biological control of weeds. Eriophyoids are often host specific even to the level of the race of the plant and the host tissue they attack. They are easily distributed by the wind. They can be used with other biological control agents. Some species are able to transmit specific plant viruses to the weed. Eriophyoids can reduce photosynthesis and cause witches' brooming and flowerbud deformities, resulting in reduced seed production. Negative aspects of eriophyoids as weed control agents include the fact that they are relatively slow acting and must often be used in conjunction with other natural enemies. Most are quite sensitive to low relative humidity. Furthermore, they are often killed by phytoseiids or other predators or fungal pathogens, which can disrupt their beneficial roles in the biological control of weeds.

Despite these negatives, approximately 13 eriophyoid species have been evaluated as biological weed control agents (Sobhian et al. 2004, Smith et al. 2009, 2010, Skoracka et al. 2010). Evaluation involves tests to confirm that the host specificity of the candidate natural enemy is strong. Laboratory evaluation under choice and no-choice situations using plants related to the target weed will confirm

Table 8.2 Eriophyoids Released to Date as Biological Control Agents of Weeds

Eriophyoid	Weed	Release Sites
Aceria chondrillae (multiple strains)	*Chrondrilla juncea* (skeletonweed)	United States, Australia, Argentina
Aceria genistae	*Cytisus scoparius* (Scotch broom)	New Zealand, Australia, United States (accidental)
Aceria hyperici	*Hypericum perforatum* (St. John's wort)	Australia
Aceria malherbae	*Convolvulus arvensis* (field bindweed)	United States
Aceria sp. (boneseed leaf buckle mite)	*Chrysanthemoides monilifera* (boneseed or bitou bush)	Australia
Cecidophyes rouhollahi	*Galium aparine, G. spurium* (Rubiaceae)	Canada
Floracarus perrepae	*Lygodium microphyllum* (Old World climbing fern)	United States (Florida)

Source: Adapted from Smith, L. et al., *Exp. Appl. Acarol.*, 51, 115–149, 2009.

host specificity. In some cases, laboratory (quarantine) results can indicate that non-target plants may be a host, but the results are a laboratory artifact because the natural behavior of the mite was inhibited. Because eriophyoid mites often leave plants by aerially dispersing, laboratory trials in which there is no wind could force the mites to feed on unsuitable plants. Thus, field trials in the country of origin may be necessary to confirm the specificity of the mite under more natural conditions in which the host plants are in a suitable physiological condition to potentially support the feeding and development of the mites, environmental conditions (especially relative humidity) are suitable for survival, and adequate time is allowed for the colonization to occur (Rosenthal 1983, Smith et al. 2009a).

To date, seven species have been permitted to be released as biological control agents of weeds, including *Aceria* spp. (e.g., *A. genistae, A. chondrillae, A. malherbae*), *Cecidophyes rouhollahi, Floracarus perrepae*, and *Aculus hyperici* (Rosenthal 1983, Smith et al. 2009b) (see Table 8.2). The results of the releases of these species are variable.

Aceria chondrillae is native to Europe and has been released for the biological control of skeleton weed *Chondrilla juncea* in Australia and the United States (Sobhian and Andres 1978, Smith et al. 2009b). Mite feeding induces gall formation in the vegetative and flower buds, causing stunting, reduction in seed formation, and weakness. Mites appear to be very specific to particular geographic races of skeleton weed, of which there are four. A strain of mite originating from Greece was most suitable against the weed in Australia, but this mite strain did not perform well in the United States. An Italian mite strain has been introduced to control the skeletonweed population in the United States. In different localities, the effectiveness of *C. juncea* varies. In some cases, the mites are preyed upon by phytoseiids or are negatively affected by winter weather. This mite has had significant impact on the weed in some localities but has been less effective in others (Smith et al. 2009b).

Aculus hyperici is native to Europe and has been used to control St. John's wort (*Hypericum perforatum*) in Australia (Smith et al. 2009b). Again, the results have been variable, although the mite does appear to reduce reproduction and spread of the weed. This mite reproduces on at least four other *Hypericum* species, including a native nontarget species in Australia.

Aceria malherbae is native to Europe and attacks field bindweed (*Convolvulus arvensis*). It produce galls and reduces shoot and root biomass (Smith et al. 2009b). Another taxonomic problem occurred with this project when *A. malherbae* was misidentified as *Eriophyes convolvuli*, which is now reassigned to the genus *Phyllocoptes. Aceria malherbae* has had a significant impact on field bindweed in some localities (Texas) but has been less successful in row crops. The mites more severely affect bindweed populations during droughts, and damage appears to be greater on water-stressed plants. Mite populations are negatively affected by late freezes in spring and by variations in the suitability of some weed populations. Overall, *A. malherbae* has had a significant impact on weed populations in some sites; however, it will also attack other *Convolvulus* species and 12 *Calystegia* (bindweed or morning glory) species and cannot be released in California over concerns about endemic *Calystegia* species.

Cecidophyes rouhollahi has been released in Canada to control *Galium aparine* and *G. spurium* (Rubiaceae) (also known as goosegrass, stickyweed) (Smith et al. 2009b). The mites cause severe stunting and completely prevent seed production by the weed in the laboratory. The mites appear to be unable to survive the winter, perhaps because the mites were introduced from southern France, where winters are milder. It is possible that populations of mites from a colder climate could survive winters in Canada.

Floracarus perrepae attacks the Old World climbing fern *Lygodium microphyllum*, producing galls on the fronds and causing necrosis. A population obtained from an area in Australia with a climate similar to that of South Florida has been approved for release (Smith et al. 2009b). Studies conducted during the exploration phase of this project determined that local strains of mites were adapted to plants at that location, but plants from other sites could be resistant to the mites. There are concerns that predators and the fungus *Hirsutella thompsonii* could reduce the impact of this natural enemy in subtropical South Florida, but it is too early to determine how effective this mite will be. In addition to the mite, insects that feed on the fern are being evaluated and released.

Additional research is ongoing with other eriophyoid mites as natural enemies of weeds, and eriophyoids accidentally introduced into new geographic areas are having an impact on some weed populations (Smith et al. 2009b). Skoracka et al. (2010) observed that "80% of eriophyoids have been reported on only one host species, 95% on one host genus, and 99% on one host family." Although the host specificity of these mites is considered an advantage in weed biological control from the point of view of risk analysis, "in such highly coevolved host–parasite systems the plants should have higher levels of resistance and/or tolerance to eriophyid mites and the eriophyids should have lower ability to injure the plant" (Smith et al. 2009b). Therefore, eriophyids may provide "relatively few species that can effectively control a target plant in an exotic location." Finally, additional research on the ability of these mites to switch hosts is required to improve the analysis of risks associated with these mites when used in the biological control of weeds.

8.8 IDENTIFICATION OF ERIOPHYOIDS

These mites are very small and require the use of keys with specialized terminology for morphological traits, careful mounting methods, and high-quality microscopes to identify them (Amrine and Stasny 1994, Lindquist et al. 1996, Amrine 1996, 2003). As noted above in the discussion on biological control of weeds, there may be cryptic species or biotypes with specializations to specific habitats and host plants. However, if you know the plant species in North America, the location of the mite on the plant, and the damage it causes, you are likely to be able to identify the North American species without having to use a key to the mites. See Keifer et al. (1982) for a list of host plants, symptoms of injury, and mite species in North America (available on the accompanying CD). Of course, this reference will not work if the eriophyoid to be identified is a new pest in the United States or if you are outside of North America.

The use of DNA-based methods for identifying species and cryptic species has been limited so far with eriophyoid mites (Navajas and Navia 2010); however, such methods could clarify suspect synonymies, test hypotheses of cryptic species, examine the occurrence of biotypes in host–plant associations, improve the study of colonization patterns of invasive species, and improve their use as

Table 8.3 Potential Integrated Mite Management Tools for Managing Eriophyoid Mites in Agriculture

1. Identify or develop cultivars or biotypes of crop plants that are resistant to or tolerant of the eriophyoid; for example, high trichome densities and glandular trichomes in tomatoes can confer resistance.
2. Maintain adequate water for crops to reduce population growth rates; for example, tomato rust mite populations increase more rapidly on water-stressed plants.
3. Eliminate weeds or volunteer plants in fields where overwintering mites may occur.
4. Isolate crops from plants likely to contain a virus and mite vector.
5. Modify the time of planting of crops to reduce disease transmission by mites.
6. Use pesticides to suppress rust, gall, blister, and bud mites. Apply the products at the correct time and in a manner that reaches these mites prior to their induction of galls and blisters or entry into the buds. Most acaricides appear to be able to suppress eriophyoid populations, although resistance to pesticides is a concern. Oils or sulfur can control eriophyoid mites, if applied correctly.
7. Use chemical control of pest insects in the crop in a manner that is least disruptive to the natural enemies of eriophyoid mites. Predators may provide excellent control of eriophyoids if not disrupted by nonselective pesticide applications (**conservation biological control**).
8. Biological control of invasive eriophyoid mites may be feasible using a **classical biological control** approach.
9. Consider **augmentative biological control** using predators or of fungal pathogens, which may be feasible in some circumstances, although none is currently sold for this purpose.
10. Improve the ability to detect and identify potentially invasive pests during commercial trade of plants.

Source: Based on information from Childers et al. (1996) and Van Leeuwen et al. (2010).

biological control agents against invasive plants (Navajas and Navia 2010). The importance of identifying these mites correctly and rapidly cannot be overemphasized. Without rapid and easy tools to identify these mites, incorrect identifications can delay biological control of weed projects, inhibit the ability of quarantine officers to prevent the accidental importation of invasive eriophyoids into new countries, and make control of eriophyoids more difficult.

8.9 CONTROL OF ERIOPHYOIDS

The control of eriophyoid mites until recently has focused on the use of pesticides (Childers et al. 1996, Van Leeuwen et al. 2010). To develop IMM programs, considerable knowledge about these pests' biology, ecology, and behavior must be obtained and used (Table 8.3). We need detailed life histories of key mite pests under natural conditions, including the effects of climate and host-plant type. Key gaps in our knowledge include what influences aerial dispersal and the ability of these mites to survive over winter, which could affect the cultural practices employed in pest management programs.

SELECTED REFERENCES

Allen, J.C. 1978. The effect of citrus rust mite damage on citrus fruit drop. *J. Econ. Entomol.* 71:746–750.

Allen, J.C. 1979. Effect of citrus rust mite damage on citrus fruit growth. *J. Econ. Entomol.* 72:195–201.

Amrine, J.W. 1996. *Keys to the World Genera of the Eriophyoidea (Acari: Prostigmata).* West Bloomfield, MI: Indira Publishing.

Amrine, J.W. 2003. *Catalog of the Eriophyoidea: A Working Catalog of the Eriophyoidea of the World*, Version 1.0. College Station: Texas A&M University (http://insects.tamu.edu/research/collection/hallan/acari/eriophyidae/).

Amrine, J.W. and T.A. Stasny. 1994. *Catalog of the Eriophyoidea (Acarina: Prostigmata) of the World.* West Bloomfield, MI: Indira Publishing.

Baker, E.W., T. Kono, J.W. Amrine, M.D. Delfinado-Baker, and T.N. Stasny. 1996. *Eriophyoid Mites of the United States.* West Bloomfield, MI: Indira Publishing.

Brown, W.M. and L.G. Skoglund. 1999. High plains disease: Virus, prion or extra-terrestrial? A review of a new disease. *Zeitsch. Pflanzenkrankh. Pflanzensch.* 106:660–663.

Carew, M., M. Schiffer, P. Umina, A. Weeks, and A. Hoffmann. 2009. Molecular markers indicate that the wheat curl mite, *Aceria tosichella* Keifer, may represent a species complex in Australia. *Bull. Entomol. Res.* 99:479–486.

Childers, C.C., M.A. Eastbrook, and M.G. Solomon. 1996. Chemical control of eriophyoid mites. In: E.E. Lindquist, M.W. Sabelis, and J. Bruin (eds.), *Eriophyoid Mites: Their Biology, Natural Enemies, and Control* (pp. 695–726). Amsterdam: Elsevier.

de Lillo, E. and A. Skoracka. 2010. What's "cool" on eriophyoid mites? *Exp. Appl. Acarol.* 51:3–30.

De Silva, P.H.P.R. and L.C.P. Fernando. 2008. Rearing of coconut mite *Aceria guerreronis* and the predatory mite *Neoseiulus baraki* in the laboratory. *Exp. Appl. Acarol.* 44:37–42.

Domingos, C.A., J.W. Das Melo, M.G.C. Gondim, G.J. De Moraes, R. Hanna, L.M. Lawson-Balagbo, and P. Schausberger. 2010. Diet-dependent life history, feeding preference and thermal requirements of the predatory mite *Neoseiulus baraki* (Acari: Phytoseiidae). *Exp. Appl. Acarol.* 50:201–215.

Duso, C., M. Castagnoli, S. Simoni, and G. Angeli. 2010. The impact of eriophyoids on crops: Recent issues on *Aculus schlechtendali*, *Calepitrimerus vitis* and *Aculops lycopersici*. *Exp. Appl. Acarol.* 51:151–168.

Gamliel-Atinsky, E., S. Freeman, M. Maymon, E. Belausov, R. Ochoa, G. Bauchan, A. Skoracka, J. Pena, and E. Palevsky. 2010. The role of eriophyoids in fungal pathogen epidemiology: Mere association or true interaction? *Exp. Appl. Acarol.* 51:191–204.

Hall, D.G., C.C. Childers, J.E. Eger, and J.C. Allen. 1997. Citrus rust mite (Acari: Eriophyidae) counts on fruit and the negative binomial distribution. *Florida Entomol.* 80:1–10.

Howard, F.W. and D. Moore. 2006. *Featured Creatures: A Coconut Mite*. Gainesville: Department of Entomology and Nematology, University of Florida Institute of Food and Agricultural Sciences (http://entnemdept.ifas.ufl.edu/creatures/orn/palms/Aceria_guerreronis.htm).

Hoyt, S.C. 1969. Integrated chemical control of insects and biological control of mites on apple in Washington. *J. Econ. Entomol.* 62:74–86.

Keifer, H.H., E.W. Baker, T. Kono, M. Delfinado, and W.E. Styer. 1982. *An Illustrated Guide to Plant Abnormalities Caused by Eriophyid Mites in North America*, USDA Handbook 573. Washington, D.C.: U.S. Department of Agriculture (www.sel.barc.usda.gov/acari/content/references/Keifer_et_al.pdf); also available on accompanying CD.

Lawson-Balagbo, L.K., M.G.C. Gondim, G.J. de Moraes, R. Hanna, and P. Schausberger. 2007. Life history of the predatory mites *Neoseiulus paspalivorus* and *Proctolaelaps bickleyi*, candidates for biological control of *Aceria guerreronis*. *Exp. Appl. Acarol.* 43:49–61.

Lawson-Balagbo, L.M., M.G.C. Gondim, G.J. de Moraes, R. Hanna, and P. Schausberger. 2008. Compatibility of *Neoseiulus paspalivorus* and *Proctolaelaps bickleyi*, candidate biocontrol agents of the coconut mite *Aceria guerreronis*: Spatial niche use and intraguild predation. *Exp. Appl. Acarol.* 45:1–13.

Lindquist, E.E., M.W. Sabelis, and J. Bruin (eds.) 1996. *Eriophyoid Mites: Their Biology, Natural Enemies, and Control*, Vol. 6. Amsterdam: Elsevier.

Michalska, K. 2003. Climbing of leaf trichomes by eriophyid mites impedes their location by predators. *J. Insect Behav.* 16:833–844.

Monfreda, R., M. Lekveishvili, R. Petanovic, and J.W. Amrine. 2010. Collection and detection of eriophyoid mites. *Exp. Appl. Acarol.* 51:273–282.

Nault, L.R. 1997. Arthropod transmission of plant viruses: A new synthesis. *Ann. Entomol. Soc. Am.* 90:521–541.

Navajas, M. and D. Navia. 2010. DNA-based methods for eriophyoid mite studies: Review, critical aspects, prospects and challenges. *Exp. Appl. Acarol.* 51:257–271.

Navia, D., G.J. de Moraes, G. Roderick, and M. Navajas. 2005. The invasive coconut mite *Aceria guerreronis* (Acari: Eriophyidae): Origin and invasion sources inferred from mitochondrial (16S) and nuclear (ITS) sequences. *Bull. Entomol. Res.* 95:505–516.

Navia, D., R. Ochoa, C. Welbourn, and F. Ferragut. 2010. Adventive eriophyoid mites: A global review of their impact, pathways, prevention and challenges. *Exp. Appl. Acarol.* 51:225–255.

Negloh, K., R. Hanna, and P. Schausberger. 2010. Season- and fruit age-dependent population dynamics of *Aceria guerreronis* and its associated predatory mite *Neoseiulus paspalivorus* on coconut in Benin. *Biol. Control* 54:349–358.

Oldfield, G.N. 1970. Mite transmission of plant viruses. *Annu. Rev. Entomol.* 15:343–380.

Peña, J.E. and H.A. Denmark. 1996. An eriophyid *Tegolophus perseaflorae* (Acari: Eriophyidae) new to Florida and the United States. *Florida Entomol.* 79:74–76.

Perez-Moreno, I. and M.L. Moraza-Zorrilla. 1998. Population dynamics and hibernation shelters of *Calepitrimerus vitis* in the vineyards of Rioja, Spain, with a description of a new eriophyid extraction technique (Acari: Eriophyidae). *Exp. Appl. Acarol.* 22:215–226.

Petanovic, R. and M. Kielkiewicz. 2010a. Plant–eriophyoid mite interactions. Part I. Cellular biochemistry and metabolic responses induced in mite-injured plants. *Exp. Appl. Acarol.* 51:61–80.

Petanovic, R. and M. Kielkiewicz. 2010b. Plant–eriophyoid mite interactions. Part II. Specific and unspecific morphological alterations. *Exp. Appl. Acarol.* 51:81–91.

Rosenthal, S.S. 1983. Current status and potential for biological control of field bindweed, *Convolvulus arvensis*, with *Aceria convolvuli.* In: M.A. Hoy, G.L. Cunningham, and L. Knutson (eds.), *Biological Control of Pests by Mites*, Special Publ. 3304. Berkeley: University of California Division of Agriculture and Natural Resources.

Siriwardena, P.H.A.P.,L.C.P. Fernando, and T.S.G. Peiris. 2005. A new method to estimate the population size of coconut mite, *Aceria guerreronis*, on a coconut. *Exp. Appl. Acarol.* 37:123–129.

Skoracka, A., L. Smith, G. Oldfield, M. Cristofaro, and J.W. Amrine. 2010. Host-plant specificity and specialization in eriophyoid mites and their importance for the use of eriophyoid mites as biocontrol agents of weeds. *Exp. Appl. Acarol.* 51:92–113.

Smith, L., M. Cristofaro, E. de Lillo, R. Monfreda, and A. Paolini. 2009a. Field assessment of host-plant specificity and potential effectiveness of a prospective biological control agent, *Aceria salsolae*, of Russian thistle, *Salsola tragus*. *Biol. Control* 48:237–243.

Smith, L., E. de Lillo, and J.W. Amrine. 2009b. Effectiveness of eriophyid mites for biological control of weedy plants and challenges for future research. *Exp. Appl. Acarol.* 51:115–149.

Sobhian, R. and L.A. Andres. 1978. The response of the skeletonweed gall midge, *Cystiphora schmidti* (Diptera: Cecidomyiidae), and gall mite, *Aceria chondrillae* (Eriophyidae), to North American strains of rush skeletonweed (*Chondrilla juncea*). *Environ. Entomol.* 7:506–508.

Sobhian, R., A. McClay, S. Hasan, M. Peterschmitt, and R.B. Hughes. 2004. Safety assessment and potential of *Cecidophyes rouhollahi* (Acari: Eriophyidae) for biological control of *Galium spurium* (Rubiaceae) in North America. *J. Appl. Entomol.* 128:258–266.

Spieser, F., B. Graf, P. Walther, and J. Noesberger. 1998. Impact of apple rust mite (Acari: Eriophyidae) feeding on apple leaf gas exchange and leaf color associated with changes in leaf tissue. *Environ. Entomol.* 27:1149–1156.

Van Leeuwen, T., J. Witters, R. Nauen, C. Duso, and L. Tirry. 2010. The control of eriophyoid mites: State of the art. *Exp. Appl. Acarol.* 51:205–224.

Waite, G.K. 1999. New evidence further incriminates honey-bees as vectors of lychee erinose mite *Aceria litchii* (Acari: Eriophyiidae). *Exp. Appl. Acarol.* 23:145–147.

Walde, S.J., J.M. Hardman, and C.N. Magagula. 1997. Direct and indirect species interactions influencing within-season dynamics of apple rust mite, *Aculus schlechtendali* (Acari: Eriophyidae). *Exp. Appl. Acarol.* 21:587–614.

[illegible]

CHAPTER 9

The Tenuipalpidae (Flat or False Spider Mites) as Pests

9.1 BIOLOGY

The Tenuipalpidae are small (200 to 300 μm, or 0.2 to 0.3 mm), slow-moving, flattened, red or green phytophagous mites that are most common in tropical or subtropical climates (see Figure S9.1A on the CD). All tenuipalpids feed on plants (Pritchard and Baker 1952, Meyer 1981, Baker and Tuttle 1987, Gerson 2008). Gerson (2008) reported that about 900 described species in approximately 30 genera are known, but many may remain undiscovered on tropical and subtropical plants. Childers et al. (2003a) reported that over 622 tenuipalpid species can be found in 30 genera, and 3 species (*Brevipalpus californicus*, *B. obovatus*, and *B. phoenicis*) attack a total of 928 plant species in 513 genera in 139 families. Mesa et al. (2009) have provided a catalog of the Tenuipalpidae of the world and a key to the genera.

Brevipalpus is one of the most important genera of tenuipalpids, with approximately 300 described species, but only a few are important as agricultural pests, including *B. californicus*, *B. obovatus*, *B. lewisi*, and *B. phoenicis* (Childers et al. 2003a, Welbourn et al. 2003) (Figure 9.1). Unfortunately, intraspecific variation in *B. phoenicis*, *B. californicus*, and *B. obovatus* has resulted in numerous synonymous species (Welbourn et al. 2003), and there is confusion as to whether the species represent a complex of cryptic species. Some confusion may occur due to artifacts induced during the slide-mounting process, due to age, or to feeding status (Welbourn et al. 2003).

Another tenuipalpid, the red palm mite (*Raoiella indica*), is in the process of colonizing tropical and subtropical areas of the Western Hemisphere and has caused serious economic injury to palms (many types), bananas, and plantains in the Caribbean (Flechtmann and Etienne 2004, Welbourn 2006, Roda et al. 2008). It is reported to occur on ornamental plants, such as gingers, but it is not clear whether all of these plants are truly hosts (Peña et al. 2006, Hoy et al. 2006, Cocco and Hoy 2009). It reduces coconut production and infests landscape palms important to the tourist industry.

Despite the economic importance of the family Tenuipalpidae (false spider mites or flat mites), this family has been studied much less than the Tetranychidae (Pritchard and Baker 1952, Jeppson et al. 1975, Childers et al. 2003a, Gerson 2008). Knowledge of the biology and integrated methods of control of tenuipalpids is relatively sparse (Jeppson et al. 1975, Gerson 2008).

The life cycle includes egg, larva, protonymph, deutonymph, and adult females. Males are present in only some species. Males of some species are very rare because the species is infected with the endosymbiont *Cardinium*, which can modify sex ratios in arthropods (Weeks et al. 2001, Weeks and Breeuwer 2003, Chigira and Miura 2005). Helle et al. (2005) reported that several *Brevipalpus* species (*B. obovatus*, *B. phoenicis*, *B. californicus*) are thelytokous and have only two chromosomes, making these mites the only known animal species in which the females are haploid.

Figure 9.1 Scanning electron micrograph of an adult *Brevipalpus phoenicis*. (Photograph provided by Ron Ochoa, U.S. Department of Agriculture, Agricultural Research Service.)

The average tenuipalpid life cycle requires 3 to 4 weeks, depending on the temperature and host plant. These mites are slow moving and less obvious than spider mites (Tetranychidae). Tenuipalpids feed on leaves, most commonly on the lower surfaces near the midrib or veins, but some feed on bark, in leaf sheaths of grass, or within the plant galls that they form by their feeding. Many tenuipalpids are not of economic importance because they occur on unimportant host plants or because their populations remain below economic injury levels.

9.2 SOME TENUIPALPID PESTS AROUND THE WORLD

The following brief descriptions highlight some of the more important tenuipalpid pest species and their hosts around the world.

9.2.1 *Dolichotetranychus floridanus*

The pineapple flat mite (*Dolichotetranychus floridanus*) commonly occurs where pineapples are grown, although it can occur on other plants, including grasses and bamboo (Jeppson et al. 1975). The females have an elongated, oval body and are red to orange. This species only rarely is of major importance; however, pineapple flat mites can cause damage to young plants when they feed on the soft tissues at the base of the plant. The feeding results in rust-like spots that can be infected by bacteria and fungi so the buds rot. Severely infested plants may be stunted and produce small or no fruit. This species can be controlled by the application of systemic acaricides, but contact poisons are not effective because the mites are well protected by the bases of old leaves and the young leaves in the bud. One *Amblyseius* phytoseiid has been found to be associated with *D. floridanus*, but its efficacy as a natural enemy is unknown.

9.2.2 *Brevipalpus californicus*

The citrus flat mite (*Brevipalpus californicus*) is infamous because it is a cosmopolitan pest of citrus and ornamental plants with a broad host range. In addition to citrus, this mite attacks cotton, tea, tobacco, and a variety of deciduous and subtropical fruits. It has a worldwide distribution. This mite can transmit leprosis of citrus and also is an important pest of tea in Sri Lanka, India, and Java. Helle et al. (2005) reported that *B. californicus* has only two chromosomes and is thelytokous (producing only females), but it is, in fact, haploid (rather than diploid). Chigira and Miura (2005) reported that *B. californicus* is infected with the endosymbiotic bacterium *Candidatus* Cardinium (Cytophaga–Flavobacterium–Bacteroides phylum); when tetracycline-treated females were cured of the symbiont, they produced male progeny. This bacterium is known to cause feminization and induction of thelytoky in insects and appears to feminize haploid males (in this case, *B. californicus*), making them functional females. *Brevipalpus californicus* can be difficult to detect because they lie flat against the leaf surface and are slow to move; however, populations can be detected by their red color and the light-colored exuviae they leave behind after they molt. All stages of *B. californicus* develop on the undersides of tea leaves, causing a reddish discoloration (Cranham 1966). Extensive damage leads to darkening of the lower foliage, a scorched appearance in the basal part of the leaf, and a reduction in the size of new foliage. Defoliation can occur, and tea bushes may die. *Brevipalpus californicus* is susceptible to sulfur and other acaricides and may also be contained by providing a host-free period or by removing infested branches of nearby infested shade trees.

9.2.3 *Brevipalpus lewisi*

The citrus flat mite (*Brevipalpus lewisi*) is a pest of citrus, pomegranates, walnuts, grapes, and ornamentals (Ebeling and Pence 1949, Michelbacher 1956, Buchanan et al. 1980, Rice and Weinberger 1981,Childers et al. 2003a). It is a serious problem in citrus in Japan and California, and it attacks grapes in Europe and Australia. The mite is found at the stem end of citrus fruit, near or under the fruit 'button.' Eggs are laid on fruit, twigs, and leaves. The mites prefer green to ripe fruit and always prefer fruit to leaves. Harvesting the fruit removes most of the population. The mites are small, flat, slow-moving, and light reddish-brown to bright red in color. In California, the overwintering stage is the adult. Peak populations occur during the warmest months, and the mites do well at very high temperatures and low relative humidity. The mites prefer to feed on areas of citrus damaged by leafhoppers or citrus thrips. Heavy populations can cause scarring of the fruit surface, and damage causes a reduction in the quality or grade of the fruit. On grapes, the mites occur on all the green parts of the vine, and feeding prevents development of the grape berries. Sulfur is an effective control material.

9.2.4 *Brevipalpus obovatus*

The ornamental flat mite or privet mite (*Brevipalpus obovatus*) is also known as *B. inornatus* (Morishita 1954, Zhang 2003). It attacks plants in more than 50 genera, including cotton, tea, coffee, strawberries, bananas, privet, anthuriums, fuchsias, and citrus. It has a nearly worldwide distribution. Adults are pale to dark red and have patterns of dark pigmentation within the body. Females reproduce by thelytoky, and males seldom occur. Helle et al. (2005) reported that two populations of *B. obovatus* had a very low frequency of males (12 out of 15,000 individuals), and all eggs examined had two chromosomes, indicating it is thelytokous. The mites overwinter as adults around the bases of plants or in sheltered places on the ventral surfaces of leaves, but they can develop year around in mild climates (Morishita 1954); under greenhouse conditions, for example, these mites can reproduce all year. Damage caused by the ornamental flat mite varies according to the plant infested.

The mites feed on the ventral leaf surface, on stems, and leaf petioles. The mites apparently kill the plant cell by injecting a toxin into the tissues. On citrus, the symptoms are referred to as **leprosis**. *Brevipalpus obovatus* often occurs with *B. phoenicis* on citrus, and the two species together can cause very severe damage.

9.2.5 *Brevipalpus oncidii*

Brevipalpus oncidii is a pest of orchids (*Oncidium* and *Odontoglossum*) in greenhouses in California and England (Pritchard 1951).

9.2.6 *Brevipalpus phoenicis*

Brevipalpus phoenicis is a pest of citrus, coffee, papaya, and tea and more than 486 other plant species around the world (Oomen 1982, Childers et al. 2003c, Hazarika et al. 2009). This mite is also known as the red and black flat mite. In mild climates, generations are continuous and overlapping. Feeding damage caused by this species on citrus is referred to as phoenicis blotch. The spots caused are similar to those in the early stages of leprosis, but there is no gumming of affected areas. The biology and appearance of this species are similar to those of *B. obovatus* and *B. californicus*, and they are often found together on the same plants. *Brevipalpus phoenicis* is a vector of coffee ringspot (a viral disease) (see Figure S9.2 on the CD). De Carvalho Mineiro et al. (2008) monitored populations of *B. phoenicis* on coffee plantations in Brazil and identified predators (Phytoseiidae and Stigmaeidae) able to affect their population dynamics. Like *B. obovatus* and *B. californicus*, *B. phoenicis* has only two chromosomes and is thelytokous (Helle et al. 2005). When Rodrigues et al. (2004) compared the genetics of populations of *B. phoenicis* from Florida and Brazil using mitochondrial DNA sequences and random amplification of polymorphic DNA by polymerase chain reaction (RAPD-PCR), they discovered that these colonies were genetically distinct, which is not unusual in thelytokous clonal lines. Groot et al. (2005) discussed the effect of thelytoky and why *B. phoenicis*, despite being asexual, has the ability to adapt to changes in its environment, including adapting to many host plants.

In principle, all of the progeny from a single, asexual female are clones of that female, with the exception of any mutations that may occur in the progeny. Two hypotheses could explain the adaptability of the *Brevipalpus* clones: the *general purpose genotype* and the *frozen niche variation* models (Groot et al. 2005). According to the general purpose genotype model, an asexual species consists of clones that can all survive and reproduce in all of the different habitats. The frozen niche variation model predicts that different clones are specialized for different niches. Groot et al. (2005) tested populations of *B. phoenicis* from three different host plants and transplanted them to their own and the two alternative host plants before evaluating mite survival and egg production. The authors concluded that the *B. phoenicis* clones evaluated were specialized for different niches, and the frozen niche variation model described the broad ecological range of this species because each mite population had low survival on at least one of the novel host plants. This finding suggests that "*B. phoenicis* consists of a collection of different specialist clones rather than a single generalist clone," which could affect pest management decisions (Groot et al. 2005). Instead of considering that all clones on all adjacent (noncrop) plants should be controlled, as suggested by Childers et al. (2003a), it may be difficult for the *B. phoenicis* clones on nearby plant species to switch to the crop of interest. Groot et al. (2005), however, concluded that more research is needed to determine on how many plants additional clones can survive. The basic question as to how this genetic variation can be maintained in the different clones also was discussed but not resolved (Groot et al. 2005). Mutations can result in genetic differences in clones, as can the possible sexual reproduction of rare males with females. An alternative hypothesis is that sexual females may become asexual when they are infected by a symbiont (such as Cardinium); if mites from a genetically diverse sexual species are converted to asexual reproduction by the symbiont, then a broad array of different clones can be produced.

These data suggest that information obtained on one clone of *Brevipalpus phoenicis* may not be appropriate for making pest management decisions on other clones found on different host plants or in different geographic areas. For example, the susceptibility or resistance status of clones to pesticides, host-plant range, and other biological traits could vary from clone to clone.

9.2.7 *Brevipalpus russulus* and *B. sayedi*

Brevipalpus russulus is a pest of cactus and succulents, both in ornamentals and rangeland plants, and *Brevipalpus sayedi* is a pest of hickory and pecan trees in Florida, Maryland, and Indiana.

9.2.8 *Tenuipalpus pacificus*

Tenuipalpus pacificus is a pest of orchids in California, Florida, Panama, Australia, Thailand, England, and Holland (Pritchard 1951, Denmark 2006) (see Figure S9.1 on the CD). Feeding causes dark spots on the leaves and eventual necrosis of the tissue.

9.2.9 *Cenopalpus pulcher*

Cenopalpus pulcher is a widespread but only occasional pest of apple, pear, prune, and walnut trees in Europe. This mite was recently discovered in apples and pears in Oregon, where it is an invasive pest.

9.2.10 *Raoiella indica*

Until recently, the red palm mite (*Raoiella indica*) was found only in India, Egypt, Israel, Mauritius, Reunion, Sudan, Iran, Oman, Pakistan, and the United Arab Emirates (Dowling et al. 2008); however, in 2004, this pest was detected in Martinique and subsequently has spread throughout the Caribbean (Flechtmann and Etienne 2004, Hoy et al. 2006, Peña et al. 2006, Welbourn 2006, Roda et al. 2008). It now is established in Florida and Central and South America and is expected to spread throughout the Western Hemisphere wherever the climate is suitable. This mite is easily distributed by wind currents and movement of infested plants.

As an invasive pest, *Raoiella indica* has been able to increase to incredible densities. Several experienced acarologists have marveled at the number of *R. indica* found on a single coconut tree; the undersurface of coconut palm fronds can be literally covered in *R. indica*, and each tree can contain millions of these mites. Population densities this high cause yellowing and browning of the foliage (see Figure S9.3 on the CD). This mite has reached greater population densities than any other tenuipalpid species this author has ever observed and can match the population densities of some tetranychid species (Tetranychidae) in outbreak conditions.

Females of *Raoiella indica*, on average, are 245 μm (0.245 mm) long and 182 μm (0.182 mm) wide; they are oval and reddish in color (Figure 9.2). Females develop dark markings on the dorsum of the body after feeding. The male is smaller, but similar to the female in shape except for having a tapering of the posterior end of the body. Adult females are larger than males and less active. Males and females are sexually mature when they emerge, and males actively seek out females, suggesting there is a sex pheromone involved. When a male locates a female deutonymph in the quiescent stage, he will settle close to it and wait for up to 2 days for her to molt. When female deutonymphs begin to molt, the male becomes active and moves under her, bending his posterior up and forward to mate.

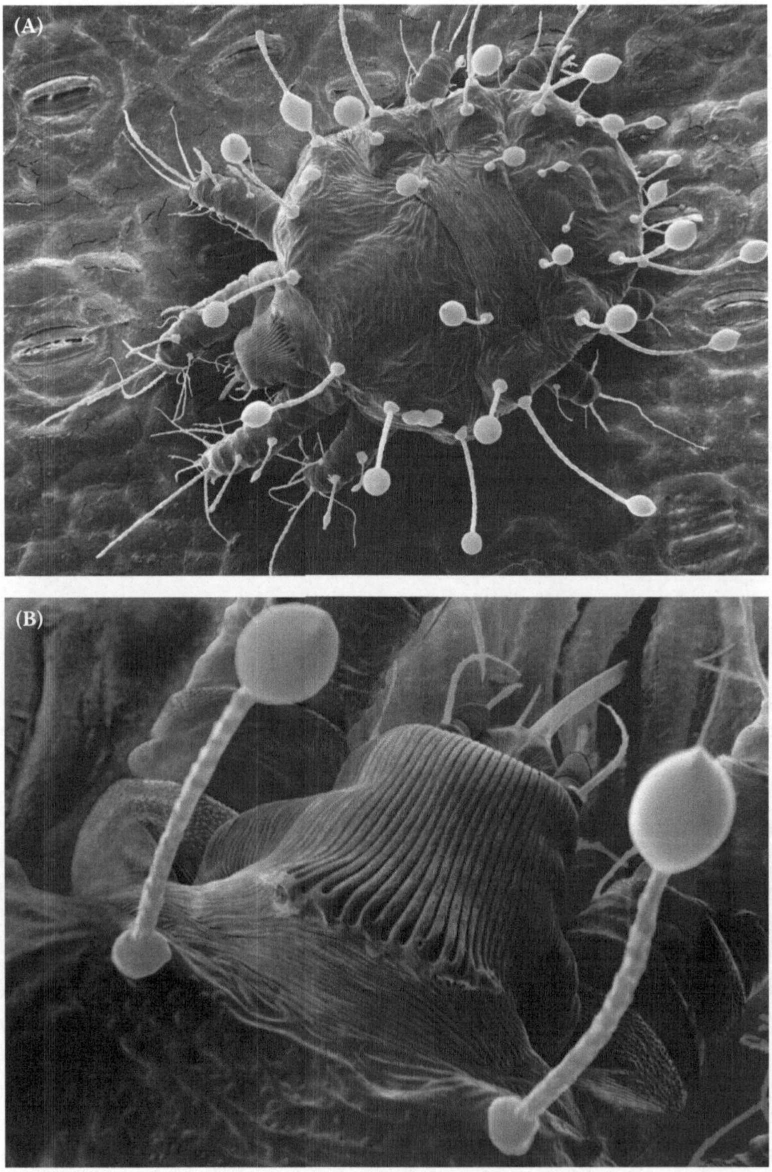

Figure 9.2 (A) Scanning electron micrograph of the red palm mite (*Raoiella indica*) showing droplets secreted from its setae. It has been speculated that these droplets are repellent to some natural enemies. (B) Close-up of mouthparts inserted into a stomata of a leaf. (Photographs by the U.S. Department of Agriculture, Agricultural Research Service.)

The life cycle from egg to adult typically requires 23 to 28 days for females and 20 to 22 days for males. Mated females have a 5- to 6-day preoviposition period and oviposit for 47 days under laboratory conditions. Unmated females deposit an average of 18 eggs after a 2-day preoviposition period, oviposit for 40 days, and live for approximately 48 days. Males, produced by unmated females, live an average of 22 days. The ovoid egg is reddish, approximately 100 μm long by 80 μm wide. The freshly laid egg is attached to the leaf surface, and a fine white stipe (slender hairlike structure) as long as or longer than the egg is present at one end. The tip of the stipe may be coiled and have a droplet clinging to it. The incubation period averages 8 days for fertilized eggs and 7.3 days for unfertilized eggs. This species has an unusual genetic system, in which all eggs laid by unmated females develop into males, and mated females produce only female progeny (Nagesha-Chandra and Channabasavanna 1984). All active stages are red with blackish marks on the dorsum. Unlike the *Brevipalpus* species, this mite is reported to have four and two chromosomes, suggesting it is arrhenotokous, with an approximate ratio of one female to one male (Helle et al. 2005).

The red palm mite is a pest of several ornamental and fruit-producing palm species, such as coconut and areca palms, and has been found attacking bananas and plantains in the Caribbean (see Figure S9.3C on the CD), although it was not possible to rear these mites on young potted bananas and plantains in a quarantine facility (Welbourn 2006, Cocco and Hoy 2009). The red palm mite has been collected from ornamental plants other than palms such as *Heliconia rostrata*, *Strelitzia reginae*, *Alpinia purpurata*, and *Etlingera elatior* in the Caribbean, but it is not clear whether these are valid host plants or whether the enormous mite populations on coconuts and other palms have temporarily moved on to substory plants under the palms. Palm nurseries, landscape palms, and horticultural gardens are affected by this new pest.

The reddish mites are easily seen against green leaves. Heavy infestations of the mites are typically on the lower surface of the leaves, and yellow speckles and blotches are visible on the leaves due to feeding damage (see Figure S9.3 on the CD). Yellowing of the leaves may be severe in very young to mature trees. Palms affected by *Raoiella indica* show scattered yellow spots on both leaflet surfaces, with most of the leaflets affected being located in the middle area of the leaf. Coconut palms severely affected by the mite show entirely yellow leaves, particularly on the lower third of the plant. The yellow color of the leaflets is followed by abortion of the flowers or small nuts in coconut palms. It is unknown if this yellowing is solely the result of mite feeding in combination with the dry season, or if this mite transmits a plant pathogen. There is no information as to whether this mite is a disease vector, but other tenuipalpid species are vectors of disease.

In less than a year after its introduction to Martinique, the mite appeared on coconut palms on nearby islands, strongly supporting the premise that the mite is dispersed by wind currents. In Dominica, where coconut is grown with bananas, the mite can be found on bananas with numbers close to 200 mites per square centimeter. The lower leaves of banana and plantain turn yellow with small patchy green-yellow areas.

In India, populations of *Raoiella indica* are negatively affected by rainfall and high relative humidity. Populations on palms are highest during hot, sunny, and dry conditions. Chemical control is difficult and expensive when the palm trees are very tall; however, chemical control may be necessary to manage these mites if they occur in palm nurseries without adequate natural enemies present. Quarantines have been implemented to prevent the further spread of this pest in the Western Hemisphere, although the ability of quarantines to keep this pest from new geographic regions appears limited.

Several natural enemies exist in India, Mauritius, and Egypt, where predators have been cited as important (Moutia 1958, Nagesha-Chandra and Channabasavanna 1984, Somchoudhury and Sarkar 1987, Zaher et al. 1996). Only a few studies have been conducted to resolve which are effective. In India, the phytoseiid *Amblyseius channabasavanni* is known to feed on *Raoiella indica* and developed rapidly and reproduced at a high rate (2.7 eggs per day) on this food (Daniel 1981). In Mauritius, the phytoseiid *A. caudatus* Berlese was considered the main predator of *R. indica* on coconut palms (Moutia 1958). Jagadish and Nagesha-Chandra (1979, 1982) evaluated *Typhlodromips tetranychivorus* as a predator of the red palm mite. Ueckermann (2004) found the phytoseiid *A. largoensis* (Muma) "in association" with *R. indica*, but did not indicate how effective it was as a natural enemy. In addition to phytoseiid predators, the lady beetles *Stethorus tetranychi* and *S. parcempunctatusi* (Coccinellidae) are known to feed on *R. indica* in India, preferring eggs followed by larvae, nymphs, and adults.

Endemic natural enemies in the Western Hemisphere provide some suppression of red palm mite populations, although it is not currently considered sufficient (Peña et al. 2009, Carrillo et al. 2010). Coccinellids endemic to the Western Hemisphere will feed on this abundant new prey. The phytoseiid *Neoseiulus longispinosus* (Evans), an Old World species probably native to the Asian region, was found attacking *Raoiella indica* in the Caribbean. *Amblyseius largoensis* feeds on *R. indica* in the Caribbean and in Florida (Roda et al. 2008). Likewise, it is likely that endemic fungal pathogens, such as *Hirsutella*, will be effective against this pest, at least during rainy seasons. However, because this mite is an invasive pest that is difficult to control with pesticides and endemic natural enemies have not been successful in reducing populations sufficiently to date, *R. indica* has become the target of a classical biological control program in Florida. Taxonomic problems with the correct identification of the phytoseiids imported from Mauritius may prevent the release of these predators, however (Bowman and Hoy, unpublished; Hoy, unpublished).

9.3 TRANSMISSION OF PLANT DISEASES

Brevipalpus species have been associated with a number of diseases in crops such as citrus, coffee, passion fruit, orchids, and ornamentals. Kitajima et al. (2003a,b) suggested that these mites may be as important as eriophyoid mites as vectors of plant disease. *Brevipalpus* mites are known to transmit citrus leprosis, passion fruit green spot virus, coffee ringspot virus, and orchid fleck virus, as well as diseases in such ornamental plants as *Annona*, *Brunfelsia*, *Clerodendron*, *Hedera*, *Hibiscus*, *Ligustrum*, *Malaviscus*, *Pelargonium*, *Pittosporum*, *Salvia*, *Schefflera*, and *Thunbergia* (Kitajima et al. 2003a). These viruses are in the family Rhabdoviridae, and it is likely that other diseases will be discovered that are vectored by tenuipalpid mites.

Citrus leprosis virus in citrus, which is transmitted by *Brevipalpus phoenicis*, *B. californicus*, and *B. obovatus*, was a serious disease of citrus in Florida prior to 1925; this disease, also called nail-head rust or scaly bark, appears to have disappeared in both Florida and Texas and has not been observed for over 40 to 50 years (Childers et al. 2003c). Reasons for the disappearance of the disease in Florida and Texas are unknown, but it could be due to winter freezes and the use of sulfur applications. There is concern that citrus leprosis could be reintroduced into the United States.

Citrus leprosis causes stem lesions and bark scaling symptoms, as well as lesions on fruits and stems (see Figure S9.4 on the CD). This disease is a serious problem in Brazil, Argentina, Paraguay, Uruguay, Venezuela, Panama, and Colombia, where it occurs on sweet orange varieties (Childers et al. 2003b). Bastianel et al. (2010) consider this disease one of the most important diseases in the Brazilian citrus industry, as it costs growers around US$80 million for control each year. Citrus leprosis is spreading through Central America and has been found as far north as Mexico. Citrus leprosis could reappear in the United States if quarantines are insufficient to prevent its invasion (Childers et al. 2003b).

According to Bastianel et al. (2010), it is not clear if the disease is present in Asian and African citrus because the disease is difficult to diagnose. There is some confusion as to whether more than one type of virus is associated with leprosis. Leprosis has been associated with "two completely distinct viruses with similar morphology and vector," and it appears that the virus associated with the cytoplasm should be classified as the type member of a new virus genus, *Cilevirus*, while the virus associated with the nucleus is, in fact, a new genus of virus (*Dichorhabdovirus*) in the Rhabdoviridae (Bastianel et al. 2010). All active stages of *Brevipalpus* mite species can acquire and transmit the viruses, which are not transovarially transmitted. It appears that the virus does not replicate in the mite, and not all mites in Brazilian citrus groves are infected with the virus. Currently, growers spray to control mites after 10% of fruits and branches are found to contain mites. This control threshold may be too high, because sampling methods to detect the mites are not accurate, and not all mites are infected.

Table 9.1 Integrated Mite Management Methods for Tenuipalpid Mites

1. Most tenuipalpid species are not serious pests on most host plants, suggesting that natural enemies (predators and pathogens) can suppress populations or that the plants are not economically important.
2. If a crop suddenly has a tenuipalpid problem, the chemical control program should be examined to determine whether pesticides used to control other pests could be disrupting tenuipalpid natural enemies and whether other (selective) products could be used that would retain such natural enemies.
3. A variety of phytoseiid species and other predators have been identified as predators of tenuipalpids. Some phytoseiids are commercially available and potentially could be released, although there are no release rates or other information regarding such augmentative releases.
4. *Hirsutella thompsonii* appears to be an effective fungal pathogen of *Brevipalpus phoenicis* in laboratory trials and in the field. It could be developed as a microbial pesticide or utilized as a natural control agent in humid environments.
5. Chemical control can be effective, but *Brevipalpus phoenicis* has developed resistance to at least one product (hexythiazox) in citrus in Brazil, so this may not be a sustainable tactic for tenuipalpids.
6. Other tactics to employ could include quarantines to reduce the spread of tenuipalpids and their host plants from country to country or from field to field and the use of less susceptible cultivars.
7. In efforts to prevent the spread of citrus leprosis disease, planting resistant citrus cultivars, pruning to reduce disease inoculum, and controlling the tenuipalpid vectors could reduce the impact of the disease.
8. Improved monitoring methods and a realistic estimate of the proportion of mites infected with citrus leprosis virus will allow the development of a more realistic economic threshold for controlling tenuipalpids in citrus in Brazil.

Source: Based on information from Bastianel et al. (2010), Childers et al. (2003), Gerson (2008), and Jeppson et al. (1975).

Brevipalpus californicus can transmit orchid fleck virus, which damages many orchid genera around the world (Kondo et al. 2003). The virus, possibly a new genus in the Rhabdoviridae, produces chlorotic or necrotic spots and rings on many genera of orchids (as well as other plants). The mites (both nymphs and adults) remained infective for a long period, but it is not clear if the virus replicates in the mite (see Figure S9.5 on the CD). Passion fruit green spot virus is vectored by *B. phoenicis* in Brazil (Kitajima et al. 2003a,b). The disease causes green spots and patches of green tissues on senescent leaves. Flowers have necrotic lesions that encircle stems and kill the plant. Control has been achieved by the use of acaricides, but this may not be sustainable if resistance develops. Coffee ringspot virus is vectored by *B. phoenicis* to coffee (Chagas et al. 2003), with a rate of transmission in the laboratory of approximately 24% (see Figure S9.2 on the CD). Transovarial transmission of the virus does not occur.

9.4 CONTROL OF TENUIPALPIDS

Gerson (2008) reviewed the natural enemies reported to affect tenuipalpids in the field and identified fungi (*Hirsutella thompsonii, Metarhizium anisopliae*) and predators that were possible control agents. The predators included cheyletids (*Cheletogenes ornatus*), cunaxids (*Cunaxa hukoschusti*), eupalopsellids (*Exothohis* sp., *Eupalopsellus olandicus*), stigmaeids (*Agistemus exsertus, A. floridanus, Zetzellia javanica*), tarsonemids (*Acaronemus destructor*), tydeids (*Pronematus elongates*), and a variety of phytoseiids (*Typhlodromus doreenae, Euseius scutalis, E. victoriensis, Typhlodromips tetranychivorus, Phytoseiulus macropilus, Amblyseius largoensis*). Finally, a coccinelid beetle, *Stethorus tetranychi*, has been reported to feed on *Raoiella indica*. Chemical control of pests such as *Brevipalpus phoenicis*, a pest of citrus and other crops, remains important especially as a method to reduce disease transmission (Table 9.1);however,resistance to at least one product, hexythiazox, has been reported in populations of *B. phoenicis* in Brazilian citrus (Campos and Omoto 2002). Rossi-Zalaf and Alves (2006) tested the pathogenicity of 52 isolates of several fungi (*Beauveria bassiana, Metarhizium anisopliae, Paecilomyces lillacinus, P. fumosoroseus, Lecanicillium lecanii, L. muscarum*, and *Hirsutella thompsonii*) against *B. phoenicis* and found that, in the laboratory, *H. thompsonii* was the most virulent.

SELECTED REFERENCES

Baker, E.W. and D.M. Tuttle. 1987. *The False Spider Mites of Mexico* (Tenuipalpidae: Acari), USDA ARS Technical Bulletin 1706. Washington, D.C.: U.S. Department of Agriculture Agricultural Research Service.

Bastianel, M., V.M. Novelli, E.W. Kitajima, K.S. Kubo, R.B. Bassanezi, M.A. Machado, and J. Freitas-Astua. 2010. Citrus leprosis: Centennial of an unusual mite–virus pathosystem. *Plant Dis.* 94(3):284–292.

Buchanan, G.A., M. Bengston, and E.M. Exley. 1980. Population growth of *Brevipalpus lewisi* McGregor (Acarina: Tenuipalpidae) on grapevines. *Aust. J. Agric. Res.* 31:957–965.

Campos, F.J. and C. Omoto. 2002. Resistance to hexythiazox in *Brevipalpus phoenicis* (Acari: Tenuipalpidae) from Brazilian citrus. *Exp. Appl. Acarol.* 26:243–251.

Carrillo, D., J.E. Pena, M.A. Hoy, and J.H. Frank. 2010. Development and reproduction of *Amblyseius largoensis* (Acari: Phytoseiidae) feeding on pollen, *Raoiella indica* (Acari: Tenuipalpidae), and other microarthropods inhabiting coconuts in Florida, USA. *Exp. Appl. Acarol.* 52:119–129.

Chagas, C.M., E.W. Kitajima, and J.C.V. Rodrigues. 2003. Coffee ringspot virus vectored by *Brevipalpus phoenicis* (Acari: Tenuipalpidae) in coffee. *Exp. Appl. Acarol.* 30:203–213.

Chigira, A. and K. Miura. 2005. Detection of "*Candidatus* Cardinium" bacteria from the haploid host *Brevipalpus californicus* (Acari: Tenuipalpidae) and effect on the host. *Exp. Appl. Acarol.* 37:107–116.

Childers, C.C., J.V. French, and J.C.V. Rodrigues. 2003a. *Brevipalpus californicus*, *B. obovatus*, *B. phoenicis*, and *B. lewisi* (Acari: Tenuipalpidae): A review of their biology, feeding injury and economic importance. *Exp. Appl. Acarol.* 30:5–28.

Childers, C.C., J.C.V. Rodrigues, K.S. Derrick, D.S. Achor, J.V. French, W.C. Welbourn, R. Ochoa, and E.W. Kitajima. 2003b. Citrus leprosis and its status in Florida and Texas: Past and present. *Exp. Appl. Acarol.* 30:181–202.

Childers, C.C., J.C.V. Rodrigues, and W.C. Welbourn. 2003c. Host plants of *Brevipalpus californicus*, *B. obovatus*, and *B. phoenicis* (Acari: Tenuipalpidae) and their potential involvement in the spread of viral diseases vectored by these mites. *Exp. Appl. Acarol.* 30:29–105.

Cocco, A. and M.A. Hoy 2009. Feeding, reproduction, and development of the red palm mite (Acari: Tenuipalpidae) on selected palms and banana cultivars in quarantine. *Florida Entomol.* 92:276–291.

Cranham, J.E. 1966. *Insect and Mite Pests of Tea in Ceylon and Their Control*, Monographs on Tea Production in Ceylon No. 6. Talawakelle, Ceylon: The Tea Research Institute.

Daniel, M. 1981. Bionomics of the predaceous mite *Amblyseius channabasavanni* (Acari: Phytoseiidae) predaceous on the palm mite. In: G.P. Channabasavanna (ed.), *Contributions to Acarology in India* (pp. 167–172). Bangalore: Acarological Society of India.

De Carvalho Mineiro, J.L., M.E. Sato, A. Raga, and V. Arthur. 2008. Population dynamics of phytophagous and predaceous mites on coffee in Brazil, with emphasis on *Brevipalpus phoenicis* (Acari: Tenuipalpidae). *Exp. Appl. Acarol.* 44:277–291.

Denmark, H.A. 2006. *Featured Creatures: Phalaenopsis Mite*. Gainesville: Department of Entomology and Nematology, University of Florida Institute of Food and Agricultural Sciences (http://entnemdept.ifas.ufl.edu/creatures/orn/flowers/Tenuipalpus_pacificus.htm).

Dowling, A.P.G., R. Ochoa, and J.J. Beard. 2008. Preliminary results on phylogeographic patterns of the invasive red palm mite, *Raoiella indica* (Prostigmata: Tenuipalpidae). In: M. Bertrand, S. Kreiter, K.D. McCoy, A. Migeon, M. Navajas, M.S. Tixier, and L. Vial (eds.), *Integrative Acarology* (pp. 147–153). Proceedings of the 6th European Congress, European Association of Acarologists, July 21–25, Montpellier France.

Ebeling, W. and R.J. Pence. 1949. New pomegranate mite. *Calif. Agric.* 3(6):11–14.

Flechtmann, C.H.W. and J. Etienne. 2004. The red palm mite *Raoiella indica* Hirst, a threat to palms in the Americas (Acari: Prostigmata: Tenuipalpidae). *Syst. Appl. Acarol.* 9:109–110.

Gerson, U. 2008. The Tenuipalpidae: An underexplored family of plant-feeding mites. *Syst. Appl. Acarol.* 13:82–101.

Groot, T.V.M., A. Janssen, A. Pallini, and J.A.J. Breeuwer. 2005. Adaptation in the asexual false spider mite *Brevipalpus phoenicis*: Evidence for frozen niche variation. *Exp. Appl. Acarol.* 36:165–176.

Hazarika, L.K., M. Bhuyan, and B.N. Hazarika. 2009. Insect pests of tea and their management. *Annu. Rev. Entomol.* 54:267–284.

Helle, W., H.R. Bolland, and J. Gutierrez. 2005. Minimal chromosome number in false spider mites (Tenuipalpidae). *Cell. Molec. Life Sci.* 28:707.

Hoy, M.A., J. Peña, and R. Nguyen. 2006. *Red palm mite,* Raoiella indica *Hirst (Arachnida: Acari: Tenuipalpidae).* Gainesville, University of Florida IFAS Extension (http://edis.ifas.ufl.edu/IN711).

Jagadish, P.S. and B.K. Nagesha-Chandra. 1979. Biology of *Typhlodromips tetranychivorus* (Acari: Phytoseiidae) on red palm mite *Raoiella indica* (Acari: Tenuipalpidae). In: *Contributions to Acarology in India*, Proceedings of the First All India Symposium on Acarology, April 23–25, Bangalore.

Jagadish, P.S. and B.K. Nagesha-Chandra. 1982. Biology of *Typhlodromips tetranychivorus* (Acari: Phytoseiidae) on red palm mite *Raoiella indica* (Acari: Tenuipalpidae) *Acarol. Newsl.* 11:9–10.

Jeppson, L.R., H.H. Keifer, and E.W. Baker. 1975. *Mites Injurious to Economic Plants.* Berkeley: University of California Press.

Kitajima, E.W., C.M. Chagas, and J.C.V. Rodrigues. 2003a. *Brevipalpus*-transmitted plant virus and virus-like diseases: Cytopathology and some recent cases. *Exp. Appl. Acarol.* 30:1345–1160.

Kitajima, E.W., J.A.M. Rezende, and J.C.V. Rodrigues. 2003b. Passion fruit green spot virus vectored by *Brevipalpus phoenicis* (Acari: Tenuipalpidae) on passion fruit in Brazil. *Exp. Appl. Acarol.* 30:225–231.

Kondo, H., T. Maeda, and T. Tamada. 2003. Orchid fleck virus: *Brevipalpus californicus* mite transmission, biological properties and genome structure. *Exp. Appl. Acarol.* 30:215–223.

Mesa, N.C., R. Ochoa, W.C. Welbourn, G.A. Evans, and G.J. De Moraes. 2009. A catalog of the Tenuipalpidae (Acari) of the world with a key to genera. *Zootaxa* 2098:1–185.

Meyer, M.K.P.S. 1981. *Mite Pests of Crops in Southern Africa*, Science Bulletin 397. Pretoria: Department of Agriculture and Fisheries, Republic of South Africa.

Michelbacher, A.E. 1956. Spider mites on walnuts. *Calif. Agric.* 10(7):4–14.

Morishita, F.S. 1954. Biology and control of *Brevipalpus inornatus* (Banks). *J. Econ. Entomol.* 47:449–456.

Moutia, L.A. 1958. Contribution to the study of some phytophagous acarina and their predators in Mauritius. *Bull. Entomol. Res.* 49:59–75.

Nagesha-Chandra, B.K. and G.P. Channabasavanna. 1984. Development and ecology of *Raoiella indica* Hirst (Acari: Tenuipalpidae) on coconut. In: D.A. Griffiths and C.E. Bowman (eds.), *Acarology VI*, Vol. 2 (pp. 785–790). Chichester: Ellis Horwood.

Oomen, P.A. 1982. Studies on population dynamics of the scarlet mite, *Brevipalpus phoenicis*, a pest of tea in Indonesia. *Arten. Mededelingen van de Landbouwhogescholl Wageningen* 82-01:1–88.

Peña, J.E., C.M. Mannion, F.W. Howard, and M.A. Hoy. 2006. Raoiella indica *(Prostigmata: Tenuipalpidae): The Red Palm Mite: An Invasive Pest of Palms and Bananas and Other Tropical Crops of Florida*, ENY 837. Gainesville: University of Florida IFAS Extension (http://edis.ifas.ufl.edu/in681).

Peña, J.E., J.C. Rodrigues, L. Osborne, and A. Roda. 2009. Predator–prey dynamics and strategies for control of the red palm mite (*Raoiella indica*) (Acari: Tenuipalpidae) in areas of invasion in the Neotropics. *IOBC/WPRS Bull.* 50:69–79.

Pritchard, A.E. 1951. Control of orchid mites: False spider mites and spider mites must be distinguished for proper control purposes. *Calif. Agric.* 5(9):11.

Pritchard, A.E. and E.W. Baker. 1952. *The False Spider Mites of California (Acarina: Phytoptipalpidae)*, University of California Publications in Entomology. Berkeley: University of California Press.

Rice, R.E. and G.B. Weinberger. 1981. Citrus flat mite on pistachios in California. *Calif. Agric.* 35(7):25–26.

Roda, A., A. Dowling, C.A. Welbourn, J. Peña, J.C.V. Rodrigues, M.A. Hoy, R. Ochoa, R.A. Duncan, and W. DeChi. 2008. Red palm mite situation in the Caribbean and Florida. *Proc. Carib. Food Crops Soc.* 44(1):80–87.

Rodrigues, J.C.V., M. Gallo-Meagher, R. Ochoa, C.C. Childers, and B.J. Adams. 2004. Mitochondrial DNA and RAPD polymorphisms in the haploid mite *Brevipalpus phoenicis* (Acari: Tenuipalpidae). *Exp. Appl. Acarol.* 34:275–290.

Rossi-Zalaf, L.S. and S.B. Alves. 2006. Susceptibility of *Brevipalpus phoenicis* to entomopathogenic fungi. *Exp. Appl. Acarol.* 40:37–47.

Somchoudhury, A.K. and P.K. Sarkar. 1987. Observations on natural enemies found in association with coconut mite, *Raoiella indica* Hirst. *Bull. Entomol.* 28:104–107.

Ueckermann, E.A. 2004. Taxonomic Research in Acarology, paper presented at the Workshop on Biodiversity Dynamics on La Reunion Island, Nov. 29–Dec. 5.

Weeks, A.R. and J.A.J. Breeuwer. 2003. A bacterium from the Cytophaga–Flavobacterium–Bacteroides phylum that causes sex-ratio distortion. In: K. Bourtzis and T.A. Miller (eds.), *Insect Symbiosis* (pp. 165–176). Boca Raton, FL: CRC Press.

Weeks, A.R., F. Marec, and J.A.J. Breeuwer. 2001. A mite species that consists entirely of haploid females. *Science* 292:2479–2482.

Welbourn, C. 2006. *Pest Alert: Red Palm Mite* Raoiella indica *Hirst (Acari: Tenuipalpidae)*. Gainesville: Florida Department of Agriculture & Consumer Services, Division of Plant Industry (http://www.doacs.state.fl.us/pi/enpp/ento/r.indica.html).

Welbourn, W.C., R. Ochoa, E.C. Kane, and E.F. Erbe. 2003. Morphological observations on *Brevipalpus phoenicis* (Acari: Tenuipalpidae) including comparisons with *B. californicus* and *B. obovatus*. *Exp. Appl. Acarol.* 20:107–133.

Zaher, M.A., A.K. Wafa, and A.A. Yousef. 1996. Biological studies on *Raoiella indica* Hirst and *Phyllotetranychus aegyptiacus* Sayed infesting date palm trees in U.A.R. (Acarina: Tenuipalpidae). *Z. Angew. Entomol.* 63:406–441.

Zhang, Z.-Q. 2003. *Mites of Greenhouses: Identification, Biology, and Control*. Wallingford, U.K.: CAB International.

CHAPTER 10

The Penthaleidae

10.1 SYSTEMATICS AND DISTRIBUTION

The Penthaleidae (Prostigmata or Actinedida: Eupodoidea) are soft-bodied mites without external peritremes (that part of the exoskeleton surrounding the tracheal opening) and with the tracheae originating at the base of the chelicerae. Several species of Penthaleidae are of economic importance as obligatory plant feeders. These include the red-legged earth mite (RLEM) (*Halotydeus destructor*), also known as the black sand mite, and the blue oat mite (*Penthaleus major*), also known as the winter grain mite. The genus *Penthaleus* is distinctive in having a dorsal anal opening on the posterior portion of the hysterosoma (Jeppson et al. 1975).

It was recently discovered in Australia that three species of *Penthaleus* are significant agricultural pests, rather than just one as previously thought (Umina et al. 2004). The blue oat mite is widely distributed around the world, but the RLEM is restricted to southern Africa, Australia, and New Zealand (Jeppson et al. 1975). The RLEM was introduced from South Africa into Australia, and the blue oat mite probably came from Europe to Australia.

Chemical control has been used extensively to control these pests, but efforts toward an IMM program are being developed in Australia. The biology of the RLEM and blue oat mite is particularly interesting and affects management tactics. Both mites have an aestival diapause that allows them to survive the very hot dry summers in Australia (Umina and Hoffman 2003, Umina 2007, 2008). Because these mites are serious pests in Australia and much of the information on their biology and ecology was developed there, this discussion focuses on the mite-management tactics developed in Australia.

10.2 RED-LEGGED EARTH MITE (RLEM)

The life cycle is egg → larva → protonymph → deutonymph → tritonymph → adult male or female. Adults are about 1 mm in length and 0.6 mm wide with red-orange legs and a black body (see Figure S10.1 on the CD). Newly hatched mites are pinkish-orange and only 0.2 mm in length. The RLEM can be distinguished from blue oat mites in the field because the blue oat mites have an oval reddish-orange mark on their dorsum (Umina 2007, 2008). Also, RLEMs often feed in groups of up to 30 mites (Figure 10.1)(also see Figure S10.2 on the CD).

Red-legged earth mite nymphs resemble the adult mite but are smaller. Each stage takes 1 to 2 weeks to complete. Between each molt, the mite is inactive, with the legs held rigidly in front of the body. Females begin to lay eggs shortly after molting to the adult. Both males and females are found, but the sex ratio is female biased. Eggs are often deposited together, and females can produce up to 100 eggs, which usually hatch in 8 to 10 days (Umina 2008).

Figure 10.1 (A) Groups of RLEMs feeding on capeweed flowers in the spring in Australia. (B) RLEMs feed in groups on clover; note silvering of leaves from feeding. (Photographs by James Ridsdill-Smith, CSIRO Australia.)

Males spin silk webbing in which small globules of waterlike fluid are spaced at intervals along the threads. These are spermatophores, erect mushroom-like structures of distinctive shape that are produced singly. A packet of sperm is supported on a stalk rather like a golf ball on a tee. The stalk has attached ramifications and excrescences that are of consistent shape within a species.

Red-legged earth mites are most active in the cool, wet part of the year (winter to spring in Australia). During this period, the RLEM may have three generations, with each lasting 6 to 8 weeks (McDonald et al. 1995). Eggs hatch in autumn following exposure for at least 2 weeks to cooler temperatures and adequate rain. The newly hatched mites attack crop seedlings and emerging

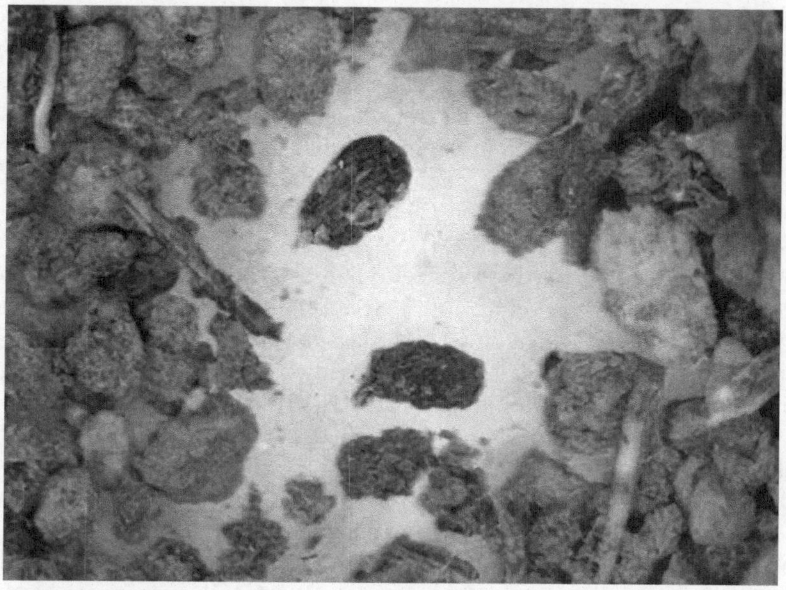

Figure 10.2 Two dead females (dark colored in center of photograph) of the RLEM are surrounded by sand grains. These dead females contain aestivating eggs that are retained within the bodies of their dead mothers during the hot dry summers in Australia. These cadavers can be blown some distance and initiate infestations in new fields. The eggs will hatch after sufficient rainfall has occurred. (Photograph by James Ridsdill-Smith, CSIRO Australia.)

pasture plants but also feed on soil microflora (algae and mosses), especially when the pasture plants are sufficiently tall to provide shade and moisture (Maclennan et al. 1998). RLEM eggs laid during the winter–spring period are orange and deposited on the undersides of leaves at the bases of plants and on debris in the soil. The eggs are often found in large clusters and can hatch in 8 to 10 days. In late spring, with the onset of hot, dry weather, longer daylengths, and physiological changes in host plants, the females produce oversummering or aestival diapause eggs. These diapause eggs are stress resistant and retained within the dead bodies of their mothers (Figure 10.2) (also see Figure S10.3 on the CD). Aestival diapause eggs can survive the extreme heat and dryness of summer and give rise to the fall generation. An exposure to 52°C for a month is optimal to break the aestival diapause, and eggs will hatch with a combination of temperatures below 20°C and adequate soil moisture. Long-distance passive dispersal probably occurs when female cadavers containing diapause eggs are blown by the wind (Ridsdill-Smith and Annells 1997).

The active stages of RLEMs spend about 90% of their time on the soil surface rather than on plant foliage. They feed for short periods and rest before settling at another feeding site (Umina 2008). Adults feed extensively, and the weight of the RLEM can increase from about 1 μg to over 100 μg while feeding on cotyledons, trifoliate leaves, and flowers. RLEMs use their sharp chelicerae to pierce the upper leaf epidermis to the palisade layer and feed on the exuded droplet of cell contents (Liu and Ridsdill-Smith 2000). The damage caused is seen as a silvering of the leaf. Feeding aggregations are common (see Figure 10.1 and Figure S10.2 on the CD). The damaged foliage is prone to desiccation and has reduced photosynthetic rates.

Red-legged earth mites are especially serious pests on newly established pastures and emerging crops. In situations where the weather is favorable and the densities are high, damage from RLEM populations can result in total crop loss. Host plants include pasture legumes, subterranean and other clovers, and alfalfa (lucerne). They also feed on ryegrass and young cereal crops, especially oats. In addition, they feed on an array of weeds (see Figure S10.4A on the CD for a photograph of a pasture damaged by RLEMs). Livestock prefer forage in pastures not damaged by the RLEM. Studies have shown that 12,000 mites per m^2 can "use as much energy as one dry sheep equivalent per hectare," resulting in a high level of competition with sheep for the pasture (Ridsdill-Smith and Pavri 2010). Long-range movement of the RLEM probably occurs when eggs in the soil adhere to farm machinery or to the legs of sheep or cattle. The diapausing eggs within their mother's dead body also are moved about by wind (see Figure S10.3 on the CD).

Control of the RLEM relies heavily on chemical control in Australia (Ridsdill-Smith et al. 2008). An integrated mite management (IMM) program is being developed and encouraged, however, because this mite has developed resistance to pesticides, so pesticide applications should be minimized when possible (Hoffman et al. 1997). See Table 10.1 for a summary of the tactics recommended. A significant improvement in RLEM management has come about through the development of a model called TIMERITE®, which is based on the onset of diapause egg production (Ridsdill-Smith et al. 2005, Ridsdill-Smith and Pavri, 2010).

10.2.1 Monitoring

Examine susceptible pastures and crops from autumn to spring for the presence of mites and damage, especially within a few weeks after planting. Mites are best seen on foliage in the morning or on overcast days. In the warmer parts of the day, the RLEMs will move to the base of plants into leaf sheaths under debris or into cracks in the ground. When disturbed, they can drop to the ground and hide. The RLEM competes with the blue oat mite for food; thus, chemical control of RLEM may not entirely remove pest pressure on the crop because the blue oat mite is a species that is generally less susceptible to chemical control and can survive to cause damage.

Red-legged earth mites can be sampled by using a soil corer, which is easy to use, inexpensive, and accurate under both wet and dry conditions. A vacuum sampler is useful for sampling mites on recently cultivated land. Samples of populations during the year have been conducted, and in Western Australia 12,000 RLEMs per m^2 were present in the spring of 1990. At least 2000 dead bodies containing 71,000 eggs per m^2 were present in the summer. This produced a fall population of 10,000 RLEM per m^2. Thus, many oversummering eggs died. Summer rainfall can have a deleterious effect on the survival of RLEM eggs during summer, especially rain that occurs early in the summer.

10.2.2 Chemical Control

Chemicals commonly have been used against the RLEM, but no currently registered pesticides can kill the diapausing eggs. Before the advent of DDT, control was neither practical nor economical. DDT was used until the early 1970s, and its main advantage was its long residual action. One annual treatment provided control, making it effective, inexpensive, and simple to use. The organophosphorus pesticides that replaced DDT and other organochlorine products provided relatively poor residual action and had to be applied a number of times through the season. These products were used for 25 to 30 years with mixed results. The inadequacies of these pesticides can be attributed to inefficient use, poor timing, or poor application methods. All treatments are highly toxic at low rates to RLEM. Modification of spray practices to manage RLEM populations is considered an important tactic in management practices. Laboratory bioassays are being conducted to determine whether any products are selective (Hoffman et al. 1997). Additional work is needed to identify products that allow natural enemies to survive yet still provide substantial control of the pest mites. Studies have indicated that broad border sprays could reduce RLEM movements from adjacent fields into favored crops (Weeks et al. 2000). Chemical control can be directed against the bare earth to kill hatching larvae and to protect young seedlings, as foliar sprays when the crop has developed, or as systemic pesticides applied as seed dressings.

Table 10.1 Integrated Mite Management (IMM) Approach to Controlling the Red-Legged Earth Mite (RLEM) in Australia Based on the TIMERITE® Model

Monitor.	Timing is key for controlling RLEM. Sprays should be applied when the mites are most susceptible to sprays so damage is reduced the following fall. This mite is active during cool wet winters, and two generations occur. The third generation produces diapausing eggs on the soil (up to 100,000 eggs per m^2). Because diapausing eggs are not killed by pesticides, the goal is to prevent females from depositing diapause eggs. This program targets a time when it is optimal to treat with a pesticide. The dates are similar at a site across years but vary between sites because diapause induction is correlated with a change in daylength.
Treat in the spring.	By treating in spring just before diapause eggs are produced, damage to new plants in the fall is greatly reduced because the number of diapause eggs produced is greatly reduced. Use the TIMERITE formula to get the correct time to spray for each area based on latitude and longitude. Spray within 2 weeks of the exact TIMERITE date.
Treat in the fall.	Sprays in fall are difficult to time because emergence of the larvae from diapause eggs is not easily predicted. The new generation of larvae can cause significant damage to young plants; however, if no spring spray was applied, then an autumn spray can be applied, although it will not be as effective. Monitor to be sure mites are present before spraying, because resistance to synthetic pyrethroids has developed in some populations.
Plant RLEM-tolerant plant varieties.	Multiple tactics are encouraged to reduce the rate of development of pesticide resistance in the RLEM. When planting new pastures or crops with a risk of damage from RLEM, try to use resistant plant varieties (lists are available).
Manage grazing.	Grazing pastures heavily in spring results in less plant material available for the mites and fewer mites in the fall.
Encourage natural enemies.	No highly effective natural enemies have been identified to control RLEM populations, but a number of natural enemies provide some suppression of RLEM populations. Broad-spectrum pesticides are detrimental, so avoid their use if possible. It appears that untreated shelter belts can increase predator numbers and allow them to reenter adjacent pastures after pesticide residues have decayed.
Limit RLEM movement.	RLEM can move within and between pastures and fields as adults and aerially when diapausing eggs are blown about. Adults can move 7 to 16 m and will move through unfavored crops to more favored plants. Borders containing plants on which the RLEM cannot reproduce can limit movements by the RLEM. Border sprays can reduce movements of RLEM but need to be larger than 10 m wide. The diapause eggs are on the soil surface during the summer and can move with soil through wind or attached to farm machinery. Avoid bare soil during the summer and use shelter belts to minimize wind erosion. Clean farm machinery thoroughly.

Source: Adapted from Australian Wool Innovation Limited, Sydney, Australia (http://www.wool.com/Grow_Timerite.htm).

A properly timed spring spray can significantly reduce populations of the RLEM the following autumn if the females are killed before they begin to produce diapause eggs in mid- to late spring. The proper timing can be predicted using the TIMERITE® model (Ridsdill-Smith and Pavri 2010). Unfortunately, RLEM populations may be resistant to some chemicals, so alternative management practices are desirable.

10.2.3 Biological Control

A variety of predators and a pathogen attack the RLEM in Australia, with the most important being predatory mites (James 1995, Ridsdill-Smith 1997). In addition, generalist predators that have been recorded as natural enemies of the RLEM include *Chrysopa* spp. (Chrysopidae); *Hemerobius* spp. (Hemerobiidae); lady beetles (*Harmonia*, *Diomus*) (Coccinellidae); several phytoseiids, including *Amblyseius* spp. (*A. victoriensis, A. messor, A. masiaka, A. dieteri*) and *Typhlodromus* (= *Metaseiulus*) *occidentalis*; a parasitid mite; anystid mites (*Walzia australica, Erythracarus* sp.); a bdellid; a cunaxid; a stigmaeid; an erythraeid (*Balaustium murorum*); and a fungus disease (*Neozygites acaracida*) (James 1995, Umina et al. 2004).

No single predator dominates the fauna associated with the RLEM. The most important predatory families are the Parasitidae, Bdellidae, Erythraeidae, Anystidae, and Cunaxidae. All predators are probably generalists and may feed on pollen and plant exudates, as well as RLEMs. The native anystid *Walzia australica* appears to be a predator equally effective as the introduced species *Anystis wallacei*. All of the predators listed above were observed either in the field or laboratory feeding on RLEM and the blue oat mite. *Neozygites*, a fungal pathogen, is commonly found in RLEM populations, with infection levels ranging from 0 to 50%. The disease seems most prevalent in blue oat mite populations. Infected mites are yellowish (RLEM) or red (blue oat mite). The infected mites become sterile before they die. The pathogen may be the cause of the population crashes frequently observed during wet winters (James 1995, Ridsdill-Smith 1997).

Natural enemies were sought in South Africa as part of a classical biological control program but to no avail. In southern France, an *Anystis* species was found preying on a close relative of the RLEM (Ridsdill-Smith 1997). This predator was thought to be *Anystis baccarum*, which was already present in Australia. Eventually it was found that a mistake in identification had been made and the French species was misidentified. The introduction program was set back nearly 30 years by this taxonomic problem (you may have noticed that taxonomic problems are not uncommon in biological control projects). The French species was introduced into Western Australian pastures during 1965. It was introduced as *Anystis salicinus*, but the name was changed to *A. wallacei*.

Monitoring after the releases suggests that *Anystis wallacei* can provide control of the RLEM, but *A. wallacei* moves very slowly. It has been spreading at a rate of only 45 meters per year, and roads or waterways serve as effective barriers. *Anystis wallacei* populations of 40 or more per m^2 can reduce RLEM populations by 50% and sometimes to less than 20% in pastures with few or no predators. Some scientists, however, have questioned the effectiveness of this predator; it will eat eggs and adults, but if other food is around it will often ignore them. The predator may have an indirect benefit, however, because it disturbs the RLEM, which makes the pest stop feeding and retreat to the crown of the plant, thus reducing damage to the crop.

A number of predators have been identified in South Africa, and some are under consideration for introduction in a classical biological control program. Work also is under way to improve methods for spreading *Anystis wallacei* throughout the RLEM-infested regions. Field-reared *A. wallacei* have been collected and released into new sites with a high rate of successful establishment. Collected *A. wallacei* must be stored under cool and moist conditions, and releases must occur under similar conditions where RLEM is abundant. The presence of adequate plant material (no overgrazing that would reduce relative humidity) increase the likelihood that *A. wallacei* will establish.

10.2.4 Cultural Controls

Rotating crops or pastures with non-host plants can reduce RLEM populations. Cultivation can reduce RLEM populations by decreasing the number of oversummering eggs. Burning dry plant material after harvesting of the crop can provide some control. Clean fallowing and control of weeds around fields and pastures also can reduce RLEM populations. When pastures have a history of high densities of RLEMs, sowing the pastures with a high-clover content should not be done. In addition, proper managing of grazing can reduce RLEM populations because pastures with shorter plants have a lower relative humidity, which increases the mortality of the pest. Other cultural methods include trap or border crops and mixed cropping.

10.3 BLUE OAT MITE (*PENTHALEUS* SPECIES)

10.3.1 Biology

Blue oat mites (BOMs) are major pests of agriculture in southern Australia and in regions in Europe, Asia, South Africa, New Zealand, and parts of North America with a Mediterranean climate (Jeppson et al. 1975, Ridsdill-Smith 1991). These mites attack pastures, vegetables, and other crops. This pest was introduced into Australia from Europe (Umina 2007). For many years, it was thought that the BOM (or winter grain mite) was a single species, *Penthaleus major*. It is now known, however, that this mite (at least in Australia) is a species complex with two additional species: *P. falcatus* and an undescribed species (Weeks and Hoffmann 1999, Umina et al. 2004). This taxonomic confusion raises questions about the validity of some of the earlier research on these mites. The following discussion focuses on mite research in Australia, where efforts have been directed to develop an IMM program.

Each *Penthaleus* species can cause damage in crops, and each species has a different distribution, pesticide tolerance, and pest status on specific crops in Australia. Studies by Weeks and Hoffmann (1999) in Victoria and New South Wales, Australia, found that *P. major* was the most common, and *P. falcatus* was somewhat less common but relatively more abundant in drier areas. The undescribed species was restricted to an area in western Victoria state. The three species can be identified morphologically and through molecular assays. The main morphological difference in the three species is the length and number of setae on the dorsal surface of the body (Umina et al. 2004). Based on laboratory studies, these three species differ in their responses to pesticides and their host-plant preferences. For example, most pest outbreaks in wheat are due to the undescribed species, whereas *P. falcatus* is most common in canola, and *P. major* and *Halotydeus destructor* are serious pests of pastures. In addition, because these *Penthaleus* species reproduce by thelytoky (females producing only females), different clones of *P. major* appear also to differ in their biology and ecology. Robinson and Hoffmann (2001) concluded that the control measures directed against the blue oat mite were most effective against *P. major*, probably because it is susceptible to the pesticides used. The more difficult species to control is *P. falcatus*, which sometimes survived after two chemical treatments. Likewise, the undescribed *Penthaleus* species also had a higher tolerance to pesticides (compared to *H. destructor*), and control failures have occurred. Once again, the accurate identification of pest species (or even clones) is demonstrated to be important in developing effective pest management programs. Umina et al. (2004) suggested that the taxonomic confusion about the blue oat mite may extend to other countries based on reexamination of illustrations of published descriptions.

These *Penthaleus* species are pests of a variety of crops (including oats, wheat, and canola) and pastures. Despite three species being identified, the generic term blue oat mite (or winter grain mite) is often used to distinguish them from the RLEM in the field. Blue oat mites have an oval reddish-orange mark on their dorsum which is the anal opening (see Figure S10.1 on the CD). A dorsal anal opening is unusual in the Acari. Blue oat mites most often are found feeding alone, whereas RLEMs often feed in groups. Adult BOMs can reach densities of up to 15,000 mites per m^2. Damage occurs when crops are newly planted, and BOM feeding can lead to the loss of entire crops (Robinson and Hoffmann 2001).

The life cycle is egg → larva → protonymph → deutonymph → adult female. Some authors have indicated that there could be a third nymphal stage. The BOM is approximately 1 mm in length and 0.7 to 0.8 mm in width, and it has red-orange legs. BOMs are active in the cool, wet part of the year, and two or three generations can develop during that time, with each generation lasting 8 to 10 weeks. BOMs spend much of their time on the soil surface and feed during the cooler parts of the day on plants. During the hotter part of the day, they move to moist soil or under foliage and may even burrow into the soil to avoid the heat and low relative humidity. Long-range dispersal probably occurs when BOM eggs on soil are moved by livestock and farm machinery. In addition, summer winds can blow diapausing BOM eggs to new sites (Umina 2007).

10.3.2 Aestival Diapause

The blue oat mite is active in winter, which is the cool, moist period of the year in Australia, and they produce drought-resistant diapause eggs in the spring; the eggs do not hatch until the following autumn (another example of an aestival diapause). Blue oat mite females deposit diapause eggs on vegetation or around plants.

10.3.3 Damage

Like the RLEM, *Penthaleus* species feed on cotyledons, trifoliate leaves, and flowers using their sharp chelicerae to pierce the upper leaf epidermis to the palisade layer. BOMs feed on the exuded droplet of cell contents (Umina 2007). The damage appears as a silvering of the leaf. *Penthaleus* species prefer to feed on small grains and grasses and can run rapidly over the surface of plants and the soil. The BOM can cause high seedling mortality in annual pastures if they emerge from diapause eggs at the time of seed germination. In newly established pastures, newly hatched BOMs can reach very high densities.

In high-rainfall areas (>500 mm/year) of Australia, such as Victoria state, significant damage is caused if either or both the RLEM and the BOM are present in 4 of 5 years on newly established pastures sown in autumn. Total failure of seedlings to establish can require reseeding of pastures. More often, reduction in seedling survival and vigor results in changes in plant species composition. Chemical control is considered necessary on most new pastures, with an estimated 60 to 80% of new sowings being treated in any one year. In high rainfall areas, the mites are less significant problems in established pastures. Chemical control is less often applied (only 10 to 20% of fields each year in dairy areas) to established pastures, because the effects of these mites are either unnoticed or underestimated.

10.3.4 Natural Biological Control in Australia

The diverse natural enemies listed above for the RLEM also cause mortality to the BOM in Australia, including predatory mites in the families Anystidae, Bdellidae, Erythraeidae, Parasitidae, and Cunaxidae. Small beetles, spiders, and ants may also play a role. The fungal pathogen *Neozygites acaracida* may be important in wet winters and could cause some population crashes.

10.3.5 Cultural Controls

Rotating crops or pastures with crops that are not hosts of these mites can reduce pest populations, decreasing the need for chemical control (Umina 2007). When *Penthaleus major* is the predominant species, canola and lentils are possible useful rotation crops. If *P. falcatus* is the most common blue oat mite, crops can be rotated with lentils. Cultivation will decrease the number of mites that survive the summer through destruction of the aestivating eggs. Burning crop stubble can also kill large numbers of diapausing summer eggs. Because broad-leaved weeds provide alternative foods for immature mites, control of weeds within crops and around pastures can help reduce mite densities. Finally, managing grazing can also reduce blue oat mite populations to below damaging levels, perhaps because the shorter pasture plants have a lower relative humidity, resulting in reduced food resources and increased mite mortality.

10.3.6 Host-Plant Resistance

Host-plant resistance is one of the easiest and more rapid ways to achieve implementation of an integrated pest management (IPM) program. It is relatively easy to distribute seeds, especially for annual crops, and achieve adoption by growers. Although considerable effort has been targeted toward developing or identifying plant varieties resistant to the RLEM, less information is available for these *Penthaleus* species in Australia due to the fact that growers and scientists were initially slow to recognize that much of the damage caused by "earth mites" was due to *Penthaleus* species (Umina et al. 2004). As a result, research to identify resistant varieties or to develop them has been delayed. A concern is that host-plant resistance to one *Penthaleus* species, once identified, may not confer resistance to the other species.

10.3.7 Chemical Control

Chemical control remains the most common tool for control of *Penthaleus* species in Australia (Umina et al. 2004); however, the pesticides used are not effective against either aestivating or active winter eggs. The most common products used are organophosphates and synthetic pyrethroids. Pesticides are used to prevent mite outbreaks by using residual chemicals as bare-earth treatments before or at the time of planting to kill newly hatched mites and to protect plants as seedlings. Pesticides with a long residual toxicity are applied as border sprays to prevent mites from moving into the crop or pasture, and systemic pesticides are applied to seeds. As noted above, the different species of *Penthaleus* have different sensitivities to pesticides, so an accurate diagnosis must be made (Umina et al. 2004). Unlike the situation with the RLEM, the use of spring sprays based on the TIMERITE® model to suppress populations in the fall does not appear to work on *Penthaleus* species in Australia. Umina et al. (2004) attributed this to the differences in timing of diapause egg production in *Penthaleus* species and concluded that spraying *Penthaleus* species within 2 to 3 weeks of emergence in the fall is likely to be a better method for timing pesticide applications.

SELECTED REFERENCES

Hoffmann, A.A., S. Porter, and I. Kovacs. 1997. The response of the major crop and pasture pest the red-legged earth mite (*Halotydeus destructor*) to pesticides: Dose–response curves and evidence for tolerance. *Exp. Appl. Acarol.* 21:151–162.

James, D.G. 1995. Biological control of earth mites in pastures using endemic natural enemies. *Plant Prot. Q.* 10:58–59.

Jeppson, L.R., H.H. Keifer, and E.W. Baker. 1975. *Mites Injurious to Economic Plants*. Berkeley: University of California Press.

Liu, A. and T.J. Ridsdill-Smith. 2000. Feeding by redlegged earth mite (*Halotydeus destructor*) on seedlings influences subsequent plant performance of different pulse crops. *Aust. J. Exp. Agric.* 40:715–723.

Maclennan, K.E., G. McDonald, and S.A. Ward. 1998. Soil microflora as hosts of redlegged earth mite (*Halotydeus destructor*). *Entomol. Exp. Appl.* 86:319–323.

McDonald, G.M., K. Moritz, E. Merton, and A.A. Hoffmann. 1995. The biology and behaviour of redlegged earth mite and blue oat mite on crop plants. *Plant Prot. Quart.* 10:52–55.

Ridsdill-Smith, T.J. (ed.) 1991. *Proceedings of the National Workshop on Redlegged Earth Mite, Lucerne Flea, and Blue Oat Mite*. Perth: Western Australia Department of Agriculture.

Ridsdill-Smith, T.J. 1997. Biology and control of *Halotydeus destructor* (Tucker) (Acarina: Penthaleidae): A review. *Exp. Appl. Acarol.* 21:195–224.

Ridsdill-Smith, T.J. and A.J. Annells. 1997. Seasonal occurrence and abundance of redlegged earth mite *Halotydeus destructor* (Acari: Penthaleidae) in annual pastures of southwestern Australia. *Bull. Entomol. Res.* 87:413–423.

Ridsdill-Smith, T.J. and C. Pavri. 2010. *TIMERITE*. CSIRO and Australian Wool Innovation Limited (http://www.wool.com/Grow_Timerite.htm).

Ridsdill-Smith, T.J., C. Pavri, E. De Boer, and D. Kriticos. 2005. Predictions of summer diapause in the redlegged earth mite *Halotydeus destructor* (Acari: Penthaleidae) in Australia. *J. Insect Physiol.* 51:717–726.

Ridsdill-Smith, T.J., A.A. Hoffmann, G.P. Mangano, J.M. Gower, C.C. Pari, and P.A. Umina. 2008. Strategies for control of the redlegged earth mite in Australia. *Aust. J. Exp. Agric.* 48:1506–1513.

Robinson, M.T. and A.A. Hoffmann. 2001. The pest status and distribution of three cryptic blue oat mite species (*Penthaleus* spp.) and redlegged earth mite (*Halotydeus destructor*) in southeastern Australia. *Exp. Appl. Acarol.* 25:699–716.

Umina, P. 2007. *Blue Oat Mite*, Agriculture Notes No. AG1300. State of Victoria Department of Primary Industries (http://www.dse.vic.gov.au/DPI/nreninf.nsf/v/B3E8A9AE92961D47CA257443007FD42A/$file/Blue_Oat_Mite.pdf).

Umina, P. 2008. *Redlegged Earth Mite*, Agriculture Notes No. AGO414. State of Victoria Department of Primary Industries (http://www.dpi.vic.gov.au/dpi/nreninf.nsf/v/14CC21106EFFA2C1CA257443007FD3EF/$file/Redlegged_Earth_Mite.pdf).

Umina, P. and A.A. Hoffmann. 2003. Diapause and implications for control of *Penthaleus* species and *Halotydeus destructor* (Acari: Penthaleidae) in southeastern Australia. *Exp. Appl. Acarol.* 31:209–223.

Umina, P.A., A.A. Hoffmann, and A.R. Weeks. 2004. Biology ecology and control of the *Penthaleus* species complex (Acari: Penthaleidae). *Exp. Appl. Acarol.* 34:211–237.

Weeks, A.R. and A.A. Hoffmann. 1999. The biology of *Penthaleus* species in south eastern Australia. *Entomol. Exp. Appl. Acarol.* 92:179–189.

Weeks, A.R., M. Turelli, and A.A. Hoffmann. 2000. Dispersal patterns of pest earth mites (Acari: Penthaleidae) in pastures and crops. *J. Econ. Entomol.* 93:1415–1423.

CHAPTER 11

Friends or Foes?

In addition to the plant-feeding mites described in Chapters 6 to 10, several other mite families are associated with plants. Not all are pests; some are beneficial, and some are rare or limited in their distribution (Evans et al. 1961, Krantz 1971). The Tetranychidae, the predatory Phytoseiidae (discussed in Chapter 12), Eriophyoidea, Tarsonemidae, and Tenuipalpidae are the most often encountered mites on plants; however, you may also see species belonging to the Anystidae, Tuckerellidae, Stigmaeidae, Tydeidae, Acaridae, Hypoaspidae, Hemisarcoptidae, and Oribatida (Gerson and Smiley 1990). A brief overview is given of these families from the point of view of their pest status in agricultural crops.

11.1 THE ANYSTIDAE: FRIENDS OF LIMITED VALUE

Anystid species (Actinedida or Prostigmata) in the genus *Anystis* have been studied relatively little, and their role as natural enemies of pest mites is not well understood (Laing and Knop 1983). All of the species, though, have been found to be predatory and very active. These distinctive, large (0.5 to 1.5 mm), reddish or orange mites are often observed on the soil surface. They are sometimes called whirligig mites due to their characteristic movements (see Figure S11.1on the CD).

Most studies have involved observations of feeding behavior or basic life history in perennial crops and, if toxic pesticides are not used, these mites may be useful generalist predators in orchards and vineyards (Laing and Knop 1983). Their slow development rate (typically, they have only two generations a year) makes them unsuitable for commercial mass rearing and releases. The life cycle consists of egg, prelarva, larva, protonymph, deutonymph, tritonymph, and adults. Sperm transfer involves the deposition of spermatophores on the substrate by males (Otto 1999).

Anystis agilis is found in California vineyards and can feed on spider mites (*Tetranychus pacificus*) and the grape leafhopper (*Erythroneura elegantula*). Sorensen et al. (1976) found that development in the laboratory averaged 49 days, and adult females deposited one to three egg clusters averaging 13 eggs per cluster. This species appears to have approximately two generations per year; however, densities of these mites averaged only one mite per 100 grape leaves during a June in California vineyards, so they were not abundant. Adults of *A. agilis* consumed an average of 39 female spider mites or 6 nymphal *E. elegantula* per day in the laboratory. Based on their slow development rate and relatively low reproductive rate, *A. agilis* is unlikely to be an important natural enemy of spider mites in vineyards (Sorensen et al. 1976). In addition, their behavior is impaired when webbing produced by *T. pacificus* is present, and *A. agilis* is sensitive to pesticides. For these reasons, *A. agilis* is unsuited for augmentative releases, although it could be a useful generalist predator in organic orchards or vineyards.

Anystis baccarum has been described as a "relatively large fast-moving, orange-red mite" that is "a voracious generalist, feeding on any prey that it can overpower" (Laurin and Bostanian 2007). Populations of *A. baccarum* in Quebec, Canada, were abundant in apple orchards and vineyards, where they appear to have two generations per year. Laurin and Bostanian (2007) evaluated the toxicity of eight insecticides to *A. baccarum* and found them highly susceptible to lambda-cyhalothrin, phosmet, and carbaryl, but five other insecticides (methoxyfenozide, acetamiprid, thiamethoxam, imidacloprid, and spinosad) were nontoxic.

In apple orchards in Northern Ireland treated only with fungicides but no acaricides, *Anystis baccarum* was the dominant predator under Northern Ireland's cool, damp climatic conditions (Cuthbertson et al. 2003a, Cuthbertson and Murchie 2004). It reduced populations of the apple rust mite (*Aculus schlechtendali*) and preyed on the apple-grass aphid (*Rhopalosiphum insertum*) (Cuthbertson et al. 2003b). In laboratory trials, *A. baccarum* fed on small instars of *Panonychus ulmi* and on *Bryobia rubrioculus* and *B. praetiosa*.

Halliday and Paull (2004) evaluated the anystid *Chaussieria capensis* as a predator of *Halotydeus destructor*, and Holm and Wallace (1989) evaluated *Anystis baccarum* in the laboratory as a potential predator of the cattle tick (*Boophilus microplus*). Anystids also have been observed feeding on *Panonychus citri*, as well as on citrus thrips (Mostafa et al. 1975).

11.2 THE HYPOASPIDAE: FRIENDS, ESPECIALLY FOR AUGMENTATIVE RELEASES

The taxonomic position of mites in the genus *Hypoaspis* varies depending upon the taxonomist consulted. Here, the genus is placed in the family Hypoaspidae (Mesostigmata or Gamasida), but others have placed these mites in the family Laelapidae.

Commercial mushroom farms may have problems with mushroom flies (Sciaridae, Diptera). Sciarids are pests because larvae of the genus *Lycoriella* feed on the mycelium, destroy the fruiting body, and they break down the structure of the compost, causing a reduction in yield. Pesticides applied to the compost, casing layer, and mushroom house interiors and exteriors often control these flies. Resistance to organophosphates has been reported in the flies, and the pesticides themselves can cause reduced mushroom yield; thus, other approaches have been investigated to reduce pesticide use. The use of nematodes and *Bacillus thuringiensis* has been considered to control this sciarid. A soil-inhabiting mite, *Hypoaspis miles*, appears able to control the glasshouse sciarid *Bradysia paupera* and the mushroom sciarids *Lycoriella solani* and *L. ingenua* (Enkegaard et al. 1997, Ydergaard et al. 1997, Ali et al. 1999, Jess and Bingham 2004, Jess and Schweizer 2009). The predatory mite *Stratiolaelaps scimitus* (Laelapidae) shows promise as a control agent for the sciarid *Bradysia matogrossensis* (Freire et al. 2007).

Sciarids also are pests of greenhouse crops, damaging young or weak plants grown in a moist, organic environment. Larvae chew roots and reduce uptake of water and nutrients but affect strong, healthy plants only at very high levels of infestation. Indirect damage is caused when sciarid larvae transmit nematodes, viruses, and fungal spores to the plant roots. Adult sciards transmit fungal spores to plant foliage. Along with sciarids, another pest in greenhouses is the western flower thrips (*Frankliniella occidentalis*), and the use of *Hypoaspis* predators, along with entomopathogenic nematodes and less-toxic pesticides, is being investigated for their control (Premachandra et al. 2003, Wiethoff et al. 2004, Thoeming and Poehling 2006).

A number of companies mass rear *Hypoaspis aculeifer* and *H. miles* to control sciarid larvae in greenhouses. These predators also feed on thrips pupae, collembola, and nematodes in the soil. Because *Hypoaspis* mites are fed on grain mites (*Tyrophagus putrescentiae*) that can damage plant foliage, the *Hypoaspis* predators should be applied to the soil surface. If grain mites remain in the product, they should not be applied directly to the plant. Care should be taken when working with predatory mites in greenhouses to reduce the likelihood that greenhouse workers will become allergic to them. Kronqvist et al. (2005) found that Swedish greenhouse workers could develop allergies to both *Phytoseiulus persimilis* and *H. miles.*

11.3 THE TUCKERELLIDAE: POTENTIAL PESTS RARELY FOUND EXCEPT IN TROPICAL AND SUBTROPICAL CLIMATES

The Tuckerellidae (Prostigmata or Actinedida: Tetranychoidea) is monogeneric and contains only a few species (approximately 20), although additional species may be discovered as additional tropical and subtropical plants are examined. A few species are occasionally pests of crops (Baker and Pritchard 1953, Jeppson et al. 1975). *Tuckerella* has been collected from roots of plants in California, but other species are found on the aerial portions of plants. *Allochaetophora californica* feeds on Bermuda grass, and *Linotetranus cylindricus* is found in moss. *Tuckerella ornata* has been found on oranges in South Africa. These mites rarely occur in large numbers and rarely are pests. Tuckerellids are notable for their beautiful setae, which have given them the common name of *peacock mites.* It is thought that these setae protect them from predators (see Figure S11.2 on the CD). These mites are usually bright red, with large leaf-shaped setae, and they have a tail of long, peacock-like plumes. The tail can be flicked over the body to brush away potential predators. Although attractive, one species has been associated with the spread of a virus-like necrosis of citrus in South America. In Australia, one species feeds on the introduced weed *Lantana*, while others have been found on rainforest trees and shrubs, and several others are associated with the roots of grasses.

11.4 THE TYDEIDAE: MOSTLY FRIENDS AS PREDATORS, ALTERNATIVE PREY, AND SANITIZING AGENTS

The Tydeidae (Prostigmata or Actinedida) is a family containing more than 300 species in more than 40 genera. Tydeids are small mites (approximately 0.25 mm in length) found around the world (Figure 11.1). Detailed knowledge of the biology of many species is unknown, but they appear to have diverse feeding habits. They have been identified as plant pests in a few cases and as useful

Figure 11.1 On the left is the head of an insect pin showing a male (bottom left) and two female tydeid mites (*Homeopronematus anconai*). These tiny mites feed on pollen and can also prey on eriophyoid mites. They are not considered plant pests, although they probably feed on foliage. The figure on the right shows an adult female under higher magnification. (Photograph by Nancy Knop, Department of Entomology, University of California–Berkeley.)

predators in others, or as important alternative prey for some phytoseiid predators (Flaherty and Hoy 1971, Laing and Knop 1983, Calis et al. 1988). Some species apparently are scavengers, while others feed on fungi, pollen, and honeydew (McGregor 1932, Brickhill 1958, Zaher and Shehata 1963, McCoy et al. 1969, Flaherty and Hoy 1971, Calvert and Huffaker 1974, Wahab et al. 1974). Some are found in stored products, where they prey on other mites (Hughes 1976). One species of tydeid in granaries is beneficial because it is the preferred prey of a predatory mite. Some tydeids are found on the bark or leaves of woody plants, as well as in bird nests and forest litter. Tydeids are found on tomatoes, citrus, grapes, and deciduous fruit trees. Tydeids are small, soft-bodied mites with needle-like chelicerae. They are smaller than spider mites but larger than eriophyoid mites. Many species of tydeids are relatively fast moving (faster than tetranychids and eriophyoids), and some are able to run backward as quickly as they move forward (not many mites do that). Note that the first two pair of legs are well separated from the posterior pairs. These mites are often found under exuviae, under insect scale covers, along midveins of leaves, and in crevices on plants.

Homeopronematus (*Pronematus*) *anconai* feeds on the tomato russet mite (*Aculops lycopersici*) in California and can reduce tomato russet mite population densities substantially (Hessein and Perring 1986). This mite also feeds on pollen, fungi, and (probably) plant tissue; it can be reared on a diet consisting solely of wind-dispersed pollens, and vigorous colonies can be maintained indefinitely in the laboratory on cattail pollen (Flaherty and Hoy 1971). When it is a predator of the tomato rust mite, it could be considered a beneficial predatory species. This same species is found in San Joaquin Valley vineyards, where it is an important prey for the phytoseiid *Metaseiulus occidentalis*. Both predator and tydeids overwinter under the bud scales of grapevines, and the tydeid serves as food for *M. occidentalis* females after they emerge from diapause late in the winter. As a result, the tydeids are considered beneficial in vineyards. By contrast, the primary spider mite pests *Tetranychus pacificus* and *T. willametti* overwinter as females in diapause under the bark of the vine trunk, so there is no overlap in the distribution of diapausing *M. occidentalis* and the diapausing tetranychids during the winter (Flaherty and Hoy 1971).

Efforts were made to build up populations of *Homeopronematus anconai* by applying cattail pollen to vines with a duster to increase populations of *Metaseiulus occidentalis* early in the growing season so they could suppress tetranychid pest mites later on (Flaherty and Hoy 1971, Calvert and Huffaker 1974). The tydeid populations did increase, but pollen applications did not become a practical pest management tool, in part because the tydeids are sensitive to sulfur, which is applied early in the growing season to control powdery mildew in California vineyards (Knop and Hoy 1983a–c). In addition to being susceptible to sulfur, this population of *H. anconai* is susceptible to benomyl, propargite, cyhexatin, and hexakis (Knop and Hoy 1983b). Kryocide (cryolite) appears to allow this tydeid to survive, but carbaryl, methomyl, and dibrom are toxic, as is permethrin (Knop and Hoy 1983b).

Homeopronematus anconai appears to have an unusual mating behavior for tydeids, with males guarding quiescent female deutonymphs (Knop 1985). Males transfer sperm directly to females, and if females are not mated within about 24 hours after emerging they do not mate at all; they produce only male progeny, suggesting that this species is arrhenotokous. By contrast, Hernandes et al. (2006) found *Lorryia formosa* to be thelytokous.

Other tydeids feed on eriophyoid mites, nematodes, and other invertebrates. *Tydeus gloveri* is found in Florida citrus groves where it is a "scavengermite commonly found aggregating in and around detritus found on leaves and fruit of citrus trees." Thus, it probably is not a plant pest. Smirnoff (1957) suggested, however, that an undescribed species of *Lorryia* caused injury to citrus trees in Morocco, and Malchenkova (1967) reported a *Tydeus* species caused damage to grape foliage in Moldavia. *Parapronematus acaciae* was reported to be a predator of citrus rust mite, but it is not a predator of rust mite, the citrus red mite, or the Texas citrus mite (McCoy et al. 1969). This species can be reared in the laboratory on *Penicillium* and *Colletotrichum*, two common leaf-inhabiting fungi. It does not feed on pollen, including citrus pollen, or on whitefly larvae or armored scale insects; thus, it is considered a scavenger (McCoy et al. 1969). *Tydeus californicus* was reported to feed on citrus bud mite (*Aceria sheldoni*), making it a predator; however, Fleschner and Arakawa (1952) indicated they could rear *T. californicus* on citrus and avocado leaves free of fungus or prey, and they directly observed the mites feeding on the foliage. Soliman et al. (1974) reared *T. californicus* on sweet potato leaves. *Homeopronematus anconai* feeds on wind-blown pollens such as cattail (*Typha* sp.) and honeydew, and this mite cannot be reared on artificial substrates. In studies, it was reared on grape (*Vitis vinifera*) or blackberry (*Rubus vitifolia*) foliage to which the pollen and other foods were added, suggesting that it also feeds on foliage (Flaherty and Hoy 1971, Knop and Hoy 1983b).

Several species feed on fungi in crops. *Tydeus caudatus* appears to feed on grape downy mildew in Italian vineyards and also preys on eriophyoid mites on the vines (Duso et al. 2005). *Lorryia formosa* can become quite abundant in citrus groves. Because it is abundant, some observers have assumed that the tydeid is a plant pest; however, *L. formosa* has been found to feed on sooty mold produced on scale honeydew and could be considered beneficial because it acts as a sanitizing agent (Mendel and Gerson 1982). *Orthotydeus lambi*, a tydeid found on wild and cultivated grapes in the northeastern United States, was able to reduce the incidence and severity of grape powdery mildew (English-Loeb et al. 1999, Melidossian et al. 2005). Leaf morphology appears to play a role in the abundance of *O. lambi*. Many plants, especially trees and vines, have domatia, which are tufts of hairs or pits located in major leaf-vein axes (O'Dowd and Willson 1989, 1991). Domatia may represent a mutualism, in which the plant provides refuge to the mites and the mites protect the plant from fungal and arthropod parasites. Most grape cultivars lack domatia, but wild grapes possess them. It could be useful to incorporate domatia into new grape cultivars through crop breeding so tydeids and phytoseiids have a protected environment (Walter and Denmark 1991). Unfortunately, two common fungicides used on grapes—mancozeb and sulfur—are toxic to *O. lambi* and suppress its populations. This tydeid is a natural enemy of an important plant pathogen, although its role in fungicide-treated vineyards is limited.

Tydeid biology is diverse: Some tydeids may act as predators, some serve as alternate prey for phytoseiids, and some clean up after honeydew producers or reduce damage due to powdery mildew or sooty mold. Although some may feed on plants, the author is unaware of any tydeids that have been truly documented as serious crop pests and is dubious that pesticide applications would be required for them. It is difficult to conclude, however, that tydeids are usually beneficial because we still know so little about the biology of most species.

11.5 THE ACARIDAE: USUALLY FOES BUT OCCASIONALLY BENEFICIAL?

Generally, the Acaridae (Astigmata or Acaridida) appear to be pests of stored products, initiators of allergies, and pests of agricultural products (see Chapter 24). Are there any beneficial or useful species among the more than 400 species in 90 genera? In greenhouses, *Rhizoglyphus robini* and *R. echinopus* are pests of bulbs (Ascerno et al. 1981, Gencsoylu et al. 1998). Others may feed on vegetables, as well, especially when the soil contains large amounts of organic matter (Santos et al. 1981). Lesna et al. (1995, 2000) investigated whether the predator *Hypoaspis aculeifer* could control *Rhizoglyphus robini* in the greenhouse, with promising results.

Tyrophagus putrescentiae was reported to be an important predator of southern corn rootworm (*Diabrotica undecimpunctata howardi*) in peanut and cornfields in North Carolina (Brust and House 1988). These mites fed on rootworm eggs, rapidly located them in the soil, and preferred them to fungi, organic debris, and dead arthropods. *Tyrophagus putrescentiae* also was reported to feed on dead and living adults and eggs of the grape phylloxera *Daktulosphaira vitifolii* (Homoptera: Phylloxeridae) in leaf galls (Rack and Rilling 1978). *Tyrophagus* species, including *T. putrescentiae*, can be reared on a diet of living nematodes, suggesting that this mite can provide some control of pest nematodes in the soil (Walter et al. 1986).

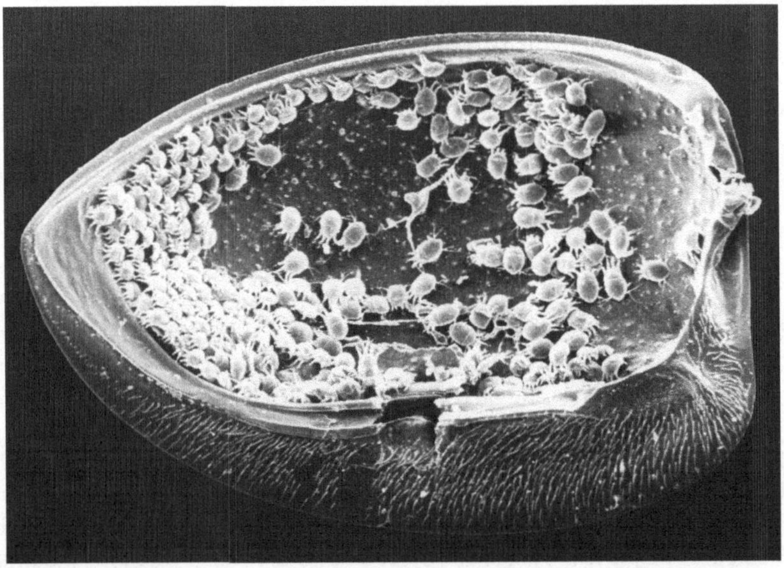

Figure 11.2 Phoretic deutonymphs of *Hemisarcoptes* on the inner cavity of the elytron of a *Chilocorus* beetle. (Photograph by Uri Gerson, Hebrew University of Jerusalem, Israel.)

11.6 THE HEMISARCOPTIDAE: FRIENDS REQUIRING MORE STUDY

All species of *Hemisarcoptes* (Acaridida or Astigmata) are predators or parasites of armored scale insects (Homoptera: Diaspididae) and have been used in classical biological control programs (Gerson et al. 1990, Hill et al. 1993). *Hemisarcoptes coccophagus*, *H. malus*, *H. dzhashii*, and *H. cooremani* are found in many parts of the world. Hemisarcoptids are whitish, soft-bodied mites that feed on adult scales or their eggs and deposit eggs on the scales (Izraylevich and Gerson 1993, 1995). Deutonymphs (hypopi) of *Hemisarcoptes* are phoretic on *Chilocorus* beetles, which are predators of scale insects and are transported from site to site by their beetle hosts (Figure 11.2).

Hemisarcoptids are nonspecific predators or parasites of armored scale insects, and their effectiveness as natural enemies varies from site to site. The mites appear to feed on scale females and their eggs. Feeding mites tend to take up the color of their prey (Gerson et al. 1990). Thus, *Hemisarcoptes* feeding on *Lepidosaphes beckii* are purple, but they are red if feeding on *Epidiaspis leperii* and yellow if feeding on *Quadraspidiotus juglansregiae*, which can make it difficult to detect the mites. The presence of fewer than five mites allows the scale to survive, but with its fecundity reduced. Scales attacked by five to ten mites do not produce progeny, and scales on which more than ten mites have fed will die (Gerson and Schneider 1981).

Unmated *Hemisarcoptes coccophagus* females deposit no eggs; after mating, the females deposit an average of 17 eggs per female with a sex ratio of 1 male to 2 females. *Hemisarcoptes coccophagus* has a facultative deutonymphal stage (hypopus) that forms when the larvae or the protonymphs are forced to subsist on insufficient food. Development requires about 16 days at 28°C if the hypopial stage does not occur. Only about 6% of the population goes into the hypopial stage (Houck and O'Connor 1990). *Chilocorus* lady beetles disseminate the hypopi, if produced, by phoresy. The hypopi move to the venter of the elytra of its beetle host, where it stays until the beetle finds new scale prey. The hypopi lack mouthparts and do not feed on their beetle hosts, although they have been observed to swell while attached, suggesting that the hypopi may be absorbing beetle hemolymph, which contains alkaloids (Houck 1994). The alkaloids do not appear to be toxic to the *Hemisarcoptes*. Mites developing on healthy host scales skip this stage and molt directly to the tritonymphal stage.

Hemisarcoptes species may be useful natural enemies of a variety of pests: *Hemisarcoptes malus* is a natural enemy of oystershell scale (*Lepidosaphes ulmi*) on apples in eastern Canada and was successfully transferred to western Canada (Tothill 1919, Gerson et al. 1990). *Hemisarcoptes malus* was introduced to Bermuda to control *L. newsteadi* on cedar trees, where it also attacked purple scale (*L. beckii*) on citrus. *Hemisarcoptes coccophagus* is important in controlling the date palm scale (*Parlatoria blanchardi*) in Africa (Kaufmann 1977), and it also attacks citrus scales. Hill et al. (1993) reported the successful establishment of *H. coccophagus* on lataniae scale (*Hemiberlesia lataniae*) in kiwifruit in New Zealand, where it used two species of New Zealand ladybirds (*Scymnus fagus* and *Halmus chalybeus*) for phoresy.

The use of these mites in biological control programs has been delayed by a lack of information on their taxonomy, biology, and ecology. Luck et al. (1999) found that *Hemisarcoptes cooremani* was not a good candidate for augmentative releases against the California red scale (*Aonidiella aurantii*) because the heavily sclerotized body of the red scale prevented the mite from getting under the scale cover to feed. Izraylevich and Gerson (1993) found that control of scales by *H. coccophagus* depends on the host scale species, its age structure, and the host plant. It is likely that many species of *Hemisarcoptes* include cryptic species, so information about different populations in different areas may be conflicting. These mites also are susceptible to many common pesticides, including sulfur and oil (Pickett and Patterson 1953, Gerson et al. 1990); however, *H. coccophagus* have been reared on potato tubers containing scale insects (Gerson 1967, Gerson and Schneider 1981), so it is possible that *Hemisarcoptes* species could be mass reared for augmentative releases into crops. Gerson et al. (1990) noted that the close association of *Hemisarcoptes* species with *Chilocorus* beetles and their restriction to diaspidid scale hosts suggest that these arthropods have coexisted for a very long time and that the *Hemisarcoptes* mites may not be as pathogenic due to this long host association.

11.7 THE STIGMAEIDAE: FRIENDS, ESPECIALLY IN UNSPRAYED ORCHARDS AND VINEYARDS

Zetzellia and *Agistemus* predators (Actinedida or Prostigmata: Stigmaeidae) have been considered of value in managing agricultural pest mites occurring on the foliage of plants (Santos 1976a,b, Bostanian et al. 2006). Stigmaeids are small to medium-sized mites (0.2 to 0.5 mm); these predators are red to yellow in color and oval or elongate. The life cycle consists of egg, larva, protonymph, deutonymph, and adults. Males and females copulate (direct sperm transfer). These species are arrhenotokous, with unfertilized eggs producing only males. They disperse by wind. They live in the soil and on plants and are usually predators of other mites. A few prey on scale insects or parasitize flies.

11.7.1 *Zetzellia*

Zetzellia mali is unable to suppress spider mite populations by itself because it has only two or three generations a year in temperate climates (White and Laing 1977); however, these predators may assist other predators in suppressing pest mites, especially in unsprayed environments. Stigmaeids are considered less effective natural enemies than phytoseiids, perhaps because stigmaeids are more sensitive to acaricides and insecticides (although different geographic populations of *Z. mali* appear to differ in their response to pesticides). The life cycle of stigmaeids appears to be longer than that of phytoseiids, which also could limit their ability to respond to spider mite population increases. *Zetzellia mali* is a predator of eggs of *Tetranychus urticae*, *Panonychus ulmi*, and *Bryobia* species and of active stages of eriophyoid mites on fruit trees in North America, Europe, and elsewhere (Santos 1976a,b). *Zetzellia mali* overwinters in the adult stage in diapause between bark scales and often is found in clusters of 150 or more females. Studies in apples suggest that *Z. mali* may practice intraguild predation (Clements and Harmsen 1990). When spider mite prey are rare, *Z. mali* is able to prey on eggs of the phytoseiid *Metaseiulus occidentalis* more frequently than on eggs of the phytoseiid *Typhlodromus pyri* because *M. occidentalis* places its eggs along the midveins more often than *T. pyri*, which is where *Z. mali* spends its time searching for prey (MacRae and Croft 1996).

Zetzellia mali appears to detect prey by contact and may not respond to prey kairomones. The encounter rate thus varies with prey density, although there also appears to be a difference in the ease with which prey types are attacked. The chorion of *Panonychus ulmi* eggs is tougher than the integument of active stages of the apple rust mite (*Aculus schlechtendali*), and *Z. mali* appears to prefer *A. schlechtendali* over *P. ulmi* eggs in apple orchards in Canada. In laboratory trials, *Z. mali* prey preference varied with the relative, but not absolute, density of its prey, as well as the ease of handling the prey (Walde et al. 1995). Contrary to the earlier statement regarding the inability of *Z. mali to* detect prey kairomones, Zahedi-Golpayegani et al. (2007) reported that *Z. mali* responded to odors in olfactometers, with a positive response to odors from the Hawthorne spider mite (*Amphitetranychus viennensis*) and a negative response to odors containing a conspecific predator (*Z. mali*), indicating that this predator may avoid competition with other *Z. mali*.

11.7.2 *Agistemus*

The genus *Agistemus* includes a number of predators of spider mites, including *A. fleschneri*, a predator of *Panonychus ulmi*, *Tetranychus cinnabarinus*, and *T. kanzawai*. *Agistemus floridanus* is a predator of *Eotetranychus sexmaculatus* and various scale insects on citrus in Florida. *Agistemus exsertus* is a predator of *P. citri* (Yue and Tsai 1995) and *Aculops lycopersici* (Osman and Zaki 1986). *Agistemus faneri* is a predator of tetranychids, and *A. longisetus* is a predator of *P. ulmi* and *Bryobia rubrioculus*.

11.8 THE ORIBATIDA (CRYPTOSTIGMATA): USUALLY BENEFICIAL IN THE SOIL BUT MAY CAUSE CROP ROOT DAMAGE AND CONTAMINATE FOODS

Oribatid mites primarily are found in the soil and in leaf litter (see Figures V.2 to V.4 in Part V). In greenhouses and in some field crops, oribatids can become minor problems when they feed on roots, foliage, or fruits (Jeppson et al. 1975, Zhang 2003). Because these mites reproduce slowly, it is rare for them to cause significant damage to agricultural crops; however, they can contaminate foods. Skubala et al. (2006) surveyed 90 different foods, including fruits, vegetables, and mushrooms, obtained from shops in Poland to determine how many were contaminated with mites. They found 53 mite species present on the foods, with oribatids being the most common (56% of total mites), and concluded that "accidental acarophagy" must occur fairly frequently. The mites were most often found on fruits "with uneven surface, covered with numerous hairs, including raspberries, strawberries and red currants. About 50% of mites still remained on vegetables and fruits after cleaning them in running water."

SELECTED REFERENCES

Ali, O., R. Dunne, and P. Brennan. 1999. Effectiveness of the predatory mite *Hypoaspis miles* (Acari: Mesostigmata: Hypoaspidae) in conjunction with pesticides for control of the mushroom fly *Lycoriella solani* (Diptera: Sciaridae). *Exp. Appl. Acarol.* 23:65–77.

Ascerno, M.E., F.L. Pfleger, and H.F. Wilkins. 1981. Effect of root rot and *Rhizoglyphus robini* on greenhouse forced Easter lily development. *Environ. Entomol.* 10:947–949.

Baker, E.W. and A.E. Pritchard. 1953. The family categories of tetranychoid mites, with a review of the new families Linotetranidae and Tuckerellidae. *Ann. Entomol. Soc. Am.* 46:243–258.

Bostanian, N.J., J.M. Hardman, G. Racette, J. Franklin, and J. Lasnier. 2006. Inventory of predacious mites in Quebec commercial apple orchards where integrated pest management programs are implemented. *Ann. Entomol. Soc. Am.* 99:536–544.

Brickhill, C.D. 1958. Biological studies on two species of tydeid mites from California. *Hilgardia* 27:601–620.

Brust, G.E. and G.J. House. 1988. A study of *Tyrophagus putrescentiae* (Acari: Acaridae) as a facultative predator of southern corn rootworm eggs. *Exp. Appl. Acarol.* 4:335–344.

Calis, J.N.M., W.P.J.Overmeer, and L.P.S. van der Geest. 1988. Tydeids as alternative prey for phytoseiid mites in apple orchards. *Med. Fac. Landbouww. Rijksuniv. Gent.* 53(2b):793–798.

Calvert, D.J. and C.B. Huffaker. 1974. Predator [*Metaseiulus occidentalis*]–prey [*Pronematus* spp.] interactions under sulfur and cattail pollen applications in a noncommercial vineyard. *Entomophaga* 19:361–369.

Clements, D.R. and R. Harmsen. 1990. Predatory behavior and prey stage preference of stigmaeid and phytoseiid mites and their potential compatibility in biological control. *Can. Entomol.* 122:321–328.

Cuthbertson, A.G.S. and A.K. Murchie. 2004. The phenology, oviposition and feeding rate of *Anystis baccarum*, a predatory mite in Bramley apple orchards in Northern Ireland. *Exp. Appl. Acarol.* 34:367–373.

Cuthbertson, A.G.S., A.C. Bell, and A.K. Murchie. 2003a. Impact of the predatory mite *Anystis baccarum* (Prostigmata: Anystidae) on apple rust mite *Aculus schlechtendali* (Prostigmata: Eriophyidae) populations in Northern Ireland Bramley orchards. *Ann. Appl. Biol.* 142:107–114.

Cuthbertson, A.G.S., C.C. Fleming, and A.K. Murchie. 2003b. Detection of *Rhopalosiphum insertum* (apple-grass aphid) predation by the predatory mite *Anystis baccarum* using molecular gut analysis. *Agric. Forest Entomol.* 5:219–225.

Duso, C., A. Pozzebon, C. Capuzzo, V. Malagnini, S. Otto, and M. Borgo. 2005. Grape downy mildew spread and mite seasonal abundance in vineyards: Effects on *Tydeus caudatus* and its predators. *Biol. Control* 32:143–154.

English-Loeb, G., A.P. Norton, D.M. Gadoury, R.C. Seem, and W.F. Wilcox. 1999. Control of powdery mildew in wild and cultivated grapes by a tydeid mite. *Biol. Control* 14:97–103.

Enkegaard, A., M.A. Sardar, and H.F. Brodsgaard. 1997. The predatory mite *Hypoaspis miles*: Biological and demographic characteristics on two prey species, the mushroom sciarid fly *Lycoriella solani* and the mould mite, *Tyrophagus putrescentiae*. *Entomol. Exp. Appl.* 82:125–146.

Evans, G.O., J.G. Sheals, and D. MacFarlane. 1961. *The Terrestrial Acari of the British Isles*, Vol. I. London: British Museum of Natural History.

Flaherty, D.L. and M.A. Hoy. 1971. Biological control of Pacific mites and Willamette mites in San Joaquin Valley vineyards. Part III. Role of tydeid mites. *Res. Popul. Ecol.* 13:80–96.

Fleschner, C.A. and K.Y. Arakawa. 1952. The mite *Tydeus californicus* on citrus and avocado leaves. *J. Econ. Entomol.* 45:1092.

Freire, R., G. Moraes, E. Silva, A. Vaz, and R. Campos Castilho. 2007. Biological control of *Bradysia matogrossensis* (Diptera: Sciaridae) in mushroom cultivation with predatory mites. *Exp. Appl. Acarol.* 42:87–93.

Gencsoylu, I., W. Liu, K.A. Usmani, and C.O. Knowles. 1998. Toxicity of acaricides to the bulb mite *Rhizoglyphus echinopus* (Acari: Acaridae). *Exp. Appl. Acarol.* 22:343–351.

Gerson, U. 1967. Observations on *Hemisarcoptes coccophagus* Meyer (Astigmata: Hemisarcoptidae), with a new synonym. *Acarologia* 9:632–638.

Gerson, U. and R. Schneider. 1981. Laboratory and field studies on the mite *Hemisarcoptes coccophagus* Meyer (Astigmata: Hemisarcoptidae), a natural enemy of armored scale insects. *Acarologia* 22:199–208.

Gerson, U. and R.L. Smiley. 1990. *Acarine Biocontrol Agents: An Illustrated Key and Manual.* London: Chapman & Hall.

Gerson, U., B.M. O'Connor, and M.A. Houck. 1990. Acari. In: D. Rosen (ed.), *Armored Scale Insects: Their Biology, Natural Enemies, and Control*, Vol. 4B (pp. 77–98). Amsterdam: Elsevier.

Halliday, R.B. and C. Paull. 2004. Assessment of *Chaussieria capensis* (Acari: Anystidae) as a predator of *Halotydeus destructor* (Acari: Penthaleidae). *African Entomol.* 12:286–290.

Hernandes, F.A., R.J. Fazzio Feres, and F. Nomura. 2006. Biological cycle of *Lorryia formosa* (Acari, Tydeidae) on rubber tree leaves: A case of thelytoky. *Exp. Appl. Acarol.* 38:237–242.

Hessein, N.A. and T.M. Perring. 1986. Feeding habits of the Tydeidae with evidence of *Homeopronematus anconai* (Acari: Tydeidae) predation of *Aculops lycopersici* (Acari: Eriophyidae). *Int. J. Acarol.* 12:215–221.

Hill, M.G., D.J. Allen, R.C. Henderson, and J.G. Charles. 1993. Introduction of armoured scale predators and establishment of the predatory mite *Hemisarcoptes coccophagus* (Acari: Hemisarcoptidae) on lantana scale, *Hemiberlesia lataniae* (Homoptera: Diaspididae) in kiwifruit shelter trees in New Zealand. *Bull. Entomol. Res.* 83:369–376.

Holm, E. and M.M.H. Wallace. 1989. Distribution of some anystid mites (Acari: Anystidae) in Australia and Indonesia and their role as possible predators of the cattle tick, *Boophilus microplus* (Acari: Ixodidae). *Exp. Appl. Acarol.* 6:77–83.

Houck, M.A. 1994. Adaptation and transition into parasitism from commensalism: A phoretic model. In: M.A. Houck (ed.), *Mites: Ecological and Evolutionary Analyses of Life-History Patterns* (pp. 252–281). London: Chapman & Hall.

Houck, M.A. and B.M. O'Connor. 1990. Ontogeny and life history of *Hemisarcoptes cooremani* (Acari: Hemisarcoptidae). *Ann. Entomol. Soc. Am.* 83:869–886.

Hughes, A.M. 1976. *The Mites of Stored Food and Houses*, 2nd ed. London: Her Majesty's Stationery Office.

Izraylevich, S. and U. Gerson. 1993. Mite parasitization on armored scale insects: Host suitability. *Exp. Appl. Acarol.* 17:861–875.

Izraylevich, S. and U. Gerson. 1995. Spatial patterns of the parasitic mite *Hemisarcoptes coccophagus* (Astigmata: Hemisarcoptidae): Host effect, density dependence of aggregation, and implications for biological control. *Bull. Entomol. Res.* 85:235–240.

Jeppson, L.R., H.H. Keifer, and E.W. Baker. 1975. *Mites Injurious to Economic Plants.* Berkeley: University of California Press.

Jess, S. and J.P.W.Bingham. 2004. Biological control of sciarid and phorid pests of mushroom with predatory mites from the genus *Hypoaspis* (Acari: Hypoaspidae) and the entomopathogenic nematode *Steinernema feltiae*. *Bull. Entomol. Res.* 94:159–167.

Jess, S. and H. Schweizer. 2009. Biological control of *Lycoriella ingenua* (Diptera: Sciaridae) in commercial mushroom (*Agaricus bisporus*) cultivation: A comparison between *Hypoaspis miles* and *Steinernema feltiae*. *Pest Manag. Sci.* 65:1195–1200.

Kaufmann, T. 1977. *Hemisarcoptes* sp. and biological control of the date palm scale, *Parlatoria blanchardi* Targioni, in the Sahel region of Niger. *Environ. Entomol.* 6:882–884.

Knop, N.F. 1985. Mating behavior in the tydeid mite *Homeopronematus anconai* (Acari: Tydeidae). *Exp. Appl. Acarol.* 1:115–125.

Knop, N.F. and M.A. Hoy. 1983a. Tydeid mites in vineyards. *Calif. Agric.* 37(11/12):16–18.

Knop, N.F. and M.A. Hoy. 1983b. Biology of a tydeid mite, *Homeopronematus anconai* (n. comb.) (Acari: Tydeidae), important in San Joaquin Valley vineyards. *Hilgardia* 51:1–30.

Knop, N.F. and M.A. Hoy. 1983c. Factors limiting the utility of *Homeopronematus anconai* (Acari: Tydeidae) in integrated pest management in San Joaquin Valley vineyards. *J. Econ. Entomol.* 76:1181–1186.

Krantz, G.W. 1971. *A Manual of Acarology.* Corvallis: Oregon State University.

Kronqvist, M., E. Johansson, B. Kolmodin-Hedman, H. Oman, M. Svartengren, and M. Hage-Hamsten. 2005. IgE sensitization to predatory mites and respiratory symptoms in Swedish greenhouse workers. *Allergy* 60:521–526.

Laing, J.E. and N.F. Knop 1983. Potential use of predaceous mites other than phytoseiids for biological control of orchard pests. In: M.A. Hoy, G.L. Cunningham, and L. Knutson (eds.), *Biological Control of Pests by Mites* (pp. 28–35), Special Publication 3304. Berkeley: University of California Press.

Laurin, M.C. and N.J. Bostanian. 2007. Laboratory studies to elucidate the residual toxicity of eight insecticides to *Anystis baccarum* (Acari: Anystidae). *J. Econ. Entomol.* 100:1210–1214.

Lesna, I., M.W. Sabelis, H.R. Bolland, and C.G.M. Conijn. 1995. Candidate natural enemies for control of *Rhizoglyphus robini* Claparede (Acari: Astigmata) in lily bulbs: Exploration in the field and preselection in the laboratory. *Exp. Appl. Acarol.* 19:655–669.

Lesna, I., C.G.M. Conijn, M.W. Sabelis, and N.M. van Straalen. 2000. Biological control of the bulb mite, *Rhizoglyphus robini*, by the predatory mite, *Hypoaspis aculeifer*, on lilies: Predator–prey dynamics in the soil, under greenhouse and field conditions. *Biol. Control Sci. Technol.* 10:179–193.

Luck, R.F., G. Jiang, and M.A. Houck. 1999. A laboratory evaluation of the astigmatid mite *Hemisarcoptes cooremani* Thomas (Acari: Hemisarcoptidae) as a potential biological control agent for an armored scale, *Aonidiella aurantii* (Maskell) (Homoptera: Diaspididae). *Biol. Control* 15:173–183.

MacRae, I.V. and B.A. Croft. 1996. Differential impact of egg predation by *Zetzellia mali* (Acari: Stigmaeidae) on *Metaseiulus occidentalis* and *Typhlodromus pyri* (Acari: Phytoseiidae). *Exp. Appl. Acarol.* 20:143–154.

Malchenkova, N.I. 1967. A mite of the genus *Tydeus* (Acariformes, Tydeidae) that is a grape pest in Moldavia. *Entomol. Rev.* 46:66–68.

McCoy, C.W., A.G. Selhime, and R.F. Kanavel. 1969. The feeding behavior and biology of *Parapronematus acaciae* (Acarina: Tydeidae). *Florida Entomol.* 52:13–19.

McGregor, E.A. 1932. The ubiquitous mite, a new species on citrus. *Proc. Entomol. Soc. Wash.* 34(4):60–64.

Melidossian, H.S., R.C. Seem, G. English-Loeb, W.F.Wilcox, and D.M. Gadoury. 2005. Suppression of grapevine powdery mildew by a mycophagous mite. *Plant Dis.* 89:1331–1338.

Mendel, Z. and U. Gerson. 1982. Is the mite *Lorryia formosa* Cooreman (Prostigmata: Tydeidae) a sanitizing agent in citrus groves? *Acta Oecol.* 3:47–51.

Mostafa, A.R., P. DeBach, and T.W. Fisher. 1975. Anystid mite: Citrus thrips predator. *Calif. Agric.* 29(3):5.

O'Dowd, D.J. and M.F. Willson. 1989. Leaf domatia and mites on Australasian plants: Ecological and evolutionary implications. *Biol. J. Linn. Soc.* 27:191–236.

O'Dowd, D.J. and M.F. Willson. 1991. Associations between mites and leaf domatia. *Trends Ecol. Evol.* 6:179–182.

Osman, A.A. and A.M. Zaki. 1986. Studies on the predation efficiency of *Agistemus exsertus* Gonzalez (Acarina, Stigmaeidae) on the eriophyid mite, *Aculops lycopersici* (Massee). *Z. Anz. Schadlingskde, Pflanzenschutz, Unweltschutz* 59:125–136.

Otto, J.C. 1999. Systematics and natural history of the genus *Chaussieria* Oudemans (Acarina: Prostigmata: Anystidae). *Zool. J. Linn. Soc.* 126:251–306.

Pickett, A.D. and N.A. Patterson. 1953. The influence of spray programs on the fauna of apple orchards in Nova Scotia. IV. A review. *Can. Entomol.* 85:472–478.

Premachandra, W.T.S.D., C. Borgemeister, O. Berndt, R.U. Ehlers, and H.M. Poehling. 2003. Combined releases of entomopathogenic nematodes and the predatory mite *Hypoaspis aculeifer* to control soil-dwelling stages of western flower thrips, *Frankliniella occidentalis*. *BioControl* 48:529–541.

Rack, G. and G. Rilling. 1978. Uber das Vorkommen der Modermilbe, *Tyrophagus putrescentiae* (Schrank) in Blattgallen der Reblaus, *Dactylosphaera vitifolii* Shimer. *Vitis* 17:54–66.

Santos, M.A. 1976a. Evaluation of *Zetzellia mali* as a predator of *Panonychus ulmi* and *Aculus schlechtendali*. *Environ. Entomol.* 5:187–191.

Santos, M.A. 1976b. Prey selectivity and switching response of *Zetzellia mali*. *Ecology* 57:390–394.

Santos, P.F., J. Phillips, and W.G. Whitford. 1981. The role of mites and nematodes in early stages of buried litter decomposition in a desert. *Ecology* 62:664–669.

Skubala, P., A. Marzec, and M. Sokolowska. 2006. Accidental acarophagy: Mites found on fruits, vegetables and mushrooms. *Biological Lett.* 43(2):249–255.

Smirnoff, W.A. 1957. An undescribed species of *Lorryia* (Acarina, Tydeidae) causing injury to citrus trees in Morocco. *J. Econ. Entomol.* 50:361–362.

Soliman, Z.R., M.A. Zaher, and G.S. ElSafi. 1974. An attempt for rearing *Tydeus californicus* (Banks) on sweet potato leaves in Egypt. *Bull. Soc. Entomol. Egypt* 58:217–219.

Sorensen, J.T., D.N. Kinn, R.L. Doutt, and J.R. Cate. 1976. Biology of the mite *Anystis agilis* (Acari: Anystidae): A California vineyard predator. *Ann. Entomol. Soc. Am.* 69:905–910.

Thoeming, G. and H.N Poehling. 2006. Integrating soil-applied azadirachtin with *Amblyseius cucumeris* (Acari: Phytoseiidae) and *Hypoaspis aculeifer* (Acari: Laelapidae) for the management of *Frankliniella occidentalis* (Thysanoptera: Thripidae). *Environ. Entomol.* 35:746–756.

Tothill, J.D. 1919. Some notes on the natural control of the oystershell scale. *Bull. Entomol. Res.* 9:183–196.

Wahab, A.E.A., A.A. Yousef, and H.M. Hemeda. 1974. Biological studies on the tydeid mite, *Tydeus californicus* (Banks). *Bull. Soc. Entomol. Egypt* 58:349–353.

Walde, S.J., C.N. Magagula, and M.L. Morton. 1995. Feeding preference of *Zetzellia mali:* Does absolute or relative abundance of prey matter more? *Exp. Appl. Acarol.* 19:307–317.

Walter D.E. and H.A. Denmark. 1991. Use of leaf domatia on wild grape (*Vitis munsoniana*) by arthropods in central Florida. *Florida Entomol.* 74:440–446.

Walter, D.E., R.A. Hudgens, and D.W. Freckman. 1986. Consumption of nematodes by fungivorous mites *Tyrophagus* spp. (Acarina: Astigmata: Acaridae). *Oecologia* 70:357–361.

White, N.D.G. and J.E. Laing. 1977. Some aspects of the biology and a laboratory life table of the acarine predator *Zetzellia mali*. *Can. Entomol.* 109:1275–1281.

Wiethoff, J., H.M. Roehling, and R. Meyhofer. 2004. Combining plant- and soil-dwelling predatory mites to optimize biological control of thrips. *Exp. Appl. Acarol.* 34:239–261.

Ydergaard, S., A. Enkegaard, and H.F. Brodsgaard. 1997. The predatory mite *Hypoaspis miles*: Temperature dependent life table characteristics on a diet of sciarid larvae, *Bradysia paupera* and *B. tritici*. *Entomol. Exp. Appl.* 85:177–187.

Yue, B.S. and J.H. Tsai. 1995. *Agistemus exertus* Gonzalez (Acari: Stigmaeidae) as a predator of citrus red mite (*Panonychus citri* [McGregor]). *J. N.Y. Entomol. Soc.* 103:107–113.

Zahedi-Golpayegani, A., A. Saboori, and M.W. Sabelis. 2007. Olfactory response of the predator *Zetzellia mali* to a prey patch occupied by a conspecific predator. *Exp. Appl. Acarol.* 43:199–204.

Zaher, M.A. and K.K. Shehata. 1963. Biological studies on *Tydeus californicus* (Banks) in Egypt (U.A.R.). *Bull. Soc. Entomol. Egypte* 47:297–300.

Zhang, Z.-Q. 2003. *Mites of Greenhouses: Identification, Biology, and Control*. Wallingford, U.K.; CAB International.

CHAPTER 12

The Phytoseiidae: Effective Natural Enemies

The family Phytoseiidae is the most important family of acarine predators of plant pest mites in agriculture (Huffaker et al. 1969, 1970, McMurtry et al. 1970, Hoy 1982, Helle and Sabelis 1985, Hoffman and Frodsham 1993, Gerson et al. 2003). Relatively little is known about the biology of many of the approximately 2000 named phytoseiid species, because they are not found on agricultural crops. Some phytoseiids are found on crops in single countries, but some premier species are used in integrated mite management (IMM) programs in multiple countries. These include *Amblyseius barkeri, A. californicus, A. cucumeris, A. degenerans, A. fallacis, A. hibisci, A. potentillae, A. swirskii, Metaseiulus occidentalis, Phytoseiulus persimilis*, and *Typhlodromus pyri* (see Figures S12.1 to S12.4 on the CD).

The names of phytoseiids can vary in the literature due to taxonomic debates about generic status; for example, *Metaseiulus occidentalis* was originally described as *Typhlodromus*, then moved to the genus *Metaseiulus*, and then moved to the genus *Galendromus*. *Amblyseius californicus* may be called *Neoseiulus californicus* in some publications. Many species originally placed in the genus *Amblyseius* have been moved to other genera. This text uses the generic names used by the authors of the papers cited.

Phytoseiids most often have been used to control pest spider mites (Tetranychidae), but some can control thrips in greenhouses and in California citrus groves (Grafton-Cardwell et al. 1999). Others are effective natural enemies of pests in the Eriophyoidea, Tenuipalpidae, and Tarsonemidae (Kostiainen and Hoy 1996). Certain phytoseiids can be important natural enemies because they consume large numbers of prey, can maintain plant-feeding mites at low densities, can be mass reared for augmentative releases, have a high reproductive rate and rapid developmental rate comparable to their prey, and have a female-biased sex ratio equivalent to their prey, which allows them to respond numerically to increased prey density.

A great deal of research has been conducted on the Phytoseiidae, and this chapter introduces the reader to some of the basics. For further details on the biology of specific phytoseiid species, consulting additional sources is recommended. For review articles, see Huffaker et al. (1969, 1970), McMurtry et al. (1970), Hoy (1982), Tanigoshi (1982), Helle and Sabelis (1985), and McMurtry and Croft (1997). Kostiainen and Hoy (1996) published a bibliography of the Phytoseiidae that covered publications between 1960 and 1994. Between 1960 and 1994, at least 4634 papers were published on the taxonomy, biology, ecology, and roles in pest management of phytoseiid mites. The indices have publications listed by subject, by the effects of pesticides on the Phytoseiidae, by species, by original descriptions of new taxa, by descriptions of previously described species, by the prey species fed upon, and by author. This publication is included on the accompanying CD. It is likely that at least as many papers have been published since 1994 on this important family because of the continuing and increasing interest in using phytoseiids in pest management programs. Because the literature is so large, this chapter will focus on the biology of a few well-known species.

12.1 GENERAL BIOLOGY

Adult females are about 500 μm (0.5 mm) long, shiny, and pear shaped, if gravid (containing mature eggs) (Figure 12.1). Males are smaller (about half the size of females), usually are oval, are often more active than females or nymphs, and can run rapidly. Males also may be seen guarding quiescent female deutonymphs and will mate with the females immediately after they molt (Figure 12.2) (also see Figures S12.5 and S12.6 on the CD.)

The life cycle is egg → larva → protonymph → deutonymph → adult males and females (Figure 12.1 to Figure 12.3) (also see Figures S12.4 to S12.7 on the CD). Larvae of some species must feed to develop to the protonymphal stage, but larvae of other species need not feed. Larvae have three pairs of legs (Figure 12.3), and a quiescent period occurs between each molt. Males typically develop more quickly than females and may mate with their siblings. As a result, inbreeding is probably common. All phytoseiids appear to be **parahaploid** (or pseudoarrhenotokous), meaning that mating is required for females to deposit eggs (Helle et al. 1978, Hoy 1979, 1985, Nelson-Rees et al. 1980, Toyoshima and Amano 1999). All eggs are fertilized (diploid) initially, but about halfway through embryogenesis half of the chromosomes are lost in those embryos that will become males (see Figure S12.8 on the CD or Figure 3.18 in Chapter 3). As a result, the male larvae (and all active stages) that hatch will have a haploid number of chromosomes. This means that a male will express any mutations present in his haploid set of chromosomes, and these mutations can be selected for in a manner similar to haplodiploid (arrhenotokous) spider mites.

Phytoseiids are relatively fast moving (compared to their spider mite prey) and typically are white, tan, red, or brown in color. Their color may change after they feed; for example, it is common for pale phytoseiids to take on reddish colors if they feed on citrus red mites or European red mites (see Figures S12.1A and S12.3 on the CD). Phytoseiids generally are predators, and some species, such as *Phytoseiulus persimilis* and *Metaseiulus occidentalis*, are obligatory predators. *Phytoseiulus persimilis* and *M. occidentalis* do not feed on pollen or other plant-based foods and will cannibalize

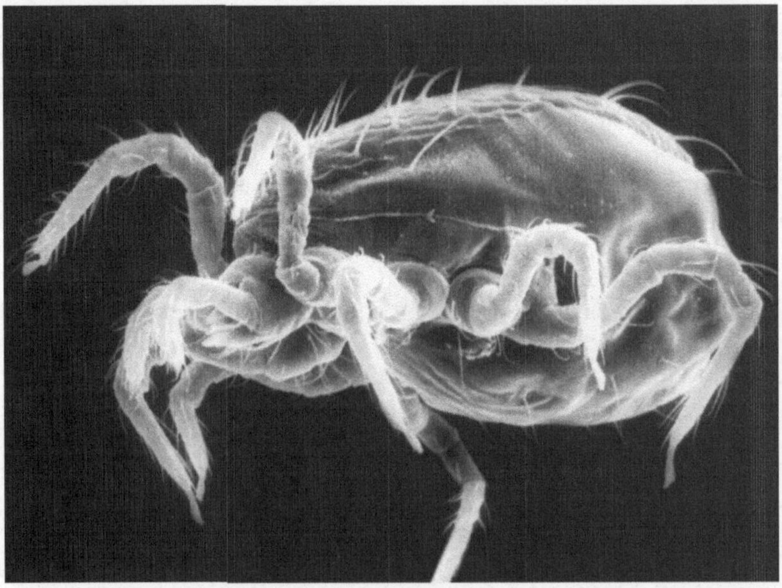

Figure 12.1 Scanning electron micrograph of a gravid female *Metaseiulus occidentalis*. Note that the palps, legs, and dorsal shield have numerous setae. This mite lacks eyes and uses setae and other receptors to perceive physical, chemical, and odorant cues. (Photograph by Ross P. Field, University of California–Berkeley.)

Figure 12.2 Male *Metaseiulus occidentalis* (bottom) guarding larger female deutonymph (top). The female is quiescent before the molt to adulthood but releases a contact sex pheromone that attracts males. When she molts to the adult stage, he will mate with her. Because males develop more rapidly than their sisters, sib mating may be common in phytoseiids. (Photograph by Jack Kelly Clark, University of California Statewide Integrated Pest Management Program.)

their own eggs or immatures, or they will feed on other phytoseiid species (**intraguild predation**). Some phytoseiid species can feed on pollen, fungal spores, and plant exudates, and some can feed, develop, and reproduce on these plant materials. Others are obligatory predators that can survive on these alternative foods but fail to deposit eggs or fail to develop on such food. Some require free water for drinking, while others are adapted to dry habitats and get all their water from their food.

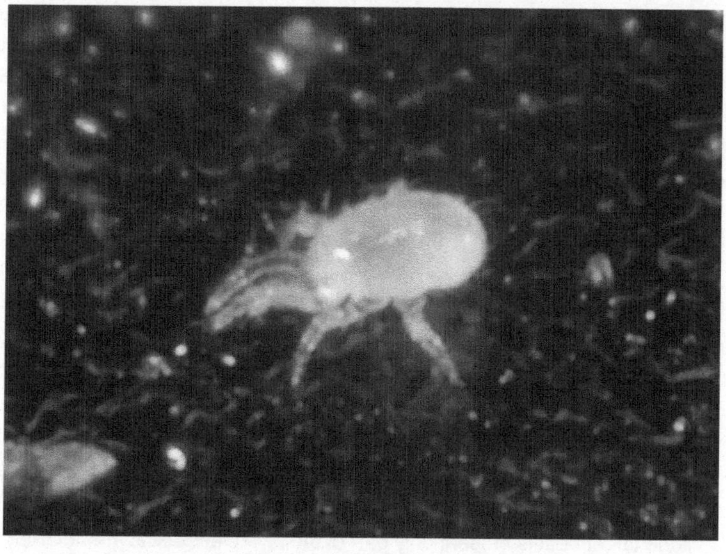

Figure 12.3 Phytoseiid larvae have three pairs of legs. This larva of *Metaseiulus occidentalis* feeds prior to becoming quiescent and molting to the protonymphal stage, which has four pairs of legs. Not all phytoseiid larvae need to feed before molting to the protonymphal stage. (Photograph by Lyle Buss, Department of Entomology and Nematology, University of Florida, Gainesville.)

An important question to answer when developing an IMM program is "Are obligatory predators (specialists) better natural enemies than those species that can survive on alternative foods (generalists) for augmentative or classical biological control releases?" This question may be resolved by considering the goals of the project and the biology and ecology of the predator and prey species involved (Table 12.1). If the target pest is present only during a limited part of the growing season, then the ability of the phytoseiid to feed on alternative foods or prey is desirable. If, however, the pest is present consistently, or you need a rapid response by the predator after augmentative releases, it probably is better to use an obligatory predator in augmentative releases or in classical biological control programs, if all else is equal (the predators are adapted to the plant architecture, cultivar, and climate). In some cases, a generalist predator that is adapted to the crop, the microclimate, and the microhabitat and is able to persist over long periods on alternative foods when the target pest is absent can be a good choice. Frischmann et al. (2006) found that generalist phytoseiids (*Typhlodromus caudiglans*, *Galendromus flumenis*) on grapevines in Washington State could slow spider mite population growth if they are present in sufficient numbers before high pest mite populations occur.

Most phytoseiids have a rapid life cycle, usually about a week at 25°C, which helps to make them effective natural enemies of spider mites, which usually have short generation times. Most phytoseiids have a female-biased sex ratio, commonly about 2.5 females to one male (about the same as spider mites); however, unlike tetranychids, phytoseiid females must mate to produce eggs, because they are parahaploid (Hoy 1979, 1985). Phytoseiid females have a brief preoviposition period during which they may disperse, then they produce approximately 2 to 4 eggs per day for a total of 20 to 50 eggs per female. Some females need to mate more than once to produce a full complement of eggs.

Mating behavior in phytoseiids is relatively complex (Figure 12.4). Quiescent female deutonymphs are often guarded by males, who hover over them until they emerge as adults (Figure 12.2) (also see Figures S12.5 and S12.6 on the CD) (Elbadry and Elbenhawy 1968, Croft 1970, Amano and Chant 1978, Hoy and Cave 1985, Tsunoda 1994). Female deutonymphs and adult females of *Metaseiulus occidentalis* produce a contact sex pheromone (Hoy and Smilanick 1979, 1981, Sonenshine 1985). Mating occurs after a ritualized courtship (Figure 12.4). During mating, the male clasps the female, and the two are venter to venter, with the male on the bottom (see Figures S12.5 and S12.6 on the CD). Males transfer the sperm packet from their genital opening to the *sperm induction pore* of the female using their modified chelicerae. The shape of the male spermatodactyl (Figure 12.5) is used to identify species. The sperm induction pore of the female, which is located laterally between legs III and IV, is used in taxonomic keys to identify species. The stored sperm are used to fertilize all oocytes. More than one spermatophore can be transferred to each storage site. Mature eggs occupy a very large proportion of the body cavity of females.

Phytoseiid females may require 20 to 25 spider mite eggs a day (or the equivalent) to deposit a full complement of eggs. Immatures and males require fewer prey. Diet preferences may be specific, and different prey may provide different nutritional content. In Figure 12.6, for example, note that *Metaseiulus occidentalis* produced more progeny (eggs) when fed a diet of *Tetranychus pacificus*

Table 12.1 Attributes of Phytoseiids and Their Use in Integrated Mite Management Programs

Type of Predator	Pros	Cons
Specialists	Specialists focus on target pests, which is a positive attribute for augmentative releases. Specialists are appropriate if the target pest is consistently present in the crop.	Specialists are not able to persist in a crop if the pest is not consistently present; they often require a specific (high) density.
Generalists	If the target pest is present only part of the year on the crop, then the ability to feed on alternative foods is a positive attribute.	Generalists are not likely to be effective in augmentative releases unless alternative foods are rare; they may not provide control of rapidly increasing pest populations.

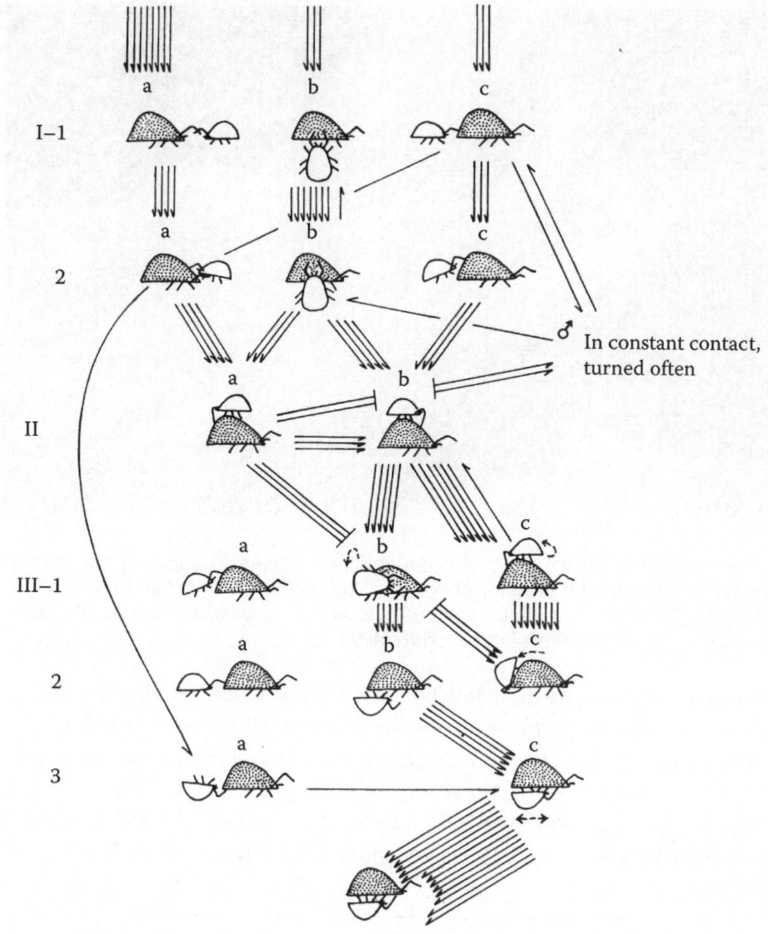

Figure 12.4 Mating behavior in phytoseiids is relatively complex and involves direct sperm transfer. Males use their modified chelicerae (spermatodactyls) to transfer sperm packets into the openings of the female located between legs III and IV. Steps in the mating behavior of *Metaseiulus occidentalis*: (1) Males and females encounter each other, and the male maintains constant contact. (2) The male climbs on top of the female. (3) The male turns and climbs under the female; while they are venter to venter, he transfers sperm. The arrows represent the number of times each behavioral option (a, b, c) occurred. (Adapted from Hoy, M.A. and Cave, F.E., *Ann. Entomol. Soc. Am.*, 78, 588–593, 1985. With permission.)

eggs than if fed a diet of active stages of the same species (Bruce-Oliver and Hoy 1990). Feeding trials were conducted using the same temperature, RH, and predator strain. The data also illustrate that laboratory life-table experiments give very specific results that depend very much on the test conditions. Laboratory data are not necessarily useful predictors of the behavior of phytoseiids under field conditions, where multiple prey species are present, and temperature, relative humidity, prey density, and other factors may vary.

Some phytoseiid species are unable to feed successfully on eggs of some spider mites because the prey eggs have a very tough chorion; for example, *Metaseiulus occidentalis* cannot feed on eggs of the European red mite, although it feeds well on eggs of *Tetranychus urticae*. Phytoseiids need approximately 30 seconds to several minutes to consume a single prey egg, but they do not always completely consume them, especially if the prey is abundant (Hoyt 1970). Incomplete consumption of prey can make phytoseiids more effective biological control agents. It is possible, but not proven,

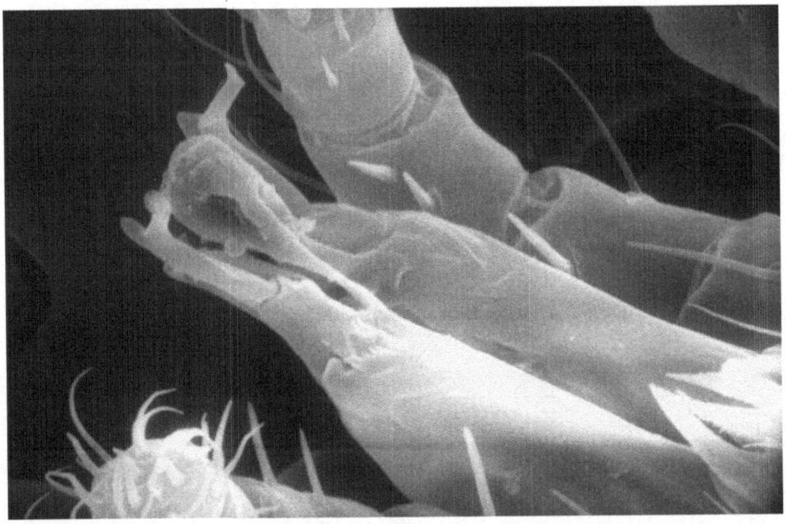

Figure 12.5 Scanning electron micrograph of male chelicerae (center) showing the spermatodactyls used by phytoseiid males for transfer of sperm to the openings between legs III and IV on the side of female phytoseiids. The shape of spermatodactyls is used in keys to species. (Photograph by Ross Field, University of California–Berkeley.)

that toxins or paralytic agents are injected by phytoseiids when feeding on active spider mites. When *M. occidentalis* attacks *T. urticae* adults, for example, the spider mite becomes less and less active within seconds after being attacked, suggesting that paralytic agents are injected.

Phytoseiids in temperate climates typically have a hibernal diapause, which involves a reproductive arrest in adult females (Sapoznikova 1964, Hoy and Flaherty 1970, 1975, Hoy 1975, Morewood and Gilkeson 1991, van der Geest et al. 1991, Veerman 1992, Morewood 1993). No aestival diapause

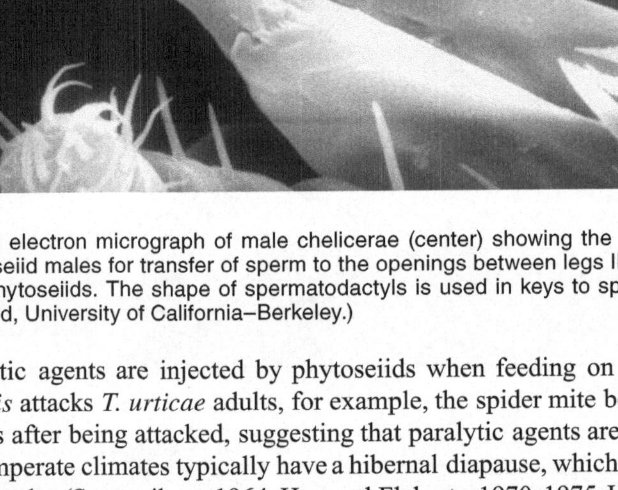

Figure 12.6 Females of *Metaseiulus occidentalis* produce more progeny when fed a diet consisting solely of eggs than when fed a diet consisting solely of active stages (larvae, nymphs, adults). Because this experiment was conducted under controlled conditions, it illustrates the point that laboratory life-table data can vary based on the conditions of the experiment (e.g., temperature, daylength, relative humidity, prey type). This makes it difficult to use such data to predict the efficacy of phytoseiids under field conditions. (Adapted from Bruce-Oliver, S.J. and Hoy, M.A., *Exp. Appl. Acarol.*, 9, 201–217, 1990. With permission.)

has been documented in the Phytoseiidae to date. Hibernally diapausing females mate before they enter hibernation sites (under bark, in leaf litter, inside bud scales) in response to decreases in temperature and daylength (Putnam 1959, Hoy and Flaherty 1975, Field and Hoy 1985). It is possible that the availability of prey affects whether females diapause; for example, Field and Hoy (1985) found that when females were reared near the critical photophase (11.2 hour of light) at a fluctuating temperature of 17.5 to 24.5°C, females of *Metaseiulus occidentalis* entered diapause more often when prey were lacking, but they did not enter diapause if prey were abundant.

Phytoseiid males and immatures do not overwinter in a hibernal diapause. Females in diapause may move about and feed, but they do not produce eggs. In the laboratory, about 80% of females of *Metaseiulus occidentalis* reared from egg to adult under a short daylength (8 hours of light and 16 hours of darkness) at 19°C enter diapause. If females are reared under a 16-hour daylength at the same temperature, females develop normally and oviposit (Hoy and Flaherty 1970, Hoy 1975). Diapausing mites are cold hardy (van der Geest et al. 1991, Morewood 1993). The African phytoseiid *Euseius fustis* failed to enter either a hibernal or aestival diapause under the conditions tested, suggesting that this tropical species does not have a diapause (Bruce-Oliver et al. 1995). Likewise, Morewood (1993) indicated that *Phytoseiulus persimilis*, a species considered to be subtropical in origin, does not have a diapause. Morewood (1993), however, observed that *P. persimilis* is cold hardy, suggesting that cold hardiness and diapause are independent genetic characteristics.

It is possible to monitor the abundance of phytoseiids in diapause in orchards and vineyards by dissecting the buds of grapevines, by examining burlap or cardboard bands placed around the trunks of deciduous orchard trees, or by examining artificial ground shelters (Putnam 1959, Hoy and Flaherty 1975, Broufas et al. 2002, Horton et al. 2002, Kawashima and Jung 2010).

12.2 PHYTOSEIID SYSTEMATICS

As with many mite groups, the taxonomy of the Phytoseiidae requires considerable work (De Moraes et al. 2004). It is likely that new species remain to be described, especially from the tropics and subtropics. Other problems stem from the fact that some taxonomists are 'splitters' and others are 'lumpers.' As a result, some species have two or three generic names. When unsure about the identity of a particular species, examine the species name carefully along with the taxonomic authority. If the species and taxonomic authority match up, they are probably the same species but have been assigned to different genera by different taxonomists. Many *Amblyseius* species have had such name changes.

Other taxonomic problems are due to the fact that species were described in different geographic areas, but the taxonomist did not compare them to species found elsewhere. Thus, **synonyms** may occur because the phytoseiid could have been introduced into the country or it has a wider geographic distribution than previously known. In other cases, some phytoseiids may actually consist of two or more **cryptic species** (Congdon and McMurtry 1985, Edwards et al. 1997, Noronha and De Moraes 2004, Tixier et al. 2006a,b, 2008a,b, 2010). It will probably take quite awhile to resolve these issues. Note that it is necessary to have both sexes for a taxonomist to identify a phytoseiid to species using morphological characters (Okassa et al. 2010). Jeyaprakash and Hoy (2002) provided molecular tools to discriminate among six phytoseiid species commonly used in augmentative biological control programs; these tools allow all life stages of *Iphiseius degenerans*, *Metaseiulus occidentalis*, *Neoseiulus californicus*, *N. cucumeris*, *N. fallacis*, and *Phytoseiulus persimilis* to be identified. A DNA extraction protocol has been developed that allows the mites to be slide mounted without loss of taxonomic characters (Jeyaprakash and Hoy 2010).

Crossing virgin males and females of different phytoseiid populations can resolve some species questions (do they produce viable progeny or not?) (Congdon and McMurtry 1985). However, partial reproductive isolation occurs between populations of some species due to the presence of

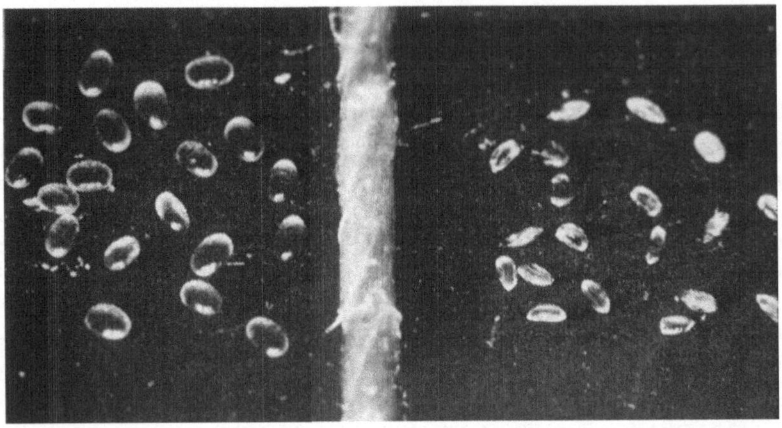

Figure 12.7 Eggs of *Metaseiulus occidentalis* (left) are oval and translucent. The eggs on the right are shriveled eggs of *M. occidentalis* obtained from incompatible crosses between different populations. The incompatibility is caused by a microbial endosymbiont. Such reproductive incompatibility can result in the development of strains (biotypes) with different characteristics.

intracellular symbionts that are transovarially transmitted (Johanowicz and Hoy 1999). Normal eggs are produced when compatible strains of *Metaseiulus occidentalis* are crossed, and shriveled eggs are produced when incompatible strains are crossed (Figure 12.7). In such cases, we may be seeing speciation in progress, because the populations with and those without the symbiont are becoming genetically isolated, even though they may not be geographically isolated. Molecular genetic methods offer hope to resolve some of the taxonomic difficulties, especially those caused by inadequate morphological differences. Okassa et al. (2010)developed morphological and molecular methods for discriminating between *Phytoseiulus persimilis* and *P. macropilis*, two important natural enemies of plant-feeding mites.

12.3 PHYTOSEIIDS IN AUGMENTATIVE BIOLOGICAL CONTROL PROGRAMS

Commonly used phytoseiids in augmentative biological control programs include *Phytoseiulus persimilis* (released to control *Tetranychus* species), *Metaseiulus occidentalis* (*Tetranychus* species), *Neoseiulus* (=*Amblyseius*) *cucumeris* and *N. degenerans* (thrips and spider mites), *Amblyseius californicus* (*Tetranychus* species), *Neoseiulus barkeri* (thrips larvae in greenhouses), *A. cucumeris* (cyclamen and broad mites in greenhouses), and *A. swirskii* (whiteflies, thrips, and other small insects in greenhouses). *Phytoseiulus persimilis*, *A. californicus*, and *M. occidentalis* have become established permanently in some locations after augmentative releases have been made (McMurtry et al. 1978, Hart et al. 2002, Walter et al. 2006, Fraulo et al. 2008). This has raised questions about the impact of these non-native species on the endemic phytoseiid populations or other mite species. Walter et al. (2006) showed, however, that *P. persimilis*, although established in Australia, did not penetrate deeply into rainforest. It is possible that the abundant endemic species of phytoseiids in natural vegetation can outcompete an introduced phytoseiid, even though the introduced species performs well in artificial agricultural ecosystems. In some cases, the introduced phytoseiid performs well in agricultural crops because it has developed resistances to key pesticides, while the endemic species may not be able to persist in sprayed environments.

Phytoseiids have different biological characteristics that make them more or less suitable for specific pest management programs. Some species are more active, such as *Phytoseiulus persimilis* (also known as an *acaricide on legs*), and they have a very high reproductive rate that allows them to clean up a pest population rapidly. The negative aspects of this behavior are that these predators must then disperse to find prey when local pest populations decline. They primarily are effective in situations when multiple releases can be made (inundative augmentation programs), such as greenhouse crops or vegetables, where it is important to get rapid control because the tolerance for damage is low.

McMurtry and Croft (1997) separated phytoseiid species into four types: Type I includes *Phytoseiulus persimilis*, *P. macropilis*, and *P. longipes*. These species have a high rate of reproduction, move rapidly, and feed voraciously on prey. Type II phytoseiids (*Metaseiulus occidentalis*, *Neoseiulus* species, and a few *Typhlodromus* species) are selective predators of tetranychids, especially species that produce dense webbing, but they have a slower rate of prey consumption and dispersal compared to Type I species. They may be able to persist even when prey densities are low. Type II species are suitable for pest management programs in deciduous orchards and vineyards where they can establish and persist for years. They do not respond as rapidly as Type I species, but they stay the course. Type III phytoseiids are generalist predators represented by some *Neoseiulus* species and most *Typhlodromus* and *Amblyseius* species; they feed on a variety of foods. Type IV phytoseiids are specialized pollen feeders or generalist predators (including *Euseius addoensis*, *E. elinae*, *E. finlandicus*, *E. fructicolus*, *E. sojaensis*, *E. stipulatus*, *E. tularensis*, and *E. victoriensis*). Another comparison of several phytoseiids in relation to their biological attributes can be found in Field and Hoy (1986).

Care should be taken by greenhouse workers when working with predatory mites in greenhouses to reduce the likelihood that they will become allergic to the mites. Kronqvist et al. (2005) found that Swedish greenhouse workers could develop allergies to both *Phytoseiulus persimilis* and *Hypoaspis miles*.

12.4 LIFE-TABLE ANALYSES OF PHYTOSEIIDS

Many life-table analyses (estimation of generation time, intrinsic rate of increase, and finite rate of increase) have been conducted on phytoseiid species under laboratory conditions. Sabelis (1985) provided a table with summaries of these data for diverse species. Unfortunately, the data are difficult to interpret with regard to predicting the effectiveness of these phytoseiids as natural enemies in specific situations in the field. Life tables are very dependent upon the conditions under which the data were obtained (e.g., Figure 12.6). Slight differences in temperature, relative humidity, food quality, or other methods can affect the data obtained, which makes it difficult to compare data from different species, or even from different laboratories when the same species are studied. Furthermore, different populations (biotypes) of the same species may vary in these important biological attributes. Thus, life tables alone may not predict the outcome of IMM programs. It is likely that information on prey specificity, prey range, and adaptation to the local environment (biotic and abiotic) is more important.

Many feeding studies have been conducted with phytoseiids under laboratory conditions in order to predict what prey or other foods they feed on in the field. Some studies are no-choice tests, which allow one to determine if the predator can feed, develop, and reproduce on this diet. No-choice tests do not indicate whether that prey item is preferred. Laboratory data allow one to conclude the phytoseiid is unlikely to be effective as a natural enemy if it is unable to feed, develop, and reproduce on a particular prey species. Unfortunately, successful feeding, development, and reproduction on a specific prey species do not predict whether the phytoseiid will, in fact, be effective as a natural enemy of that pest in a specific cropping system.

It was assumed, until relatively recently, that mixtures of phytoseiids were beneficial or at least not detrimental. Some cropping systems have several phytoseiid species, and detrimental effects from these mixtures may occur. For example, phytoseiids that are obligatory predators can become cannibalistic if prey densities are low. **Cannibalism** of their own species (Schausberger 2003) or intraguild predation on another phytoseiid species could be beneficial if it allows the predator to persist during periods of low prey density. Alternatively, cannibalism by phytoseiids on their own immatures could interfere with effective biological control in the field. Studies are being conducted to determine which phytoseiid species work well together and when intraguild predation could disrupt effective biological control (Rosenheim et al. 1995, Schausberger and Croft 2000a,b, Hatherly et al. 2005). At present, there are no clear rules to recommend augmentative releases of specific phytoseiid-species combinations; however, environmental conditions also influence intraguild predation. Roda et al. (2000) showed that leaf trichomes and spider mite webbing protected eggs of the phytoseiid *Typhlodromus pyri* from intraguild predation by the western flowerthrips *Frankliniella occidentalis.*

12.5 PREY-LOCATION BEHAVIOR

How do phytoseiids, mites that lack eyes, find their prey? Can they detect their prey at a distance, or do they have to encounter it by chance? The answer is that they use cues and sensory receptors on legs I and on their palps to detect prey at both short and relatively long distances (Hislop and Prokopy 1981, Hoy and Smilanick 1981, Sabelis and van de Baan 1983, Sabelis and Dicke 1985, Dicke and Sabelis 1988, Dicke et al. 1988, 1990, Janssen 1999).

Phytoseiids can detect webbing and chemical cues (feces, kairomones, mite-induced plant volatiles) deposited by spider mite prey. Figure 12.8 shows the movement patterns of hungry females of *Metaseiulus occidentalis* under controlled laboratory conditions in which spider mites were allowed to feed, defecate, and spin silk on leaf discs; they were then removed, leaving behind only the silk and chemical cues produced by either *Tetranychus urticae* or the European red mite (*Panonychus ulmi*). In Figure 12.8A, the females move over the (control) uninfested leaf discs rapidly and turn infrequently; they are not arrested. In Figure 12.8B, females behave differently because silk and chemical cues from *T. urticae* are present. These hungry females move more slowly and turn more frequently, behavior consistent with searching for a favored prey. In Figure 12.8C, females move rapidly, even though there are chemical cues (although no silk) deposited by *Panonychus ulmi*; this may be due to the fact that *P. ulmi* is not a favored prey of *M. occidentalis.* In Figure 12.8D, the leaf disc was divided into halves, with the left half having no cues (control) from *T. urticae*, while the right side did. Note the behavior changes when females are on different halves. The behavior shown in Figure 12.8E and Figure 12.8F is consistent with the conclusion that *T. urticae* is a more appropriate prey than *P. ulmi* for *M. occidentalis.*

Responses to these chemical cues can vary in phytoseiid species depending on their physiological status. Foraging behavior can be altered after the phytoseiids are exposed to dead conspecifics, perhaps due to transmission of a disease (Schutte et al. 1998). Dicke et al. (1986) found that *Amblyseius potentillae* that was deficient in vitamin A modified its searching behavior in response to chemical cues emitted by different spider mite species. It searched for a prey species that was inferior but that had the carotenoids needed by the predators for diapause induction. Krips et al. (1999) showed that *Phytoseiulus persimilis* reared on *Tetranychus urticae* that were reared on lima beans were not attracted to spider-mite-induced volatiles from *Gerbera.* Starvation did not affect this response; however, when *P. persimilis* was reared on *T. urticae* reared on *Gerbera*, they responded to the mite-induced volatiles from *Gerbera.* Kappers et al. (2010) found that 15 different cucumber accessions produced significantly different amounts of volatiles and different odor profiles, indicating that it might be possible to breed for cucumber varieties that are more attractive to predatory mites.

The effectiveness of some phytoseiid species as natural enemies of spider mites may depend, in part, on the structure of the foliage on which they are searching (Walter 1996). Several smaller species of phytoseiids respond to leaf surface structure, showing a preference for leaves with abundant leaf hairs, especially when prey is scarce and the weather is hot and dry. It is thought that the hairy

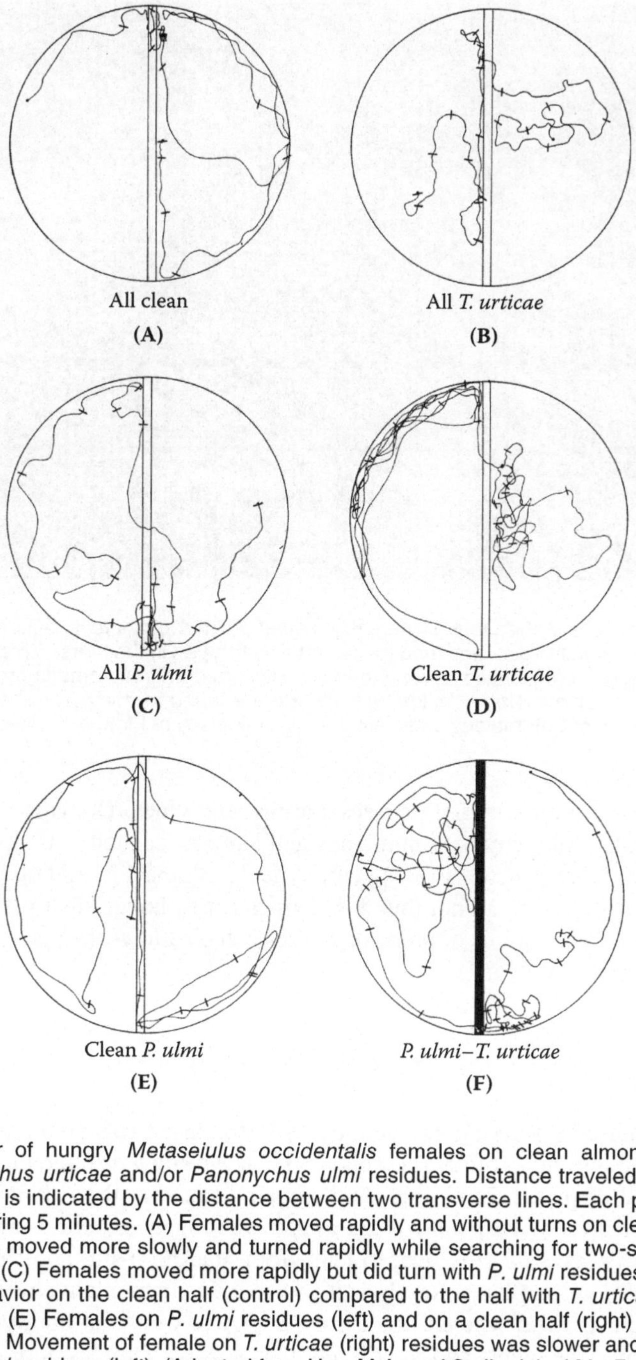

Figure 12.8 Behavior of hungry *Metaseiulus occidentalis* females on clean almond leaf discs or with *Tetranychus urticae* and/or *Panonychus ulmi* residues. Distance traveled on the disc within 15 seconds is indicated by the distance between two transverse lines. Each pattern records movement during 5 minutes. (A) Females moved rapidly and without turns on clean discs (control). (B) Females moved more slowly and turned rapidly while searching for two-spotted spider mite (*T. urticae*). (C) Females moved more rapidly but did turn with *P. ulmi* residues, a less-favored prey. (D) Behavior on the clean half (control) compared to the half with *T. urticae* residues was very different. (E) Females on *P. ulmi* residues (left) and on a clean half (right) did not behave differently. (F) Movement of female on *T. urticae* (right) residues was slower and with more turns than on *P. ulmi* residues (left). (Adapted from Hoy, M.A. and Smilanick, J.M., *Entomol. Exp. Appl.*, 29, 241–253, 1981. With permission.)

leaves provide a higher relative humidity for these predators; however, phytoseiids may do poorly on foliage with glandular hairs that can trap them (Walter 1996). Some plants have **extrafloral nectaries** (glands that produce sugary secretions containing, in some cases, amino acids), and these may provide moisture and nutrients for some phytoseiids. Structures in the junction of veins may shelter mites (Figure 12.9) (also see Figure S12.10A–C on the CD). These acarodomatia, or **domatia**, vary

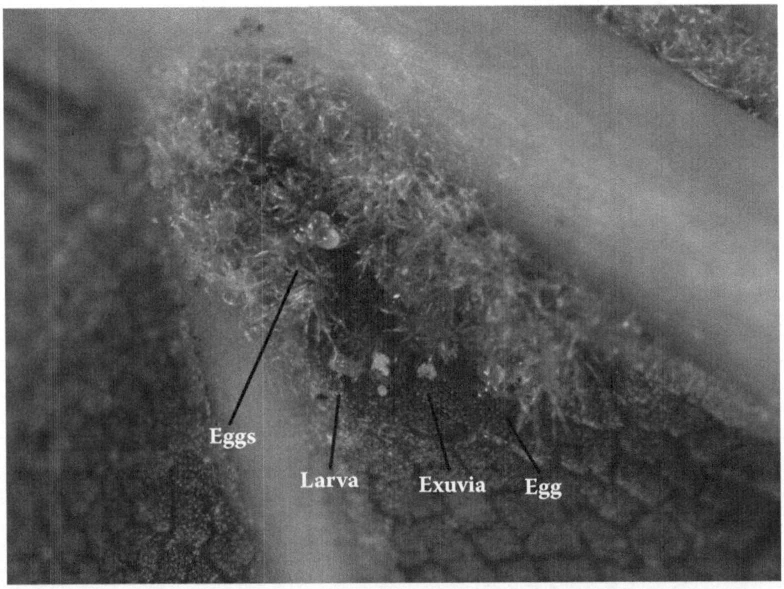

Figure 12.9 Leaf domatia are often found where veins meet the midvein of foliage. Leaf hairs provide shelter, places to deposit eggs, and food (pollen often is trapped in the hairs). This domatia contains phytoseiid eggs, a phytoseiid larva, and exuvia. It is thought that domatia provide a home for "body guards" for plants. Spider mites typically are not found in domatia. (Photograph by Lyle Buss, Department of Entomology and Nematology, University of Florida, Gainesville.)

in shape and size and are often found on trees, shrubs, and vines. Domatia may consist of raised domes, pouches or pits, and many have numerous leaf hairs associated with them. Phytoseiid mites often are abundant in these structures (along with tydeids, stigmaeids, and other small arthropods), although spider mites typically are not found in domatia. It is thought that plants with domatia are able to maintain larger populations of predatory mites, which allows the plants they inhabit to have fewer plant-feeding mites (Walter 1996).

12.6 DISPERSAL

Phytoseiids disperse by walking and aerially (Johnson and Croft 1976, Hoy et al. 1984, 1985, Field and Hoy 1985, Jung and Croft 2001). None (as far as we know) is phoretic on insects. Certainly they lack specialized dispersal forms such as the hypopi found in the Acaridida. As noted above, some phytoseiids, such as *Phytoseiulus persimilis*, have longer legs and move more rapidly than other species. Phytoseiids walk from leaf to leaf, from plant to plant, and can walk some meters over soil from weeds to crop plants. Rarely is it possible, in augmentative releases, to release phytoseiids on each and every leaf where there is prey. So, we rely on their ability to walk to distribute themselves evenly over the plants and to adjacent plants. Most females do not move far if prey at the release site is abundant and the plants are healthy and vigorous. Immatures rarely move from the leaf upon which they hatch, unless there is no food; thus, movements from leaf to leaf or plant to plant tend to be made by newly emerged and mated females during their preoviposition period. As a result, there may be a lag before released predators move to adjacent plants after augmentative releases because such releases typically involve only adult females. This is another reason to make augmentative releases early, when the pest population is low.

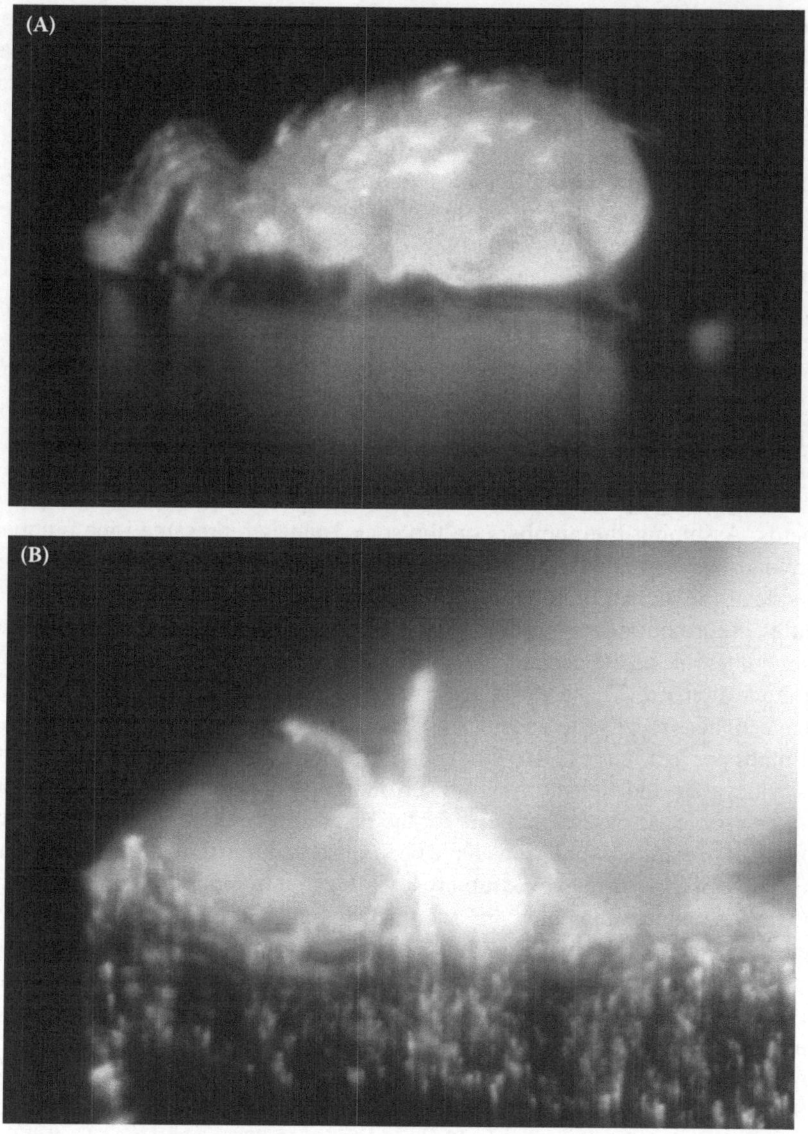

Figure 12.10 (A) A gravid adult female of *Metaseiulus occidentalis* remained low on the substrate when placed in an air stream. (B) By contrast, a newly emerged adult female of *M. occidentalis* raised her body off the substrate in a typical posture before being blown off. This behavior allows aerial dispersal of this phytoseiid. (From Hoy, M.A. et al., *Calif. Agric.*, 38(9), 21–23, 1984. Photographs by Ross Field and reprinted with permission of the University of California.)

Aerial dispersal of adult phytoseiids typically involves newly emerged, mated females (Johnson and Croft 1976, Hoy et al. 1984, 1985, Sabelis and Dicke 1985). There is a specific behavior associated with aerial dispersal of *Metaseiulus occidentalis*, and females disperse deliberately. Newly emerged, nongravid *M. occidentalis* females stand on their posterior legs and release themselves to be blown off the leaf at specific wind speeds (Figure 12.10) (also see Figure S12.9A,B on the CD). They hunker down at higher wind speeds (see Figure S12.9C on the CD) (Hoy et al. 1984).

Aerial dispersal of *Metaseiulus occidentalis* (and spider mites) within and out of almond orchards in California was studied by placing greased clear-plastic panels between trees using a pulley and rope system so the panels could be changed weekly (Hoy et al. 1984,1985)(see Figure 4.3 in Chapter 4). The panels outside the two almond orchards were at the cardinal directions (N, S, E, W) at three heights. The results of sampling through two growing seasons in two orchards indicated aerial movements of adult females of the spider mites *Tetranychus urticae* and *T. pacificus*, and the phytoseiid *M. occidentalis* occurred more or less at random with respect to wind direction within the orchard. Most mites, however, moved out of the orchards downwind (in the Central Valley of California prevailing winds are typically from the northwest). Aerial dispersal was not random during the day. Most females dispersed between 1600 and 2200, when relative humidity and wind speeds increased and temperatures decreased (Hoy et al. 1985). Sticky panels located 200 meters downwind from the orchard trapped both spider mites and *M. occidentalis*, indicating that mites can move at least that far, although the numbers were reduced compared to the number on the panels just outside the orchard.

Aerial dispersal was not random during the growing season but peaked shortly after the populations on the almond foliage peaked (Hoy et al. 1985). Huge numbers of mites dispersed from the almond orchards. Assuming the numbers on the sticky panels represented the numbers in the air volume downwind, we estimated that 170 million spider mites left a 5.7-hectare almond orchard on the south side of the orchard and another 30 million left on the east side for a total of 200 million spider mites. Approximately 8 million *Metaseiulus occidentalis* left the 5.7-ha almond orchard during the same interval. In the second orchard, approximately 1330 million spider mites left the 18-hectare almond orchard. These estimates could be low, because it is not unusual for almond orchards to have higher spider mite densities on leaves (reaching 30 to 40 mites per leaf in contrast to the 9 to 11 mites per leaf in these two studies). It has long been known that spider mites can blow into orchards, vineyards, and field crops from surrounding areas, creating an instant pest problem. Mite hot spots can serve as a source of pests for the rest of the orchard, vineyard, or field.

Explosions of spider mite populations have been observed to occur in almond orchards that, in the previous week, appeared to have extremely low (or no) mite populations. The best explanation appears to be that the populations blew in, established, and multiplied rapidly in the very hot, dry climate (a generation can be completed in 4 to 5 days). Such aerial movements can make it difficult to predict mite population densities, even when sampling a site once a week. As a result, mite management can fail, because predators cannot keep up, and pesticides may not be applied in a timely manner under these circumstances.

Aerial dispersal must be hazardous to both spider mites and predatory mites (Johnson and Croft 1976, Hoy 1982, Hoy et al. 1985, Sabelis and Dicke 1985). There is no evidence that either predators or prey can direct their landing onto a specific suitable plant. They appear to be "aerial plankton," landing at random. Furthermore, if a mite lands on an unsuitable plant, we have no information as to whether they relaunch themselves into the air or if they can walk to nearby host plants. Most likely, there is a high rate of mortality in the dispersants; however, once a mated spider mite female lands on a suitable host plant, she is able to initiate a colony. A phytoseiid female must land on a suitable host plant that is also supplied with suitable food.

Loughner et al. (2009) mixed grape plants with and without trichomes and found that the presence of nonglandular trichomes on plants can influence phytoseiid (*Typhlodromus pyri*) dispersal. Plants with an abundance of nonglandular trichomes had more phytoseiids. Hypotheses for this include the fact that trichomes capture alternative food such as pollen, they provide microhabitats that are favorable for the phytoseiids, or they provide refuge from predation. In the case of *T. pyri*, Loughner et al. (2009) found that this predator will disperse by walking and aerially when trichomes were lacking; they suggested that in crops where plants have no trichomes, this predator will not establish. What is not known is whether this is true for other phytoseiid species.

12.7 PLANT-EMITTED VOLATILES AND BIOLOGICAL CONTROL

Plants produce a variety of volatile chemicals and release them from flowers, fruits, and foliage. The controlled release of volatiles during attacks by plant-feeding pests is thought to help plants deter pests or attract their predators (Cortesero et al. 2000, Farmer 2001). Plants infested with spider mites produce volatiles that increase the resistance of uninfested leaves to attack by spider mites. These volatiles also induce the expression of defense genes in neighboring plants. The production of mite-induced plant volatiles apparently can vary based on the genetic characteristics of the spider mite species, the plant cultivar, and the genetics of the predators. Kappers et al. (2010) found that different cucumber accessions varied in the jasmonic-acid- and spider-mite-induced volatiles perceived by *Phytoseiulus persimilis*. The Kanzawa spider mite (*Tetranychus kanzawai*) is polyphagous, and different populations produce either red or white feeding scars (Matsushima et al. 2006). The spider mite strain producing the red damage caused significantly more salicylates to be produced in leaves than the white strain, and the volatile blend emitted from leaves also differed. Predatory mites also vary in their responses to the plant-emitted volatiles. Shimoda et al. (2005) found that the generalist phytoseiid *Neoseiulus californicus* responded to five synthetic volatiles produced by spider-mite-infected plants, including linalool, methyl salicylate, (*Z*)-3-hexen-1-ol, (*E*)-2-hexenal, and (*Z*)-3-hexenyl acetate. Ishiwari et al. (2007) found that *N. womersleyi* responded to mixtures of three synthetic compounds produced by *T. kanazwai*-induced tea plant volatiles. When the same predator was reared on *T. urticae*-infested bean plants, they showed a significant preference for a mixture of four synthetic compounds mimicking the bean volatiles: DMNT, methyl salicylate, beta-caryophyllene, and (*E,E*)-4,8,12-trimethyl-1,3,7,11-tridecatentraene.

Predatory mites are able to perceive these volatile chemical cues and even learn about them. It is not yet clear how distant the phytoseiid predators can perceive the cues in the field. It is unknown whether the predators can colonize the infested plants from long distances and reduce plant damage in agricultural cropping systems. Very interesting laboratory and greenhouse studies have been conducted on the role of these chemical cues and the responses by phytoseiids (Dicke and Sabelis 1988, Dicke et al. 1990, Janssen 1998, Mayland et al. 2000, Maeda et al. 2001, De Boer and Dicke 2004, De Boer et al. 2005), but this information has not yet been applied in IMM programs.

12.8 PESTICIDE RESISTANCES IN PHYTOSEIIDS

Several phytoseiid species have developed resistances to pesticides through field selection; for example, *Metaseiulus occidentalis* developed resistances to organophosphate (OP) insecticides in apple orchards in Washington (Hoyt 1969) and to sulfur in vineyards in California (Hoy and Standow 1982), allowing it to become an important predator in deciduous orchards and vineyards, respectively (Table 12.2). The case study in Chapter 16 on apples highlights the role these field-selected resistant strains have played in an IMM program. The apple pest management program developed for Washington apple orchards in the early 1970s was based on OP resistance because the codling moth could be controlled with azinphosmethyl, without disrupting biological control of spider mites (Hoyt 1969). In fact, *M. occidentalis* probably developed resistance to DDT in apple orchards, although this was not documented. The OP-resistant strain of *M. occidentalis* eventually was imported into Australia, where it established and was used in apple and peach pest management there (Readshaw 1975, Field 1978). Other phytoseiids have become resistant to OPs, allowing them to become important predators of mites in apple orchards in Michigan, Canada, and New Zealand (Table 12.2). As a general rule, the prey had to become resistant prior to the development of resistance in the predators (obligatory predators have to have food to persist long enough to become resistant) (Hoy 1998). Unfortunately, different populations of the same species of phytoseiid vary considerably

Table 12.2 Resistances to Pesticides in Some Phytoseiids Developed in the Field or by Laboratory Selection

Species	Pesticide	Lab or Field	Refs.
Amblyseius andersoni	Organophosphates, carbaryl	Field	Dunley et al. (1991)
Amblyseius fallacis	Azinphosmethyl, carbaryl	Both	Croft and Meyer (1973)
	Carbaryl	Both	Meyer (1975)
	Organophosphates	Field	Motoyama et al. (1970)
	Permethrin	Lab	Strickler and Croft (1982)
	Pyrethroids and DDT	Lab	Croft et al. (1982)
Amblyseius finlandicus	Azinphosmethyl, dimethoate	Lab	Kostiainen and Hoy (1994b)
Amblyseius hibisci	Parathion	Field	Kennett (1970)
Amblyseius nicholsi	Phosmet	Lab	Huang et al. (1987), Du and Zhong (1989), Du et al. (1991)
Amblyseius potentillae	Organophosphates	Field	Anber and Oppenoorth (1989)
Amblyseius womersleyi	Permethrin	Lab	Mochizuki (1997, 2003)
	Methidathion	Lab	Sato et al. (2000)
Metaseiulus occidentalis	Sulfur	Field	Hoy and Standow (1982)
	Permethrin	Lab	Hoy and Knop (1981)
	Dimethoate, carbaryl	Lab	Roush and Hoy (1981a,b)
	Parathion	Field	Huffaker and Kennett (1953), Morgan and Anderson (1958)
Phytoseiulus persimilis	Methidathion	Lab	Fournier et al. (1988a,b)
Typhlodromus arboreus	Azinphosmethyl, carbaryl	Field	Croft and AliNiazee (1983)
Typhlodromus pyri	Parathion, carbaryl	Field	Overmeer and van Zon (1983)
	Organophosphates, carbaryl	Field	Kapetanakis and Cranham (1983), Hadam et al. (1986)
	Pyrethroids	Field	Solomon and Fitzgerald (1993)
	Pyrethroids	Lab	Markwick (1986), Hardman et al. (2000)

Note: This is not a complete list of pesticide-resistant species.

in their tolerances to pesticides, making it difficult to predict the results of specific pesticide applications without a good history of prior results of sprays (Figure 12.11).Differences in resistance levels apparently are due to differences in pesticide treatment histories. Although phytoseiids disperse aerially, it appears that phytoseiids may move relatively slowly from orchard to orchard or from vineyard to vineyard, so few pesticide-resistance alleles are introduced into the endemic populations, which are very large. As a result, as a pest manager, you will need local pesticide histories, and you cannot be certain that a particular pesticide is safe to use unless you know the history of that site or that a phytoseiid population with a known level of resistance was introduced and established.

12.8.1 Genetic Improvement of Phytoseiids

Several phytoseiids have been selected in the laboratory for resistance to carbaryl and synthetic pyrethroids (Table 12.2) (Hoy 1985). The implementation of a strain of *Metaseiulus occidentalis* resistant to OPs, carbaryl, and sulfur in an almond IMM is described in Chapter 17.Hardman et al. (2000) evaluated a pyrethroid-resistant strain of *Typhlodromus pyri* in Canadian apple orchards and found that it could perform well. In addition to selection for resistances to pesticides, phytoseiids have been selected in the laboratory for an inability to diapause under greenhouse conditions during the winter (Hoy 1984, Field and Hoy 1985, 1986, Morewood and Gilkeson 1991, van Houten et al. 1995), and for improved high-temperature tolerance (Voroshilov 1979).

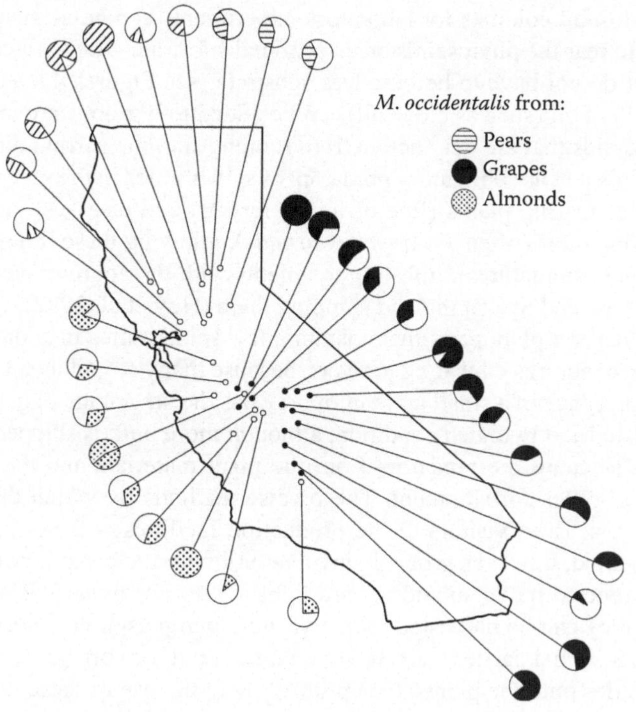

Figure 12.11 Different populations of *Metaseiulus occidentalis* show difference levels of resistance to organophosphorus pesticides. Populations were sampled from grapes, almonds, and pears in California. The relative level of resistance in each population and crop is indicated by the degree and type of shading. The most resistant population in each crop has a completely shaded circle, and the least resistant has proportionately less shading. Very local selection pressures result in these different levels of resistance to pesticides. (From Hoy, M.A., *Annu. Rev. Entomol.*, 30, 345–370, 1985. With permission.)

12.9 REARING METHODS FOR SPIDER MITES AND PHYTOSEIIDS

At present, most spider mites and phytoseiids must be reared on natural foods (plants and prey or pollen, respectively) (McMurtry and Scriven 1965, Kennett and Hamai 1980, Overmeer 1985, Ochieng et al. 1987, James 1993, Megevand et al. 1993). This is expensive and time consuming and requires substantial amounts of greenhouse or other space if large-scale releases are to be made (millions of mites). Some mass-rearing methods are useful only for producing predators for relatively small-scale field tests.

Several artificial diets have been developed, but these are primarily used for the temporary feeding, shipping, and holding of phytoseiids that are obligatory predators (McMurtry and Scriven 1962, Shehata and Weismann 1972, Kennett and Hamai 1980, Ochieng et al. 1987, Rasmy et al. 1987, Shih et al. 1993). Effective and inexpensive artificial diets suitable for long-term rearing are needed either for the spider mite prey (which would be a plant substitute) or for the phytoseiids that are obligatory predators. Obligatory predators, such as *Phytoseiulus persimilis*, often are the most effective natural enemies for augmentative biological control, but they are the most difficult to rear because they may not accept alternative foods such as pollen or grain mites. If phytoseiids (or their prey) could be reared on high-quality artificial diets over multiple generations they could be produced more inexpensively, and augmentative biological control of spider mites in glasshouses, strawberries, and vegetables could be more cost effective.

Maintaining phytoseiid colonies for laboratory experiments can be achieved in multiple ways. One of the easiest is to rear the phytoseiids on an artificial substrate with a water moat, which works with phytoseiids that do not have to be on a leaf substrate (see Figures 4.6 and 4.8 in Chapter 4). Alternatively, a repellent oil (such as clove oil) can be added to a groove around a plastic substrate that may keep phytoseiids that are very active from running into the surrounding water moat.

Rearing phytoseiids that are obligatory predators of spider mites in the greenhouse is a tritrophic endeavor that involves rearing plants (free of mites, insects, and diseases), infesting them with a known number of prey mites (often *Tetranychus urticae*), allowing these mites to multiply (predators often prefer eggs or immatures), infesting the plants with the appropriate ratio of phytoseiids, harvesting the predators, and packaging and shipping them (Hoy et al. 1982).

Some producers harvest plant parts that contain a few spider mites in addition to the predators. This method of release requires educating growers, because many are reluctant to release any spider mites; however, the presence of a small amount of prey has the advantage of getting predators to the release site in a well-fed and hydrated condition, although more bulk is shipped.

Many producers separate the predators from the plant materials and the prey (spider mites), packaging them into bottles for shipment. The precise methods by which the predators are harvested are trade secrets. Even visitors to the production facilities will not be shown these trade secrets. The basic method, however, probably involves utilizing the behavior of phytoseiid females. Female phytoseiids tend to move up and toward light if they are hungry. They may aggregate in specific sites where they can be harvested with a vacuum pump aspirator. Such predators could be placed into containers with a carrier such as vermiculite, bran, or corn grits, and the mites can be shaken out of the bottles onto the plants. One problem with the use of these dry carriers is that the predators can arrive partially desiccated and debilitated; however, the carrier does make it easier to shake the predators out evenly on the plants.

Several phytoseiids, such as *Amblyseius cucumeris*, are reared in containers with grain mites (*Tyrophagus putrescentiae*) reared on grain or bran. If care is taken to avoid excess moisture buildup in the containers that could result in fungal growth, large numbers of *A. cucumeris* can be produced relatively easily and inexpensively. These predators may be sold in sachets containing bran or grain and the grain mites so the predators continue to reproduce at the release site and disperse out into the crop over time. The grain mites, however, may cause damage to some plants, so caution should be exercised in such releases.

One predatory mite, *Amblyseius finlandicus*, could not be reared in the laboratory because egg cannibalism was high. An egg-harvesting method was used to start colonies with uniform age classes (Kostiainen and Hoy 1994a). Because many phytoseiids like to deposit their eggs into cotton fibers or artificial leaf hairs, the cotton tufts can be moved after a day or two to new rearing containers, leaving behind the larger stages that would have cannibalized the eggs and larvae.

Some predators are highly mobile and difficult to contain on leaf discs resting on water-soaked cotton (see Figure 4.6 in Chapter 4). The Huffaker cell is a useful method for observing the biology of such species (Huffaker 1948). Argov et al. (2002) described additional rearing methods for several phytoseiids used in a classical biological control project directed against the citrus rust mite in Israel. The *Euseius* species could not be reared on any substrate except beeswax-coated paper, perhaps due to the need for these mites to imbibe liquids from leaves. Some phytoseiids can be reared on factitious (not natural) prey such as fungivorous astigmatid mites. Rearing of fungivorous astimatid mites can be conducted on fungi reared on potato dextrose agar (Okabe and O'Connor 2001).

It is possible to mass rear phytoseiids in field plots, although this is not done commercially, as far as the author is aware (Hoy et al. 1982). The field plot has the disadvantage of producing predators relatively late in the season, usually after spider mite populations have increased, so it is not likely to be useful for augmentative releases in which the effect is desired immediately. It is unclear whether augmentative releases made late in a growing season would be successful in suppressing mites during the following growing season; however, such releases could result in their establishment in perennial crops (unless the crop is treated with pesticides to which they are susceptible). The method used by Hoy et al. (1982) involved planting soybeans, infesting them with spider mites (reared in the greenhouse) when the beans were in the dicotyledon stage, and then infesting the beans with predators after monitoring indicated that each leaf had one to two spider mites. The inoculation ratio was about one *Metaseiulus occidentalis* for every 50 to 100 spider mites, which means the predators had abundant food for a long time. The soybeans were sprayed with carbaryl, which enhanced spider mite reproduction and eliminated contamination by other predators. The carbaryl application did not kill *M. occidentalis* because this predator strain was resistant to carbaryl. As a result, a nearly pure culture of phytoseiids was possible. When the predators had nearly eliminated the spider mites, the soybeans were cut and the plants placed in almond trees. At that point, most of the spider mites had been devoured so they did not create a pest problem, and the predators moved off the foliage and into the almond tree. This method can be used for experiments on a field scale but probably is not practical for commercial production.

Research on the effects of nutrition during mass rearing of phytoseiids for augmentative releases has been limited (Dicke et al. 1989). The goal of commercial producers is to produce large numbers of predators at low cost, but little attention has been given to the quality of these mass-produced phytoseiids. Carotenoids, for example, are essential for *Amblyseius potentillae* to enter diapause, but a diet consisting solely of *Vicia faba* pollen does not produce adequate provitamin A to allow these predators to enter diapause even though they can feed and develop on this pollen (Dicke et al. 1989). Interestingly, *A. potentillae* and *Typhlodromus pyri* reared on *V. faba* pollen responded to volatile kairomones of more prey species than when reared on the two-spotted spider mite *Tetranychus urticae*. Predation rates by *A. potentillae* females reared on two-spotted spider mite or on *V. faba* pollen were different, with a lower predation rate for females reared on the pollen. All of these dietary changes affected predator performance, and it is unknown what other variables could occur in mass-rearing facilities that could reduce predator performance. Ouyang et al. (1992) showed that reproduction, development, and survivorship of *Euseius tularensis* varied when they were fed 17 different plant pollens. *Euseius tularensis* readily feeds and reproduces on pollen, but some pollens, although suitable for feeding to the first generation, were not suitable for subsequent generations, resulting in reduced survival, reproduction, and longevity. Other important aspects of mass-reared phytoseiids could be their ability to find prey, and this apparently can vary based on the source of the phytoseiid colony and the species of prey on which the phytoseiid is fed. Maeda et al. (2001) found that *Amblyseius womersleyi* collected from 13 different sites in Japan behaved differently in laboratory assays toward volatiles of several species of spider mites on infested bean leaves, suggesting that both genetics and prey species affect predatory behavior. Disease can affect the performance of phytoseiids mass reared for augmentative releases (Bjornson et al. 2000) (also see Chapter 14 for a discussion of pathogens of mites).

SELECTED REFERENCES

Amano, H. and D.A. Chant. 1978. Mating behaviour and reproductive mechanisms of two species of predacious mites, *Phytoseiulus persimilis* Athias-Henriot and *Amblyseius andersoni* (Chant) (Acarina: Phytoseiidae). *Acarologia* 20:196–213.

Anber, H.A.I. and F.J. Oppenoorth. 1989. A mutant esterase degrading organophosphates in a resistant strain of the predaceous mite *Amblyseius potentillae* (Garman). *Pestic. Biochem. Physiol.* 35:283–297.

Argov, Y., S. Amitai, G.A.C. Beattie, and U. Gerson. 2002. Rearing, release and establishment of imported predatory mites to control citrus rust mite in Israel. *BioControl* 47:399–409.

Bjornson, S., D.A. Raworth, and C. Bedard. 2000. Abdominal discoloration and the predatory mite *Phytoseiulus persimilis* Athias-Henriot: Prevalence of symptoms and their correlation with short-term performance. *Biol. Control* 19:17–27.

Broufas, G.D., D.S. Koveos, and D.I. Georgatsis. 2002. Overwintering sites and winter mortality of *Euseius finlandicus* (Acari: Phytoseiidae) in a peach orchard in southern Greece. *Exp. Appl. Acarol.* 26:1–12.

Bruce-Oliver, S.J. and M.A. Hoy. 1990. Effect of prey stage on life table attributes of a genetically manipulated strain of *Metaseiulus occidentalis* (Acari: Phytoseiidae). *Exp. Appl. Acarol.* 9:201–217.

Bruce-Oliver, S.J., J.S. Yaninek, and M.A. Hoy 1995. Photoperiod and temperature studies to determine whether diapause is found in successive generations of the African phytoseiid, *Euseius fustis* (Pritchard and Baker) (Acari: Phytoseiidae). *Exp. Appl. Acarol.* 19:565–572.

Congdon, B.A. and J.A. McMurtry. 1985. Biosystematics of *Euseius* on California citrus and avocado with the description of a new species (Phytoseiidae). *Int. J. Acarol.* 11:23–30.

Cortesero, A.M., J.O. Stapel, and W.J. Lewis. 2000. Understanding and manipulating plant attributes to enhance biological control. *Biol. Control* 17:35–49.

Croft, B.A. 1970. Comparative studies on four strains of *Typhlodromus occidentalis* (Acarina: Phytoseiidae). I. Hybridization and reproductive isolation studies. *Ann. Entomol. Soc. Am.* 63:696–699.

Croft, B.A. and M.T. AliNiazee. 1983. Differential resistance to insecticides in *Typhlodromus arboreus* Chant and associated phytoseiid mites of apple in the Willamette Valley, Oregon. *Environ. Entomol.* 12:1420–1423.

Croft, B.A. and R.H. Meyer. 1973. Carbamate and organophosphorus resistance patterns in populations of *Amblyseius fallacis*. *Environ. Entomol.* 2:691–695.

Croft, B.A., S.W. Wagner, and J.G. Scott. 1982. Multiple and cross-resistances to insecticides in pyrethroid-resistant strains of the predatory mite, *Amblyseius fallacis*. *Environ. Entomol.* 11:161–164.

De Boer, J.G. and M. Dicke. 2004. The role of methyl salicylate in prey searching behavior of the predatory mite *Phytoseiulus persimilis*. *J. Chem. Ecol.* 30:255–271.

De Boer, J.G., T.A.L. Snoeren, and M. Dicke. 2005. Predatory mites learn to discriminate between plant volatiles induced by prey and nonprey herbivores. *Animal Behav.* 69:869–679.

De Moraes, G.J., J.A. McMurtry, H.A. Denmark, and C.B. Campos. 2004. *A Revised Catalog of the Mite Family Phytoseiidae*, Zootaxa 454. Auckland, New Zealand: Magnolia Press.

Dicke, M. and M.W. Sabelis. 1988. How plants obtain predatory mites as bodyguards. *Nethl. J. Zool.* 38:148–165.

Dicke, M., M.W. Sabelis, and A. Groeneveld. 1986. Vitamin A deficiency modifies response of predatory mite *Amblyseius potentillae* to volatile kairomone of two-spotted spider mite, *Tetranychus urticae*. *J. Chem. Ecol.* 12:1389–1396.

Dicke, M., M.W. Sabelis, and M. DeJong. 1988. Analysis of prey preference in phytoseiid mites by using an olfactometer, predation models and electrophoresis. *Exp. Appl. Acarol.* 5:225–241.

Dicke, M., M. DeJong, M.P.T. Alers, F.C.T. Stelder, R. Wunderink, and J. Post. 1989. Quality control of mass-reared arthropods: Nutritional effects on performance of predatory mites. *J. Appl. Entomol.* 108:462–475.

Dicke, M., T.A. van Beek, M.A. Posthumus, N. Ben Dorn, H. van Bokhoven, and A.E. de Groot. 1990. Isolation and identification of volatile kairomone that affects acarine predator–prey interactions. *J. Chem. Ecol.* 16:3381–396.

Du, T.Y. and J.J. Zhong. 1989. The selection for and genetical analyses of phosmet–dimethol resistance in *Amblyseius nicholsi* Ehara et Lee. In: D.J. Lin (ed.), *Studies on the Integrated Management of Citrus Insect Pests* (pp. 56–62). Guangzhou, China: Academic Book and Periodical Press (in Chinese).

Du, T.Y., Z.D. Tian, J.J. Xiong, and H.P. Lu. 1991. Study on native organophosphorus-resistant population of *Amblyseius nicholsi* Ehara et Lee. *Nat. Enem. Insects* 13(2):61–65.

Dunley, J.E., R.H. Messing, and B.A. Croft. 1991. Levels and genetics of organophosphate resistance in Italian and Oregon biotypes of *Amblyseius andersoni* (Acari: Phytoseiidae). *J. Econ. Entomol.* 84:750–755.

Edwards, O.R., E.L. Melo, L. Smith, and M.A. Hoy. 1997. Discrimination of three *Typhlodromalus* species (Acari: Phytoseiidae) using random amplified polymorphic DNA markers. *Exp. Appl. Acarol.* 21:101–109.

Elbadry, E.A. and E.M. Elbenhawy 1968. Studies on the mating behaviour of the predaceous mite *Amblyseius gossipi* (Acarina, Phytoseiidae). *Entomophaga* 13:159–162.

Farmer, E.E. 2001. Surface-to-air signals. *Nature* 411:854–586.

Field, R.P. 1978. Control of the two spotted mite in a Victorian peach orchard with an introduced insecticide resistant strain of the predatory mite *Typhlodromus occidentalis* Nesbitt (Acarina: Phytoseiidae). *Aust. J. Zool.* 26:519–527.

Field, R.P. and M.A. Hoy 1985. Diapause behavior of genetically improved strains of the spider mite predator *Metaseiulus occidentalis* (Acarina: Phytoseiidae). *Entomol. Exp. Appl.* 38:113–120.

Field, R.P. and M.A. Hoy. 1986. Evaluation of genetically improved strains of *Metaseiulus occidentalis* (Nesbitt) (Acarina: Phytoseiidae) for integrated control of spider mites on roses in greenhouses. *Hilgardia* 54(2):1–32.

Fournier, D., M. Pralavorio, A. Cuany, and J.B. Berge. 1988a. Genetic analysis of methidathion resistance in *Phytoseiulus persimilis*. *J. Econ. Entomol.* 81:1008–1013.

Fournier, D., M. Pralavorio, J. Coulon, and J.B. Berge. 1988b. Fitness comparison in *Phytoseiulus persimilis* strains resistant and susceptible to methidathion. *Exp. Appl. Acarol.* 5:55–64.

Fraulo, A.B., R. McSorley, and O.E. Liburd. 2008. Effect of the biological control agent *Neoseiulus californicus* (Acari: Phytoseiidae) on arthropod community structure in north Florida strawberry fields. *Florida Entomol.* 91:436–445.

Frischmann, D.A., D.G. James, L.C. Wright, and W.E. Synder. 2006. Effects of generalist phytoseiid mites and grapevine canopy structure on spider mite (Acari: Tetranychidae) biocontrol. *Environ. Entomol.* 35:56–67.

Gerson, U., R.L. Smiley, and R. Ochoa. 2003. *Mites (Acari) for Pest Control*. Oxford, U.K.: Blackwell Scientific.

Grafton-Cardwell, E.E., Y. Ouyang, and R.A. Striggow. 1999. Predacious mites for control of citrus thrips, *Scirtothrips citri* (Thysanoptera: Thripidae) in nursery citrus. *Biol. Control* 14:29–36.

Hadam, J.J., M.T. AliNiazee, and B.A. Croft. 1986. Phytoseiid mites (Parasitiformes: Phytoseiidae) of major crops in Willamette Valley, Oregon, and pesticide resistance in *Typhlodromus pyri* Scheuten. *Environ. Entomol.* 15:1255–1263.

Hardman, J.M., D.L. Moreau, M. Snyder, S.O. Gaul, and E.D. Bent. 2000. Performance of a pyrethroid-resistant strain of the predator mite *Typhlodromus pyri* (Acari: Phytoseiidae) under different insecticide regimes. *J. Econ. Entomol.* 93:590–604.

Hart, A.J., J.S. Bale, A.G. Tullett, M.R. Worland, and K.F.A. Walters. 2002. Effects of temperature on the establishment potential of the predatory mite *Amblyseius californicus* McGregor (Acari: Phytoseiidae) in the U.K. *J. Insect Physiol.* 48:592–599.

Hatherly, I.S., J.S. Bale, and K.F. Walters. 2005. Intraguild predation and feeding preferences in three species of phytoseiid mite used for biological control. *Exp. Appl. Acarol.* 37:43–55.

Helle, W. and M.W. Sabelis (eds.) 1985. *Spider Mites: Their Biology, Natural Enemies, and Control*, Vol. 1B. Amsterdam: Elsevier.

Helle, W., H.R. Bolland, R. van Arendonk, R. De Boer, and G.G.M. Schulten. 1978. Genetic evidence for biparental males in haplo-diploid predator mites (Acarina: Phytoseiidae). *Genetica* 49:165–171.

Hislop, R.G. and R.J. Prokopy. 1981. Mite predator responses to prey and predator-emitted stimuli. *J. Chem. Ecol.* 7:895–904.

Hoffman, M.P. and A.C. Frodsham. 1993. *Natural Enemies of Vegetable Insect Pests*. Ithaca, NY: Cornell University Cooperative Extension.

Horton, D.R., D.A. Broers, T. Hinojosa, T.M. Lewis, E.R. Miliczky, and R.R. Lewis. 2002. Diversity and phenology of predatory arthropods overwintering in cardboard bands placed in pear and apple orchards of central Washington State. *Ann. Entomol. Soc. Am.* 95:469–480.

Hoy, M.A. 1975. Diapause in the mite *Metaseiulus occidentalis*: Stages sensitive to photoperiodic induction. *J. Insect Physiol.* 21:745–751.

Hoy, M.A. 1979. Parahaploidy of the "arrhenotokous" predator, *Metaseiulus occidentalis* (Acarina: Phytoseiidae), demonstrated by x-irradiation of males. *Entomol. Exp. Appl.* 26:97–104.

Hoy, M.A. (ed.) 1982. *Recent Advances in Knowledge of the Phytoseiidae*, Publication No. 3284. Berkeley: Division of Agricultural Sciences, University of California.

Hoy, M.A. 1984. Genetic improvement of a biological control agent: Multiple pesticide resistances and non-diapause in *Metaseiulus occidentalis* (Nesbitt) (Phytoseiidae). In: D.A. Griffiths (ed.), *Acarology 6*, Vol. 2 (pp. 673–679). Chichester, U.K.: Ellis Horwood.

Hoy, M.A. 1985. Recent advances in genetics and genetic improvement of the Phytoseiidae. *Annu. Rev. Entomol.* 30:345–370.

Hoy, M.A. 1998. Myths, models and mitigation of resistance to pesticides. *Phil. Trans. R. Soc. Lond. B* 353:1718–1795.

Hoy, M.A. and F.E. Cave. 1985. Mating behavior in four strains of *Metaseiulus occidentalis* (Acari: Phytoseiidae). *Ann. Entomol. Soc. Am.* 78:588–593.

Hoy, M. and D. Flaherty. 1970. Photoperiodic induction of diapause in a predaceous mite, *Metaseiulus occidentalis*. *Ann. Entomol. Soc. Am.* 63:960–963.

Hoy, M.A. and D.L. Flaherty. 1975. Diapause induction and duration in vineyard-collected *Metaseiulus occidentalis*. *Environ. Entomol.* 4:262–264.

Hoy, M.A. and N.F. Knop. 1981. Selection for and genetic analysis of permethrin resistance in *Metaseiulus occidentalis*: Genetic improvement of a biological control agent. *Entomol. Exp. Appl.* 30:10–18.

Hoy, M.A. and J.M. Smilanick. 1979. A sex pheromone produced by immature and adult females of the predatory mite, *Metaseiulus occidentalis*, Acarina: Phytoseiidae. *Entomol. Exp. Appl.* 26:291–300.

Hoy, M.A. and J.M. Smilanick. 1981. Non-random prey location by the phytoseiid predator *Metaseiulus occidentalis*: Differential responses to several spider mite species. *Entomol. Exp. Appl.* 29:241–253.

Hoy, M.A. and K.A. Standow. 1982. Inheritance of resistance to sulfur in the spider mite predator *Metaseiulus occidentalis*. *Entomol. Exp. Appl.* 31:316–323.

Hoy, M.A., D. Castro, and D. Cahn. 1982. Two methods for large-scale production of pesticide–resistant strains of the spider mite predator *Metaseiulus occidentalis* (Nesbitt) (Acarina, Phytoseiidae). *Zeitsch. Angew. Entomol.* 94:1–9.

Hoy, M.A., H.E. van de Baan, J.J.R. Groot, and R.P. Field. 1984. Aerial movements of mites in almonds: Implications for pest management. *Calif. Agric.* 38(9):21–23.

Hoy, M.A., J.J.R. Groot, and H.E. van de Baan. 1985. Influence of aerial dispersal on persistence and spread of pesticide-resistant *Metaseiulus occidentalis* in California almond orchards. *Entomol. Exp. Appl.* 37:17–31.

Hoyt, S.C. 1969. Integrated chemical control of insects and biological control of mites on apple in Washington. *J. Econ. Entomol.* 62:74–86.

Hoyt, S.C. 1970. Effect of short feeding periods by *Metaseiulus occidentalis* on fecundity and mortality of *Tetranychus mcdanieli*. *Ann. Entomol. Soc. Am.* 63:1382–1384.

Huang, M.D., J.J. Xiong, and T.Y. Du. 1987. The selection for and genetical analysis of phosmet resistance in *Amblyseius nicholsi*. *Acta Entomol. Sin.* 30(2):133–139.

Huffaker, C.B. 1948. An improved cage for work with small insects. *J. Econ. Entomol.* 41:648–649.

Huffaker, C.B. and C.E. Kennett. 1953. Differential tolerance to parathion to two *Typhlodromus* predatory on cyclamen mite. *J. Econ. Entomol.* 46:707–708.

Huffaker, C.B., M. van de Vrie, and J.A. McMurtry. 1969. The ecology of tetranychid mites and their natural control. *Annu. Rev. Entomol.* 14:125–274.

Huffaker, C.B., M. van de Vrie, and J.A. McMurtry. 1970. Ecology of tetranychid mites and their natural enemies: A review. II. Tetranychid populations and their possible control by predators: An evaluation. *Hilgardia* 40(11):391–458.

Ishiwari, H., T. Suzuki, and T. Maeda. 2007. Essential compounds in herbivore-induced plant volatiles that attract the predatory mite *Neoseiulus womersleyi*. *J. Chem. Ecol.* 33:1670–1681.

James, D.G. 1993. Pollen, mould mites and fungi: Improvements to mass rearing of *Typhlodromus doreenae* and *Amblyseius victoriensis*. *Exp. Appl. Acarol.* 17:271–276.

Janssen, A. 1999. Plants with spider-mite prey attract more predatory mites than clean plants under greenhouse conditions. *Entomol. Exp. Appl.* 90:191–198.

Jeyaprakash A. and M.A. Hoy. 2002. Mitochondrial 12S rRNA sequences used to design a molecular ladder assay to identify six commercially available phytoseiids (Acari: Phytoseiidae). *Biol. Control* 25:136–142.

Jeyaprakash, A. and M.A. Hoy. 2010. A DNA extraction procedure that allows mite specimens to be slide mounted: Phytoseiid species evaluated as a model. *Exp. Appl. Acarol.* 52(2):131–140.

Johanowicz, D.L. and M.A. Hoy. 1999. *Wolbachia* infection dynamics in experimental laboratory populations of *Metaseiulus occidentalis*. *Entomol. Exp. Appl.* 93:257–266.

Johnson, D.T. and B.A. Croft 1976. Laboratory study of the dispersal behavior of *Amblyseius fallacis* (Acarina: Phytoseiidae). *Ann. Entomol. Soc. Am.* 69:1019–1023.

Jung, C. and B.A. Croft. 2001. Aerial dispersal of phytoseiid mites (Acari: Phytoseiidae): Estimating falling speed and dispersal distance of adult females. *Oikos* 94:182–190.

Kapetanakis, E.G. and J.E. Cranham. 1983. Laboratory evaluation of resistance to pesticides in the phytoseiid predator *Typhlodromus pyri* from English apple orchards. *Ann. Appl. Biol.* 103(3):389–400.

Kappers, I.F., F.W.A.Verstappen, L.L.P. Luckerhoff, H.J. Bouwmeester, and M. Dicke. 2010. Genetic variation in jasmonic acid- and spider mite-induced plant volatile emission of cucumber accessions and attraction of the predator *Phytoseiulus persimilis*. *J. Chem. Ecol.* 36:500–512.

Kawashima, M. and C. Jung. 2010. Artificial ground shelters for overwintering phytoseiid mites in orchards. *Exp. Appl. Acarol.* 52(1):35–47.

Kennett, C.E. 1970. Resistance to parathion in the phytoseiid mite *Amblyseius hibisci*. *J. Econ. Entomol.* 63:1999–2000.

Kennett, C.E. and J. Hamai. 1980. Oviposition and development in predaceous mites fed with artificial and natural diets (Acari: Phytoseiidae). *Entomol. Exp. Appl.* 28:116–122.

Kostiainen, T. and M.A. Hoy. 1994a. Egg harvesting allows large-scale rearing of *Amblyseius finlandicus* (Acari: Phytoseiidae) in the laboratory. *Exp. Appl. Acarol.* 18:155–165.

Kostiainen, T. and M.A. Hoy 1994b. Genetic improvement of *Amblyseius finlandicus* (Acari: Phytoseiidae): Laboratory selection for resistance to azinphosmethyl and dimethoate. *Exp. Appl. Acarol.* 18:469–484.

Kostiainen, T.S. and M.A. Hoy. 1996. *The Phytoseiidae as Biological Control Agents of Pest Mites and Insects: A Bibliography*, IFAS Monograph 17. Gainesville: Institute of Food and Agricultural Sciences, University of Florida (available on the CD accompanying this book).

Krips, O.E., P.E.L. Willems, R. Gols, M.A. Posthumus, and M. Dicke. 1999. The response of *Phytoseiulus persimilis* to spider mite-induced volatiles from *Gerbera*: Influence of starvation and experience. *J. Chem. Ecol.* 25:2623–2641.

Kronqvist, M., E. Johansson, B. Kolmodin-Hedman, H. Oman, M. Svartengren, and M. Hage-Hamsten. 2005. IgE sensitization to predatory mites and respiratory symptoms in Swedish greenhouse workers. *Allergy* 60:521–526.

Loughner, R., K. Wentworth, G. Loeb, and J. Nyrop. 2009. Leaf trichomes influence predatory mite densities through dispersal behavior. *Entomol. Exp. Appl.* 134:78–88.

Maeda, T., J. Takabayashi, S. Yano, and A. Takafuji. 2001. Variation in the olfactory response of 13 populations of the predatory mite *Amblyseius womersleyi* to *Tetranychus urticae*-infested plant volatiles (Acari: Phytoseiidae, Tetranychidae). *Exp. Appl. Acarol.* 25:55–64.

Markwick, N.P. 1986. Detecting variability and selecting for pesticide resistance in two species of phytoseiid mites. *Entomophaga* 31:225–236.

Matsushima, R., R. Ozawa, M. Uefune, T. Gotoh, and J. Takabayashi. 2006. Intraspecies variation in the Kanzawa spider mite differentially affects induced defensive response in lima bean plants. *J. Chem. Ecol.* 22:2501–2512.

Mayland, H., D.C. Margolies, and R.E. Charlton. 2000. Local and distant prey-related cues influence when an acarine predator leaves a prey patch. *Entomol. Exp. Appl.* 96:245–252.

McMurtry, J.A. and B.A. Croft. 1997. Life-styles of phytoseiid mites and their roles in biological control. *Annu. Rev. Entomol.* 42:291–321.

McMurtry, J.A. and G.T. Scriven. 1962. The use of agar media in transporting and rearing phytoseiid mites. *J. Econ. Entomol.* 55:412–414.

McMurtry, J.A. and G.T. Scriven. 1965. Insectary production of phytoseiid mites. *J. Econ. Entomol.* 58:282–284.

McMurtry, J.A., C.B. Huffaker, and M. van de Vrie. 1970. Ecology of tetranychid mites and their natural enemies: A review. I. Tetranychid enemies: Their biological characters and the impact of spray practices. *Hilgardia* 40:331–390.

McMurtry, J.A., E.R. Oatman, P.A. Phillips, and C.W. Wood. 1978. Establishment of *Phytoseiulus persimilis* (Acari: Phytoseiidae) in southern California. *BioControl* 23:175–179.

Megevand, B., A. Klay, D. Gnanvossou, and G. Paraiso. 1993. Maintenance and mass rearing of phytoseiid predators of the cassava green mite. *Exp. Appl. Acarol.* 17:115–128.

Meyer, R.H. 1975. Release of carbaryl-resistant predatory mites in apple orchards. *Environ. Entomol.* 4:49–51.

Mochizuki, M. 1997. Permethrin resistance and stability in the predatory mite, *Amblyseius womersleyi* Schicha (Acari: Phytoseiidae). *Jpn. J. Appl. Entomol. Zool.* 41:1–5.

Mochizuki, M. 2003. Studies on use of the pesticide resistant predatory mite *Amblyseius womersleyi* Schicha (Acari: Phytoseiidae) for integrated pest management on tea plants. *Bull. Natl. Inst. Veg. Tea Sci.* 2:93–138.

Morewood, W.D. 1993. Diapause and cold hardiness of phytoseiid mites (Acarina: Phytoseiidae). *Eur. J. Entomol.* 90:3–10.

Morewood, W.D. and L.A. Gilkeson 1991. Diapause induction in the thrips predator *Amblyseius cucumeris* (Acarina: Phytoseiidae) under greenhouse conditions. *Entomophaga* 36:253–263.

Morgan, C.V.G. and N.H. Anderson. 1958. Notes on parathion-resistant strains of two phytophagous mites and a predacious mite in British Columbia. *Can. Entomol.* 90:92–97.

Motoyama, N., G.C. Rock, and W.C. Dauterman. 1970. Organophosphorus resistance in an apple orchard population of *Typhlodromus (Amblyseius) fallacis*. *J. Econ. Entomol.* 63:1139–1142.

Nelson-Rees, W.A., M.A. Hoy, and R.T. Roush. 1980. Heterochromatinization, chromatin elimination and haploidization in the parahaploid mite *Metaseiulus occidentalis* (Nesbitt) (Acarina: Phytoseiidae). *Chromosoma (Berl.)* 77:263–276.

Noronha, A.C.D.S. and G.J. De Moraes. 2004. Reproductive compatibility between mite populations previously identified as *Euseius concordis* (Acari: Phytoseiidae). *Exp. Appl. Acarol.* 32:271–279.

Ochieng, R.S., G.W. Oloo, and E.O. Amboga. 1987. An artificial diet for rearing the phytoseiid mite, *Amblyseius teke* Pritchard and Baker. *Exp. Appl. Acarol.* 3:169–173.

Okabe, K. and B.M. O'Connor. 2001. A method for both mass and individual rearing of fungivorous astimatid mites (Acari). *Exp. Appl. Acarol.* 25:493–504.

Okassa, M., M.S. Tixier, and S. Kreiter. 2010. Morphological and molecular diagnostics of *Phytoseiulus persimilis* and *Phytoseiulus macropilis* (Acari: Phytoseiidae). *Exp. Appl. Acarol.* 52(3):291–303.

Ouyang, Y., E.E. Grafton-Cardwell, and R.L. Bugg. 1992. Effects of different pollens on development, survivorship, and reproduction of *Euseius tularensis* (Acari: Phytoseiidae). *Environ. Entomol.* 21:1371–1376.

Overmeer, W.P.J. 1985. Rearing and handling. In: W. Helle and M.W. Sabelis (eds.), *Spider Mites: Their Biology, Natural Enemies, and Control*, Vol. 1B (pp. 161–170). Amsterdam: Elsevier.

Overmeer, W.P.J. and A.Q. van Zon. 1983. Resistance to parathion in the predaceous mite *Typhlodromus pyri* Scheuten (Acarina: Phytoseiidae). *Meded. Fac. Landbouwwet. Rijksuniv. Gent* 48(2):247–251.

Putnam, W.L. 1959. Hibernation sites of phytoseiids (Acarina: Phytoseiidae) in Ontario peach orchards. *Can. Entomol.* 91:735–741.

Rasmy, A.H., M.E. Elbagoury, and A.S. Reda. 1987. A new diet for reproduction of two predaceous mites *Amblyseius gossipi* and *Agistemus exsertus* (Acari: Phytoseiidae, Stigmaeidae). *Entomophaga* 32:277–280.

Readshaw, J.L. 1975. Biological control of orchard mites in Australia with an insecticide-resistant predator. *J. Aust. Inst. Agric. Sci.* 41:213–214.

Roda, A., J. Nyrop, M. Dicke, and G. English-Loeb. 2000. Trichomes and spider-mite webbing protect predatory mite eggs from intraguild predation. *Oecologia* 125:428–435.

Rosenheim, J.A., H.K. Kaya, L.E. Ehler, H.J. Marois, and B.A. Jaffee. 1995. Intraguild predation among biological-control agents: Theory and evidence. *Biol. Control* 5:303–335.

Roush, R.T. and M.A. Hoy 1981a. Genetic improvement of *Metaseiulus occidentalis*: Selection with methomyl, dimethoate, and carbaryl and genetic analysis of carbaryl resistance. *J. Econ. Entomol.* 74:138–141.

Roush, R.T. and M.A. Hoy. 1981b. Laboratory, glasshouse, and field studies of artificially selected carbaryl resistance in *Metaseiulus occidentalis*. *J. Econ. Entomol.* 74:142–147.

Sabelis, M.W. 1985. Capacity for population increase. In: W. Helle and M.W. Sabelis (eds.), *Spider Mites: Their Biology, Natural Enemies, and Control*, Vol. 1B (pp. 25–42). Amsterdam: Elsevier.

Sabelis, M.W. and M. Dicke. 1985. Long-range dispersal and searching behaviour. In: W. Helle and M.W. Sabelis (eds.), *Spider Mites: Their Biology, Natural Enemies, and Control*, Vol. 1B (pp. 141–160). Amsterdam: Elsevier.

Sabelis, M.W. and van de Baan. 1983. Location of distant spider mite colonies by phytoseiid predators: Demonstration of specific kairomones emitted by *Tetranychus urticae* and *Panonychus ulmi*. *Entomol. Exp. Appl.* 33:303–314.

Sapoznikova, F.D. 1964. The photoperiodic response of the mite *Typhlodromus* (*Amblyseius*) *similis* (C.L. Koch) (Acarina: Phytoseiidae). *Zoologicheskii Zhurnal* 43:1140–1144.

Sato, M.E., T. Miyata, A. Kawai, and O. Nakano. 2000. Selection for resistance and susceptibility to methidathion and cross resistance in *Amblyseius womersleyi* Schicha (Acari: Phytoseiidae). *Appl. Entomol. Zool.* 35:393–399.

Schausberger, P. 2003. Cannibalism among phytoseiid mites: A review. *Exp. Appl. Acarol.* 29:173–191.

Schausberger, P. and B.A. Croft. 2000a. Cannibalism and intraguild predation among phytoseiid mites: Are aggressiveness and prey preference related to diet specialization? *Exp. Appl. Acarol.* 24:709–725.

Schausberger, P. and B.A. Croft. 2000b. Nutritional benefits of intraguild predation and cannibalism among generalist and specialist phytoseiid mites. *Ecol. Entomol.* 25:473–480.

Schutte, C., P. van Baarlen, H. Dijkman, and M. Dicke. 1998. Change in foraging behaviour of the predatory mite *Phytoseiulus persimilis* after exposure to dead conspecifics and their products. *Entomol. Exp. Appl.* 88:295–300.

Shehata, K.K. and L. Weismann. 1972. Rearing the predacious mite *Phytoseiulus persimilis* Athias-Henriot on artificial diet (Acarina: Phytoseiidae). *Biologia (Bratislava)* 27(8):609–615.

Shih, C.I., H.Y. Chang, P.H. Husu, and Y.F. Hwang. 1993. Responses of *Amblyseius ovalis* (Evans) (Acarina: Phytoseiidae) to natural food resources and two artificial diets. *Exp. Appl. Acarol.* 17:503–519.

Shimoda, T., R. Ozawa, K. Sano, E. Yano, and J. Takabayashi. 2005. The involvement of volatile infochemicals from spider mites and from food-plants in prey location of the generalist predatory mite *Neoseiulus californicus*. *J. Chem. Ecol.* 31:2019–2032.

Solomon, M.G. and J.D. Fitzgerald. 1993. Orchard selection for resistance to a synthetic pyrethroid in organophosphate-resistant *Typhlodromus pyri* in the U.K. *Biocontrol Sci. Technol.* 3:127–132.

Sonenshine, D.E. 1985. Pheromones and other semiochemicals of the Acari. *Annu. Rev. Entomol.* 30:1–28.

Strickler, K. and B.A. Croft. 1982. Selection for permethrin resistance in the predatory mite *Amblyseius fallacis*. *Entomol. Exp. Appl.* 31:339–345.

Tanigoshi, L.K. 1982. Advances in knowledge of the biology of the Phytoseiidae. In: M.A. Hoy (ed.), *Recent Advances in Knowledge of the Phytoseiidae* (pp. 1–22), ANR Publication No. 3284. Berkeley: Division of Agriculture and Natural Resources, University of California.

Tixier, M.S., S. Kreiter, F. Ferragut, and B. Cheval. 2006a. The suspected synonymy of *Kampimodromus hmiminai* and *Kampimodromus adrianae* (Acari: Phytoseiidae): Morphological and molecular investigations. *Can. J. Zool.* 84:1216–1222.

Tixier, M.S., S. Kreiter, Z. Barbar, S. Ragusa, and B. Cheval. 2006b. Status of two cryptic species, *Typhlodromus exhilaratus* Ragusa and *Typhlodromus phialatus* Athias-Henriot (Acari: Phytoseiidae): Consequences for taxonomy. *Zool. Scripta* 35:115–122.

Tixier, M.S., S. Guichou, and S. Kreiter. 2008a. Morphological variation in the biological control agent *Neoseiulus californicus* (McGregor) (Acari: Phytoseiidae): Consequences for diagnostic reliability and synonymies. *Invertebr. Syst.* 22:453–469.

Tixier, M.S, S. Kreiter, B.A. Croft, and B. Cheval. 2008b. *Kampimodromus aberrans* (Acari: Phytoseiidae) from the USA: Morphological and molecular assessment of its density. *Bull. Entomol. Res.* 98:125–134.

Tixier, M.S., M. Ferrero, M. Okassa, S. Guichou, and S. Kreiter. 2010. On the specific identity of specimens of *Phytoseiulus longipes* Evans (Mesostigmata: Phytoseiidae) showing different feeding behaviours: Morphological and molecular analyses. *Bull. Entomol. Res.* 100(5):569–579.

Toyoshima, S. and H. Amano. 1999. Cytological evidence of pseudo-arrhenotoky in two phytoseiid mites, *Phytoseiulus persimilis* Athias-Henriot and *Amblyseius womersleyi* Schicha. *J. Acarol. Soc. Jpn.* 8:125–142.

Tsunoda, T. 1994. Mating behavior of the predacious mite, *Amblyseius womersleyi* Schicha (Acari: Phytoseiidae). *Appl. Entomol. Zool.* 29:141–147.

van der Geest, L.P.S., W.P.J. Overmeer, and A.Q. van Zon. 1991. Cold-hardiness in the predatory mite *Amblyseius potentillae* (Acari: Phytoseiidae). *Exp. Appl. Acarol.* 11:167–176.

van Houten, Y.M., P. van Startum, J. Bruin, and A. Veerman. 1995. Selection for non-diapause in *Amblyseius cucumeris* and *Amblyseius barkeri* and exploration of the effectiveness of selected strains for thrips control. *Entomol. Exp. Appl.* 77:289–295.

Veerman, A. 1992. Diapause in phytoseiid mites: A review. *Exp. Appl. Acarol.* 14:1–60.

Voroshilov, H.V. 1979. Heat-resistant lines of the mite *Phytoseiulus persimilis* A.H. *Genetika* 15(1):70–76.

Walter, D.E. 1996. Living on leaves: Mites, tomenta, and leaf domatia. *Annu. Rev. Entomol.* 41:101–114.

Walter, D.E., G.N. Azam, G. Waite, and J. Hargreaves. 2006. Risk assessment of an exotic biocontrol agent: *Phytoseiulus persimilis* (Acari: Phytoseiidae) does not establish in rainforest in southeast Queensland. *Aust. J. Ecol.* 23:587–592.

CHAPTER 13

Predatory Insects and Plant-Feeding Mites

13.1 INSECTS AS PREDATORS OF PLANT-FEEDING MITES: PROS AND CONS

Insects in the orders Coleoptera, Thysanoptera, Hemiptera (Heteroptera), Diptera, and Neuroptera may feed on mites that are pests in agricultural crops. These predatory insects can be used in two ways in integrated mite management (IMM) programs: conservation and augmentation. Because of concerns about unintended consequences, the release of generalist predators probably is not appropriate in classical biological control programs.

Unfortunately, we know less than we would like about the population dynamics of these insect predators within the complex of phytophagous, predatory, and parasitic arthropods in crops. In specific situations, predatory insects have been documented to provide effective control of a variety of pests (Gerson and Smiley 1990, Rosenheim 1998, Symondson et al. 2002, Letourneau et al. 2009).

Many factors appear to affect the status of predatory insects in agricultural crops, including the crop cultivar planted, the pesticide type and patterns of use, pest density, and intraguild predation. Parrella et al. (1980) evaluated the compatibility of the predatory thrips *Leptothrips mali,* the coccinellid *Stethorus punctum*, and the heteropteran *Orius insidiosus* in the laboratory as predators of spider mites and concluded that *L. mali* and *O. insidiosus* were not compatible because *Orius* killed and fed on *Leptothrips* (intraguild predation). As acknowledged, the results are only "suggestive of what could happen in the apple orchard" (Parrella et al. 1980). It is becoming increasingly clear that complex interactions can occur between pests and their diverse natural enemies (parasites, pathogens, predators) in crop ecosystems and that it is difficult to tease out the interrelationships under field conditions and to predict specific outcomes (Chazeau 1985, Rosenheim 1998, Briggs and Borer 2005, Chow et al. 2009, Letourneau et al. 2009). Despite this concern, as a general rule, *these natural enemies should be conserved*, if at all possible, through modifications in cultural practices and pesticide applications.

Predatory insects by themselves often are less effective than phytoseiid predators because most are generalists. At present, we do not understand the complex dynamics of the interactions of generalist predators and their many pest and nonpest prey in agricultural agroecosystems (Blommers 1994, Symondson et al. 2002, Letourneau et al. 2009). Because predatory insects are larger than predatory mites, predatory insects require larger amounts of prey on which to develop and reproduce; however, if the generalist predator has fed on other arthropods or other foods (pollen or honeydew) in the crop and has built up in numbers so that it can suppress pest mites, then these generalist predators can be effective in reducing mite populations. The role of generalist predators in IMM depends on the crop, the geographic area, economic injury levels, and a variety of logistical issues. In the future, it may be possible to manipulate predator numbers by habitat manipulation (Landis et al. 2000, James 2003). Certainly, modifying spray practices to further enhance their effectiveness (Hull and Beers 1985, Desneux et al. 2007) should be a component of any integrated pest management (IPM) program.

Table 13.1 What Is the Take-Home Message Regarding Predatory Insects and Integrated Mite Management in Agricultural Crops?

1. Predatory insects (Coleoptera, Thysanoptera, Diptera, Neuroptera, and Hemiptera) and spiders may be found in agricultural crops. Many are generalist predators, but several (such as six-spotted thrips and *Stethorus* species) specialize on spider mites as prey.
2. Generalist predatory insects and spiders should be **conserved** in agricultural crops through modification of pesticide applications; despite their role as intraguild predators, they usually are more beneficial than detrimental.
3. Generalist predatory insects should not be released in **classical biological control** programs due to concerns about nontarget effects.
4. Predatory insects, such as six-spotted thrips or *Stethorus*, may sometimes be useful in **augmentative biological control**, but they are expensive.
5. More information on the biology, ecology, and behavior of predatory insects under field conditions is required to understand and appropriately deploy generalist predatory insects and spiders in agricultural cropping systems.

Several predatory insects are commercially reared and sold for augmentative releases, but they are relatively more expensive than phytoseiids, and research-based guidelines on their release rates and timing may be lacking for specific crops or environments. Cannibalism by these insects makes it difficult, and expensive, to produce them in large quantities. Highly effective artificial diets are rare, and most predatory insects must be reared on live prey, which can be expensive to produce. As a result, the costs of predators remain high relative to a pesticide spray. The availability of high-quality artificial diets could increase the role of predatory insects in augmentative releases (Table 13.1).

The advantages to predatory insects over phytoseiid predators include the fact that they are larger than phytoseiids and thus can consume more prey per unit time (higher **functional response**). Insect predators also have a greater longevity and higher fecundity than phytoseiids and thus have a potential for a greater **numerical response**. In spite of these advantages, relying on predatory insects in some IMM programs may be ineffective unless the plant can tolerate relatively high densities of pest mites or abundant alternative prey foods are available.

Deterrents to the use of predatory insects as effective natural enemies of mites may include a lack of specialization on mites so they only feed on mites when the mites are very abundant. Some predatory insects are repelled by heavy webbing. If the predatory insects arrive in a crop after mite populations are very high, it may be too late to prevent crop damage or losses. Generalist predatory insects have a slower development time compared to spider mites or other pest mites and, as a result, the predatory insects exhibit a very **delayed numerical response** to the prey population. Predatory insects may be unable to persist when prey populations are low (especially if no alternative foods are available), so rebounds of the pest can occur. Finally, many predatory insects are susceptible to nonselective pesticides and are disrupted by applications made to control insect pests.

Predatory insects are important natural enemies of spider mites in certain geographic regions at specific times under natural conditions. Each situation should be examined carefully to avoid prejudice. Unfortunately, predatory insects compete with phytoseiids for prey and can also prey on them (intraguild predation) so sorting out the relative costs and benefits may not be simple (Table 13.2).

Should growers be advised to make augmentative releases of predatory insects to control mites in greenhouses or other crops? The answer, in many cases, is no. These predatory insects do not remain in a field crop unless adequate food is present and will disperse to find food. Mite populations that are sufficiently high to allow predatory insects to reproduce can result in substantial damage to the crop, especially if the economic injury level is low. Crop species and cultivars vary in their tolerance to feeding by mites, so it is difficult to say that augmentative releases are never justified, especially in contained environments such as greenhouses, where they are less likely to disperse. It will be appropriate to conduct augmentative releases as experiments until release rates, timing, and benefits can be quantified for your specific crop and pest mite.

Table 13.2 A Comparison of Positive and Negative Aspects of Generalist Predatory Insects Compared to Phytoseiids for Integrated Mite Management in Agricultural Crops

Positive	Negative
1. Predatory insects can consume larger numbers of prey per unit time compared to phytoseiids.	1. They have a long generation time compared to pest spider mites.
2. They live longer than phytoseiids.	2. They require large amounts of prey to develop, so pest populations must be high to allow successful development to adulthood; because they will leave when pest densities decline, rebounds of spider mites can occur.
3. Adults can fly rapidly from site to site.	3. Many will cannibalize or practice intraguild predation on other natural enemies in the crop, including phytoseiids.
4. They have a high rate of reproduction compared to phytoseiids.	4. Many (but not all) are sensitive to pesticides.
5. They may be able to feed on nonprey foods (pollen or honeydew) which can result in control of even low-density populations of spider mites.	5. Generalist predatory insects are not suitable for releases in classical biological control programs due to concerns about nontarget effects.
6. Some predatory insects can be mass reared for augmentative releases.	6. Mass-reared predatory insects may be more expensive to produce than phytoseiids because of their high rate of consumption of prey, their propensity to cannibalize, and their long generation time; highly effective artificial diets are needed.

Growers should be encouraged to modify their spray practices to maintain predatory insects (and predatory mites) in the cropping system (**conservation biological control**). Predatory insects will consume mites that are not under adequate control by phytoseiids or have not been adequately controlled, perhaps due to resistance in the pest species. Their presence could help to delay the development of pesticide resistance in the pest mites.

13.2 FAMILY COCCINELLIDAE (ORDER COLEOPTERA): *STETHORUS* ARE MITE SPECIALISTS

Coccinellids (ladybirds, ladybugs, lady beetles) are small, ranging from 1 to 10 mm in length, and are found around the world (Hagen 1962, Chazeau 1985, Obrycki and Kring 1998). Over 5000 coccinellid species have been described, and most are considered beneficial predators of aphids, scale insects, and mealybugs, although a few are plant pests (such as species in the subfamily Epilachninae, which includes the Mexican bean beetle). In temperate regions, these beetles enter diapause and may gather in large groups to overwinter. Most coccinellids are brightly colored, perhaps as a signal to ward off predators. Many coccinellid species, such as *Coleomegilla maculata*, *Hippodamia convergens*, and *Harmonia axyridis,* will feed on mites, but the mites are not sufficient for reproduction; however, the tribe Stethorini includes specialized mite predators.

The Stethorini includes the genera *Stethorus* and *Parastethorus* and consists of about 90 described species that are black and known as 'spider mite destroyers.' These beetles also feed on false spider mites, or Tenuipalpidae (Chazeau 1985, Biddinger et al. 2009). The Stethorini are considered by some to be highly effective natural enemies of tetranychids second only to the Phytoseiidae, whereas others consider them as unable to control spider mite populations at low enough levels (Biddinger et al. 2009). According to several authors, adult *Stethorus punctum* can find apple trees infested with low-density populations of *Panonychus ulmi* even though the prey population was less than one mite per leaf. Similar results have been seen in small patches of spider mites in avocados, citrus, and raspberries (McMurtry and Johnson 1966, Hull et al. 1977a,b, Haney et al. 1987, Chen 1993, Congdon et al. 1993, Chen and Zhao 1994).

Figure 13.1 Adult of *Stethorus picipes* (Coleoptera: Coccinellidae). *Stethorus* species specialize as predators of spider mites. (Photograph by Jack Kelly Clark, University of California Statewide Integrated Pest Management Program.)

Stethorus species are very small (1 to 1.5 mm), round, black beetles with brown or yellow legs and antennae (Figure 13.1)(also see Figure S13.1 on the CD). Most look similar, and identification to species is difficult because male genitalia must be examined. Most species cannot be identified using only females (Biddinger et al. 2009). Males have a flattened or notched posterior margin of their last abdominal segment, while females do not. As a result of the difficulty in accurate identifications, the literature can be very confusing, because taxonomic errors may be common.

Stethorus species are found in many geographic areas and climates, including temperate and tropical, and they are especially important in orchards. A number of species, including *S. punctillum*, have been widely distributed around the world in classical biological control programs. One species, *Parastethorus nigripes* (formerly placed in the genus *Stethorus*), was introduced from Australia (Field 1979) into California almond orchards (Hoy and Smith 1982), where it apparently failed to establish. It subsequently showed up in Texas, where it feeds on the Banks grass mite *Oligonychus pratensis* (Pollock and Michaels 2002).

The developmental stages of *Stethorus* species include egg, four larval stages, pupa, and adult males and females (see Figure S13.1 on the CD). *Stethorus* eggs are elongated ovals and are pale cream or pink. They are laid singly on the leaf or on bark, often in the midst of a mite colony, or nearby. Larvae can crawl among the silk webbing produced by *Tetranychus* mites, unlike many lady beetles that do not perform well in dense webbing. *Stethorus* species can feed on extrafloral nectaries, aphid honeydew, and pollen, as well as a variety of mite species (Wheeler et al. 1973). In temperate climates, *Stethorus* species overwinter as adults in diapause and have two or three generations per year (McMurtry et al. 1974, Chazeau 1985). *Stethorus picipes* has a facultative reproductive diapause that is induced by short daylengths.

The life cycle is relatively long compared to that of spider mites or phytoseiids (Putnam 1955, Raros and Haramoto 1974, Walters 1974, Daniel 1976, Tanigoshi and McMurtry 1977, Houck 1991, Rott and Ponsonby 2000a,b, Roy et al. 2002, Kishimoto 2003, Mori et al. 2005). Development of *Stethorus punctillum* fed on *Tetranychus urticae* at 21°C requires about 21 days; however, the fecundity of *S. punctillum* is high, as females produce 197 to 1290 eggs over 106 to 786 days. The consumption rate of prey also is high, as immature *S. punctillum* can consume a total of 239 *T. urticae*, and adult females consume 52 to 87 spider mite eggs per day. If food is scarce, *Stethorus* can become cannibalistic, thus making them difficult to rear and increasing the expense of mass production.

Adults actively fly and will aggregate near mite colonies, which is an advantage over the Phytoseiidae. They consume the entire mite; as a result, it may be difficult to confirm that spider mite populations declined due to predation by these beetles if sampling is done infrequently and the adult beetles have left. *Stethorus* species may feed on several tetranychid species, but some *Stethorus* species may not prefer some spider mites; for example, S. *punctillum*, *S. punctum*, and *S. gilvifrons* do not readily feed on *Bryobia* species (Putnam 1955). *Stethorus punctillum* has been reported to have a broader prey range than *S. punctum* and was reported to prey on spruce spider mites (*Oligonychus ununguis*) on ornamentals (Wheeler et al. 1973).

Finding abundant *Stethorus* on a crop usually indicates that spider mite populations are high. If *Stethorus* eggs and immatures, as well as adults, are on many leaves, they should be able to clean up the pest population in a few days. At this point, most of the mite feeding damage has been done, so it probably is not worthwhile to spray for mites. So far, *Stethorus* species have been most effective in "woody perennial systems … where overwintering sites are available and they are protected from toxic pesticides" (Biddinger et al. 2009).

One of the best documented examples of *Stethorus* as an important natural enemy is that of *S. punctum* in Pennsylvania apple and peach orchards (Mowery et al. 1975, Hull et al. 1976, 1977a,b, Asquith and Hull 1979, Hull and Beers 1985). This example also demonstrates that changes in production practices can alter an IPM program dramatically. During the 1970s, *S. punctum* was found to have developed resistance to organophosphate insecticides that were used to control insect pests in apple orchards. In addition, effective miticides were not available for the European red mite, so an IMM program was promoted once *S. punctum* was recognized to be an important predator of the European red mite (*Panonychus ulmi*). The use of alternate row middle spraying of pesticides (spraying only half of the trees on each side of the row middle and not spraying alternate rows) was conducive to maintaining these predators in apple orchards because it provided unsprayed refuges for the beetles. By the mid-1990s, new miticides became available, and *S. punctum* began to disappear from apple orchards because the lower European red mite populations resulting from these miticide sprays prevented the beetles from reproducing. In addition, the lepidopterous pests developed resistance to organophosphate insecticides, and new pesticides began to be used that were toxic to *S. punctum*. By 2005, the phytoseiid *Typhlodromus pyri* had become more important as a predator of *P. ulmi* because it could feed on alternative foods such as rust mites, pollen, and fungi when spider mite densities were low. As a result, *S. punctum* no longer is considered essential to the apple IMM program in Pennsylvania. A parallel example was observed in Italian apple orchards, where *S. punctillum* developed resistance to azinphosmethyl and became important as a mite predator (Pasqualini and Malavolta 1985, Pasqualini and Antropoli 1994). However, alternative pesticides are now being used to control the azinphosmethyl-resistant insect pests and biological control of mites is now achieved by the phytoseiid mites *T. pyri* and *Amblyseius andersoni*.

Biological control of the two-spotted spider mite with *Stethorus punctillum*, the phytoseiid *Amblyseius californicus*, and other endemic predators has been documented in strawberries in Spain (Garcia-Mari and Gonzalez-Zamora 1999). Although the phytoseiid was an important predator, *S. punctillum* and the coniopterygid *Conwentzia psociformis* were found to be the main predators during some periods of the year. Garcia-Mari and Gonzalez-Zamora (1999) concluded that in strawberry crops near Valencia, "naturally occurring predators are able to control infestations of spider mites and maintain them below damaging levels" and that augmentative releases of phytoseiids were not necessary if disruptive pesticides were not applied. They developed a presence–absence sampling method and found that, when the spider mite density ranged between 0.1 and 1 mite per leaflet, the prey-to-predator ratio should be between 1 and 5 for an immediate decrease in the pest population (Garcia-Mari and Gonzalez-Zamora 1999). This combination of naturally occurring phytoseiid and insect predators achieved control as good as that achieved by the more expensive augmentative releases of *Phytoseiulus persimilis* as long as the pesticides used were not toxic to *A. californicus* and *S. punctillum*. Rott and Ponsonby (2000a) evaluated *S. punctillum* and *A. californicus* as biological control agents of two-spotted spider mites in greenhouses and found that combinations of predators can sometimes result in improved pest control.

McMurtry et al. (1970) evaluated the effects of pesticides on *Stethorus* species. In general, pesticides are highly detrimental to coccinellid populations. Roy et al. (1999) found that *S. punctillum* was abundant in untreated red raspberries in Quebec, Canada, while commercial blocks treated with pesticides had few. Obrycki and Kring (1998) pointed out that pesticides have the "greatest effect on local populations of coccinellids," and "the greatest gains may be attained through reduction of toxic pesticides in coccinellid habitats." Pesticides affect coccinellids both directly and indirectly (through consumption of treated prey). Even if not killed, pesticides can reduce fecundity or longevity of coccinellids or eliminate their food supply. The International Organization of Biological Control has developed standardized test methods for evaluating the effects of pesticides on natural enemies (Hassan 1985), and the modification of spray practices is expected to result in enhanced biological control of mite pests by *Stethorus* species.

Other methods for conserving coccinellids include providing refuges from toxic sprays, providing suitable sites for overwintering of diapausing beetles, using crop cultivars (containing waxes, trichomes, secondary plant compounds) that are suitable for coccinellid searching behavior, and providing food sprays containing sugars and proteins to crops (Obrycki and Kring 1998). James (2003) showed that adding synthetic **herbivore-induced plant volatiles (HIPVs)** to Washington State hops attracted a variety of predatory insects, including *Stethorus punctum picipes*. HIPVs are chemicals produced by plants in response to damage caused by pest spider mites and insects (Hunter 2002, Frost et al. 2008). These chemicals may prime the plant (and its plant neighbors) to respond with direct and indirect plant defenses (Frost et al. 2008). The chemicals also may be perceived by natural enemies and attract them to the damaged plant. As an example, *S. punctum* responded to the herbivore-induced plant volatiles methyl salicylate and (*Z*)-3-hexenyl acetate in hops, and James (2003) concluded that the results "are a small, but significant first step indicating the potential of HIPVs as field attractants for beneficial arthropods." In pest management programs, it is often important to reduce pest populations as early in the season as possible to prevent economic injury. If HIPVs can be developed to provide such early-season pest control by insect and mite predators, then improved mite management (and insect pest management) may be developed. Methods for rearing *Stethorus* species have been developed by Scriven and Fleschner (1960) and Walters (1974).

13.3 FAMILY STAPHYLINIDAE (ORDER COLEOPTERA): *OLIGOTA* SPECIES MAY BE USEFUL PREDATORS OF SPIDER MITES

Oligota (recently designated as in the genus *Holobus*) is a cosmopolitan genus with more than 170 species (Chazeau 1985, Frank and Thomas 1999). Most live in decaying plants or in fungi, in stored products, under tree bark, or in bird or ant nests where they may prey on mites and small insects, or feed on dead arthropods. *Oligota* species that are predators of spider mites feed on all life stages of the mites, although some instars of the beetles perform better on different stages of mites. These are small beetles, about 1 to 2 mm in length (Figure 13.2). Relatively few species have been studied in detail, but the biology of several has been summarized (Badgley and Fleschner 1956, Chazeau 1985, Shimoda et al. 1997, Kishimoto 2003, Perumalsamy et al. 2009). The developmental

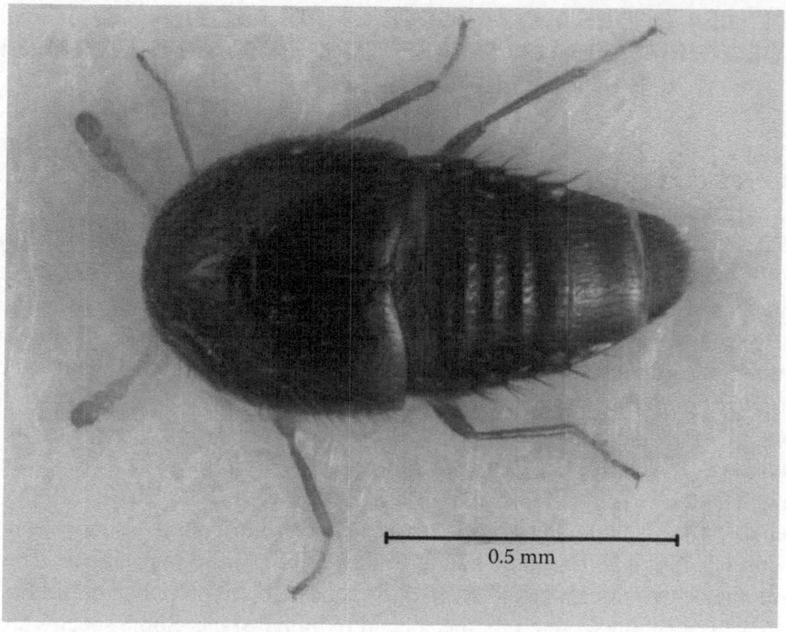

Figure 13.2 Adult of *Oligota minuta* (Coleoptera: Staphylinidae). This predator may prey on a variety of small arthropods, including spider mites. (Photograph by Lyle Buss, Department of Entomology and Nematology, University of Florida, Gainesville.)

time for *Oligota oviformis* and *O. flavicornis*, for example, averaged 21 to 30 days. Adults lived 30 to 35 days, and females produced 40 to 50 eggs (*O. flavicornis* reared on *Panonychus ulmi*) or over 300 eggs (*O. oviformis* reared on tetranychids). The life cycle includes an egg, three larval stages, prepupa, and pupa, with pupation taking place in the soil. This pupation site may be an important consideration in modifying cultural practices to enhance the ability of these beetles to persist in crops. In the case of *O. oviformis*, adults live about 72 days, with females depositing approximately 300 eggs. In tropical areas, development may be continuous, but in temperate areas there may be only one or two generations per year.

Oligota pygmaea is considered an important predator of *Oligonychus coffeae* (Tetranychidae) infesting tea in India (Perumalsamy et al. 2009). Badgley and Fleschner (1956) noted that *Oligota oviformis* occasionally can be found in large numbers in California citrus and avocado trees. They considered this species to be third in importance in controlling mites in avocados and citrus, behind phytoseiids and *Stethorus*. During development, *O. oviformis* larvae can consume 200 to 300 mites, while adults consume about 10 mites per day over their adult life of 32 days for a total of about 320 mites (Badgley and Fleschner 1956). Badgley and Fleschner (1956) considered that pupation in the soil "may have some bearing on the rather sporadic occurrence of large populations of the beetles." Because California citrus groves are not cultivated, the soil becomes very hard, and Badgley and Fleschner (1956) speculated that the larvae were not able to enter the soil to pupate under the citrus trees. By contrast, avocado groves may provide a better situation for the larvae to enter the soil because there is more leaf litter under avocado trees that allow the prepupae to survive.

Frank et al. (1992) reviewed the distribution and prey records for *Oligota minuta* and found that it fed on "various tetranychid mites on various plants," including the cassava green mite in Colombia. They noted that *Oligota pygmaea* is a predator of *Bryobia arborea*, *Tetranychus urticae*, *Oligonychus yothersi*, and *Panonychus ulmi* (see Figure S13.2 on the CD).

13.4 ORDER THYSANOPTERA (PHLAEOTHRIPIDAE, ASOLOTHRIPIDAE, THRIPIDAE): THRIPS MAY BE GENERALISTS OR SPECIALISTS (SIX-SPOTTED THRIPS)

Thrips (fringed-winged insects) are small (0.2 to 5 mm), elongated insects representing approximately 4500 species (Lewis 1973, Chazeau 1985). Thrips have two pair of wings that are held over the back when at rest, and many species are plant feeders or feed on pollen or fungi. Three families (Phlaeothripidae, Aeolothripidae, and Thripidae) contain species that prey on spider mites or other small arthropods such as thrips, coccids, and whiteflies (Ananthakrishnan 1993, Mound 2005). The Aeolothripidae consists of at least 190 species in 23 genera; some genera are **facultative** predators that use plant tissues as part of their diet. One genus in the Aeolothripidae (*Franklinothrips*) has 14 species that are ant mimics, and all are obligate predators, with several being predators of pest thrips such as *Thrips palmi* (Mound and Reynaud 2005).

The biology of thrips is variable (Mound 2005), and relatively few species of predatory thrips have been studied (Bailey 1940, Gilstrap and Oatman 1976, Parrella et al. 1982, Milne and Walter 1997, Araraki and Okajima 1998, Watson et al. 1998, Hoddle et al. 2000, Hoddle 2003a,b, Kishimoto 2003, Gotoh et al. 2004, Li et al. 2006). Eggs are laid either within plant tissues or exposed, depending upon the species. Nymphs resemble adults in their biology and appearance. They have a prepupal stage and one or two pupal stages. Wing pads appear in the prepupae or the first pupal stage. Some pupate in a silk cocoon secreted by the second nymphal stage. Some species are thelytokous, and others are arrhenotokous. In the case of *Franklinothrips vespiformis*, Araraki et al. (2001) showed that *Wolbachia* endosymbionts cause thelytoky.

The biology of several acariphagous thrips has been studied (Putman 1955, Gilstrap and Oatman 1976, Bournier et al. 1978, 1979, Parella et al. 1982). The complete life cycle can be completed in as little as 2 weeks, and adults are relatively long lived. All species of *Scolothrips* are obligate predators of spider mites, while *Frankliniella occidentalis* and *F. schultzei* are phytophagous, but they can be important mite predators in some cases (facultative predators) (Mound 2005). Mound (2005) indicated that most "tropical Aeolothripidae seem to be obligate predators, although most related temperate species in the genus *Aeolothrips* are facultative predators in flowers."

One of the most common predatory thrips in temperate climates is the six-spotted thrips *Frankliniella sexmaculatus* (Thripidae) (Figure 13.3) (also see Figure S13.3 on the CD). All active stages of *F. sexmaculatus* prey on *Eotetranycus sexmaculatus* (in citrus), *Oligonychus punicae* (avocado), *Panonychus citri* (citrus), *P. ulmi* (apple), *Tetranychus urticae* (strawberries), and spider mites in melons, walnuts, and cotton in the United States. This thrips seems well adapted to spider mite prey and is reported to use the silk threads deposited by mites to locate prey. It can be difficult to distinguish predatory thrips immatures, such as *F. sexmaculatus*, from plant-feeding immatures, so pest managers may not recognize these predators. Eggs of spider mites are completely consumed, but other instars of the mite may be only partially consumed. Because adults of *F. sexmaculatus* can fly to spider mite outbreaks, they can sometimes eliminate a spider mite outbreak in a short time.

In Canada, *Haplothrips faurei* has been recorded preying on *Panonychus ulmi* and *Bryobia arborea* in apples and peaches. In the United States, *Haplothrips subtilissimus* is a predator of *P. ulmi* in apple, as are *Leptothrips mali* and *Aeolothrips melaleucus*.

Some flower thrips, such as western flower thrips (*Frankliniella occidentalis*), are considered pests of cotton and other plants (Trichilo and Leigh 1986, Pickett et al. 1988); however, *F. occidentalis* preys on spider mites in cotton, where they are considered opportunistic predators (Trichilo and Leigh 1986). Agrawal et al. (1999) found that *F. occidentalis* fed on more prey when the plant tissues had been induced by prior feeding by spider mites, suggesting that induced host-plant resistance may affect feeding preferences by omnivores. Milne and Walter (1997) found that the cotton-leaf-feeding thrips *Frankliniella schultzei* in Australia also feeds on spider mites and that this addition to their diet shortened their development time, increased fecundity, and extended their life

Figure 13.3 Adult of six-spotted thrips, *Frankliniella sexmaculatus* (Insecta: Thripidae). These are important predators of spider mites and are distinctive as adults with their six spots on their wings. Unfortunately, the larvae resemble plant-feeding thrips species. (Photograph by Jack Kelly Clark, University of California Statewide Integrated Pest Management Program.)

span. Predation by *F. schultzi* and by *F. occidentalis* was reduced, however, when the spider mite eggs were covered with silk, indicating that these thrips are not well adapted to mites as prey. Milne and Walter (1997) determined that cotton was not the primary host plant for *F. schultzei* and that *F. schultzei* performed better on its primary host, *Malvaviscus arboreus*, where it fed on pollen. Milne and Walter (1997) concluded that "it is probable that predation by flower thrips makes a difference to the thrips only under circumstances that otherwise affect performance negatively, i.e. when there is no pollen available, as in seedling cotton." Wilson et al. (1996) found that three thrips species (*Thrips imaginis, T. tabaci*, and *F. schultzei*) in Australia fed on two-spotted spider mites and were considered "potentially important predators of spider mites in the field, especially given their abundance on young cotton and preference for inhabiting situations in which mite colonies are found."

Pesticides can disrupt predatory thrips in crops (Li et al. 2006, Bosco and Tavella 2010) so efforts should be made to conserve them. Mass production of predatory thrips for augmentative releases currently is expensive, so few are made.

13.5 HETEROPTERA (HEMIPTERA: TRUE BUGS): GENERALIST PREDATORS OF SMALL ARTHROPODS, INCLUDING MITES

Predatory Heteroptera (Hemiptera) also are, relatively speaking, less studied than phytoseiids. Yeargan (1998) reviewed the information available for predatory Heteroptera in North America and concluded that the "predatory heteropteran faunas of major North American agroecosystems are well known. We know far less, however, about the ecological factors that affect their choice of habitats and their abundance in various agroecosystems and the roles these predators play." He further suggested that, "It may be unrealistic to expect predatory Heteroptera to keep some types of pest populations in check (e.g., aphids, spider mites), [and] it is likely that these predators often play an important role in reducing the intragenerational survivorship of many other pests."

It may be that heteropteran predators are more common in row crops (such as corn, soybeans, alfalfa wheat, peanuts, sorghum, and cotton) than in orchards and vineyards. Heteropteran predators, with a relatively long developmental time, are limited as predators of rapidly reproducing spider mites, which may develop from egg to adult in less than a week. It also is unclear whether these predators will seek out high-density spider mite populations or be able to respond to the rapid population growth that is often seen in spider mite populations.

Coll (1998) reported that a number of predatory Heteroptera also feed on plants, including several predatory anthocorid species in the genus *Orius* (*O. insidiosus*, *O. pallidicornis*, *O. tristicolor*, and *O. vicinus*). *Geocoris* (Lygaeidae) species (*G. bullatus*, *G. punctipes*, *G. pallens*, and *G. uliginosus*) also are able to feed on plants. Species of the Anthocoridae and Miridae do feed on spider mites, but the relationships between these facultative predators and other insect and mite species in agricultural crops is not easy to resolve. *Geocoris* species (Lygaeidae) are facultative predators of spider mites, but may be of secondary importance.

13.5.1 Family Anthocoridae

Lattin (1999) reviewed the bionomics of the 400 to 600 species of Anthocoridae, small insects (1.4 to 4.5 mm in length) found in a variety of habitats. The genera *Anthocoris*, *Orius*, and *Tetraphleps* are predators of small arthropods, although some feed on plants (Lattin 1999). Most information about the family is based on studies of species found in agricultural crops. Herring (1966) reviewed the taxonomy of the genus *Orius* in the Western Hemisphere. At that time, 21 species were known, and Herring (1966) provided keys to them. Ryerson and Stone (1979) provided a selected bibliography on *O. tristicolor* and *O. insidiosus*.

Orius species (often called minute pirate bugs) are generalist predators, feeding on a variety of small arthropods, including spider mites, thrips, aphids, and insect eggs and larvae (Askari and Stern 1972, Isenhour and Yeargan 1981, Alauzet et al. 1994, Lariviere and Wearing 1994, Chyzik et al. 1995, Venzon et al. 2002, Tommasini et al. 2004). Adult *Orius* are oval and black with white patches on the wings (Figure 13.4) (also see Figure S13.4 on the CD). Development includes eggs and five nymphal stadia. Immatures resemble adults, but they cannot fly. Some are univoltine, but others have two generations per year. Females deposit eggs in the leaf or petiole tissues (Lundgren and Fergen 2006). When the eggs hatch, the young nymphs will begin to feed. Adults and nymphs have sucking mouthparts and remove the contents from their prey. Feeding involves ramming "its beak into the body of the mite," and *Orius* then sucks the fluid from its prey (Iglinsky and Rainwater 1950).

Under laboratory conditions *Orius insidiosis* developed from egg to adult in 12.1 days on spider mite prey (Iglinsky and Rainwater 1950). *Orius insidiosis* is a predator of *Tetranychus urticae*, *Panonychus ulmi*, and *P. citri* in melons, hops, strawberries, fruit trees, and citrus (Iglinsky and Rainwater 1950, Isenhour and Yeargan 1981). Ruberson et al. (2000) found that *O. insidiosus* in Arkansas entered adult reproductive diapause in response to short photoperiods occurring during the fifth instar and during the early adult stage.

Biological studies on four *Orius* species using *Ephestia* eggs and *Frankliniella occidentalis* adults as prey were conducted by Tommasini et al. (2004). Wearing and Colhoun (1999) studied the biology of *Orius vicinus* as predators of the mites *Aculus schlechtendali*, *Panonychus ulmi*, and *Tetranychus urticae*. Although *O. vicinus* could develop on all prey within 22 to 26 days at 20°C, when *O. vicinus* was reared on the smaller apple rust mite (*Aculus schlechtendali*) prey development was slowest, and the smallest adults were produced. Wearing and Colhoun (1999) provided a long list of the insects and mites on which *O. vicinus* has been recorded feeding, indicating that this is a generalist predator. It also has been found to feed on pollen. *Orius* species are commercially produced, and efforts are being made to rear *O. insidiosus* on an artificial diet (Ferkovich and Shapiro 2004a,b, 2005).

Figure 13.4 (A) Adult of *Orius insidiosus* (Insecta: Hemiptera: Anthocoridae). *Orius* species are generalist predators that may feed on spider mites. (Photograph by Jim Castner, Department of Entomology and Nematology, University of Florida, Gainesville.) (B) Immature of *Orius tristicolor*. (Photograph by Lyle Buss, Department of Entomology and Nematology, University of Florida, Gainesville.)

These generalist predators are often involved in intraguild predation; for example, Rosenheim (2005) found that *Orius tristicolor* is often preyed on by *Geocoris* species in California cotton fields. Cloutier and Johnson (1993) found that *O. tristicolor* fed on the phytoseiid *Phytoseiulus persimilis* "in significant numbers even when thrips were available"; however, the addition of spider mites and thrips to the two-predator system significantly reduced *O. tristicolor* predation on *P. persimilis* and thrips. Norton et al. (2001) found that acarodomatia on wild grape (*Vitis riparia*) increased survival of a tydeid and a phytoseiid (*Amblyseius andersoni*) when *Orius insidiosus* and coccinellid predators were present, suggesting that *Orius* is a predator of these beneficial mites. Clearly, the interactions of multiple generalist predator and prey species and their host plants can be complex. See Chazeau (1985) for a method for distinguishing mirids and anthocorids based on wing venation of the hemelytron.

13.5.2 Family Miridae

The Miridae is a large and diverse insect family that includes plant bugs, leaf bugs, and grass bugs, with over 10,000 known species (Wheeler 2001). They are oval or elongate and are usually less than 12 mm long. Most mirids are studied because they are important agricultural pests (such as *Lygus* bug), and all mirids were once thought to be phytophagous, but we now know that some species are predatory (Wheeler 2001). In fact, many phytophagous mirids, perhaps 25%, are facultative predators.

Most mirids overwinter as eggs that are laid in soft plant tissue. Nymphs of most species emerge in the spring when there is sufficient new tender foliage on which to feed. The five nymphal instars are each 5 to 7 days in length. Adults feed continuously, and males are short lived after mating. Eggs typically require 10 days to 2 weeks before hatching. Mirids are known to be predators of aphids, whiteflies, and mites; however, they will feed on syrphid larvae (another example of intraguild predation) (Frechette et al. 2007). Mirids are considered important predators of European red mites in apple orchards in England (Collyer 1952, 1953), but the mirid *Campylomma verbasci* is a damaging pest of fruits of some apple cultivars (Reding et al. 2001). The nymphs of the overwintering generation cause damage to young fruits in the brief interval from full bloom until about one week after petal fall. Mirids are important predators of mites in cotton (Butler 1965).

The mirid *Macrolophus caliginosus* was shown to respond to herbivore-induced plant volatiles produced by plants infested with the two-spotted spider mite or the green peach aphid *Myzus persicae* (Moayeri et al. 2007). Interestingly, the mirids were more attracted to HIPVs emitted by plants containing both aphids and mites than to plants containing only one potential prey. This mirid is being used in European greenhouses to control multiple pests, including aphids, whiteflies, and mites. Hansen et al. (1999) studied the biology of *M. caliginosus* using two-spotted spider mites as prey on tomatoes. Females lived 29 days, producing 0.7 eggs per female per day. *Macrolophus caliginosus* had a 6-day preoviposition period and an 18-day oviposition period, and development from egg to adult required 27 days; however, life-table data indicated that a spider mite diet was inferior compared to a diet of insect prey. Lucas and Alomar (2001) found that *M. caliginosus* was a suitable prey for another mirid, *Dicyphus tamaninii*, when the plant diet available was of poor quality, indicating that mirids will eat mirids (intraguild predation), especially when other prey is scarce.

Chouinard et al. (2006) studied the effectiveness of *Hyaliodes vitripennis* as a predator in apple orchards and found that it fed on European red mites and aphids in Quebec, Canada, but these mirids reduced the effectiveness of naturally occurring phytoseiid mites "either by resource competition or by direct predation." Fitzgerald et al. (2007) pointed out that, because broad-spectrum pesticide use is declining in a variety of crops, more species of predatory insects and predatory mites will be able to survive on a long-term basis. As a result, the role of intraguild predation is likely to become a more important area of research regarding managing and predicting pest and predator interactions in these crops. Even sublethal doses of pesticides can affect intraguild predation among predatory insects and mites (Provost et al. 2003). In pesticide-treated crops, it is often the phytoseiids that have been able to develop resistance to pesticides, leading to their dominance. This leads us to the question: Will mirids and other insect predators displace phytoseiids in organic agriculture or in crops using reduced-pesticide inputs in the future?

13.6 CECIDOMYIIDAE (ORDER DIPTERA): *FELTIELLA* SPECIES CAN BE EFFECTIVE PREDATORS OF SPIDER MITES

The Cecidomyiidae (known as gall midges) includes phytophagous, saprophagous, or zoophagous species (Chazeau 1985). These are very small flies (about 2 mm in length) and very delicate in appearance (Figure 13.5) (also see Figure S13.5 on the CD). The body and wings have long hairs, which are easily rubbed off. The larvae usually are plant feeders, often producing galls in which they live. A number of species are predators of mites, aphids, scales, and other small arthropods. Gagne (1995) revised the genus *Feltiella* and found that there were eight species, rather than the 24 previously described. He noted that errors in identification had been made previously because the color of larvae was considered a useful taxonomic character. Gagne (1995) observed that larval color varies with diet and, as a result, synonymized the genus *Feltiella* and *Therodiplosis*.

The biology of the cosmopolitan *Feltiella acarisuga* (known also as *Therodiplosis*) is better understood than any other species. *Feltiella acarisuga* is the most widely distributed species (Gagne 1995, Opit et al. 1997, Gillespie et al. 2000, Mo and Liu 2006, 2007, Osborne et al. 2008). Eggs are deposited on plant tissue infested with tetranychid mites. Upon hatching the larvae crawl to spider mites and feed on all life stages. Larvae are relatively slow moving and are yellow, orange,

Figure 13.5 (A) Larva of *Feltiella* (Diptera: Cecidomyiidae), a predator of spider mites. (B) Adult *Aphidoletes aphidimyza* (Cecidomyiidae). The adults are not predatory. (Photographs by Jack Kelly Clark, University of California Statewide Integrated Pest Management Program.)

or red, depending on the body contents. There appear to be three instars and larvae reach about 2 mm in length before pupating. The full-grown larva spins a cocoon on the leaf. Adults are not predaceous. Mo and Liu (2007) found that larval *F. acarisuga* fed on 38, 60, and 87 *Tetranychus urticae* eggs as first-, second- and third-instar larvae, respectively, under laboratory conditions, and females deposited 33 eggs during their lifespan of approximately 13 days at 26.7°C and 85% relative humidity.

Gillespie et al. (2000) evaluated *Feltiella acarisuga* as a natural enemy of the two-spotted spider mite on greenhouse vegetables in Canada. They found that the development time of the population they tested ranged from 10 days at 27°C to 34 days at 15°C. Development at 20°C was shorter at 96% relative humidity than at 84%, and high mortality of larvae occurred at lower relative humidities. As a general rule, this predator requires relative humidities above 50%. At 25°C, females lived for about 5 days and produced about 46 eggs. Releases of 1000 individuals per hectare have been effective in controlling spider mites on tomato, pepper, and cucumber.

In addition to feeding on tetranychids, the immatures of one species of cecidomyiid are predators of eriophyoid mites on tea in Indonesia, and two undescribed species have been found feeding on citrus rust mite (*Phyllocoptruta oleivora*) on citrus in Florida (Villanueva et al. 2006).

13.7 ORDER NEUROPTERA (CHRYSOPIDAE, CONIOPTERIGIDAE, HEMEROBIIDAE): GENERALIST PREDATORS THAT MAY SOMETIMES FEED ON MITES

The Order Neuroptera includes about 6000 species that are easily identified by their conspicuous wing venation and their larval piercing mouthparts (Chazeau 1985). Three families (Chrysopidae, Coniopterygidae, and Hemerobiidae) are considered important natural enemies of a variety of pests. All three families contain species that can feed on spider mites, although none is a specialized mite predator (McMurtry et al. 1970). Additional information on the Neuroptera can be found in an online digital lacewing bibliography (http://lacewing.tamu.edu/bibliography/index.html) that contains literature on the insect super order Neuropterida (orders Neuroptera, Megaloptera, and Raphidioptera), as well as references to systematics, biology, ecology, and biological control.

The **Chrysopidae** (green lacewings or golden eyes) contains over 1200 species around the world (Winterton and de Freitas 2006). Chrysopids are mainly predators of aphids, although they also feed on pollen, honey, and other arthropods (Miller et al. 2004, Oswald 2004). There are at least 80 genera, although some are rare, and the most important genus for biological control of pests in agriculture is the genus *Chrysoperla* (Winterton and de Freitas 2006). Adults are 10 to 15 mm in length and often fly at night. They may be collected easily because they are attracted to lights. During the day, adults remain at rest on the undersides of leaves. Adults fly rather slowly and become active at dusk (Balduf 1974).

The stalked eggs are distinctive and may be deposited in batches on foliage (Figure 13.6) (also see Figure S13.6 on the CD). The production of egg stalks is thought to protect the eggs from cannibalism by chrysopid larvae and other predators. Species prefer particular habitats; some are found primarily on trees, others on low vegetation (field crops, grasses, and weeds), and others occur in tall herbs, shrubs, or trees (Balduf 1974).

Predatory adults have chewing mouthparts, so they ingest solid as well as liquid components of their prey. Adults provided with water, honeydew, or prey can survive a comparatively long time (60 to 80 days), and females can produce up to several hundred eggs (Balduf 1974). Larvae are elongate, and their surface is tuberculate with numerous setae; some describe the larvae as "alligator-like" in appearance. Some chrysopid species carry debris on their abdomens. Three larval instars are followed by a pupal instar, which takes place within a cocoon. Larvae feed by sucking fluids from their prey. Various chrysopids have been reported to feed on planthoppers, aphids, coccids, pyralid

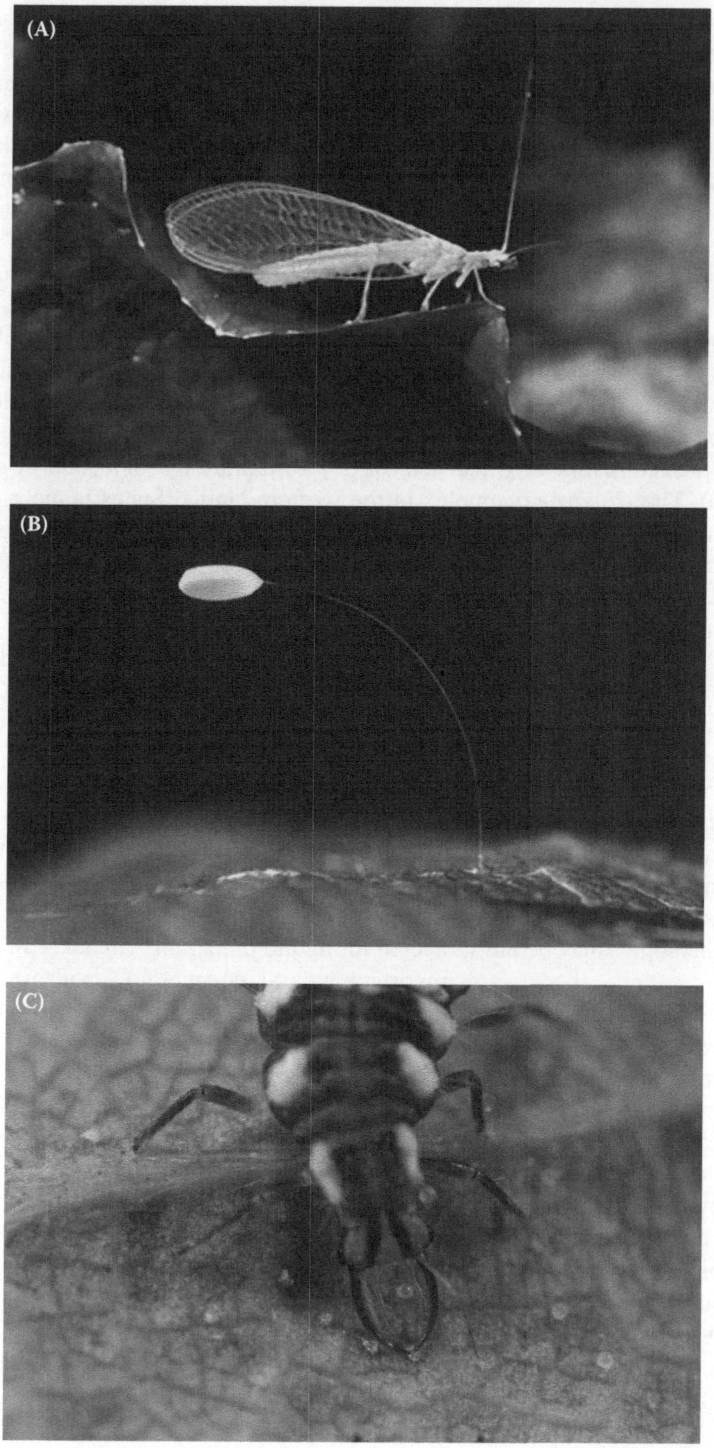

Figure 13.6 (A) Adult and (B) egg of a green lacewing (Neuroptera: Chrysopidae). (Photographs by Jack Kelly Clark, University of California Statewide Integrated Pest Management Program.) (C) The green lacewing larva is feeding on spider mite eggs, but these insects are generalist predators. (Photograph by Lyle Buss, Department of Entomology and Nematology, University of Florida, Gainesville.)

larvae, thrips, mites, eggs of coccinellids, eggs of papillionids and other Lepidoptera, spider eggs, syrphid larvae, larval coccinellids, and phylloxera (Balduf 1974). Larvae are cannibalistic on eggs or larvae, and some feed on nectar. Although chrysopids feed on mites, it is not clear that they can develop and reproduce solely on a diet of spider mites (Chazeau 1985). Chrysopids also can feed on other natural enemies, including coccinellids. Chrysopids in temperate climates overwinter as adults and may undergo a conspicuous color change from green during the summer to reddish in fall. During winter, adults do not feed. For additional information on the Chrysopidae, see Canard et al. (1984).

Inundative and augmentative releases are often made with *Chrysoperla* species in home gardens, row crops, orchards, and greenhouses using commercially reared insects, and *Chrysoperla* species are important worldwide as commercially available natural enemies (Tauber et al. 2000). Two species in North America are *C. carnea* and *C. rufilabris*; *C. nipponensis* and *C. externa* are used in Latin America and Asia (Tauber et al. 2000). There are taxonomic difficulties, however, and what is considered to be *C. carnea* in the midwestern and eastern United States appears to actually be a species complex consisting of *C. carnea* (or *C. plorabunda* by some) and *C. downsei* (Tauber and Tauber 2000). The *C. carnea* complex in the western United States is more diverse, and the species status is unclear; multiple biotypes or cryptic species are possible, based on differences in courtship songs and reproductive isolation (Wells and Henry 1998, Tauber et al. 2000, Henry et al. 2002, Henry and Wells 2004).

Taxonomic difficulties may be resolved when additional molecular analyses are done. Winterton and de Freitas (2006) provided a molecular analysis of the Chrysopidae in which they evaluated subfamilies and tribes. Haruyama et al. (2008) analyzed the cryptic species of the *Chrysoperla nipponensis* group and discovered that at least three cryptic species occur in Japan, although all had been described as *C. carnea* at one time. Because these different biotypes or cryptic species may have different biological characteristics, voucher specimens should be preserved in 95% ethanol for future DNA analyses when conducting field trials for efficacy as natural enemies or augmentative releases. Different commercial producers of "green lacewings" may be producing different species or biotypes for augmentative releases, so using the results from published trials to predict the outcome of future releases could be problematic. In open cropping systems, where other generalist predators occur, released chrysopids may be the subject of intraguild predation (Tauber et al. 2000).

Although *Chrysoperla* species are important natural enemies of small arthropods, none is known to specialize on spider mites, so augmentative releases are rarely made for this purpose. An exception may be the work conducted by Reddy (2001, 2002) in which he reported that *C. carnea* was useful in controlling *Tetranychus ludeni* on vegetable crops, such as okra, in India. (Whether the species used was truly *C. carnea* or a different biotype or cryptic species is unknown.) An IPM program that used a resistant eggplant variety, *C. carnea*, and neem oil provided better mite control than a different eggplant variety and an acaricide (dicofol), indicating that host-plant resistance complemented by natural enemies and neem oil could provide effective control. When *C. carnea* releases were tested in tomatoes, however, *C. carnea* failed, and a bioassay using plant volatiles from eggplant, okra, peppers, and tomatoes showed that the *C. carnea* population responded positively to damaged plants and volatiles produced by all crops, except tomatoes.

Another chrysopid species may be useful predators of spider mites. Ho (2000) reported that *Mallada basalis* was released in the 1990s on citrus, tea, papaya, and ornamentals in Taiwan for control of spider mites with promising results. Although this species is considered a generalist predator as larvae, and adults feed on nectar, honeydew, and pollen, *M. basalis* appears to have potential as a natural enemy of *Panonychus citri* on citrus and Indian jujube, and of *Tetranychus kanazwai* and *T. urticae* on strawberry (Cheng et al. 2010). Artificial diets for *M. basalis* have been tested (Lo et al. 2002), a mass rearing method has been developed, cold storage methods have been established, and the basic biology has been studied under various temperatures (Chang 2000, Cheng et al. 2010). This species has some tolerance to insecticides, fungicides, and acaricides, which enhances its ability to be used in IPM programs (Cheng et al. 2010). Cheng et al. (2009, 2010) found that mass-reared *M. basalis* fed equally well on two spider mites common in papaya plantations (*T. kanzawai* and *P. citri*) and concluded that augmentative releases could be useful.

Augmentative releases of *Chrysoperla* adults may be made, although adults are not predaceous (unlike adults in the genus *Chrysopa*) and may disperse from the release site. All larval stages feed on small arthropods; thus, eggs are often released because they are less expensive to produce than adults. Silvers et al. (2002) evaluated the quality of mass-reared *Chrysoperla rufilabris* eggs from three commercial insectaries and found a dramatic improvement in their quality over that of predator eggs evaluated previously by O'Neil et al. (1998). Because chrysopids are important natural enemies of a variety of arthropod pests, they should be conserved in agricultural crops, although they can be the subject of intraguild predation by ants, assassin bugs, earwigs, and other predatory arthropods in open cropping systems. Ridgway and Murphy (1984) reviewed methods for increasing their numbers in the field and concluded that potential manipulations to increase chrysopid populations have been limited by our lack of knowledge of the population dynamics of the pests and natural enemies in the crops in which the manipulations have been attempted. Most of the studies reviewed by Ridgway and Murphy (1984) involved only insect pests.

Conservation of chrysopids in field crops can be enhanced by their ability to tolerate pesticide applications. A number of populations vary in their tolerance of pesticides, probably due to field selection for resistance in at least some cases (Grafton-Cardwell and Hoy 1985, Croft 1990, Mizell and Schiffhauer 1990, Vogt 1994, Schuster and Stansly 2000, Pathan et al. 2008). Schuster and Stansly (2000) found that different life stages of two lacewing species were better able to tolerate field-rate applications of azadirachtin, insecticidal soap, and paraffinic oil, suggesting that these "biorational" pesticides might be useful in IPM programs. In addition, a heterogeneous population of *Chrysoperla carnea* from California alfalfa fields responded to selection in the laboratory for resistance to carbaryl (Grafton-Cardwell and Hoy 1986). It appears that different species and different populations vary in their responses to pesticides, so it is difficult to transfer information on pesticide resistance or tolerance to a specific crop or chrysopid population.

The family **Coniopterygidae** (known also as mealy-wings, dusky-wings, or dusty-wings) contains small insects (2 to 6 mm in length) that resemble aphids; the body and wings are covered with a white powdery exudate (Chazeau 1985). Adults hold the wings over the body. Larvae are small and resemble chrysopid larvae. These predators are most active at dawn and dusk and thus may be overlooked. Coniopterygids in the genus *Conwentzia* may prey on spider mites, but dense webbing can be detrimental to survival of larvae. Muma (1967) described the biology of *C. vicina* as a predator of citrus rust mites (*Phyllocoptruta oleivora*), coccid crawlers, and six-spotted mites (*Eotetranychus sexmaculatus*) in Florida citrus groves. This insect is blue-gray in color and about 3 mm long; they "fly with a darting, fluttering pattern when disturbed and come to rest primarily on the lower surfaces of leaves" (Muma 1967). Eggs of *C. vicina* are laid along the margins or midribs of leaves, and there are four larval instars, a pupal stage, and adult males and females. This predator fed, developed, and reproduced most rapidly on diets of whitefly eggs and crawlers, or six-spotted mites; however, larvae often became trapped in spider mite webbing and died, indicating that this predator may prefer whiteflies to mites. The life cycle is completed in 29 to 46 days, and 29 to 83 citrus red mite eggs, larvae, or nymphs are consumed per day. Muma (1967) reported this species to be widely distributed in citrus groves in Florida. The biology of coniopterygids has been studied by Arrow (1917), Badgley et al. (1955), Collyer (1951), Fleschner and Ricker (1953), Narayanan (1942), and Quayle (1913).

Hemerobius species adults (brown lacewings or **Hemerobiidae**) are usually brown in color and are considered important predators of small, soft-bodied insect pests, although they will also feed on honeydew (Balduf 1974). The larvae of hemerobiids resemble those of chrysopids, and they may prey on spider mites, although they are better known as aphid predators. Adults are rather slow fliers and are most active at dusk or on cloudy days. Relatively little is known about their effectiveness as predators of spider mites (Figure 13.7) (also see Figure S13.7 on the CD).

Figure 13.7 Adult of *Hemerobius* (Insecta: Neuroptera: Hemerobiidae). Hemerobiids are generalist predators, especially of aphids, but they occasionally feed on spider mites. (Photograph by Jack Kelly Clark, University of California Statewide Integrated Pest Management Program.)

Hemeriobiid eggs are oval and light in color, lack stalks, and are deposited on the lower sides of the leaves and in bark crevices (Balduf 1974). One report indicates that females can produce from 80 to as many as 460 eggs. Larvae are small and tend to hide in curled leaves or in leaf and flower clusters, appearing to prefer shaded areas. The mouthparts are used for sucking, and the larvae walk or run rapidly. Larvae of various species are reported to feed on coccids, aphids, leafhoppers, mites, thrips, scales, and mealybugs (Balduf 1974). Once larvae are mature, they pupate in sheltered sites such as bark crevices, rolled up leaves, abandoned insect galls, or hollow stems, and some may deposit their silken cocoons under stones or in the soil. Once larvae produce a cocoon, they pupate. For a key to the Hemerobiidae of Florida and references to the biology of several species, see MacLeod and Stange (2009) and Slater and Baranowski (1978).

13.8 ANTS AS PREDATORS OF *T. URTICAE*

Although ants typically are considered pests, Osborne et al. (1995) reported that *Tapinoma melanocephalum* is a predator of *Tetranychus urticae* in greenhouses in Florida (see Figure S13.8 on the CD). This ant feeds on sweets, dead or living insects, and *T. urticae*. This ant is considered a pest in greenhouses, so its potential role as an augmented natural enemy of *T. urticae* in greenhouses is questionable.

13.9 SPIDERS AS PREDATORS OF MITES AND TICKS: LESS WELL STUDIED

Spiders are important predators in many agroecosystems (Turnbull 1973, Riechert and Lockley 1984, Maloney et al. 2003). There are at least 30,000 described species in approximately 2400 genera and about 54 families. Some predict that there could be as many as 50,000 species of spiders (Turnbull 1973). Nearly all spiders are carnivores, and insects are their primary prey. Many species will capture and kill more prey than they can consume, which is referred to as **wasteful killing** (Maloney et al. 2003). As natural enemies of insect pests of agriculture they have received relatively little study, in part because they are generalist predators. Reichert and Lockley (1984) suggest that:

> … conservation of the diverse spider fauna that is characteristic of most natural systems must be emphasized rather than the life histories and foraging behavior of individual spider species. What a particular spider taxon does in a system would only be important if it, by itself, could effect control. It cannot, but the spider community as a whole apparently can. In natural habitats, it is a stable assemblage, which consists of species representatives that remove significant numbers of insects from the system.

Are there any negative aspects to spiders? Yes. The answer is that spiders that forage on plants can consume predatory insects such as *Geocoris*, *Chrysoperla*, *Hippodamia*, and pollinating bees (Maloney et al. 2003); however, despite this intraguild predation, spiders are considered beneficial. Furthermore, because spiders prey on so many species, generalist spiders can maintain pest populations at low levels, even if they are not able to control pest outbreaks. Spiders have been used as biological control agents of pest insects in apple orchards and rice paddies (Maloney et al. 2003). In Japan, spider populations can be maintained and enhanced by the release of *Drosophila* flies into fields when pest insects are not abundant. Spiders are known to be long lived, tolerant of desiccation, tolerant of starvation, and able to persist over long periods if not disrupted by pesticides or by crop-production practices such as plowing, planting, crop rotations, and harvesting. As a result, spider densities tend to be much higher in organic fields than in fields that use conventional pesticides and higher in perennial crops than in annual crops.

Will spiders be effective predators of mites and ticks? Guarisco (1991)reported that two common house spiders preyed on ticks and the venomous brown recluse spider *Loxosceles reclusa*. Spiders in South Africa have been shown to feed on spider mites in the laboratory and to reduce pest mites in "certain crops to such a level that a substantial increase in yield may be expected" (ARC 2010). Mansour et al. (1995) studied the potential of four species of spiders as predators of *Tetranychus cinnabarinus* in Israel and found that *Chiracanthium mildei* could feed on 27.5 mites per day in the laboratory, indicating that spiders play a role as natural enemies of spider mites.

SELECTED REFERENCES

Agrawal, A.A., C. Kobayashi, and J.F. Thaler. 1999. Influence of prey availability and induced host-plant resistance on omnivory by western flower thrips. *Ecology* 80:518–523.

Alauzet, C., D. Dargagnon, and J.C. Malausa. 1994. Bionomics of a polyphagous predator: *Orius laevigatus* (Het.: Anthocoridae). *Entomophaga* 39:33–40.

Ananthakrishnan, T.N. 1993. Bionomics of thrips. *Annu. Rev. Entomol.* 38:71–92.

Araraki, N. and S. Okajima. 1998. Notes on the biology and morphology of a predatory thrips, *Franklinothrips vespiformis* (Crawford) (Thysanoptera: Aeolothripidae): First record from Japan. *Entomol. Sci.* 1:359–363.

Araraki, N., T. Miyoshi, and H. Noda. 2001. *Wolbachia*-mediated parthenogenesis in the predatory thrips *Franklinothrips vespiformis* (Thysanoptera: Insecta). *Proc. R. Soc. Lond. B* 268:1011–1016.

ARC. 2010. *Spiders as Predators of Mites*. Pretoria, South Africa: Agricultural Research Council (www.arc.agric.za/home.asp?pid=4206).

Arrow, G.J. 1917. The life-history of *Conwentzia psociformis* Curt. *Entomol. Monthly Mag.* 53:254–257.

Askari, A. and V.M. Stern. 1972. Biology and feeding habits of *Orius tristicolor* (Hemiptera: Anthocoridae). *Ann. Entomol. Soc. Am.* 65:96–100.

Asquith, D. and L.A. Hull. 1979. Integrated pest management system in Pennsylvania apple orchards. In: D.J. Boethel and R.D. Eikenbary (eds.), *Pest Management Programs for Deciduous Tree Fruits and Nuts* (pp. 203–220). New York: Plenum Press.

Badgley, M.A. and C.A. Fleschner. 1956. Biology of *Oligota oviformis* Casey (Coleoptera: Staphylinidae). *Ann. Entomol. Soc. Am.* 44:501–502.

Badgley, M.E., C.A. Fleschner, and J.C. Hall. 1955. The biology of *Spiloconis picticornis* Banks (Neuroptera: Coniopterygidae). *Psyche* 62:75–81.

Bailey, S.P. 1940. The black hunter, *Leptothrips mali* (Fitch). *J. Econ. Entomol.* 33:539–544.

Balduf, W.V. 1974. *The Bionomics of Entomophagous Insects*, Part II. Faringdon, U.K.: E.W. Classey.

Biddinger, D.J., D.C. Weber, and L.A. Hull. 2009. Coccinellidae as predators of mites: Stethorini in biological control. *Biol. Control* 51:268–283.

Blommers, L.H.M. 1994. Integrated pest management in European apple orchards. *Annu. Rev. Entomol.* 39:213–241.

Bosco, L. and L. Tavella. 2010. Population dynamics and integrated pest management of *Thrips tabaci* on leek under field conditions in northwest Italy. *Entomol. Exp. Appl.* 125:276–287.

Bournier, A., A. Lacasa, and Y. Pivot. 1978. Biologie d'un thrips predateur *Aeolothrips intermedius* (Thys.: Aeolothripidae). *Entomophaga* 23:403–410.

Bournier, A., A. Lacasa, and Y. Pivot. 1979. Regime alimentaire d'un thrips predateur *Aeolothrips intermedius* Bagnall (Thys.: Aeolothripidae). *Entomophaga* 24:353–361.

Briggs, C.J. and E.T. Borer. 2005. Why short-term experiments may not allow long-term predictions about intraguild predation. *Ecol. Appl.* 15:1111–1117.

Butler, G.D. 1965. *Spanogonicus albofasciatus* as an insect and mite predator (Hemiptera: Miridae). *J. Kansas Entomol. Soc.* 38:70–75.

Canard, M., Y. Semeria, and T.R. New (eds.) 1984. *Biology of Chrysopidae*. Series Entomologica 27. The Hague: Dr. W. Junk Publishers.

Chang, C.P. 2000. Investigation on the life history of *Mallada basalis* (Walker) (Neuroptera: Chrysopidae) and the effects of temperatures on its development. *Chin. J. Entomol.* 20:73–87.

Chazeau, J. 1985. Predaceous insects. In: W. Helle and M.W. Sabelis (eds.), *Spider Mites: Their Biology, Natural Enemies, and Control*, Vol. 1B (pp. 211–246). Amsterdam: Elsevier.

Chen, W. 1993. Predation behavior of *Stethorus chengi* Sasaji on patchily distributed preys. *J. Shanghai Agric. Coll.* 11:209–213.

Chen, W. and Z. Zhao. 1994. Study on predation of *Stethorus chengi* Sasaji to *Panonychus citri* McGregor. *J. Southwest Agric. Univ.* 16:27–31.

Cheng, L.L., J.R. Nechols, D.C. Margolies, J.F. Campbell, P.S. Yang, C.C. Chen, and C.T. Lu. 2009. Foraging on and consumption of two species of papaya pest mites, *Tetranychus kanzawai* and *Panonychus citri* (Acari: Tetranychidae) by *Mallada basalis* (Neuroptera: Chrysopidae). *Environ. Entomol.* 38:715–722.

Cheng, L.L., J.R. Nechols, D.C. Margolies, J.F. Campbell, and P.S. Yang. 2010. Assessment of prey preference by the mass-produced generalist predator, *Mallada basalis* (Walker) (Neuroptera: Chrysopidae), when offered two species of spider mites, *Tetranychus kanzawai* Kishida and *Panonychus citri* (McGregor) (Acari: Tetranychidae), on papaya. *Biol. Control* 53:267–272.

Chouinard, G., S. Bellerose, C. Brodeur, and Y. Morin. 2006. Effectiveness of *Hyaliodes vitripennis* (Say) (Heteroptera: Miridae) predation in apple orchards. *Crop Protect.* 25:705–711.

Chow, A., A. Chau, and K.M. Heinz. 2009. Compatibility of *Amblyseius* (*Typhlodromips*) *swirskii* (Athias-Henriot) (Acari: Phytoseiidae) and *Orius insidiosus* (Hemiptera: Anthocoridae) for biological control of *Frankliniella occidentalis* (Thysanoptera: Thripidae) on roses. *Biol. Control* 42:188–196.

Chyzik, R., M. Klein, and Y. Ben-Dov. 1995. Reproduction and survival of the predatory bug *Orius albidipennis* on various arthropod prey. *Entomol. Exp. Appl.* 75:27–31.

Cloutier, C. and S.G. Johnson. 1993. Predation by *Orius tristicolor* (Hemiptera: Anthocoridae) on *Phytoseiulus persimilis* (Acarina: Phytoseiidae): Testing for compatibility between biocontrol agents. *Environ. Entomol.* 22:477–482.

Coll, M. 1998. Living and feeding on plants in predatory Heteroptera. In: M. Coll and J.R. Ruberson (eds.), *Predatory Heteroptera: Their Ecology and Use in Biological Control* (pp. 89–129). Lanham, MD: Entomological Society of America.

Collyer, E. 1951. The separation of *Conwentzia pineticola* End. from *Conwentzia psociformis* (Curt.) and notes on their biology. *Bull. Entomol. Res.* 42:555–564.

Collyer, E. 1952. Biology of some predatory insects and mites associated with the fruit tree red spider mite (*Metatetranychus ulmi* (Koch)) in south-eastern England. I. Biology of *Blepharidopterus angulatus* (Fall.) (Hemiptera-Heteroptera, Miridae). *J. Hort. Sci.* 27:117–129.

Collyer, E. 1953. The biology of some predatory insects and mites associated with the fruit tree red spider mite (*Metatetranychus ulmi* (Koch)) in south-eastern England. II. Some important predators of the mite. *J. Hort. Sci.* 28:85–97.

Congdon, B.D., C.H. Shanks, and A.L. Antonelli. 1993. Population interactions between *Stethorus punctum picipes* (Coleoptera: Coccinellidae) and *Tetranychus urticae* (Acari: Tetranychidae) in red raspberries at low predator and prey densities. *Environ. Entomol.* 22:1302–1307.

Croft, B.A. 1990. *Arthropod Biological Control Agents and Pesticides*. New York: John Wiley & Sons.

Daniel, M. 1976. Biology and predatory habits of the lady bird beetle *Stethorus keralicus* Kapur (Coleoptera: Coccinellidae), predatory on the palm mite. *J. Plantation Crops* 4:7–9.

Desneux, N., A. Decourtye, and J.M. Delpuech. 2007. The sublethal effects of pesticides on beneficial arthropods. *Annu. Rev. Entomol.* 52:81–106.

Ferkovich, S.M. and J.P. Shapiro. 2004a. Comparison of prey-derived and non-insect supplements on egg-laying of *Orius insidiosus* maintained on artificial diet as adults. *Biol. Control* 31:57–64.

Ferkovich, S.M. and J.P. Shapiro. 2004b. Increased egg-laying in *Orius insidiosus* (Hemiptera: Anthocoridae) fed artificial diet supplemented with an embryonic cell line. *Biol. Control* 31:11–15.

Ferkovich, S.M. and J.P. Shapiro. 2005. Enhanced oviposition in the insidious flower bug, *Orius insidiosus* (Hemiptera: Anthocoridae), with a partially purified nutritional factor from prey eggs. *Florida Entomol.* 88:253–257.

Field, R.P. 1979. Integrated pest control in Victorian peach orchards: The role of *Stethorus* spp. (Coleoptera: Coccinellidae). *J. Aust. Entomol. Soc.* 18:315–322.

Fitzgerald, J.D., N. Pepper, and M.G. Solomon. 2007. Interactions among predators and phytophagous mites on apple; possible impact on biocontrol of *Panonychus ulmi* by *Typhlodromus pyri* in orchards. *Biocontrol Sci. Technol.* 17:1009–1019.

Fleschner, C.A. and D.W. Ricker. 1953. Food habits of coniopterygids on citrus in southern California. *J. Econ. Entomol.* 46:458–461.

Frank, J.H. and M.C. Thomas. 1999. *Featured Creatures: Rove Beetles (of the World)*. Gainesville: Department of Entomology and Nematology, University of Florida Institute of Food and Agricultural Sciences (http://entnemdept.ufl.edu/creatures/misc/beetles/rove_beetles.htm).

Frank, J.H., F.D. Bennett, and H.L. Cromroy. 1992. Distribution and prey records for *Oligota minuta* (Coleoptera: Staphylinidae), a predator of mites. *Florida Entomol.* 75:376–380.

Frechette, B., S. Rojo, O. Alomar, and E. Lucas. 2007. Intraguild predation between syrphids and mirids: Who is the prey? Who is the predator? *BioControl* 52:175–191.

Frost, C.J., M.C. Mescher, J.E. Carlson, and C.M. De Moraes. 2008. Plant defense priming against herbivores: Getting ready for a different battle. *Plant Physiol.* 146:818–824.

Gagne, R.J. 1995. Revision of tetranychid (Acarina) mite predators of the genus *Feltiella* (Diptera: Cecidomyiidae). *Ann. Entomol. Soc. Am.* 88:16–30.

Garcia-Mari, F. and J.E. Gonzalez-Zamora. 1999. Biological control of *Tetranychus urticae* (Acari: Tetranychidae) with naturally occurring predators in strawberry plantings in Valencia, Spain. *Exp. Appl. Acarol.* 23:487–495.

Gerson, U. and R.L. Smiley.1990. *Acarine Biocontrol Agents: An Illustrated Key and Manual.* London: Chapman & Hall.

Gillespie, D.R., G. Opit, and B. Roitberg. 2000. Effects of temperature and relative humidity on development, reproduction, and predation in *Feltiella acarisuga* (Vallot) (Diptera: Cecidomyiidae). *Biol. Control* 17:132–138.

Gilstrap, F.E. and E.R. Oatman. 1976. The bionomics of *Scolothrips sexmaculatus* (Perg.) (Thysanoptera: Thripidae), an insect predator of spider mites (Acari: Tetranychidae). *Hilgardia* 44(2):27–59.

Gotoh, T., K. Yamaguchi, M. Fukazawa, and K. Mori. 2004. Effect of temperature on life history traits of the predatory thrips *Scolothrips takahashii* Priesner (Thysanoptera: Thripidae). *Appl. Entomol. Zool.* 39:511–519.

Grafton-Cardwell, E.E. and M.A. Hoy. 1985. Intraspecific variability in response to pesticides in the common green lacewing, *Chrysoperla carnea* (Stephens) (Neuroptera: Chrysopidae). *Hilgardia* 53(6):1–32.

Grafton-Cardwell, E.E. and M.A. Hoy. 1986. Genetic improvement of common green lacewing, *Chrysoperla carnea* (Neuroptera: Chrysopidae): Selection for carbaryl resistance. *Environ. Entomol.* 15:1130–1136.

Guarisco, H. 1991. Predation of two common house spiders upon medically significant pests. *Trans. Kansas Acad. Sci.* 94:79–81.

Hagen, K.S. 1962. Biology and ecology of predaceous Coccinellidae. *Annu. Rev. Entomol.* 7:289–326.

Haney, P.B., R.F. Luck, and S.S. Moreno. 1987. Increases in densities of the citrus red mite *Panonychus citri* (Acarina: Tetranychidae) in association with the argentine ant *Iridomyrmex humilis* (Hymenoptera: Formicidae) in southern California citrus. *Entomophaga* 21:49–57.

Hansen, D.L., H.F. Brodsgaard, and A. Enkegaard. 1999. Life table characteristics of *Macrolophus caliginosus* preying upon *Tetranychus urticae*. *Entomol. Exp. Appl.* 93:267–273.

Haruyama, N., H. Naka, A. Mochizuki, and M. Nomura. 2008. Mitochondrial phylogeny of cryptic species of the lacewing *Chrysoperla nipponensis* (Neuroptera: Chrysopidae) in Japan. *Ann. Entomol. Soc. Am.* 101:971–977.

Hassan, S.A. 1985. Standard methods to test side-effects of pesticides on natural enemies of insects and mites developed by the IOBC/WPRS Working Group "Pesticides and Beneficial Organisms." *Bull. OEPP/EPO* 15:214–255.

Henry, C.S. and M.M. Wells. 2004. Adaptation or random change? The evolutionary response of songs to substrate properties in lacewings (Neuroptera: Chrysopidae: *Chrysoperla*). *Anim. Behav.* 68:879–895.

Henry, C.S., S.J. Brooks, P. Duelli, and J.B. Johnson. 2002. Discovering the true *Chrysoperla carnea* (Insecta: Neuroptera: Chrysopidae) using song analysis, morphology, and ecology. *Ann. Entomol. Soc. Am.* 95:172–191.

Herring, J.L. 1966. The genus *Orius* of the Western Hemisphere (Hemiptera: Anthocoridae). *Ann. Entomol. Soc. Am.* 59:1093–1109.

Ho, C.C. 2000. Spider–mite problems and control in Taiwan. *Exp. Appl. Acarol.* 24:453–462.

Hoddle, M.S. 2003a. Predation behaviours of *Franklinothrips orizabensis* (Thysanoptera: Aelurothripidae) towards *Scirtothrips perseae* and *Heliothrips haemorroidalis* (Thysanoptera: Thripidae). *Biol. Control* 27:323–328.

Hoddle, M.S. 2003b. The effect of prey species and environmental complexity on the functional response of *Franklinothrips orizabensis*: A test of the fractal foraging model. *Ecol. Entomol.* 28:309–318.

Hoddle, M.S., L. Robinson, K. Drescher, and J. Jones. 2000. Developmental and reproductive biology of a predatory *Franklinothrips* n.sp. (Thysanoptera: Aeolothripidae). *Biol. Control* 18:27–38.

Houck, M.A. 1991. Time and resource partitioning in *Stethorus punctum* (Coleoptera: Coccinellidae). *Environ. Entomol.* 20:494–497.

Hoy, M.A. and K.B. Smith. 1982. Evaluation of *Stethorus nigripes* (Col: Coccinellidae) for biological control of spider mites in California almond orchards. *Entomophaga* 27:301–310.

Hull, L.A. and E.H. Beers. 1985. Ecological selectivity: Modifying chemical control practices to preserve natural enemies. In: M.A. Hoy and D.C. Herzog (eds.), *Biological Control in Agricultural IPM Systems* (pp. 103–122). Orlando, FL: Academic Press.

Hull, L.A., D. Asquith, and P.D. Mowery. 1976. Distribution of *Stethorus punctum* in relation to densities of the European red mite. *Environ. Entomol.* 5:337–342.

Hull, L.A., D. Asquith, and P.D. Mowery. 1977a. The functional responses of *Stethorus punctum* in relation to densities of the European red mite. *Environ. Entomol.* 6:85–90.

Hull, L.A., D. Asquith, and P.D. Mowery. 1977b. The mite searching ability of *Stethorus punctum* within an apple orchard. *Environ. Entomol.* 5:684–688.

Hunter, M.D. 2002. A breath of fresh air: Beyond laboratory studies of plant volatile–natural enemy interactions. *Agric. Forest Entomol.* 4:81–86.

Iglinsky, W. and C.F. Rainwater. 1950. *Orius insidiosus*, an enemy of a spider mite on cotton. *J. Econ. Entomol.* 43:567–568.

Isenhour, D.J. and K.V. Yeargan. 1981. Effect of temperature on the development of *Orius insidiosus*, with notes on laboratory rearing. *Ann. Entomol. Soc. Am.* 74:114–116.

James, D.G. 2003. Synthetic herbivore-induced plant volatiles as field attractants for beneficial insects. *Environ. Entomol.* 32:977–992.

Kishimoto, H. 2003. Development and oviposition of predacious insects, *Stethorus japonicus* (Coleoptera: Coccinellidae), *Oligota kashmirica benefica* (Coleoptera: Staphylinidae) and *Scolothrips takahashii* (Thysanoptera: Thripidae) reared on different spider mite species (Acari: Tetranychidae). *Appl. Entomol. Zool.* 38:15–22.

Landis, D.A., S.D. Wratten, and G.M. Gurr. 2000. Habitat management to conserve natural enemies of arthropod pests in agriculture. *Annu. Rev. Entomol.* 45:175–201.

Lariviere, M.C. and C.H. Wearing. 1994. *Orius vicinus* (Ribaut) (Heteroptera: Anthocoridae), a predator of orchard pests new to New Zealand. *N.Z. Entomol.* 17:17–21.

Lattin, J.D. 1999. Bionomics of the Anthocoridae. *Annu. Rev. Entomol.* 44:207–231.

Letourneau, D.K., J.A. Jedlicka, S.G. Bothwell, and C.R. Moreno. 2009. Effects of natural enemy biodiversity on the suppression of arthropod herbivores in terrestrial ecosystems. *Annu. Rev. Ecol. Evol. Syst.* 40:573–592.

Lewis, T. 1973. *Thrips: Their Biology, Ecology, and Economic Importance*. London: Academic Press.

Li, D.X., J. Tian, and Z.R. Shen. 2006. Effects of pesticides on the functional response of predatory thrips, *Scolothrips takahashii* to *Tetranychus viennensis*. *J. Appl. Entomol.* 130:314–322.

Lo, P.K.-C., W.-T. Lee, C.-T. Lu, and J.-Z. Yu. 2002. Artificial diet for mass production of *Mallada basalis* (Walker) for control of agricultural pests in screen houses. In: S. Chen and T.T. Lin (eds.), *International Symposium on Design and Environmental Control of Tropical and Subtropical Greenhouses* (pp. 201–205), ISHS Acta Horticulturae 578. Belgium: International Society for Horticultural Science.

Lucas, E. and O. Alomar. 2001. *Macrolophus caliginosus* (Wagner) as an intraguild prey for the zoophytophagous *Dicyphus tamaninii* Wagner (Heteroptera: Miridae). *Biol. Control* 20:147–152.

Lundgren, J.G. and J.K. Fergen. 2006. The oviposition behavior of the predator *Orius insidiosus*: Acceptability and preference for different plants. *BioControl* 51:217–227.

MacLeod, E.G. and L.A. Stange. 2009. *Brown Lacewings (of Florida) (Insecta: Neuroptera: Hemerobiidae)*, IFAS Publication No. EENY225. Gainesville: University of Florida IFAS Extension (http://edis.ifas.ufl.edu/in382).

Maloney, D., F.A. Drummond, and R. Alford. 2003. *Spider Predation in Agroecosystems: Can Spiders Effectively Control Pest Populations?* Maine Agricultural and Forestry Experiment Station Technical Bulletin 190. Orono: University of Maine.

Mansour, F., E. Bernstein, and F. Abo-Moch. 1995. The potential of spiders of different taxa and a predacious mite to feed on the carmine spider mite, a laboratory study. *Phytoparasitica* 23:217–220.

McMurtry, J.A. and H.G. Johnson. 1966. An ecological study of the spider mite *Oligonychus punicae* (Hirst) and its natural enemies. *Hilgardia* 37:363–402.

McMurtry, J.A., C.B. Huffaker, and M. van de Vrie. 1970. Ecology of tetranychid mites and their natural enemies: A review. I. Tetranychid enemies: Their biological characters and the impact of spray practices. *Hilgardia* 40(11):331–390.

McMurtry, J.A., G.T. Scriven, and R.S. Malone. 1974. Factors affecting oviposition of *Stethorus picipes* (Coleoptera: Coccinellidae), with special reference to photoperiod. *Environ. Entomol.* 3:123–127.

Miller, G.L., J.D. Oswald, and D.R. Miller. 2004. Lacewings and scale insects: A review of predator/prey associations between the Neuropterida and Coccoidea (Insecta: Neuroptera, Raphidioptera, Hemiptera). *Ann. Entomol. Soc. Am.* 97:1103–1125.

Milne, M. and G.H. Walter. 1997. The significance of prey in the diet of the phytophagous thrips, *Frankliniella schultzei*. *Ecol. Entomol.* 22:74–81.

Mizell, R.F. and D.E. Schiffhauer. 1990. Effects of pesticides on pecan aphid predators *Chrysoperla rufilabris* (Neuroptera: Chrysopidae), *Hippodamia convergens, Cycloneda sanguinea* (L.), *Olla v-nigrum* (Coleoptera: Coccinellidae) and *Aphelinus perpallidus* (Hymenoptera: Encyrtidae). *J. Econ. Entomol.* 83:1806–1812.

Mo, T.L. and T.X. Liu. 2006. Biology, life table and predation of *Feltiella acarisuga* (Diptera: Cecidomyiidae) feeding on *Tetranychus cinnabarinus* eggs (Acari: Tetranychidae). *Biol. Control* 39:418–426.

Mo, T.L. and T.X. Liu. 2007. Predation and life table of *Feltiella acarisuga* (Diptera: Cecidomyiidae) preying on eggs of *Tetranychus urticae* (Acari: Tetranychidae). *Environ. Entomol.* 36:369–375.

Moayeri, H.R.S., A. Ashouri, L. Poll, and A. Enkegaard. 2007. Olfactory response of a predatory mirid to herbivore induced plant volatiles: Multiple herbivory vs. single herbivory. *J. Appl. Entomol.* 131:326–332.

Mori, K., M. Nozawa, K. Arai, and T. Gotoh. 2005. Life-history traits of the acariphagous lady beetle *Stethorus japonicus* at three constant temperatures. *BioControl* 50:35–51.

Mound, L.A. 2005. Thysanoptera: Diversity and interactions. *Annu. Rev. Entomol.* 50:247–269.

Mound, L.A. and P. Reynaud. 2005. *Franklinothrips*: A pantropical Thysanoptera genus of ant-mimicking obligate predators (Aeolothripidae). *Zootaxa* 864:1–16.

Mowery, P.D., D. Asquith, and W.M. Bode. 1975. Computer simulation for predicting the number of *Stethorus punctum* needed to control the European red mite in Pennsylvania apple trees. *J. Econ. Entomol.* 68:250–254.

Muma, M.M. 1967. Biological notes on *Coniopteryx vicina* (Neuroptera: Coniopterygidae). *Florida Entomol.* 50:285–293.

Narayanan, E.S. 1942. On the bionomics and life-history of *Coniopteryx pusana* Withycombe (Coniopterygidae: Neuroptera). *Indian J. Entomol.* 4(1):1–4.

Norton, A.P., G. English-Loeb, and E. Belden. 2001. Host plant manipulation of natural enemies: Leaf domatia protect beneficial mites from insect predators. *Oecologia* 126:535–542.

Obrycki, J.J. and T.J. Kring. 1998. Predaceous Coccinellidae in biological control. *Annu. Rev. Entomol.* 43:295–321.

O'Neil, R.J., K.L. Giles, J.J. Obrycki, D.L. Mahr, J.C. Legaspi, and K. Katovich. 1998. Evaluation of the quality of four commercially available natural enemies. *Biol. Control* 11:1–8.

Opit, G.P., B. Roitberg, and D.R. Gillespie. 1997. The functional response and prey preference of *Feltiella acarisuga* (Vallot) (Diptera: Cecidomyiidae) for two of its prey: Male and female twospotted spider mites, *Tetranychus urticae* Koch (Acari: Tetranychidae). *Can. Entomol.* 129:221–227.

Osborne, L.S., J.E. Peña, and D.H. Oi. 1995. Predation by *Tapinoma melanocephalum* (Hymenoptera: Formicidae) on twospotted spider mites (Acari: Tetranychidae) in Florida greenhouses. *Florida Entomol.* 78:565–570.

Osborne, R.S., N.C. Leppla, and L.S. Osborne. 2008. *Featured Creatures: Predatory Gall Midge.* Gainesville: Department of Entomology and Nematology, University of Florida Institute of Food and Agricultural Sciences (http://entnemdept.ifas.ufl.edu/creatures/beneficial/f_acarisuga.htm).

Oswald, J.D. 2004. *Bibliography of the Neuropterida.* College Station: Department of Entomology, Texas A&M University (http://insects.tamu.edu/research/neuropterida/neur_bibliography/bibhome.html).

Parrella, M.P., J.P. McCaffrey, and R.I. Horsburgh. 1980. Compatibility of *Leptothrips mali* with *Stethorus punctum* and *Orius insidiosis*, predators of *Panonychus ulmi*. *Environ. Entomol.* 9:694–696.

Parrella, M.P., D.J. Rowe, and R.L. Horsburgh. 1982. Biology of *Leptothrips mali*, a common predator in Virginia apple orchards. *Ann. Entomol. Soc. Am.* 75:130–135.

Pasqualini, E. and A. Antropoli. 1994. *Stethorus punctillum. Informa. Fitopathol.* 44(5):33–36.

Pasqualini, E. and C. Malavolta. 1985. Possibilita di contenimento naturale di *Panonychus ulmi* (Koch) (Acarina: Tetranychidae) su melo in Emilia-Romagna. *Boll. Ist. Ent. Univ. Bologna* 39:221–230.

Pathan, A.K., A.H. Sayyed, M. Aslam, M. Razaq, G. Jilani, and M.A. Saleem. 2008. Evidence of field-evolved resistance to organophosphates and pyrethroids in *Chrysoperla carnea* (Neuroptera: Chrysopidae). *J. Econ. Entomol.* 101:1676–1684.

Perumalsamy, K., R. Selvasundaram, A. Roobakkumar, V.J. Rahman, A. Babu, and N.N. Muraleedharan. 2009. Life table and predation of *Oligota pygmaea* (Coleoptera: Staphylinidae), a major predator of the red spider mite *Oligonychus coffeae* (Acarina: Tetranychidae) infesting tea. *Biol. Control* 51:96–101.

Pickett, C.H., L.T. Wilson, and D. Gonzalez. 1988. Population dynamics and within-plant distribution of the western flower thrips (Thysanoptera: Thripidae), an early-season predator of spider mites infesting cotton. *Environ. Entomol.* 17:551–559.

Pollock, D.A. and G.J. Michaels. 2002. Distributions of *Stethorus nigripes* Kapur (Coleoptera: Coccinellidae), a predator of Banks grass mite (*Oligonychus pratensis* Banks) in the southern United States. *Southwest. Entomol.* 27:217–220.

Provost, C., D. Coderre, E. Lucas, and N.J. Bostanian. 2003. Impact of lambda-cyhalothrin on intraguild predation among three mite predators. *Environ. Entomol.* 32:256–263.

Putnam, W.L. 1955. Bionomics of *Stethorus punctillum* Weise (Coleoptera: Coccinellidae) in Ontario. *Can. Entomol.* 87:9–22.

Quayle, H.J. 1913. Some natural enemies of spiders and mites. *J. Econ. Entomol.* 6:85–88.

Raros, E.S. and F.H. Haramoto. 1974. Biology of *Stethorus siphonulus* Kapur (Coccinellidae: Coleoptera), a predator of spider mites, in Hawaii. *Proc. Hawaiian Entomol. Soc.* 21(3):457–465.

Reddy, G.V.P. 2001. Comparative effectiveness of an integrated pest management system and other control tactics for managing the spider mite *Tetranychus ludeni* (Acari: Tetranychidae) on eggplant. *Exp. Appl. Acarol.* 25:985–992.

Reddy, G.V.P. 2002. Plant volatiles mediate orientation and plant preference by the predator *Chrysoperla carnea* Stephens (Neuroptera: Chrysopidae). *Biol. Control* 25:49–55.

Reding, M.E., E.H. Beers, J.F. Brunner, and J.E. Dunley. 2001. Influence of timing and prey availability on fruit damage to apple by *Campylomma verbasci* (Hemiptera: Miridae). *J. Econ. Entomol.* 94:33–38.

Riechert, S.E. and T. Lockley. 1984. Spiders as biological control agents. *Annu. Rev. Entomol.* 29:299–320.

Ridgway, R.L. and W.L. Murphy. 1984. Biological control in the field. In: M. Canard, Y. Semeria, and T.R. New (eds.), *Biology of Chrysopidae* (pp. 220–228). The Hague: Dr. W. Junk Publishers.

Rosenheim, J.A. 1998. Higher-order predators and the regulation of insect herbivore populations. *Annu. Rev. Entomol.* 43:421–447.

Rosenheim, J.A. 2005. Intraguild predation of *Orius tristicolor* by *Geocoris* spp. and the paradox of irruptive spider mite dynamics in California cotton. *Biol. Control* 32:172–179.

Rott, A.S. and D.J. Ponsonby. 2000a. Improving the control of *Tetranychus urticae* on edible glasshouse crops using a specialist coccinellid (*Stethorus punctillum* Weise) and a generalist mite (*Amblyseius californicus* McGregor) as biocontrol agents. *Biocontrol Sci. Technol.* 10:487–498.

Rott, A.S. and D.J. Ponsonby. 2000b. The effects of temperature, relative humidity and host plant on the behavior of *Stethorus punctillum* Weise (Coleoptera: Coccinellidae) as a predator of the twospotted spider mite, *Tetranychus urticae* Koch (Acari: Tetranychidae). *BioControl* 45:155–164.

Roy, M., J. Brodeur, and C. Cloutier. 1999. Seasonal abundance of spider mites and their predators on red raspberry in Quebec, Canada. *Environ. Entomol.* 28:735–747.

Roy, M., J. Brodeur, and C. Cloutier. 2002. Relationship between temperature and developmental rate of *Stethorus punctillum* (Coleoptera: Coccinellidae) and its prey *Tetranychus mcdanieli* (Acarina: Tetranychidae). *Environ. Entomol.* 31:177–187.

Ruberson, J.R., Y.J. Shen, and T.J. Kring. 2000. Photoperiodic sensitivity and diapause in the predator *Orius insidiosus* (Heteroptera: Anthocoridae). *Ann. Entomol. Soc. Am.* 93:1123–1130.

Ryerson, S.A. and J.D. Stone. 1979. A selected bibliography of two species of *Orius*: The minute pirate bug, *Orius tristicolor*, and *Orius insidiosus* (Heteroptera: Anthocoridae). *Bull. Entomol. Soc. Am.* 25(2):131–135.

Schuster, D.J. and P.A. Stansly. 2000. Response of two lacewing species to biorational and broad-spectrum insecticides. *Phytoparasitica* 28:297–304.

Scriven, G.T. and C.A. Fleschner. 1960. Insectary production of *Stethorus* species. *J. Econ. Entomol.* 53:982–985.

Shimoda, T., N. Shinkaji, and H. Amano. 1997. Prey stage preference and feeding behaviour of *Oligota kashmirica benefica* (Coleoptera: Staphylinidae), an insect predator of the spider mite *Tetranychus urticae* (Acari: Tetranychidae). *Exp. Appl. Acarol.* 21:665–675.

Silvers, C.S., J.G. Morse, and E.E. Grafton-Cardwell. 2002. Quality assessment of *Chrysoperla rufilabris* (Neuroptera: Chrysopidae) producers in California. *Florida Entomol.* 85:594–598.

Slater, J.A. and R.M. Baranowski. 1978. *How to Know the True Bugs (Hemiptera: Heteroptera)*. Dubuque, IA: W.C. Brown.

Symondson, W.O.C., K.D. Sunderland, and M.H. Greenstone. 2002. Can generalist predators be effective biocontrol agents? *Annu. Rev. Entomol.* 47:561–594.

Tanigoshi, L.K. and J.A. McMurtry. 1977. The dynamics of predation of *Stethorus picipes* (Coleoptera: Coccinellidae) and *Typhlodromus floridanus* on the prey *Oligonychus punicae* (Acarina: Phytoseiidae, Tetranychidae). Part I. Comparative life history and life table studies. Part II. Effects of initial prey–predator ratios and prey distribution. *Hilgardia* 45(8):237–288.

Tauber, M.J., C.A. Tauber, K.M. Daane, and K.S. Hagen. 2000. Commercialization of predators, recent lessons from green lacewings (Neuroptera: *Chrysoperla*). *Am. Entomol.* 46:26–38.

Tommasini, M.G., J.C. van Lenteren, and G. Burgio. 2004. Biological traits and predation capacity of four *Orius* species on two prey species. *Bull. Insectol.* 57:79–93.

Trichilo, P.J. and T.F. Leigh. 1986. Predation on spider mite eggs by the western flower thrips, *Frankliniella occidentalis* (Thysanoptera: Thripidae), an opportunist in a cotton agroecosystem. *Environ. Entomol.* 15:821–825.

Turnbull, A.L. 1973. Ecology of the true spiders (Araneomorphae). *Annu. Rev. Entomol.* 18:305–348.

Venzon, M., A. Janssen, and M.W. Sabelis. 2002. Prey preference and reproductive success of the generalist predator *Orius laevigatus*. *Oikos* 97:116–124.

Vogt, H. 1994. Effects of pesticides on *Chrysoperla carnea* Stephens (Neuroptera: Chrysopidae) in the field and comparison with laboratory and semi-field results. *OILB/WPRS Bull.* 17(10):71–82.

Villanueva, R.T., R. Gagne, and C.C. Childers. 2006. Two species of Cecidomyiidae predacious on citrus rust mite, *Phyllocoptruta oleivora*, on Florida citrus. *Florida Entomol.* 89:161–167.

Walters, P.J. 1974. A method for culturing *Stethorus* spp. (Coleoptera: Coccinellidae) on *Tetranychus urticae* (Koch) (Acarina: Tetranychidae). *J. Aust. Entomol. Soc.* 14:245–246.

Watson, D.M., T.Y. Du, M. Li, J.J. Xiong, D.G. Liu, M.D. Huang, D.J. Rae, and G.A.C. Beattie. 1998. Life history and feeding biology of the predatory thrips, *Aleurodothrips fasciapennis* (Thysanoptera: Phlaeothripidae). *Bull. Entomol. Res.* 88:351–357.

Wearing, C.H. and K.J. Colhoun. 1999. Development of *Orius vicinus* (Ribaut) (Heteroptera: Anthocoridae) on different prey. *Biocontrol Sci. Technol.* 9:327–334.

Wells, M.N. and C.S. Henry. 1998. Songs, reproductive isolation and speciation in cryptic species of insects: A case study using green lacewings. In: D. Howard and S. Berlocher (eds.), *Endless Forms: Species and Speciation* (pp. 217–233). New York: Oxford University Press.

Wheeler, A.G. 2001. *Biology of the Plant Bugs (Hemiptera: Miridae): Pests, Predators, Opportunists.* Ithaca, NY: Cornell University Press.

Wheeler, A.G., R.B. Colburn, and R.D. Lehman. 1973. *Stethorus punctillum* associated with spruce spider mite on ornamentals. *Environ. Entomol.* 4:718–720.

Wilson, L.J., L.R. Bauer, and G.H. Walter. 1996. "Phytophagous" thrips are facultative predators of twospotted spider mites (Acari: Tetranychidae) on cotton in Australia. *Bull. Entomol. Res.* 86:297–305.

Winterton, S. and S. de Freitas. 2006. Molecular phylogeny of the green lacewings (Neuroptera: Chrysopidae). *Aust. J. Entomol.* 45:235–243.

Yeargan, K.V. 1998. Predatory Heteroptera in North American agroecosystems: An overview. In: M. Coll and J.R. Ruberson (eds.), *Predatory Heteroptera: Their Ecology and Use in Biological Control* (pp. 7–19). Lanham MD: Entomological Society of America.

CHAPTER 14

Pathogens and Symbionts of Mites and Ticks

14.1 MICROBIAL SYMBIONTS AND PATHOGENS

All arthropods have microorganisms associated with them (van der Geest 1985, Tanada and Kaya 1993, Boucias and Pendland 1998, Poinar and Poinar 1998, van der Geest et al. 2000). Some microorganisms are *external contaminants* of the arthropod; therefore, the types present will vary with the local environment (Lacey and Kaya 2007). Other microorganisms have a more intimate association with their acarine host. The importance of disease transmission by mites and ticks to plants or vertebrates has long been known, but the recognition that microbes have other effects on the biology of mites and ticks is becoming more clear now that molecular tools are available with which to identify the microorganisms (Hoy and Jeyaprakash 2009).

Viruses, fungi, bacteria, and microsporidia may affect mites and ticks negatively and are considered to be **pathogens**. In addition, mites and ticks are hosts to a variety of microorganisms that may have a positive effect on them (in which case, it may be considered a mutualistic symbiont). Alternatively, some mites and ticks transmit microbial disease agents to their vertebrate or plant hosts and are vectors of human, plant, or animal diseases. Transmission of plant pathogens by plant-feeding mites is discussed in Chapters 6, 8, and 9. Transmission of pathogens by mites and ticks to their vertebrate hosts is discussed in Chapters 22 and 23.

Broadly speaking, a **symbiont** is any organism living in or on another organism, without regard to whether it is detrimental (pathogenic), beneficial (mutualistic), or neutral; however, this definition of symbiont is not universally accepted, and some restrict the meaning to mutualism. Often microbial symbionts are assumed to provide a beneficial service to their arthropod host, such as nutrition or improved fitness traits (Douglas 1994, O'Neill et al. 1997, Bourtzis and Miller 2003, 2006, 2009). Microbial symbionts can be extracellular or intracellular (microbial **endosymbiont**).

Microbial symbionts of mites and ticks are found primarily in the reproductive tract and in the gut. For example, bacterial *endosymbionts* (living within cells) such as ***Wolbachia*** and ***Cardinium*** are found in the reproductive tracts of many species of Acari, where they may reduce or enhance fitness, alter sex ratios, or cause reproductive incompatibilities. If populations containing *Wolbachia* are crossed to populations lacking *Wolbachia* or to populations with a different type of *Wolbachia*, partial to complete reproductive incompatibility may occur (Johanowicz and Hoy 1998, Weeks and Stouthamer 2003, Enigl and Schausberger 2007, Kitajima et al. 2007, Hoy and Jeyaprakash 2009). The microbial endosymbiont *Cardinium* appears to cause the tenuipalpid mite *Brevipalpus phoenicis* to consist of only haploid females (Weeks et al. 2001). Three other species of *Brevipalpus* also are primarily female and infected with *Cardinium*. It is believed that *B. phoenicis* historically was arrhenotokous, but infection with *Cardinium* caused the production of haploid females only. Curing the mite of *Cardinium* with antibiotics restores the production of haploid males.

Several species of putative gut symbionts have been identified, and gut symbionts may live either in the lumen of the gut or in specialized organs or structures. Putative gut bacteria were identified in the phytoseiid *Metaseiulus occidentalis* and the two-spotted spider mite (*Tetranychus urticae*), but information on the roles these gut bacteria play in the biology of their hosts remains limited (Hoy and Jeyaprakash 2009). Although *Wolbachia*, *Cardinium*, and the gut bacteria are considered symbionts, it is possible that these microbial symbionts could become pathogens if the relationship between the mite host and the microorganism becomes disrupted due to stresses placed on the mites. Symbionts do not typically increase their densities beyond a certain level, which appears to be a result of signaling between the microorganism and the host so the ideal number of symbionts is maintained. Generally, we know very little about the roles of symbionts in pest or beneficial mite species because, until molecular tools became available, it was difficult to study them.

Microbial pathogens of mites and ticks have been studied as potential control agents for plant-feeding mites (tetranychids, eriophyoids, tarsonemids) and ticks. Pathogens have been found to reduce the fitness of phytoseiid predators and have been studied as controls for the *Varroa* mite parasitic on honey bees (Davidson et al. 2002, Kanga et al. 2002) and for the control of grain mites (Mwangi et al. 1991, Bruin and van der Geest 2009). The focus of this chapter is to describe microbial pathogens as possible tools for integrated mite management (IMM) and the requirements for their development. The most common pathogens of mites and ticks are fungi, microsporidia (which are now known to be fungi), and viruses, although some bacteria also are pathogens.

When a pathogen is discovered to be important as a natural enemy of plant-feeding mites, it can be used in classical, conservation, or augmentative biological control programs. If augmentative releases are considered, then large quantities of high-quality pathogens must be commercially available. Commercializing a pathogen, although less time consuming and expensive than developing a chemical pesticide, requires multiple steps (Table 14.1).

Unfortunately, pathogens also may infect important natural enemies of mites used in augmentative biological control programs, such as species of the Phytoseiidae. Pathogens of phytoseiids include bacteria, viruses, microsporidia, and fungi (Bjornson et al. 2000, Becnel et al. 2002, Olsen and Hoy 2002, Bjornson 2009, Hoy and Jeyaprakash 2009, Schutte and Dicke 2009). Only a few bacteria are known to be pathogens of mites, but Gols et al. (2007) found the pathogenic bacterium *Acaricomes phytoseiuli* in mass-reared cultures of *Phytoseiulus persimilis*. Mass-reared phytoseiids are especially likely to become infected with one or more microorganism in mass-rearing facilities, where crowding is common and contamination of rearing units can occur.

14.2 VIRUSES OF MITES AND TICKS

Viruses have been studied as potential control agents of insect pests, especially the Lepidoptera (Hunter-Fujita et al. 1998, Miller and Ball 1998); however, much less is known about viruses that infect mites and ticks. Naturally occurring viruses have been found in some pest mites and ticks that are highly pathogenic to their host under certain environmental conditions (Samish and Rehacek 1999). No viruses are commercially produced and used as pesticides of mites and ticks at present. Producing such a commercial product would require rearing the virus in cell culture or in living mites. Typically, viruses are ingested and their speed of infection is relatively slow. Viruses also are sensitive to ultraviolet (UV) light, so their development as a pesticide would require a specialized formulation to preserve the activity of the viral pesticide in the environment.

Citrus red mite populations in California citrus commonly are infected with a non-included virus (McMurtry 1985, McCoy et al. 2009). The citrus red mites excrete the virus and, when dead, are often found stuck to the leaf surface by a smear of black material excreted from the anus (see Figure S14.1 on the CD). It appears that healthy mites encounter the virus when feeding in an area previously infected with diseased mites (Reed et al. 1975). The disease caused by this virus can destroy

Table 14.1 Steps Involved in Commercializing a Microbial Pesticide

1. *Isolation of a suitable pathogen*—Screen a large number of isolates to determine which show activity against the target pest.
2. *Laboratory and greenhouse tests*—Both are required to evaluate efficacy and to determine what environmental issues (e.g., temperature, relative humidity, daylength) could affect efficacy.
3. *Production*—Fermentation is the growth or production of a microorganism in a large-scale system. Maintaining virulence during large-scale production is essential. If cell cultures are required for endoparasites, production will be more expensive.
4. *Formulation*—This is a critical issue that has implications for storage, handling, and effectiveness. The goal is to protect the microorganism for as long as possible under challenging outdoor conditions (temperature, relative humidity, ultraviolet light).
5. *Efficacy trials*—Large-scale trials examine the effectiveness of commercial formulations.
6. *Registration trials*—These trials begin only after a final formulation is developed.
7. *Toxicity testing*—This testing is not as demanding as that for synthetic organic pesticides. Toxicity test costs are lower and may take about 6 months to meet Environmental Protection Agency standards in the United States.
8. *Registration*—Regulation is based on the product's biological and ecological traits and intended areas of application.
9. *Commercial-scale production and marketing*—Microbial pest control products are expected to increase in popularity, but they are not expected to replace synthetic organic pesticides.

Source: Based on information from Falcon (1985), Lacey (2001), and Montesinos (2003).

laboratory colonies and can reduce high field populations in California and Arizona. Unfortunately, high mortality may occur in the field only after significant plant injury has been caused. In experimental plots, virus applied in an aqueous suspension resulted in satisfactory control, but the pH of the solution was important. Good results were obtained by brushing diseased mites onto trees. To do so, large numbers of infected mites were collected in an orchard with a vacuum machine, and the mites were then held on green lemons for about a week to allow the titer of the virus to increase. The virus is rapidly inactivated by sunlight, but the activity could be extended by proper formulation methods, especially by adding a UV protectant. That may explain the greater rate of success achieved by releasing living infected mites as compared to spraying virus in a water solution.

Temperature influences the effectiveness of the virus; citrus red mite control is much more difficult to achieve in spring and fall when the infected mites live longer at low temperatures. Very hot weather also negatively influences the virus. The host range of the virus appears limited. The phytoseiid species tested were not susceptible, and of seven tetranychid species tested, only *Tetranychus cinnabarinus* was infected in the laboratory, and the mortality rate was low (Shaw et al. 1967).

A spherical noninclusion virus was found infecting the European red spider mite *Panonychus ulmi* in Ontario, Canada (Putman and Herne 1966). Many of the symptoms are similar to those in the citrus red mite. The infected mites typically are a darker red color as immatures. The mites contain crystalline and spheroidal viral protein inclusions, and their legs are rigidly extended from the body. Infected European red spider mites deposit virus on the leaves, perhaps in their excrement or in oral secretions at feeding sites. The virus on the leaves is UV sensitive and seldom remains infective for more than a few days. It is inactivated after exposure to water. Introduction of the virus into orchard populations of European red mite rapidly induced **epizootics** that reduced populations. Natural epizootics occur primarily in dense populations.

An unidentified virus was detected in greenhouse populations of *Tetranychus urticae* after releasing a field-collected population of *T. urticae* into a greenhouse in Berkeley, California (M.A. Hoy, unpubl.). The virus was highly lethal under laboratory and greenhouse conditions, and the greenhouse and its contents had to be treated with a sodium hypochlorite solution to sanitize the greenhouse before a new colony could be started from another source. The disease might be of interest to commercial producers, but, unfortunately, the frozen specimens were lost. The point is that other lethal viruses of spider mites might be found if surveys were conducted.

Two viruses were observed in the predatory mite *Metaseiulus occidentalis* (Phytoseiidae) by transmission electron microscopy, but it is not clear whether these viruses were pathogenic (Poinar and Poinar 1998). A few observations have been made that suggest that ticks can be infected with viruses, but little information is available about their importance and impact on the ticks (Samish and Rehajeck 1999). Information about the role of viruses in the biology of mites and ticks is very limited, and additional investigation is needed.

14.3 FUNGAL PATHOGENS

Chandler et al. (2000) reviewed what is known about fungal pathogens of mites and ticks, noting that 58 species of fungi have been found infecting at least 73 species of mites in the Acaridida, Oribatida, Actinedida, Gamasida, or Ixodida in either the field or the laboratory. Most fungal pathogens are in the Entomophthorales (Entomophthoromycotina, formerly Zygomycota), Ascomycota, and the imperfect (or Deuteromycota) fungi, which are all transmitted horizontally. A number of species, such as *Hirsutella thompsonii* (Deuteromycota or Mitosporic fungi) (Figure 14.1) and *Neozygites floridana* (Entomophthorales), are specific to mites, while other fungal species may kill both mites and insects (Gerson et al. 1979, Chandler et al. 2000). Many different isolates of these fungi can be detected using molecular tools, and the isolates may have different levels of specificity or virulence.

Acaropathogenic fungi infect their hosts through specialized spores that attach to, then germinate on and penetrate, the integument (Figure 14.2). Once inside the host, the fungus grows, resulting in death within 3 to 10 days. Under favorable conditions (high relative humidity or the presence of free water), spores are produced on the outside of the cadaver (Figure 14.1). Under less-favorable environmental conditions, spores can be produced inside the mite that allow the fungus to survive until favorable conditions return (Figure 14.3).

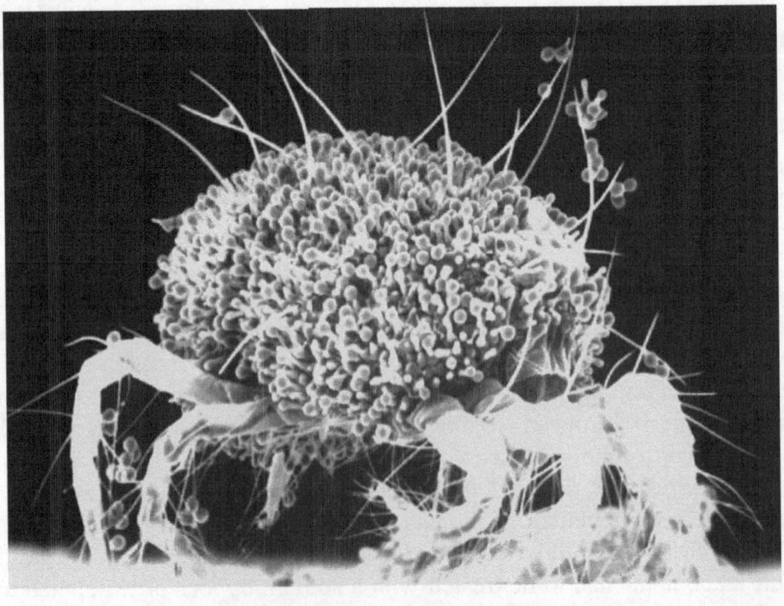

Figure 14.1 The two-spotted spider mite (*Tetranychus urticae*) with externally borne primary conidiospores of *Hirsutella thompsonii*. (Photograph by Drion Boucias, Department of Entomology and Nematology, University of Florida, Gainesville.)

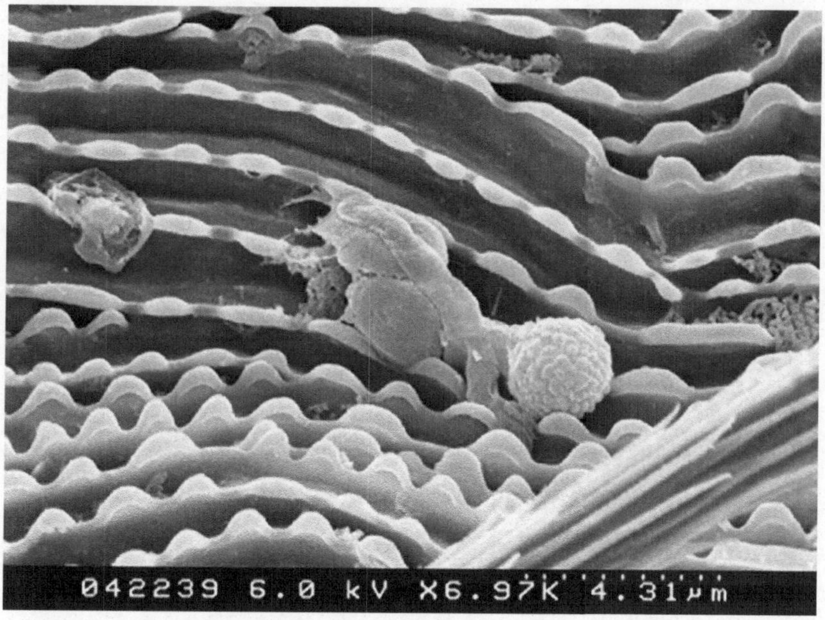

Figure 14.2 Close-up view of a germ tube emerging from a conidiospores of *Hirsutella thompsonii* on the exoskeleton of the two-spotted spider mite (*Tetranychus urticae*). Infection by the pathogen involves penetration of the exoskeleton by the germ tube. (Photograph by Drion Boucias, Department of Entomology and Nematology, University of Florida, Gainesville.)

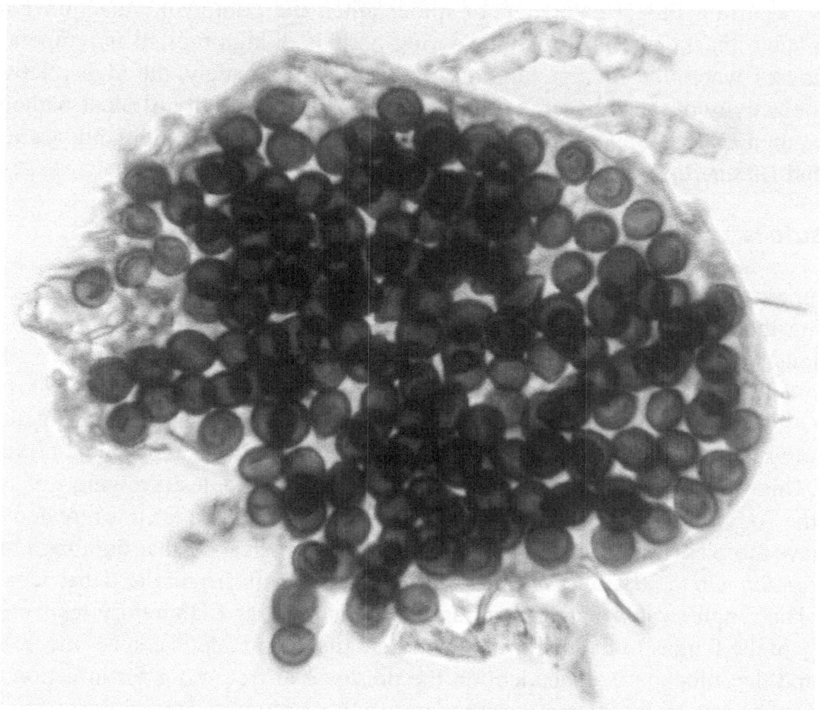

Figure 14.3 Internal resting spores of *Hirsutella thompsonii* (Entomopththorales: Entomophthoraceae) in the two-spotted spider mite (*Tetranychus urticae*) viewed under the light microscope. (Photograph by Drion Boucias, Department of Entomology and Nematology, University of Florida, Gainesville.)

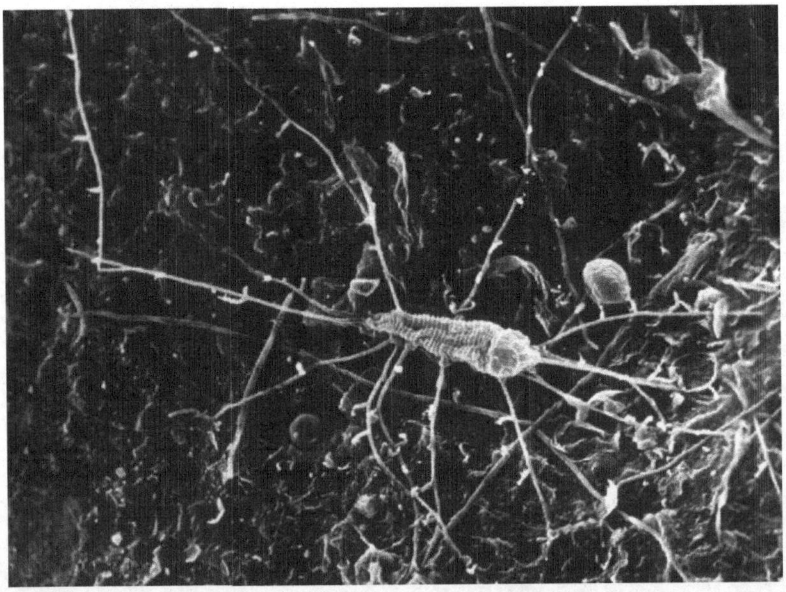

Figure 14.4 *Hirsutella thompsonii* has killed this citrus rust mite (*Phyllocoptruta oleivora*) (Eriophyoidea). During the 1970s, Mycar®, a commercial product containing an isolate of *H. thompsonii*, was available for use in Florida citrus groves but the use of fungicides to control citrus pathogens made the product ineffective. (Photograph by Drion Boucias, Department of Entomology and Nematology, University of Florida, Gainesville.)

Naturally occurring fungal pathogens of spider mites and eriophyoid mites have been found, especially in subtropical or tropical regions, during periods of high rainfall in temperate climates, or in greenhouses where the relative humidity is high. Unfortunately, the high relative humidity that favors the activity of these beneficial fungi also favors the activity of plant-pathogenic fungi. The most common acaropathogenic fungi infecting tetranychids and erioiphyoids are in the genera *Neozygites* and *Hirsutella* (Chandler et al. 2000).

14.3.1 *Hirsutella*

One commercial product (Mycar®) containing an isolate of *Hirsutella thompsonii* was produced for controlling the citrus rust mite (*Phyllocoptruta oleivora*) in Florida during the 1970s; under ideal conditions, the product was as effective as chemical acaricides (Chandler et al. 2000) (Figure 14.4). *Hirsutella thompsonii* is a fungal pathogen of eriophyoid and tetranychid mites and occurs naturally in Florida citrus groves. It can cause decreases in citrus rust mite population densities when the rainy season begins at the end of June or early July in this subtropical climate (McCoy 1985, 1996). One way to detect infected mites is to find hyphal bodies (growing fungal fragments that float in the body cavity) in dark-colored, sluggish rust mites; however, trees treated with copper fungicides have more citrus rust mites than untreated trees, indicating that the fungicide interferes with the *Hirsutella*. Up to 70% of rust mites can be infested with *Hirsutella* under ideal conditions in the field. The fungus was isolated on an artificial medium, and laboratory tests confirmed the pathogenicity of the fungus to the rust mite. Both immatures and adults can be infected under field conditions, and the infection is dependent on the presence of free water for infection to occur. In Florida, epizootics caused by interactions among weather, mites, and fungi occur regularly in the summer. Diseased mites can be found on fruit and foliage throughout the year. Because fungicides applied to control plant pathogens of citrus also kill *H. thompsonii*, commercial use of this product did not persist in Florida.

Strains of *Hirsutella thompsonii* have been found to infect many mite species, including *Tetranychus cinnabarinus*, *T. urticae*, *Eutetranychus orientalis*, *E. banksi*, *Eotetranychus sexmaculatus*, and *Varroa jacobsoni* (Figure 14.1) (also see Figure S14.2 on the CD). Under laboratory conditions, the fungus has not infected a tick or species of Astigmata, Cryptostigmata, or Mesostigmata.

Hirsutella thompsonii has been used to control *Tetranychus cinnabarinus* under glasshouse conditions but concerns about unintended negative effects limit its use. Daytime temperatures in glasshouses may rise above 30°C in the summer, and relative humidities are too low to allow the fungus to persist. At night, relative humidity does increase, but growers are unlikely to allow it to increase to 100%, which would allow the fungus to infect mites, because fungal plant pathogens also can germinate under these conditions. This fungus could only be used in glasshouses with special relative humidity regimes and with crops that are tolerant of plant diseases. Samson and McCoy (1982) found a new species of *Hirsutella* (*H. tydeicola*) causing epizootics in populations of *Tydeus gloveri* (Tydeidae) in citrus groves in Florida. *Tydeus gloveri* was described as being a scavenger on citrus.

14.3.2 *Neozygites*

A fungus called *Neozygites* (or *Triplosporium* or *Entomophthora*) also shows promise as a mite control product. It has been found around the world, and at least six species appear to be specific to mites (Chandler et al. 2000). It has been identified from a variety of spider mites, and different isolates, or biotypes, or species of *Neozygites* have different levels of pathogenicity, thus creating taxonomic confusion. The taxonomy of this group remains unresolved as to the number of species and their host ranges.

Neozygites floridana appears to be specific to tetranychid mites and can cause epizootics in spider mite populations in corn, peanuts, cotton, soybeans, and lima beans (Brandenburg and Kennedy 1982, Klubertanz et al. 1991). Epizootics can occur when light, humidity, and temperature conditions are conducive, but a full understanding of the biology of these different isolates remains unknown and is key to their deployment in IMM programs. The pathogen can overwinter as resting spores in dead mites in close proximity to overwintering mites. About 3% of the mites that dispersed aerially from a North Carolina cornfield on wind currents were infected with the pathogen and transported it to new sites (Smitley et al. 1986).

Neozygites tanajoae, a pathogen of the cassava green mite (CGM) in Brazil, was considered for use in Africa as a biological control agent of the CGM. It appears to be specific to the CGM and was called *Neozygites floridana* in prior studies; however, it appears to have several differences from that species. These differences include 18S ribosomal DNA sequences, host ranges, nutritional requirements for growth *in vitro*, tolerances to cold, and ability to withstand cryopreservation techniques (Delalibera et al. 2004). Hountondji et al. (2005) found that *N. tanajoae* sporulated in response to volatiles produced by CGM-damaged leaves but not from undamaged leaves, indicating that these chemical cues modulate the sporulation of this fungus.

Neozygites adjarica was found infecting the Banks grass mite (*Oligonychus pratensis*) and *Tetranychus urticae* in field corn in Kansas (Dick and Buschman 1995). Epizootics were associated with high mite populations and extended periods of high relative humidity (>80%). All *N. adjarica* infections in Kansas occurred late in the field season (middle to late August), which was too late for it to serve as a dependable tool in an IMM program.

Fungi have been evaluated as potential control agents for the honey bee mite (*Varroa jacobsoni*) (see Figure S14.2 on the CD). Additional control efforts directed against this pest are discussed in Chapter 20. Fungi in the genera *Metarhizium* and *Beauveria*, as well as *Aspergillus*, *Fusarium*, *Paecilomyces*, and *Verticillium*, are associated with ticks (Samish and Rehacek 1999). It is thought that naturally occurring fungal populations can reduce tick populations by as much as 50% under humid environmental conditions. *Metarhizium* and *Beauveria* are commercially produced because they are used to suppress insects, but these fungi can also kill ticks. Relatively few field experiments have been conducted using these products for tick management (Samish and Rehacek 1999).

Utilizing endemic fungal pathogens in IMM programs clearly will require detailed understanding of the fungal–acarine interactions and the environmental conditions that favor mite population suppression. In addition, the use of chemical pesticides may have to be modified to incorporate fungal pathogens into IMM programs. As already noted, copper fungicides can disrupt the role of *Hirsutella thompsonii* in Florida citrus groves, and Morjan et al. (2002) indicated that glyphosate products used for weed control also have fungicidal effects on the acaropathogenic fungus *Neozygites floridana*.

14.4 MICROSPORIDIAL PATHOGENS

Microsporidia, once considered to be protozoans, are now known to be fungi and are spore-forming, intracellular pathogens that can be transmitted both vertically and horizontally. Microsporidia can parasitize both vertebrates and invertebrates and can cause diseases in immune-compromised humans (Wittner and Weiss 1999). Many are somewhat host specific, so they could be potentially useful in IMM programs; however, because they are intracellular parasites and cannot be mass produced, their main role may be as naturally occurring infections.

Microsporidia typically are found in acarine (or insect) colonies that have been stressed, which can result in natural enemies that are less effective in augmentative releases; for example, microsporidia have been found in mass-rearing facilities where phytoseiids such as *Phytoseiulus persimilis*, *Neoseiulus cucumeris*, and *Neoseiulus barkeri* are reared (Bjornson and Keddie 1999, Bjornson 2009). Olsen and Hoy (2002) found a new microsporidium, *Oligosporidium occidentalis* (Becnel et al. 2002), in the phytoseiid *Metaseiulus occidentalis* and demonstrated that it reduced fecundity and longevity and altered sex ratios.

Microsporidia may be underreported as parasites of mites. Larsson et al. (1997) described a microsporidian parasite of the acarid *Tyrophagus putrescentiae*, and eight new microsporidian parasites of oribatid mites from forest soils have recently been described (Purrini and Weiser 2009). Some ticks have been reported to be infected with microsporidia (Rehacek and Weiser 1978; Tokarev and Movile 2004).

14.5 COMMERCIALIZATION OF MICROBIAL PESTICIDES

The development of pesticide resistance in mites and ticks, the increasing costs of developing new agricultural chemicals and registering them, and concerns over the environmental hazards of synthetic organic pesticides have led to a search for other management options (Lacey et al. 2001). Only a few microbial pesticide products have shown promise of being useful in IMM programs in agriculture, although pathogens probably are important agents of natural mortality in some cases but are not recognized as such. Ticks are serious pests of livestock (and humans) due to debilitation of their hosts by their feeding, the ability of ticks to transmit a variety of disease-causing pathogens, and their ability to develop resistance to pesticides (see Chapter 22). There is a growing recognition that fungal pathogens may be agents of natural mortality of tick populations, and some research has been conducted to determine whether fungal pathogens can be used to suppress tick populations, in part because ticks have developed resistances to pesticides (Bruin and van der Geest 2009).

Pathogens of mites (or ticks) could be used in IMM programs in at least two ways: augmentation and conservation. We can rely on naturally occurring pathogens to reduce mite and tick populations and possibly enhance their efficacy by modifying environmental conditions or eliminating the

Table 14.2 Potential Barriers to Commercialization of Microbial Products

1. Products currently are not highly reliable under field conditions, often due to limitations in formulation and lack of information about the basic biology of the microorganism.
2. The activity of microorganisms may be *too specific*, which is a disadvantage when the cropping system has multiple pests. This specificity is an advantage for reducing nontarget effects; often the pathogens do not affect other natural enemies.
3. Because most microbial products have a short residual activity in the field, multiple applications are necessary, thus increasing labor and product costs.
4. Microbial pesticides are expensive because it is difficult to develop large-scale production methods.
5. Microbial products may act slowly, making them appear ineffective or allowing excessive damage to occur in the crop. Growers often want to see lots of dead pests immediately after application.

Source: Based on information from Falcon (1985), Lacey (2001), and Montesinos (2003).

application of pesticides that are toxic to the acaropathogen (conservation). Or, we can mass produce effective acaropathogens, formulate them, and apply them as commercially available microbial pesticides (augmentation).

Microbial pesticides usually are less toxic than conventional pesticides to humans and nontarget organisms, are often effective in relatively small quantities, and can decompose quickly, thereby resulting in lower exposures and other pollution problems (Lacey et al. 2001). Usually fewer data are required to register a microbial pesticide than a synthetic organic product. Microbial pesticides may be registered in less than a year in the United States (compared to more than 3 years for conventional chemical pesticides) and at substantially lower costs; however, the U.S. Environmental Protection Agency does review microbial products to ensure that they have no negative effects on human health or the environment. Data required for registration of new microbial products include their composition, toxicity, and degradation rate and products.

Table 14.1 described the steps involved in developing and registering a microbial pesticide. The registration of pathogens formulated for use against agricultural mite pests will require substantial research effort if acaropathogens are to be registered at a rate comparable to the registration of chemical control agents (acaricides).

Negative aspects of microbial acaricides can be their relative specificity. The specificity requires accurate identification of pests and could limit the size of the market and increase their cost relative to chemical pesticides, which are often toxic to diverse arthropod species (Table 14.2). A grower might have to treat with several products to control multiple arthropod pests, unless multiple pathogens are formulated together. Another concern is that environmental conditions have to be conducive to infection. Typically, acaropathogenic organisms require high relative humidity to achieve infection or transmission. In addition, fungal, viral, and bacterial pathogens usually require careful formulation so they are not inactivated by UV light. For many growers, microbial pesticides take too long to achieve control, which can have a negative psychological effect on growers hoping for a rapid kill of the pests. As a result, there are several potential barriers to the commercialization of microbial pest control products (see Table 14.2).

SELECTED REFERENCES

Becnel, J.J., A. Jeyaprakash, M.A. Hoy, and A. Shapiro. 2002. Morphological and molecular characterization of a new microsporidian species from the predatory mite *Metaseiulus occidentalis* (Nesbitt) (Acari, Phytoseiidae). *J. Invertebr. Pathol.* 79:163–172.

Bjornson, S. 2009. Natural enemies of mass-reared predatory mites (family Phytoseiidae) used for biological pest control. In: S. Bruin and L.P.S. van der Geest (eds.), *Diseases of Mites and Ticks* (pp. 299–306). Heidelberg: Springer.

Bjornson, S. and B.A. Keddie. 1999. Effects of *Microsporidium phytoseiuli* (Microsporidia) on the performance of the predatory mite *Phytoseiulus persimilis* (Acari: Phytoseiidae). *Biol. Control* 15:153–161.

Bjornson, S., D.A. Raworth, and C. Bedard. 2000. Abdominal discoloration and the predatory mite *Phytoseiulus persimilis* Athias-Henriot: Prevalence of symptoms and their correlation with short-term performance. *Biol. Control* 19:17–27.

Boucias, D.G. and J.C. Pendland. 1998. *Principles of Insect Pathology*. Dordrecht: Kluwer Academic.

Bourtzis, K. and T.A. Miller (eds.) 2003. *Insect Symbiosis*, Vol. 1. Boca Raton, FL: CRC Press.

Bourtzis, K. and T.A. Miller (eds.) 2006. *Insect Symbiosis*, Vol. 2. Boca Raton, FL: CRC Press.

Bourtzis, K. and T.A. Miller (eds.) 2009. *Insect Symbiosis*, Vol. 3. Boca Raton, FL: CRC Press.

Brandenburg, R.L. and G.G. Kennedy. 1982. Relationship of *Neozygites floridana* (Entomophthorales: Entomophthoraceae) to twospotted spider mite (Acari: Tetranychidae) populations in field corn. *J. Econ. Entomol.* 75:691–694.

Bruin, J. and L.P.S. van der Geest (eds.) 2009. *Diseases of Mites and Ticks*. Heidelberg: Springer.

Chandler, D., G. Davidson, J.L. Pell, B.V. Ball, K. Shaw, and K.D. Sunderland. 2000. Fungal biocontrol of Acari. *Biocontrol Sci. Technol.* 10:357–384.

Davidson, G., K. Phelps, K.D. Sjuderland, J.K. Pell, B.V. Ball, K.E. Shaw, and D. Chandler. 2002. Study of temperature–growth interactions of entomopathogenic fungi with potential for control of *Varroa destructor* (Acari: Mesostigmata) using a nonlinear model of poikilotherm development. *J. Appl. Microbiol.* 94:816–825.

Douglas, A.E. 1994. *Symbiotic Interactions*. Oxford, U.K.: Oxford University Press.

Delalibera, I., A.E. Hajek, and R.A. Humber. 2004. *Neozygites tanajoae* sp. nov., a pathogen of the cassava green mite. *Mycologia* 96:1002–1009.

Dick, G.L. and L.L. Buschman. 1995. Seasonal occurrence of a fungal pathogen, *Neozygites adjarica* (Entomophthorales:Neozygitaceae),infecting banks grass mites, *Oligonychus pratensis*, and twospotted spider mites, *Tetranychus urticae* (Acari: Tetranychidae),in field corn. *J. Kansas Entomol. Soc.* 68:425–436.

Enigl, M. and P. Schausberger. 2007. Incidence of the endosymbionts *Wolbachia*, *Cardinium* and *Spiroplasma* in phytoseiid mites and associated prey. *Exp. Appl. Acarol.* 42:75–85.

Falcon, L.A. 1985. Development and use of microbial insecticides. In: M.A. Hoy and D.C. Herzog (eds.), *Biological Control in Agricultural IPM Systems* (pp. 229–242). Orlando, FL: Academic Press.

Gerson, U., R. Kenneth, and T.I. Muttath. 1979. *Hirsutella thompsonii*, a fungal pathogen of mites. II. Host–pathogen interactions. *Ann. Appl. Biol.* 91:29–40.

Gols, R., C. Schutte, R. Stouthamer, and M. Dicke. 2007. PCR-based identification of the pathogenic bacterium, *Acaricomes phytoseiuli*, in the biological control agent *Phytoseiulus persimilis* (Acari: Phytoseiidae). *Biol. Control* 42:316–325.

Hountondji, F.C.C., M.W. Sabelis, R. Hanna, and A. Janssen. 2005. Herbivore-induced plant volatiles trigger sporulation in entomopathogenic fungi: The case of *Neozygites tanajoae* infecting the cassava green mite. *J. Chem. Ecol.* 31:1003–1021.

Hoy, M.A. and A. Jeyaprakash. 2009. Symbionts, including pathogens, of the predatory mite *Metaseiulus occidentalis*: Current and future analysis methods. In: S. Bruin and L.P.S. van der Geest (eds.), *Diseases of Mites and Ticks* (pp. 329–347). Heidelberg: Springer.

Hunter-Fujita, F.R., P.F. Entwistle, H.F. Evans, and N.E. Cook. 1998. *Insect Viruses and Pest Management*. New York: Wiley.

Johanowicz, D.L. and M.A. Hoy. 1998. Experimental induction and termination of non-reciprocal reproductive incompatibilities in a parahaploid mite. *Entomol. Exp. Appl.* 87:51–58.

Kanga, L.H.B., R.R. James, and D.G. Boucias. 2002. *Hirsutella thompsonii* and *Metarhizium anisopliae* as potential microbial control agents of *Varroa destructor*, a honey bee parasite. *J. Invertebr. Pathol.* 81:175–184.

Kitajima, E.W., T.V.M. Groot, V.N. Novelli, J. Freitas-Astua, G. Alterti, and G.J. De Moraes. 2007. *In situ* observation of the *Cardinium* symbionts of *Brevipalpus* (Acari: Tenuipalpidae) by electron microscopy. *Exp. Appl. Acarol.* 42:263–271.

Klubertanz, T.H., L.P. Pedigo, and R.E. Carlson. 1991. Impact of fungal epizootics on the biology and management of the twospotted spider mite (Acari: Tetranychidae) in soybean. *Environ. Entomol.* 20:731–735.

Lacey, L.A. and H.K. Kaya (eds.) 2007. *Field Manual of Techniques in Invertebrate Pathology*, 2nd ed. Dordrecht: Springer.

Lacey, L.A., R. Frutos, H.K. Kaya, and P. Vail. 2001. Insect pathogens as biological control agents: Do they have a future? *Biol. Control* 21:230–248.

Larsson, J.I.R., M.Y. Steiner, and S. Bjornson. 1997. *Intexta acarivora* gen. et sp. n. (Microspora: Chytridiopsidae): Ultrastructural study and description of a new microsporidian parasite of the forage mite *Tyrophagus putrescentiae* (Acari: Acaridae). *Acta Protozool.* 36:295–304.

McCoy, C.W. 1985. Citrus: Current status of biological control in Florida. In: M.A. Hoy and D.C. Herzog (eds.), *Biological Control in Agricultural IPM Systems* (pp. 481–499). Orlando, FL: Academic Press.

McCoy, C.W. 1996. Pathogens of eriophyoids. In: E.E. Lindquist, M.W. Sabelis, and J. Bruin (eds.), *Eriophyoid Mites: Their Biology, Natural Enemies, and Control* (pp. 481–490). Amsterdam: Elsevier.

McCoy, C.W., R.A. Sanson, D.G. Boucias, L.S. Osborne, J.E. Peña, and L.J. Buss. 2009. *Pathogens Affecting Insects and Mites of Citrus*. Winter Park, FL: Friends of Microbes (www.insectpathogens.com).

McMurtry, J.A. 1985. Citrus. In: W. Helle and M.W. Sabelis (eds.), *Spider Mites: Their Biology, Natural Enemies, and Control*, Vol. 1B (pp. 339–347). Amsterdam: Elsevier.

Miller, L.K.J. and L.A. Ball (eds.) 1998. *The Insect Viruses*. New York: Plenum Press.

Mwangi, E.N., O.O. Dipeolu, R.M. Newson, G.P. Kaaya, and S.M. Hassan. 1991. Predators, parasitoids and pathogens of ticks: A review. *Biocontrol Sci. Technol.* 1:147–156.

Montesinos, E. 2003. Development, registration and commercialization of microbial pesticides for plant protection. *Int. Microbiol.* 6:245–252.

Morjan, W.E., L.P. Pedigo, and L.C. Lewis. 2002. Fungicidal effects of glyphosate and glyphosate formulations on four species of entomopathogenic fungi. *Environ. Entomol.* 31:1206–1212.

Olsen, L.E. and M.A. Hoy. 2002. Heat curing *Metaseiulus occidentalis* (Nesbitt) (Acari: Phytoseiidae) of a fitness-reducing microsporidium. *J. Invertebr. Pathol.* 79:173–178.

O'Neill, S.L., A.A. Hoffmann, and J.H. Werren. 1997. *Influential Passengers: Inherited Microorganisms and Arthropod Reproduction*. Oxford, U.K.: Oxford University Press.

Poinar, G., Jr. and R. Poinar. 1998. Parasites and pathogens of mites. *Annu. Rev. Entomol.* 43:449–470.

Purrini, K. and J. Weiser. 2009. Eight new microsporidian parasites of moss-mites (Oribatei, Acarina) in forest soils. *Zeitsch. Angew. Entomol.* 91(1–5):217–224.

Putman, W.L. and D.H.C. Herne. 1966. The role of predators and other biotic agents in regulating the population density of phytophagous mites in Ontario peach orchards. *Can. Entomol.* 98:808–820.

Reed, D.K., H. Tashiro, and J.B. Beavers. 1975. Determination of mode of transmission of the citrus red mite virus. *J. Invertebr. Pathol.* 26:239–246.

Rehacek, H. and H. Weiser. 1978. Natural infection of the tick *Dermacentor reticulatus* (Fabr.) with the microsporidian *Nosema slovaca* Weiser et Rehacek in Slovakia. *Folia Parasitol. (Praha)* 25:165–170.

Samish, M. and J. Rehacek. 1999. Pathogens and predators of ticks and their potential in biological control. *Annu. Rev. Entomol.* 44:159–182.

Samson, R.A. and C.W. McCoy. 1982. A new fungal pathogen of the scavenger mite, *Tydeus gloveri*. *J. Invertebr. Pathol.* 40:216–220.

Schutte, C. and M. Dicke. 2009. Verified and potential pathogens of predatory mites (Acari: Phytoseiidae). In: S. Bruin and L.P.S. van der Geest (eds.), *Diseases of Mites and Ticks* (pp. 307–328). Heidelberg: Springer.

Shaw, J.G., C. Moffitt, and G.T. Scriven. 1967. Biotic potential of phytoseiid mites fed on virus-infected citrus red mites. *J. Econ. Entomol.* 60:1751–1752.

Smitley, D.R., G.G. Kennedy, and W.M. Brooks. 1986. Role of the entomogenous fungus, *Neozygites floridana*, in population declines of the twospotted spider mite, *Tetranychus urticae*, on field corn. *Entomol. Exp. Appl.* 41:255–264.

Tanada, Y. and H.K. Kaya. 1993. *Insect Pathology*. San Diego, CA: Academic Press.

Tokarev, I.S. and A.A. Moville. 2004. A first record of microsporidia in the ixodid tick *Ixodes ricinus* L. (Ixodidae) in the territory of the Commonwealth of Independent States Republic Moldova. *Parazitologiya* 38:552–556.

van der Geest, L.P.S. 1985. Pathogens of spider mites. In W. Helle and M.W. Sabelis (eds.), *Spider Mites: Their Biology, Natural Enemies, and Control*, Vol. 1B (pp. 247–258). Amsterdam: Elsevier.

van der Geest, L.P.S., S.L. Elliot, J.A.J. Breeuwer, and E.A.M. Beerling. 2000. Diseases of mites. *Exp. Appl. Acarol.* 24:497–560.

Weeks, A.R. and R. Stouthamer. 2003. Increased fecundity associated with infection by a *Cytophaga*-like intracellular bacterium in the predatory mite, *Metaseiulus occidentalis*. *Proc. R. Soc. Biol. Lett.* 271(Suppl. 4):193–195.

Weeks, A.R., F. Marec, and J.A.J. Breeuwer. 2001. A mite species that consists entirely of haploid females. *Science* 292:2479–2482.

Wittner, M. and L.M. Weiss (eds.) 1999. *The Microsporidia and Microsporidiosis*. Washington, D.C.: American Society for Microbiology.

PART IV

Exemplars of Integrated Mite Management Programs for Plant-Feeding Mites

Part IV provides an overview of integrated mite management (IMM) programs for plant-feeding mites where different tactics are deployed, as appropriate. In the first example (Chapter 15), involving the cassava green mite in Africa, a classical biological control program was determined to be the most effective tactic because the cassava green mite was an invasive pest, chemical control was too expensive for local growers, cultural practices (hand pruning) failed to work, and host-plant resistance was not available.

The second example (Chapter 16) is based on the pioneering work conducted by Stan Hoyt in apple orchards in Washington State. It illustrates how integrating biological and chemical control can result in improved mite management. This work served as a model for subsequent IMM programs in other areas, including apples produced in Michigan, Southern California, and New Zealand and peaches in Australia. In the third example (Chapter 17), multiple tactics were deployed in an IMM program in California almond orchards. Although a predatory mite was found to be effective in suppressing spider mites in unsprayed almond orchards, the trees could not tolerate high levels of spider mites without defoliating. Chemical control of a key insect pest, the navel orangeworm, with a carbamate insecticide killed the predatory mites, allowing secondary outbreaks of the spider mites. This IMM program involved: (1) releasing a laboratory-selected, carbaryl-organophosphate-resistant strain of the predatory mite; (2) monitoring predator–spider mite interactions; (3) applying spot treatments early in the growing season to assist the predator; (4) improving water and dust management; and (5) using lower-than-label rates of acaricides to suppress the spider mites at key times during the growing season.

The fourth example (Chapter 18) compares the mite species and their damage in citrus groves in Florida and in California, illustrating that different climates and cultivars result in different pest mite problems. Even within California, climatic differences can result in dramatically different pest problems in the different growing regions; thus, developing IMM programs is often very site specific. The final example (Chapter 19) discusses the difficulties of developing a biological control-based pest management program for ornamentals grown in greenhouses due to the low economic injury levels accepted; however, progress is being made in obtaining the data needed for IMM programs in these crops.

Note that each IMM program requires consideration of a variety of components, including the crop cultivars, climate, other pests (insects, mites, diseases) in the crop, availability of selective pesticides that are not disruptive to natural enemies, effective monitoring methods, and considerable knowledge of the biology and ecology of both the pest and beneficial species in each crop in each region. IMM is an information-intensive approach to pest management.

[illegible]

[illegible]

[illegible]

[illegible]

[illegible]

CHAPTER 15

Classical Biological Control of the Cassava Green Mite in Africa

15.1 STEPS IN A CLASSICAL BIOLOGICAL CONTROL PROGRAM

Although it is rare that an integrated mite management (IMM) program can rely on a single tactic, in some cases classical biological control may be the sole appropriate management tactic. The introduction of natural enemies in classical biological programs is often effective when an invasive pest has invaded a new country without its specific natural enemies (Herren 1987, Herren and Neuenschwander 1991, Bale et al. 2008). In many cases, endemic natural enemies (predators, parasitoids, and pathogens) may provide some suppression of the invasive pest populations but lack the ability to provide adequate control. Classical biological control, although taking longer to implement than chemical control, is especially appropriate when other management tools are ineffective or difficult to implement (Van Driesche and Bellows 1993, 1995). Classical biological control has a long history of success around the world, although it has been less successful with pest mites than with pest insects or weeds (McMurtry 1984, Kauffman and Nechols 1992, Van Driesche and Bellows 1993, 1995, McMurtry and Croft 1997), probably because it has been attempted less often. This chapter describes a successful classical biological control program for the cassava green mite (CGM) (Acari: Tetranychidae) in Africa. Classical biological control programs usually require a number of years (10 or more); in the case of the CGM, it took 15 years to complete the project (Yaninek 2007). Time is required to conduct the background literature research, to deal with taxonomic issues, and to identify, screen, and introduce the natural enemies (Table 15.1). Mass-rearing methods may have to be developed (Haug et al. 1987, Hanna and Toko 2001), and release methods designed. After releases are made, it may take several years for the natural enemies to establish, disperse, and have a measurable impact on the target pest. Evaluation of efficacy, and of costs and benefits, cannot occur until the pest and its natural enemies are in an equilibrium, which may take years. However, if effective natural enemies become established, they can provide inexpensive and long-term control of the pest. Although initial costs may be high and the development of the program may be lengthy, the potential benefits can be long term. On average, classical biological control programs return $250 for each $1 invested *each year* (Bale et al. 2008).

15.2 CASSAVA GREEN MITE (*MONONYCHELLUS TANAJOA*) IN AFRICA

Cassava, which is native to Latin America, has become the most important root crop in the lowland tropics throughout the world. It supplies between 300 and 500 million people with a major portion of their caloric intake (Yaninek and Herren 1988, Yaninek et al. 1989). Cassava was

Table 15.1 Key Steps in Classical Biological Control Programs

1. *Determine that the pest is foreign.* Appropriately curated material should be identified by an expert taxonomist so an effective exploration program can be carried out in the native range of the pest for effective (and specific) natural enemies.
2. *Confirm that the pest problem justifies the effort.* The nature and amount of damage should justify the cost of a classical biological control program. An analysis of losses and a comparison of the costs and expected efficacy of other control methods will help to justify the program.
3. *Determine that the pest is a good candidate for classical biological control.* Many pests cannot be effectively controlled through host-plant resistance, cultural controls, genetic control techniques, or pesticides. Costs and environmental hazards of pesticides and the risks of pesticide resistance may preclude the use of chemical control as a sustainable approach.
4. *Conduct a survey for the pest and its natural enemies in its area of origin.* The pest's original range may be difficult to determine because the pest may be scarce there, usually because it is under the control of its natural enemies. If surveys are made in the area where the pest is a problem and in areas where it is not, it will be possible to identify the natural enemies available and those that are suitable for a classical biological control program. The area of origin of the pest is the likely site of the greatest diversity of natural enemies.
5. *Match the climate and other ecological conditions.* This is extremely important, as the natural enemy must be able to establish and persist in the new environment. Historically, natural enemy establishment rates have averaged only 24%; this rate is improved if climate matching and other care is taken to achieve establishment.
6. *Obtain information on the biology of the candidate natural enemies.* These studies can be conducted in the country of origin, in quarantine facilities, or from literature reviews.
7. *Obtain permits to import.* Importation of natural enemies into countries is regulated and requires permits; for example, in the United States, importation and release are regulated by both state and federal agencies.
8. *Conduct risk evaluations in quarantine.* The natural enemy should not pose a risk to agriculture, to the environment, or to non-target species. Quarantine screening ensures that arthropod pests, plant diseases, hyperparasites (parasites of parasites), or pathogens are not introduced accidentally. Parasitoids, predators, or pathogens of beneficial species such as honey bees, silkworms, or lac insects should not be introduced. Introduced agents should not attack native species, especially threatened and endangered ones. The equivalent of an environmental assessment is performed prior to releases. Time, funding, and logistical issues (space in quarantine) preclude the importation, evaluation, and release of all natural enemies associated with a target pest. Evaluations should include: (a) host or prey specificity, (b) searching efficiency, (c) climate matching, and (d) ability to suppress and maintain a pest population at low densities.
9. *Develop mass rearing methods.* Large-scale rearing using the target pest as prey is ideal but can be expensive. Quality controls should be implemented to ensure that natural enemies are vigorous, healthy, and as genetically diverse as possible to make establishment and adaptation to the new environment feasible.
10. *Obtain permission to release.* State, national, and international agencies may have to be consulted before permission to release can be obtained. Because natural enemies may not recognize national borders, consultations with neighboring countries may be required.
11. *Carefully consider release methods.* Choose sites where the pest is present in good numbers and no pesticides are applied. Release healthy, young, and well-fed natural enemies at a time when the environment is most permissive. Minimize the time the natural enemy is left in hot locations in transit. Release adequate numbers so it is easy for them to find mates. Generally, about 100 to 1000 should be released multiple times in as many appropriate sites as possible.
12. *Evaluate for establishment.* Monitor to determine if the natural enemy has persisted and reproduced. Establishment means that it is able to survive adverse conditions, such as winter or the dry season, throughout the year. There is an unofficial '3-year rule' for monitoring because it is fairly common for a natural enemy to establish but be present in such low numbers that it cannot be detected easily. Sometimes natural enemies appear to establish but adverse weather conditions subsequently eliminate the population.
13. *Evaluate the impact.* This is difficult and often requires a number of years after the pest and natural enemies have come to some type of equilibrium; it is usually impossible to predict the impact of natural enemies immediately after releases. Can the natural enemy reduce the pest population below the economic injury level in a stable manner throughout the range over several years with different weather patterns? It is rare that a single natural enemy species is able to control a pest; there are few such panaceas. Sometimes, establishment of two or more natural enemies is required, and the assistance of endemic generalist predators and pathogens may be required.
14. *Conduct an economic assessment.* Engage the help of an economist early in the program to calculate the costs and benefits of the program so they are accurately estimated. Economic analyses are rarely conducted but are important in obtaining funding for future programs.
15. *Document the entire program.* We need to learn from the experiences of both successful and unsuccessful classical biological control programs. Reference specimens and voucher specimens should be submitted to museums in 70% ethanol (for slide mounting) and in 95% ethanol for future DNA analyses so the identity of the species can be confirmed. Name changes may occur along with taxonomic revisions, and cryptic species may be discovered within an apparent "single" species using DNA analysis methods.

brought to Africa 300 years ago free of its arthropod pests and became a subsistence food throughout much of Africa (Bellotti and Schoonhoven 1978, Norgaard 1988). Traditionally, cassava was considered a hardy plant because it yielded useful roots even under drought conditions (Bellotti and van Schoonhoven 1978, Herren and Neuenschwander 1991). Cassava is considered especially valuable for survival during famine periods in Africa because it can be left in the ground for up to 24 months; it provides a sustained food supply when other crops have failed due to drought or other problems.

The cassava mealybug, *Phenacoccus manihoti* (Pseudococcidae), was accidentally introduced into Africa from South America in the 1970s and was the subject of one of the largest classical biological control programs ever undertaken (Herren 1987, Neuenschwander et al. 1989). The introduction of a parasitoid wasp, *Epidinocarsis lopezi*, was spectacularly successful in controlling the cassava mealybug in Africa, resulting in dramatic control of the pest with a benefit:cost ratio of at least $149:1 each year on a continent-wide basis by 1988. This success made the development of the large-scale classical biological control program directed against the cassava green mite and its subsequent release into 20 countries easier (Herren 1987, Neuenschwander et al. 1989, Yaninek 2007).

Several spider mites are found on cassava in Latin America, including the two-spotted spider mite complex (*Tetranychus cinnabarinus* and *T. urticae*) and species of *Mononychellus* (Byrne et al. 1983, Bellotti et al. 1999). Taxonomic confusion arose regarding the species called the cassava green mite, which may include *Mononychellus tanajoa*, *M. caribbeaneae*, and *M. progresivus*. In Africa, the cassava green mite, probably *M. tanajoa*, was first reported in Uganda in 1971 and rapidly spread throughout the cassava-growing areas of Africa at a rate of 375 km per year (Yaninek and Herren 1988). By 1982, the mite had spread throughout the entire cassava-growing region of Africa. At 27°C, it takes *M. tanajoa* 12.5 days to develop from egg to adult; it has a life span of 24 days, during which time females can deposit about 60 eggs (Figure 15.1) (also see Figure S15.1 on the CD). Severe damage to cassava occurs during hot dry weather, due to chlorosis of leaves, reductions in leaf area (reducing yield in the roots), and defoliation (Figure 15.2) (also see Figure S15.2 on the CD). Large populations of CGM can even kill the entire apical shoot.

15.3 CONTROL MEASURES ATTEMPTED

Chemical control of *Mononychellus tanajoa* is expensive and difficult for subsistence farmers in Africa (Yaninek and Herren 1988). Few have the training or the equipment necessary to apply chemical pesticides safely, and few understand the possible health and environmental risks of chemical control. Chemical control can cause secondary pest outbreaks and resistance to pesticides in the pest. Cultural control practices for the CGM were limited (Yaninek and Herren 1988). Cassava plants that are 2 to 9 months old are most vulnerable to attack, so altering planting time was the principal form of cultural control attempted; however, because subsistence farmers need to plant cassava soon after the first rains of the long wet season, altering the time of planting was difficult to do. Traditional practices of intercropping or mixed cropping can help reduce some pests, but this was not true for the CGM. Hand removal of new leaves infested with the CGM was not effective, because it was labor intensive and resulted in new growth on lateral shoots, which soon provided suitable new leaves for the CGM. African farmers often develop pest-resistant cultivars by selecting plants that are relatively tolerant of local pests and diseases, but the CGM had not been present in Africa long enough for much local selection to be effective. Neotropical cultivars that might confer some resistance to the CGM are susceptible to a cassava mosaic virus that was present in Africa; thus, these cultivars were not useful. Genetic selection for resistant cultivars was attempted in a number of cassava improvement programs throughout the African cassava-growing area, but it is a long-term endeavor.

Figure 15.1 Cassava green mite females, eggs, immatures, and males. This species invaded Africa and caused serious damage to cassava there prior to the conclusion of the classical biological control program. (Photograph by Steve Yaninek, International Institute for Tropical Agriculture, Cotonou, Bénin.)

Classical biological control was a suitable tactic because the pest is not native to Africa, it cannot be controlled readily any other way, and chemical control is too expensive for subsistence farmers. There was no history of prolonged and severe mite attacks on cassava in Latin America, lending support to the hypothesis that effective natural enemies might exist in South America (Yaninek and Herren 1988, Hanna and Toko 2001).

A drawback to pursuing a classical biological control project against the CGM was the fact that almost nothing was known about the biology of the CGM or its natural enemies in Latin America (Bellotti et al. 1987, Yaninek 2007). It was suggested that plant resistance in neotropical cultivars, control by unknown natural enemies, and weather were factors that combined to suppress CGM populations below economically important levels in Latin America.

Exploration for natural enemies of the CGM began in 1977 by the International Institute of Tropical Agriculture (IITA), the Centro Internacional de Agricultura Tropical (CIAT), and the Empresa Brasileira de Pesquisa Agropecuária (EMBRAPA) in Brazil. International donors provided funding to the Consultative Group on International Agricultural Research (CGIAR). Surveys throughout the cassava-growing area in South America identified a variety of natural enemies, including species in the Chrysopidae, Cecidomyiidae (*Arthrocnodax* sp.), Thysanoptera (*Scolothrips* spp.), Syrphidae, Anthocoridae, Lygaeidae, Staphylinidae (*Oligota minutus*), Coccinellidae (*Stethorus*), and Phytoseiidae. Found in lesser numbers were spiders, other predatory mite families, and some pathogens, including the fungal pathogen *Neozygites*; however, insect predators were considered to be opportunists or effective only at high population densities.

The international project focused on phytoseiids, because phytoseiids are known to have a long association with tetranychids, some have relatively specific prey requirements, and some are superior predators when mite populations are low (McMurtry 1984, McMurtry and Croft 1997). Phytoseiids are known to be effective natural enemies of tetranychids in a large variety of crops (Kostiainen and Hoy 1996).

Figure 15.2 (A) Cassava foliage undamaged by the cassava green mite. (B) Damage to cassava foliage by CGM feeding. Serious yield losses occurred in cassava crops after the invasion of this tetranychid into Africa. (Photographs by Steven Yaninek, International Institute of Tropical Agriculture, Cotonou, Bénin.)

The CGM project is the largest coordinated classical biological control program initiated against a spider mite pest to date, and it has had a continent-wide impact. Explorations in the neotropics were made in areas that corresponded ecologically to the very large and climatically diverse areas in Africa where the CGM was a pest. It was important to identify natural enemies that could survive in Africa that also were successful predators of the CGM. As phytoseiid species were identified, several biological traits were studied, and the data were used to rank the predators for potential release in Africa.

15.4 RISK EVALUATION FOR CLASSICAL BIOLOGICAL CONTROL

About 50 species of Phytoseiidae were found on cassava in South America (de Moraes et al. 1989, Yaninek 2007). Some were imported into quarantine facilities in London or the Netherlands for identification and evaluation, including *Amblyseius aerialis, Euseius concordis, Galendromus annectens, Neoseiulus idaeus, N. anonymus, N. californicus, Phytoseiulus mexicanus, Typhlodromalus aripo, T. limonicus, T. tenuiscutus*, and *T. manihoti*. Once released from quarantine, they were sent to Cotonou, Bénin, in West Africa for additional evaluation and possible mass rearing and release in appropriate climatic zones in Africa. Prior to release, mass-rearing methods had to be developed (Haug et al. 1987) so a total of 11.6 million phytoseiids could be shipped to 16 countries in 847 locations (Yaninek 2007) (Figure 15.3) (also see Figure S15.3 on the CD). *Galendromus annectens, N. anonymus, N. californicus, N. ideaeus, T. aripo*, and *T. manihoti* were recovered after releases, and three are considered to be established (Yaninek et al. 1998) (Figure 15.4) (also see Figure S15.4 on the CD). *Typhlodromalus aripo* has been the most successful, having "become established in more than 20 countries, where it has reduced *M. tanajoa* populations and significantly increased cassava yields by a third" (Yaninek 2007).

Typhlodromalus aripo was released directly onto cassava plants beginning in 1993 and became widely established in Africa by the end of 1995 (Hanna and Toko 2001) (Figure 15.5) (also see Figure S15.5 on the CD). This phytoseiid spread over an area of more than 150,000 km^2 within 2 years. *Typhlodromalus aripo* appears to be relatively specific to the CGM, although it can feed and develop on other foods, including *Tetranychus urticae, Oligonychus gossypii*, several types of pollens (maize, castor bean and leucaena), whiteflies, cassava-plant exudate, and cassava-leaf substrate (Onzo 2003, Onzo et al. 2009). The use of these alternative foods is important because it allows *T. aripo* to persist even when CGM densities are low. The use of the plant apex between the leaf primordia as a refuge by *T. aripo* also is an important factor in its effectiveness. The predators rest in the apices of the plant during the day but move to the young leaves during the night to feed. They then return to the apices the next morning. This behavior results in the protection of young foliage, which is especially important in photosynthesis.

15.5 TAXONOMIC PROBLEMS

The CGM classical biological control program was hampered by taxonomic difficulties (note that taxonomic problems are common in most acarine groups). The Phytoseiidae of South America were not well known, and over 25 new species were detected during the exploration for natural enemies of the CGM on cassava. One, *Typhlodromalus limonicus*, originally was reported to be present throughout the Western Hemisphere from California to Brazil. It was later reclassified as two species (*T. limonicus* sensu strictu and *T. manihoti*) after it was recognized that populations from Colombia were not as easy to rear as those from California. Species status was established on the basis of reproductive incompatibility and morphological differences, but this took almost 10 years, exemplifying the difficulty of detecting, verifying, and describing a previously unknown species.

Among the many species of phytoseiids collected from cassava in South America, it is likely that some are synonyms and others represent groups of unrecognized sibling species. Molecular markers, such as random amplification of polymorphic DNA by polymerase chain reaction (RAPD-PCR), offer a possibility of speeding up the process of discriminating between cryptic species. RAPD-PCR markers were used to compare three morphologically similar species collected from cassava; the data suggested that the *Typhlodromalus manihoti* populations from Colombia and the *T. manihoti* populations from Brazil might be different species (Edwards et al. 1997).

Figure 15.3 (A) Cassava trees were used to rear the cassava green mite to feed phytoseiids being evaluated for release. (B) Rearing trays for producing phytoseiids for laboratory evaluation and release into the field. (Photographs by Steve Yaninek, International Institute for Tropical Agriculture, Cotonou, Bénin.)

Figure 15.4 The phytoseiid *Neoseiulus ideaus* feeding on a cassava green mite female. This phytoseiid was released in the classical biological control project. (Photograph by Steve Yaninek, International Institute for Tropical Agriculture, Cotonou, Bénin.)

The fungus *Neozygites* (Entomophthorales) was found infecting the CGM in Brazil (de Moraes and Delalibera 1992, Delalibera et al. 2004). Laboratory tests were conducted to determine how specific the pathogen is and especially whether it would attack phytoseiids. Tests indicated that the fungus would not attack two phytoseiids that are predators of CGM, nor would it attack *Tetranychus urticae* and *T. bastosi*, indicating that it exhibits some degree of host specificity and should be studied further as a potential component of the classical biological control program in Africa. The fungus subsequently was described as a new species (Delilibera et al. 2004).

Figure 15.5 Releases of *Typhlodromalus aripo* and the other phytoseiid species involved placing foliage containing the predators directly on the cassava plants. This field has been seriously damaged by the cassava green mite. (Photograph by Steven Yaninek, International Institute for Tropical Agriculture, Cotonou, Bénin.)

15.6 WHY NOT USE AFRICAN PHYTOSEIIDS AS NATURAL ENEMIES?

Initially, governments in Africa resisted the concept of classical biological control of the CGM. Some officials were suspicious and concerned that these new introductions would become pests. They also argued that native natural enemies should be able to suppress the pest. As part of the project, African phytoseiids were collected that included 18 genera and 103 known and 26 undescribed species from 402 host plants, which led to descriptions of 20 new species and 63 redescribed species in Africa (Yaninek and Hanna 2003). Among the African species, a doctorate student supported by the IITA evaluated one. Bruce-Oliver (1993) studied the African phytoseiid *Euseius fustis* to determine why it was not effective as a natural enemy of the CGM. The results illustrate some of the problems that can occur when relying on native generalist natural enemies and emphasize that detailed knowledge of biology and behavior is necessary to understand predator–prey interactions (Table 15.2).

One of the first questions considered was whether *Euseius fustis* could feed and reproduce on a diet consisting solely of the CGM (Bruce-Oliver et al. 1996). When *E. fustis* was given a diet of all stages of CGM or castor bean pollen, it survived less well on the mite than on the pollen over seven generations. A diet consisting solely of CGM did not allow this predator population to increase over time, indicating that it was unlikely to be an effective predator of CGM in the field. Furthermore, *E. fustis* did not exhibit the same phenology on the cassava plant as the CGM (Bruce-Oliver 1993). *Euseius fustis* populations declined during the dry seasons and increased during the wet seasons in Bénin, West Africa. This phenology is the opposite of that exhibited by the CGM, which does very well during the dry season. Because pollen counts increased during the long wet season, *E. fustis* densities increased. Finally, *E. fustis* did not inhabit the same part of the cassava plant as the CGM; it was more abundant on lower leaves (at least at the time of day when the samples were taken), but the CGM was more abundant on the terminal leaves.

Zannou et al. (2007) evaluated the effects of *Typhlodromalus aripo* on native phytoseiids in Malawi and Mozambique to determine whether this biological control program had unintended non-target effects. They found through field surveys on cassava that the abundance of the two most common native phytoseiids were apparently facilitated, or enhanced, by the introduction of *T. aripo*, and a third species was unaffected. The overall abundance of phytoseiids found on non-cassava plants also was not affected in these studies.

15.7 PROGRAM COSTS AND BENEFITS

Yaninek (2007) reported that US$9.5 million were raised for the program, which ran from 1983 to 1997. The funds were expended to operate the laboratories in Cotonou, Bénin, to support a network of collaborators in 20 countries, and to pay for foreign exploration services, international quarantine work, and training. National programs in Africa were provided basic laboratory and

Table 15.2 Problems Encountered with the African Phytoseiid *Euseius fustis* as a Natural Enemy of the Cassava Green Mite (CGM) in Africa

1. The phenology of the pest and phytoseiid was not congruent; *Euseius fustis* populations increased during the wet seasons in Bénin, West Africa, while CGM populations increased during the dry seasons.
2. The predator and prey should occupy the same area on the plant, at least when they are feeding; however, *E. fustis* was more abundant on lower leaves, while the CGM was most abundant on the top leaves.
3. The predator should be able to feed and develop well on the CGM. *Euseius fustis* performed better on a diet of pollen than it did on a diet consisting solely of the CGM.
4. The predator should be able to reproduce on a diet consisting solely of the CGM. Reproduction was poor on a diet of CGM alone.

Source: Based on information from Bruce-Oliver (1993) and Bruce-Oliver et al. (1996).

field equipment, funds for personnel, and supplies. Funds also were provided for training, including training workshops for local scientists and staff, as well as training for scientists from Africa to obtain doctorate (15), master of science (9), and bachelor of science (9) degrees in universities in Africa, South America, North America, and Europe. Yaninek (2007) attributed the successful program to creating "a legacy of biological control training, infrastructure and experiences that is still evident today" in African subsistence agriculture.

SELECTED REFERENCES

Bale, J.S., J.C. van Lenteren, and F. Bigler. 2008. Biological control and sustainable food production. *Phil. Trans. R. Soc. Lond. B* 363:761–777.

Bellotti, A. and A. van Schoonhoven. 1978. Mite and insect pests of cassava. *Annu. Rev. Entomol.* 23:39–67.

Bellotti, A.C., N. Mesa, M. Serrano, J.M. Guerrero, and C.J. Herrera. 1987. Taxonomic inventory and survey activity for natural enemies of cassava green mites in the Americas. *Insect Sci. Appl.* 8:845–849.

Bellotti, A.C., L. Smith, and S.L. Lapointe. 1999. Recent advances in cassava pest management. *Annu. Rev. Entomol.* 44:343–370.

Bruce-Oliver, S.J. 1993. Evaluation of the Indigenous African Phytoseiid, *Euseius fustis* (Pritchard and Baker) (Acari: Phytoseiidae), as a Potential Control Agent of the Cassava Green Mite, *Mononychellus tanajoa* (Bondar) (Acari: Tetranychidae), on Cassava, *Manihot esculenta* Crantz, in West Africa, PhD thesis, University of California–Berkeley.

Bruce-Oliver, S.J., M.A. Hoy, and J.S. Yaninek. 1996. Effect of some food sources associated with cassava in Africa on the development, fecundity and longevity of *Euseius fustis* (Pritchard and Baker) (Acari: Phytoseiidae). *Exp. Appl. Acarol.* 20:73–85.

Byrne, D.H., A.C. Bellotti, and J.M. Guerrero. 1983. The cassava mites. *Trop. Pest Manag.* 29:378–394.

Delalibera, I., A.E. Hajek, and R.A. Humber. 2004. *Neozygites tanajoae* sp. nov., a pathogen of the cassava green mite. *Mycologia* 96:1002–1009.

de Moraes, G.J. and I. Delalibera, Jr. 1992. Specificity of a strain of *Neozygites* sp. (Zygomycetes: Entomophthorales) to *Mononychellus tanajoa* (Acari: Tetranychidae). *Exp. Appl. Acarol.* 14:89–94.

de Moraes, G.J., J.A. McMurtry, and J.S. Yaninek. 1989. Some phytoseiid mites (Acari: Phytoseiidae) from tropical Africa with description of a new species. *Int. J. Acarol.* 15:95–102.

Edwards, O.R., E.L. Melo, L. Smith, and M.A. Hoy. 1997. Discrimination of three *Typhlodromalus* species (Acari: Phytoseiidae) using random amplified polymorphic DNA markers. *Exp. Appl. Acarol.* 21:101–109.

Hanna, R. and M. Toko. 2001. Cassava green mite biological control in Africa: project overview and summary of progress. In: R. Hanna and M. Toko (eds.), *Proceedings of the Regional Meeting of the Cassava Green Mite Biocontrol Project, Dar es Saloam, Tanzania, November 15–17, 2000* (pp. 4–22). Cotonou, Bénin: International Institute for Tropical Agriculture.

Haug, T., H.R. Herren, D.J. Nadel, and J.B. Akinwumi. 1987. Technologies for the mass-rearing of cassava mealybugs, cassava green mites and their natural enemies. *Insect Sci. Applic.* 8:879–881.

Herren, H.R. 1987. Africa-wide biological control project of cassava pests: a review of objectives and achievements. *Insect Sci. Applic.* 8:837–840.

Herren, H.R. and P. Neuenschwander. 1991. Biological control of cassava pests in Africa. *Annu. Rev. Entomol.* 36:257–283.

Kauffman, W.C. and J.E. Nechols (eds.) 1992. *Selection Criteria and Ecological Consequences of Importing Natural Enemies*. Lanham, MD: Entomological Society of America.

Kostiainen, T.S. and M.A. Hoy. 1996. *The Phytoseiidae as Biological Control Agents of Pest Mites and Insects: A Bibliography*, IFAS Monograph 17. Gainesville: Institute of Food and Agricultural Sciences, University of Florida.

McMurtry, J.A. 1984. A consideration of the role of predators in the control of acarine pests. In: D.A. Griffiths and C.E. Bowman (eds.), *Acarology VI*, Vol. 1 (pp. 109–121). Chichester, U.K.: Ellis Horwood.

McMurtry, J.A. and B.A. Croft. 1997. Life-styles of phytoseiid mites and their role in biological control. *Annu. Rev. Entomol.* 42:291–321.

Neuenschwander, P., W.N.O. Hammond, A.P. Gutierrez, A.R. Cudjoe, R. Adjakloe, J.U. Baumgartner, and U. Regev. 1989. Impact assessment of the biological control of the cassava mealybug *Phenacoccus manihoti* Matile-Ferrero (Hemiptera: Pseudococcidae), by the introduced parasitoid *Epidinocarsis lopezi* (De Santis) (Hymenoptera: Encyrtidae). *Bull. Entomol. Res.* 79:579–594.

Norgaard, R.B. 1988. The biological control of cassava mealybug in Africa. *Am. J. Agric. Econ.* 70:366–371.

Onzo, A. 2003. Population Dynamics of Cassava Green Mite and Its Predator *Typhlodromalus aripo* in Bénin, West Africa, PhD dissertation, University of Amsterdam, the Netherlands.

Onzo, A., R. Hanna, and M.W. Sabelis. 2009. Within-plant migration of the predatory mite *Typhlodromalus aripo* from the apex to the leaves of cassava: response to day–night cycle, prey location and prey density. *J. Insect Behav.* 22:186–195.

Van Driesche, R.G. and T.S. Bellows (eds.) 1993. *Steps in Classical Biological Control.* Lanham, MD: Entomological Society of America.

Van Driesche, R.G. and T.S. Bellows. 1995. *Biological Control.* New York: Chapman & Hall.

Yaninek, J.S. 2007. Biological control of the cassava green mite in Africa: overcoming challenges to implementation. In: C. Vincent, M.S. Goettel, and G. Lazarovits (eds.), *Biological Control: A Global Perspective* (pp. 28–37). Wallingford, U.K.: CAB International.

Yaninek, J.S. and R. Hanna. 2003. Cassava green mite in Africa: a unique example of successful classical biological control of a mite pest on a continental scale. In: P. Neuenschwander, C. Borgemeister, and J. Langewald (eds.), *Biological Control in IPM Systems in Africa* (pp. 61–75). Wallingford, U.K.: CAB International.

Yaninek, J.S. and H.R. Herren. 1988. Introduction and spread of the cassava green mite, *Mononychellus tanajoa* (Bondar) (Acari: Tetranychidae), an exotic pest in Africa and the search for appropriate control methods: a review. *Bull. Entomol. Res.* 78:1–13.

Yaninek, J.S., G.J. de Moraes, and R.H. Markham. 1989. *Handbook on the Cassava Green Mite (*Mononychellus tanajoa*) in Africa: A Guide to Its Biology and Procedures for Implementing Classical Biological Control.* Cotonou, Bénin: International Institute for Tropical Agriculture.

Yaninek, J.S., B. Megevand, B. Ojo, A.R. Cudjoe, E. Abole, A. Onzo, and I. Zannou. 1998. Establishment and spread of *Typhlodromalus manihoti* (Acari: Phytoseiidae), an introduced phytoseiid predator of *Mononychellus tanajoa* (Acari: Tetranychidae) in Africa. *Environ. Entomol.* 276:1496–1505.

Zannou, I.D., R. Hanna, B. Agboton, G.J. de Moraes, S. Kreiter, G. Phiri, and A. Jones. 2007. Native phytoseiid mites as indicators of non-target effects of the introduction of *Typhlodromalus aripo* for the biological control of cassava green mite in Africa. *Biol. Control* 41:190–198.

[illegible]

CHAPTER 16

Integrated Mite Management in Washington Apple Orchards

16.1 THE APPLE ECOSYSTEM

Apples originated from wild species endemic to Asia (van de Vrie 1985). When apple trees were domesticated, they were moved throughout Europe and subsequently to the other continents. Insects, mites, and pathogens such as scab, mildew, and canker attack the cultivars grown today. Apple orchards are a stable agroecosystem, offering overwintering sites and habitats for both pests and natural enemies for many years. Trees begin to produce a crop within 2 to 6 years after planting and may live for 20 to 30 years. Because blemish-free, large fruits are desired, apples historically have required multiple pesticide applications to manage the arthropod pests and diseases. Chemical or mechanical fruit thinning is used to enhance the size of the fruits (Hoyt and Burts 1974); however, intensive use of pesticides is expensive, harmful to the environment, and disruptive to natural enemies. Furthermore, consumers are demanding fruits of high quality that lack spray residues. As a result, the development of integrated pest management (IPM) programs for apples has been a high priority (van de Vrie 1985, Prokopy and Croft 1994, Jones et al. 2006, 2009, Martinez-Rocha et al. 2008). It long has been recognized that spider mites are induced pests in apple orchards because their natural enemies have been disrupted by pesticides (MacPhee and Sanford 1954, Morgan and Anderson 1958, van de Vrie 1962, Sanford 1967, Hoyt and Burts 1974). The tetranychid species found on apples vary by geographic location, which may be due to both climatic differences and cultural practices (van de Vrie 1985). Some consider spider mites to be induced pests caused by the increased use of fertilizers and pruning, as well as the use of pesticides that disrupt their natural enemies.

16.2 MITES ON APPLES

16.2.1 *Panonychus ulmi*

The European red mite *Panonychus ulmi* (or *Metatetranychus ulmi*) is common in northern Europe, the northeastern United States and Washington State, eastern Canada, South America, and New Zealand (van de Vrie 1985). It is likely that *P. ulmi* has moved around the world on nursery stock as diapausing eggs (see Figures S16.1 and S16.2 on the CD). *Panonychus ulmi* overwinters as eggs at the base of buds, on spurs, or in crevices on branches. Diapausing eggs are deposited as early as August, and egg deposition may continue into November, depending on the weather. Photoperiod and temperature induce the production of diapause eggs (Lees 1953). In spring, the larvae hatch and move onto foliage. Approximately 3 to 6 weeks are required for them to develop and begin another generation. The number of generations during the growing season varies by location, ranging from 3 to 5, or up to 9 to 10. Nondiapause eggs are deposited on lower leaf surfaces. European red mite females may disperse from tree to tree by spinning a silk thread from which they are blown by the wind; however, on foliage this species produces little webbing.

Feeding by European red mites causes bronzing of leaves, and defoliation may occur when densities are high. Reduced crop yield may occur in the subsequent year, but effects vary with apple cultivar, tree vigor, time when damage occurs, crop load, and weather (Hoyt et al. 1979). Beers and Hull (1990) evaluated the effect of European red mites on apples in Pennsylvania, allowing 1000 cumulative mite days (one spider mite feeding for one day = a mite day) to occur in the early season (early May to mid-June), mid-season (mid-June to August 1), or late season (August 1 to mid-October). Mid-season injury resulted in the greatest reduction in mean fruit weight at harvest, as well as reduced bloom and fruit load the following season. Late-season injury resulted in reduced bloom the next year; however, Hoyt et al. (1979) observed that, in Washington State, "improved red color and higher soluble solids were positively correlated with *P. ulmi* populations in one test. No other significant effects from mite feeding damage were observed on fruit quality or yield." Hoyt (1967) reported that interspecific competition with *Tetranychus mcdanieli* limits the populations of *Panonychus ulmi* in the high, central area of apple trees, and *P. ulmi* was not considered a serious pest in Washington apple orchards at that time.

Vigorous trees grown with adequate fertilizer may show little effect from feeding by mites (either *Panonychus ulmi* or *Tetranychus mcdanieli*). Hoyt et al. (1979) noted that, "In Washington, apples are thinned extensively, both chemically and by hand, and this thinning may overcome any potential for reduced yields in the second year." Yield losses caused by mites on apple are "more likely to be due to reduced fruit growth the year the population occurs rather than to reduced bloom and fruit set the following year."

16.2.2 *Tetranychus* and *Eotetranychus* Mites on Apples

In southern and eastern Europe, central and western United States, Australia, South Africa, and parts of South America, *Tetranychus urticae, T. mcdanieli, T. turkestani, T. viennensis*, and *Eotetranychus pruni* are the common mite species found on apples. *Tetranychus urticae, T. mcdanieli*, and *Eotetranychus pruni* have a similar biology and behavior. The relative importance of each species varies geographically, but ***Tetranychus mcdanieli*** was the most serious mite pest in Washington apple orchards during the 1960s (Hoyt 1969) (see Figure S16.3 on the CD). It initially causes stippling of foliage, but high populations can result in defoliation and sunburned fruits (see Figures S16.4 and S16.5 on the CD). Adult females overwinter in diapause under bark, in crevices, or in litter in the soil. Diapausing *T. mcdanieli* females also may occur on fruits, which is undesirable (see Figures S16.6 and S16.7 on the CD). In spring, *T. mcdanieli* females move onto newly opened leaves and begin to oviposit. Early in the spring, *T. mcdanieli* tends to be located in the lower central portion of the tree, and, if an acaricide is applied early in the season, good coverage of foliage in the center of the tree is important. Multiple, overlapping generations of *T. mcdanieli* develop later during the growing season and are more uniformly distributed throughout the tree, if not controlled by pesticides or predation. *Tetranychus mcdanieli* produces heavy webbing and is found on the underside of the foliage; it disperses as an adult by wind or by walking.

16.2.3 *Aculus schlechtendali*

Aculus schlechtendali, the apple rust mite (Eriophyoidea), is a common mite in Washington apple orchards (Hoyt 1969). This mite overwinters as an adult within buds or in crevices on twigs. The overwintering mites emerge in late March and early April and aggregate on the green tissue of the buds. They also feed on the underside of leaves but will colonize the upper surfaces when populations increase. These mites tend to be found on the newer foliage during the summer; as many as 600 to 2000 *A. schlechtendali* per leaf were observed by Hoyt (1967), but the distribution of the predator *Metaseiulus occidentalis* was closely related to that of *A. schlechtendali*, which served as an important prey for *M. occidentalis*. It also may serve as a prey for the stigmaeid *Zetzellia mali*; however, *Z. mali* was not considered an important predator of *Tetranychus mcdanieli* in Washington apple orchards at that time (Hoyt 1967).

16.3 IMM FROM THE 1960S TO THE EARLY 1990S

Integrated mite management in Washington apple orchards was developed in the 1960s, primarily by the efforts of S.C. Hoyt (1967, 1969), who acknowledged the importance of papers on integrated control published by Ripper (1956), Stern et al. (1959), van den Bosch and Stern (1962), Chant (1964), and van den Bosch (1965). During the 1950s and 1960s, *Tetranychus mcdanieli* developed resistance to organophosphate (OP) pesticides, dicofol, tetradifon, and binapacryl. As a result, high populations of *T. mcdanieli* occurred in orchards treated with a standard pesticide regime. Clearly, something had to change: Either a new pesticide-intensive approach or an IMM program had to be developed.

Several key components were required for an IMM program. Hoyt (1969) showed that the phytoseiid *Metaseiulus* (= *Typhlodromus* or *Galendromus*) *occidentalis* "could regulate the density of *Tetranychus mcdanieli* populations at very low levels in unsprayed orchards" (see Figure S16.8 on the CD). The distribution of *T. mcdanieli* females and its key predator, *Metaseiulus occidentalis*, was found to be relevant for managing *T. mcdanieli* in Washington apple orchards. *Tetranychus mcdanieli* females emerging from diapause in the spring deposit their first-generation eggs on flower parts or on the foliage in early spring (Hoyt 1967, 1969). Adult females of *M. occidentalis* overwinter in diapause on the trunk of the apple trees and also move onto the flowers and foliage early in the spring to feed. Hoyt (1967) noted that *M. occidentalis* is present during this first generation on these flower parts and foliage and is so "efficient at destroying colonies of *T. mcdanieli* that its relationship to the distribution of *T. mcdanieli* is evident for only short periods before the prey population is destroyed."

In fact, Hoyt (1969) showed that *Metaseiulus occidentalis* is an important predator of both *Tetranychus mcdanieli* and *Aculus schlechtendali*; he did not consider *M. occidentalis* to be an effective predator of *Panonychus ulmi* in Washington apple orchards, although this conclusion has been modified recently. (*Metaseiulus occidentalis* does feed on active stages of *P. ulmi* and now is considered an effective natural enemy of *P. ulmi*.) This predator tolerates the profuse webbing deposited by *T. mcdanieli* and responds to increases in populations of *T. mcdanieli*. Diapausing *M. occidentalis* females leave the foliage in September and seek overwintering sites in crevices on the tree trunk. One method to sample these predators is to place cardboard bands on tree trunks such that the aggregated females in diapause can be counted and removed (Horton et al. 2002).

Another component of the IMM program was the ability to use pesticides in a selective manner. Hoyt (1969) speculated that "if selective insecticides could be found, integrated chemical control of insects and biological control of mites would offer a solution to the problem." He concluded, "The extensive resistance problems with *Tetranychus mcdanieli* can be attributed in part to unnecessary acaricide treatments applied as a result of a lack of knowledge of economic thresholds." Once it was clear that an effective predator of *T. mcdanieli* was available, it was crucial to determine whether pesticides applied to control insect pests (especially the codling moth, *Cydia pomonella*) could be made selective. Selectivity can be achieved by the choice of product, timing of applications, rates at which the product is applied, and placement of the sprays. All of these approaches were employed in the IMM program. The development of moderate levels of resistance to azinphosmethyl in field populations of *Metaseiulus occidentalis* was a crucial component in developing the IMM program; it allowed azinphosmethyl to be applied (at moderate rates only) to control codling moth, the key insect pest of Washington apples. In addition, Hoyt (1969) was able to identify other products that could be applied to suppress disease agents and other insect pests without disrupting the efficacy of *M. occidentalis* populations.

Hoyt (1969) recognized that *Metaseiulus occidentalis* is an obligatory predator that required a consistent source of prey in apple orchards, which could be provided by *Aculus schlechtendali*, *Tetranychus mcdanieli*, and *Panonychus ulmi*. Rust mite populations were found to be crucial in maintaining adequate numbers of *M. occidentalis* at key times during the growing season so it remained an effective natural enemy of *T. mcdanieli*. The extensive body of research contributing to the IMM program is outlined in Table 16.1. Hoyt (1969) suggested that refinements to the program should be developed in the future, including improved cultural controls, the use of other species of predators, and the development of information on the economic injury levels of mites.

Table 16.1 Key Information and Tools Used in the Integrated Mite Management (IMM) Program during the 1960s until the Early 1990s in Washington Apple Orchards

1. *Tetranychus mcdanieli* developed resistance to organophosphates, dicofol, tetradifon, and binapacryl by 1969.
2. No effective spray program was available for *T. mcdanieli* in many orchards.
3. There are relatively few disease problems, and most insects can be controlled with relatively few sprays.
4. *Metaseiulus occidentalis* was shown to be the key predator of *T. mcdanieli* in unsprayed orchards but was killed by azinphosmethyl and other organophosphates prior to the 1960s.
5. Strains of *M. occidentalis* were discovered in the field in the 1960s that were resistant to or tolerant of some pesticides (DDT, azinphosmethyl, parathion, diazinon) used to control insects such as codling moth. Field populations of *M. occidentalis*, however, were susceptible to carbaryl, used for the thinning of apples.
6. The presence of more than one species of phytophagous mite provided prey for predators, especially *M. occidentalis*, which is an obligate predator.
7. *Aculus schlechtendali* did not become resistant to pesticides, and their presence was considered beneficial to maintaining *M. occidentalis* populations.
8. When azinphosmethyl-resistant *M. occidentalis* was discovered, applications of azinphosmethyl at low rates became selective and could be used to control codling moth, yet did not disrupt the control of *T. mcdanieli* by *M. occidentalis*.
9. Additional selectivity was obtained by applying carbaryl for fruit thinning only to the tree periphery during mid- to late May. During this time, *M. occidentalis* primarily is located in the center of the tree, so application methods were important in gaining selectivity.
10. Pesticides (Morestan®, dormant application of ethion + oil, application of dinocap at pink and petal fall, full-bloom applications of dinitrocresol, dormant applications of zinc sulfate and postbloom applications of 2,4,5,-trichloropropionic acid, naphthalene acetic acid, and naphthylacetamide) were identified that had limited toxicity to *M. occidentalis* so sprays could be applied to control pests without disrupting biological control of *T. mcdanieli*.
11. Sprays that could not be used included high rates of carbaryl and azinphosmethyl, as well as binapacryl and dicofol.
12. *Aculus schlechtendali* was recognized as being a key prey for *M. occidentalis*, especially early in the growing season, so adequate predator populations were maintained to suppress *Panonychus ulmi* and *T. mcdanieli*. Populations of *A. schlechtendali* of up to 300 per leaf caused no apparent crop damage, but densities of 20 rust mites per leaf were sufficient to maintain effective populations of *M. occidentalis*.
13. *Metaseiulus occidentalis* is effective in Washington apple orchards due to its similar distribution in time and space and its ability to destroy large populations of *T. mcdanieli* during April and May. If *A. schlechtendali* are present, *M. occidentalis* can persist and suppress *T. mcdanieli* during July and August.
14. Populations of the stigmaeid predator *Zetzellia mali* were thought to be a possible detriment to the IMM program because they fed on *A. schlechtendali* and prevented *M. occidentalis* populations from increasing early in the growing season.
15. Other phytoseiid species, anthocorids, and mirids were found in the orchards under IMM but were not considered sufficiently abundant or widely distributed to be important. *Stethorus picipes* appeared when high populations of *T. mcdanieli* occurred but were susceptible to pesticides used in the IMM program.
16. Other natural enemies increased in abundance in the orchards under IMM, including chrysopids, syrphids, anthocorids, coccinellids, mirids, and the parasitoid *Aphelinus mali*. In some cases, they were sufficiently numerous that treatments for apple aphid and wooly apple aphid were not needed.
17. Adoption of the IMM program increased from 900 acres in 1966 to 40,000 acres in 1967.

Source: Based on information from Hoyt (1967, 1969).

16.4 PROBLEMS WITH IMM IN THE 1990S

Pest management programs typically change over time. Pests develop resistances to pesticides, pesticides are eliminated due to regulatory changes, new pesticides are registered, new pests invade, consumers exert increased pressure to reduce pesticide residues on food, and new pest management tools (such as mating disruption, sterile insect control methods) are developed (Arthurs et al. 2005, Jones et al. 2006, 2009, Brunner et al. 2007a,b). All of the above have occurred to modify the apple pest management program in Washington (Table 16.2). The use of azinphosmethyl (an OP) is being phased out under the 1966 Food Quality Protection Act (Calkins and Faust 2003), and resistance to azinphosmethyl was detected in the codling moth (Knight et al. 1994, Dunley and Welter 2000).

Another significant change in apple IPM occurred with the development of a **mating disruption** protocol for codling moth (Calkins and Faust 2003). Mating disruption works by releasing a synthetic, species-specific pheromone that prevents or delays males from finding and mating with females. Codling moth overwinters as mature larvae in hibernacula under loose bark on the tree or under leaf litter at the base of the tree. Dispensers of the pheromone are placed in orchards before the first moths fly and mate in the spring so fewer codling moth eggs are deposited in the orchard. Mating disruption reduces the number of codling moths that must be controlled with insecticides. The use of mating disruption has been adopted on an estimated 75% of the apple acreage in Washington (Brunner et al. 2007b). In addition, the codling moth granulovirus can be used to suppress codling moth populations (Arthurs et al. 2005). Once alternative methods to suppress the codling moth were deployed, insects previously controlled by azinphosmethyl became significant pests and required the application of pesticides to suppress them. This led to the use of new pesticides, many of which were toxic to *Metaseiulus occidentalis*.

Tests were conducted to evaluate the new insecticides, fruit thinners, and fungicides with regard to their impact on *Metaseiulus occidentalis*, including neonicotinoids (Beers et al. 2005), novaluron, acetamiprid, thiacloprid, calcium polysulfide, ammonium thiosulfate, mineral oil, and dry flowable

Table 16.2 A Brief History of Mite Management Programs in Washington Apple Orchards

1. Codling moth (*Cydia pomonella*) is a key pest of apples in Washington State.
2. When DDT was introduced after World War II to control the codling moth, large outbreaks of spider mites occurred. Acaricide applications were made to control spider mites, and their resistance to the acaricides became a serious problem.
3. Organophosphate (OP) insecticides, especially azinphosmethyl, replaced DDT, and the OPs were toxic to both spider mites and predatory mites.
4. Spider mites soon became resistant to OPs, and their primary predator *Metaseiulus occidentalis* later became resistant to azinphosmethyl, as well.
5. Once resistance to OPs in *M. occidentalis* was recognized in the early 1970s, an integrated mite management (IMM) program developed by S.C. Hoyt was implemented on approximately 90% of the apple acreage. This IMM program remained effective through the 1990s.
6. Changes in pesticide usage occurred to control pests other than the codling moth in the 2000s. The prior use of OPs had suppressed aphids, leafhoppers, leafminers, and leafrollers, but new materials, such as carbaryl, were substituted due to resistance in these pests.
7. Carbaryl also was used for fruit thinning and initially was toxic to the predatory mite *M. occidentalis*; however, moderate levels of resistance were found in *M. occidentalis* in the field. Pyrethroids were not used in Washington apple orchards because they were disruptive to IMM.
8. After about 40 years of using azinphosmethyl to control the codling moth, new management tools were developed in the early 2000s, triggered by the development of OP resistance in the codling moth and by regulatory issues.
9. Mating disruption is now the basis of codling moth control in about 80 to 90% of Washington's apple orchards.
10. Neonicotinoids were shown to disrupt IMM, as was calcium polysulfide (blossom thinner), ammonium thiosulfate (plant nutrient), and dry flowable sulfur (fungicide for mildew control).
11. IMM programs in Washington apples have been disrupted, and acaricide use has increased since the 1990s, increasing production costs and the likelihood of resistance to acaricides developing in spider mites.

Source: Based on information from Beers et al. (2005, 2009), Brunner et al. (2007a), Hoyt (1967, 1969), Hoyt and Burts (1974), and Martinez-Rocha et al. (2008).

sulfur (Fernandez et al. 2005, Martinez-Rocha et al. 2008, Beers et al. 2009). It was recognized that spider mites would become resistant to the new acaricides (and insecticides) that became available for use, so baseline data were obtained to document any resistance that might develop (Beers et al. 1990, Knight et al. 1990, Rathman et al. 1990, Beers et al. 1997, 1998).

Jones et al. (2009) observed that, "Management systems based solely on pesticides have proven to be unstable, and the success of IPM systems in western orchards has been driven by conservation of natural enemies to control secondary pests, combined with pesticides and mating disruption to suppress the key lepidopteran pests." Furthermore, the changes mandated by the Food Quality Protection Act of 1996 "made it necessary to focus efforts on enhancing biological control not only of secondary pests but also of primary lepidopteran pests to help augment new pesticides and mating disruption tactics. The new management programs envisioned will be information extensive as well as time sensitive and will require redesign of educational and outreach programs to be successful" (Jones et al. 2009).

16.5 CURRENT AND FUTURE CHANGES TO IMM IN WASHINGTON APPLE ORCHARDS

The Apple Pest Management Transition Project was initiated to assist growers in adopting a new IPM program for apples. The project goals include: (1) Reduce the use of OP insecticides for control of pests in Washington apple orchards. (2) Increase the use of the Washington State University–Decision Aid System (DAS), which provides real-time pest management decision-making tools and assists with the adoption of the new IPM protocols. (3) Increase knowledge about reduced-risk pesticides and IPM practices among farm workers to improve the work environment. Much useful information is available on the DAS website (http://das.wsu.edu),including monitoring methods, the biology of pests, how to deploy mating disruption, season-specific recommendations for pest management, weather data, phenology models for insects and diseases, a list of pesticides available for pests and their effects on natural enemies, and management recommendations. The decision aids are based on weather data obtained from a network of weather stations located throughout the state. The model predicts the near future (10 days out) based on weather predictions (Jones et al. 2006). It will be interesting to see how the Washington apple IPM program changes in the next few years. Prokopy and Croft (1994) pointed out that IPM programs evolve, and they described four levels. First-level IPM is biologically based, using host-plant resistance, biological control, cultural control, behavioral control, and autocidal control as substitutes for pesticides in managing a *single class* of pests (such as arthropods). Second-level IPM involves the integration of multiple management practices across *all classes of pests* and is more challenging than first-level IPM. Plant pathologists, weed scientists, entomologists, and others must cooperate and work together to develop second-level IPM program, and only a few programs have incorporated second-level IPM concepts. Third-level IPM integrates combined pest-management approaches with the *entire system of crop production*. Fourth-level IPM provides the greatest challenge because it "aims at integrating third-level IPM or sustainable horticulture within a larger off-farm perspective of greater social, psychological, cultural, and political acceptability" (Prokopy and Croft 1994).

SELECTED REFERENCES

Arthurs, S.P., L.A. Lacey, and R. Fritts. 2005. Optimizing use of codling moth granulovirus: Effects of application rate and spraying frequency on control of codling moth larvae in Pacific Northwest apple orchards. *J. Econ. Entomol.* 98:1459–1468.

Beers, E.H. and L.A. Hull. 1990. Timing of mite injury affects the bloom and fruit development of apple. *J. Econ. Entomol.* 85:547–551.

Beers, E.H., S.C. Hoyt, and E.C. Burts. 1990. Effect of tree fruit species on residual activity of avermectin B1 to *Tetranychus urticae* and *Panonychus ulmi*. *J. Econ. Entomol.* 83:961–964.

Beers, E.H., A. Andersen, and R.D. Brown. 1997. Absorption and translaminar activity of abamectin in apple and pear foliage as determined by spider mite (Acari: Tetranychidae) mortality. *J. Econ. Entomol.* 90:566–573.

Beers, E.H., H. Riedl, and J.E. Dunley. 1998. Resistance to abamectin and reversion to susceptibility to fenbutatin oxide in the spider mite (Acari: Tetranychidae) populations in the Pacific Northwest. *J. Econ. Entomol.* 91:352–360.

Beers, E.H., J.F. Brunner, J.E. Dunley, M. Doerr, and K. Granger. 2005. Role of neonicotinyl insecticides in Washington apple integrated pest management. Part II. Nontarget effects on integrated mite control. *J. Insect Sci.* 5:16 (www.insectscience.org/5.16).

Beers, E.H., L. Martinez-Rocha, R.R. Talley, and J.E. Dunley. 2009. Lethal, sublethal, and behavioral effects of sulfur-containing products in bioassays of three species of orchard mites. *J. Econ. Entomol.* 102:324–335.

Brunner, J.F., J.E. Dunley, E.H. Beers, and V.P.Jones. 2007a. Building a multi-tactic biologically intensive pest management system for Washington orchards. In: A.S. Felsot and K.D. Racke (eds.), *Crop Protection Products for Organic Agriculture: Environmental, Health, and Efficacy Assessment* (pp. 131–143). Washington, D.C.: American Chemical Society.

Brunner, J.F., K.R. Granger, and M.D. Doerr. 2007b. *Implementing OP-Alternative Pest Management Programs in Washington Apple.* Wenatchee: Washington State University Tree Fruit Research & Extension Center (http://entomology.tfrec.wsu.edu/op-alternative/).

Calkins, C.O. and R.J. Faust. 2003. Overview of areawide programs and the program for suppression of codling moth in the western United States directed by the United States Department of Agriculture–Agricultural Research Service. *Pest Manag. Sci.* 59:601–604.

Chant, D.A. 1964. Strategy and tactics of insect control. *Can. Entomol.* 96:182–201.

Dunley, J.E. and S.C. Welter. 2000. Correlated insecticide cross-resistance in azinphosmethyl resistant codling moth (Lepidoptera: Tortricidae). *J. Econ. Entomol.* 93:955–962.

Fernandez, D.E., E.H. Beers, J.F. Brunner, M.D. Doerr, and J.E. Dunley. 2005. Effects of seasonal mineral oil applications on the pest and natural enemy complexes of apple. *J. Econ. Entomol.* 98:1639–1640.

Horton, D.R., D.A. Broers, T. Hinojosa, T.M. Lewis, E.R. Miliczky, and R.R. Lewis. 2002. Diversity and phenology of predatory arthropods overwintering in cardboard bands placed in pear and apple orchards of central Washington State. *Ann. Entomol. Soc. Am.* 95:469–480.

Hoyt, S.C. 1967. Population studies of five mite species on apple in Washington. In: G.O. Evans (ed.), *Proceedings of the 2nd International Congress of Acarology, Sutton Bonington, England, July 19–25, 1967* (pp. 117–133). Budapest: Akadémiai Kiadó.

Hoyt, S.C. 1969. Integrated chemical control of insects and biological control of mites on apple in Washington. *J. Econ. Entomol.* 62:74–86.

Hoyt, S.C. and E.C. Burts. 1974. Integrated control of fruit pests. *Annu. Rev. Entomol.* 19:231–252.

Hoyt, S.C., L.K. Tanigoshi, and R.W. Browne. 1979. Economic injury level studies in relation to mites on apple. In: J.G. Rodriguez (ed.), *Recent Advances in Acarology*, Vol. 1 (pp. 3–12). New York: Academic Press.

Jones, V.P., T.R. Unruh, D.R. Horton, and J.F. Brunner. 2006. Improving apple IPM by maximizing opportunities for biological control. *Good Fruit Grower* 57(1):1–8.

Jones, V.P., T.R. Unruh, D.R. Horton, N.J. Mills, J.R. Brunner, E.H. Beers, and P.W. Shearer. 2009. Treefruit IPM programs in the western United States: The challenge of enhancing biological control through intensive management. *Pest Manag. Sci.* 65:1305–1310.

Knight, A.L., E.H. Beers, S.C. Hoyt, and H. Riedl. 1990. Acaricide bioassays with spider mites (Acari: Tetranychidae) on pome fruits: Evaluation of methods and selection of discriminating concentrations for resistance monitoring. *J. Econ. Entomol.* 83:1752–1760.

Knight, A.L., J.F. Brunner, and D. Alston. 1994. Survey of azinphosmethyl resistance in codling moth (Lepidoptera: Tortricidae) in Washington and Utah. *J. Econ. Entomol.* 87:285–292.

Lees, A.D. 1953. Environmental factors controlling the evocation and termination of diapause in the fruit tree red spider mite, *Metatetranychus ulmi* (Koch) (Acarina: Tetranychidae). *Ann. Appl. Biol.* 40:449–486.

MacPhee, A.W. and K.H. Sanford. 1954. The influence of spray programs on the fauna of apple orchards in Nova Scotia. VII. Effects on some beneficial arthropods. *Can. Entomol.* 86:128–135.

Martinez-Rocha, L., E.H. Beers, and J.E. Dunley. 2008. Effect of pesticides on integrated mite management in Washington State. *J. Entomol. Soc. British Columbia* 105:97–107.

Morgan, C.V.G. and N.H. Anderson. 1958. Notes on parathion-resistant strains of two phytophagous mites and a predacious mite in British Columbia. *Can. Entomol.* 90:92–97.

Prokopy, R.J. and B.A. Croft. 1994. Apple insect pest management. In: R.L. Metcalf and W.H. Luckmann (eds.), *Introduction to Insect Pest Management*, 3rd ed. (pp. 543–585). New York: John Wiley & Sons.

Rathman, R.J., E.H. Beers, J.L. Flexner, H. Riedl, S.C. Hoyt, P.H. Westigard, and A.L. Knight. 1990. Baseline bioassays with hexythiazox and clofentezine of three mite species (Acari: Tetranychidae) occurring on Washington and Oregon tree fruits. *J. Econ. Entomol.* 83:1711–1714.

Ripper, W.E. 1956. Effect of pesticides on balance of arthropod populations. *Annu. Rev. Entomol.* 1:403–438.

Sanford, K.H. 1967. The influence of spray programs on the fauna of apple orchards in Nova Scotia. XVII. Effects on some predacious mites. *Can. Entomol.* 99:197–201.

Stern, V.M., R.F. Smith, R. van den Bosch, and K.S. Hagen. 1959. The integration of chemical and biological control of the spotted alfalfa aphid. Part I. The integrated control concept. *Hilgardia* 29:81–101.

van den Bosch, R. 1965. Man and his habitat: Problems of pollution. Integrated pest control in California. *Bull. Atomic Sci.* March:22–27.

van den Bosch, R. and V.M. Stern. 1962. The integration of chemical and biological control of arthropod pests. *Annu. Rev. Entomol.* 7:367–386.

van de Vrie, M. 1962. The influence of spray chemicals on predatory and phytophagous mites on apple trees in laboratory and field trials in the Netherlands. *Entomophaga* 7:215–250.

van de Vrie, M. 1985. Apple. In: W. Helle and M.W. Sabelis (eds.), *Spider Mites: Their Biology, Natural Enemies, and Control*, Vol. 1B (pp. 311–325). Amsterdam: Elsevier.

CHAPTER 17

Integrated Mite Management in California Almonds

17.1 ALMONDS IN CALIFORNIA

Almonds (*Prunus amygdalus*) are a very important deciduous nut crop in California, which produces 67 to 75% of the world's almonds, and the amount of acreage planted is still increasing (USDA 1999). California became the world's largest almond producer in the 1990s, with Spain coming in second (Murua et al. 1993). By 2005, approximately 680,000 acres had been planted, and approximately 50 different varieties are now grown, primarily in the Great Central Valley (which consists of both the northern Sacramento Valley and the warmer southern San Joaquin Valley). Both areas have hot, dry summers and wet, cold winters (freezes occur). Almonds are deep rooted and thrive on deep, well-drained soil. In California, all commercial almond orchards are grown using irrigation (UC IPM 1985). Almonds begin blooming in mid-February, and almonds are harvested during July and August (see Figure S17.1on the CD).

Because fruit-bud differentiation for the following year begins in summer and continues through the autumn and winter, management of spider mites is important throughout the growing season. Spider mites should be managed to reduce leaf injury, maintain photosynthesis levels, and prevent defoliation, which could reduce crop yield. Irrigation practices during the growing season can reduce or enhance the effect of spider mites feeding on almonds. Foliage on water-stressed trees is warmer than foliage on well-watered trees, allowing spider mite developmental rates to increase (see Figure S17.2on the CD). Relatively speaking, almonds are more susceptible than apples or grapes to spider mite feeding damage in California's Central Valley, and pears (especially the Bartlett variety) are more susceptible to mites than almonds. These differences in susceptibility determine which tools are appropriate for integrated mite management (IMM) programs in each crop.

The major insect pest of almonds is the navel orangeworm (NOW), *Amyelois transitella* (Lepidoptera: Pyralidae) (Rice 1978, Rice et al. 1978). Infestations of nuts by *Aspergillus* spp., which produce carcinogenic aflatoxins, are highly correlated with NOW infestations of nuts. The NOW can be controlled with pesticides and by sanitation of orchards during the winter. Removal and shredding of the "mummy" nuts that contain overwintering NOW are an important cultural control practice (Rice et al. 1978, Sibbett and Van Steenwyk 1993).

Other pests of almonds include the peach twig borer (*Anarsia lineatella*), the San Jose scale (*Quadraspidiotus perniciousus*), and six species of spider mites, including the European red mite (*Panonychus ulmi*), the citrus red mite (*P. citri*), the brown almond mite (*Bryobia rubrioculus*), the two-spotted spider mite (*Tetranychus urticae*), the Pacific spider mite (*T. pacificus*), and the strawberry mite (*T. turkestani*) (Rice and Jones 1978, Hoy 1985a, UC IPM 1985). The NOW caused annual damages estimated at US$35 million per year between 1978 and 1980, when the IMM program described below was being developed, and management primarily was achieved at that time by applications of an organophosphorus (azinphosmethyl or diazinon) or a carbamate (carbaryl) insecticide, as well as by removal of "mummy" nuts during the winter (sanitation).

When research for the IMM program in almonds began, California almond growers considered spider mites to be primary, rather than secondary, pests. Most growers sprayed their orchards one to three times a year to suppress *Tetranychus* spider mites. It is difficult to discriminate among *T. urticae*, *T. pacificus*, and *T. turkestani* in the field, so they are treated as general *Tetranychus* species because they appear to cause similar damage and have a similar **phenology** (timing of emergence and development). The *Panonychus* species and *Bryobia rubrioculus* have a different phenology and typically are less serious pests (see Figures S17.3to S17.6on the CD).

17.1.1 Economic Injury Levels

Only a few studies have been conducted to determine the effects of spider mites on almond tree growth and yield (Barnes and Andrews 1978, Andrews and LaPre 1979, Andrews and Barnes 1981, Welter et al. 1984) and economic injury levels (EILs) were not established in 1977. One study found that "high" levels of *Tetranychus* species did not affect almond-tree growth and yield during the year in which the infestations occurred (Barnes and Andrews 1978). However, the yields of trees that were not treated with acaricides were 13 to 19% lower compared to sprayed trees, and terminal shoot length was reduced by 11 and 48% in two orchards during the year *after* the mite infestations. Trunk growth was reduced by 19% in the unprotected control trees. This delay in response to mite feeding is to be expected in trees that have extensive stored nutrients. The delay occurs because the fruit buds for the subsequent year are formed during the current growing season.

Another study indicated that chlorophyll content was negatively correlated with the number of *Tetranychus* species mite days per leaf (Andrews and La Pre 1979). A **mite day** is a spider mite feeding for one day, and the effects of mite days can vary with the season, the time of year, and the cultivar. Although the peak densities of spider mites per leaf in an orchard may be similar, the number of mite days can vary and determine the amount of crop loss. If one mite feeds on a leaf for 20 weeks, then 140 mite days would accumulate. If 20 spider mites feed for 1 week on a leaf, then 140 mite days of damage have accumulated. The mite day concept is used in a variety of crops to estimate potential damage. (During feeding, spider mites remove chloroplasts from the foliage, reducing the amount of photosynthesis that can occur.)

Another estimate suggested that *Tetranychus* mites reduced development, growth, and yield in almonds in the subsequent growing season by 24 and 9%, respectively, after 517 and 424 mite days (Welter et al. 1984). The number of fruit that set was reduced by 20% after 178 mite days during the preceding growing season in trees that were 6 years old. Thus, relative to apples and grapes in California, almonds are more susceptible to spider mite damage.

To prevent such losses, growers during the 1970s and 1980s applied propargite or cyhexatin to suppress spider mites on a preventative basis, or whenever webbing or leaf stippling was seen. Often acaricides were applied without examining trees to determine whether natural enemies were present. One to three sprays were typically applied per growing season, and an economic analysis indicated that a single acaricide application cost US$50.50 per acre at that time (Headley 1983).

17.2 PEST MITES IN CALIFORNIA ALMOND ORCHARDS

The most serious mite pests in California almond orchards during the development of the IMM program were *Tetranychus urticae*, *T. pacificus*, and *T. turkestani* because the feeding associated with these pests caused stippling and defoliation at relatively low densities. Damage caused by *Panonychus ulmi*, *P. citri*, or *Bryobia rubrioculus* rarely resulted in defoliation.

17.2.1 Brown Almond Mite

Bryobia rubrioculus was observed primarily in the late winter and early spring in almond orchards while this IMM program was being developed (see Figure S17.3on the CD). Active stages rest on the woody portions of the tree and move out to feed on the upper leaf surfaces (Summers 1950, Summers and Stocking 1972, Rice and Jones 1978, Hoy 1985a). Because this mite moves off the foliage to rest, sampling must take this into account. This mite can cause stippling and defoliation, but damage usually is limited because there typically are only two generations a year. Damage primarily is found in the lower portion of the tree, which the mites first colonize in the spring. Populations decline in June because the eggs, deposited on twigs, are in diapause and do not hatch until the following spring. Overwintering eggs were destroyed by applications of dormant sprays containing oil and an organophosphate (OP) insecticide. If dormant sprays were made, *B. rubrioculus* populations were limited to orchards in the cooler growing areas of the Central Valley and to orchards with few pesticide applications.

17.2.2 European Red Mite

The European red mite (***Panonychus ulmi***) occurs throughout the California almond-growing area but usually was more abundant in the cooler areas during the time the IMM program was developed. It overwinters as an egg in diapause. Dormant sprays, if properly timed, gave good control. This species could persist all season long and reach densities of several hundred per leaf early in the growing season. It feeds on both upper and lower leaf surfaces and causes stippled foliage (see Figure S17.4on the CD). Defoliation occurred, in rare cases, after about 400 mite days. No phytoseiids known to be specialist predators of *P. ulmi* were found during the 1970s, but *Amblyseius hibisci* (Chant) and *Metaseiulus occidentalis* were observed to feed on *P. ulmi* in several orchards during 1983. Interestingly, *M. occidentalis* could feed on active stages but could not penetrate the chorion of *P. ulmi* eggs. Occasionally, high densities of brown lacewings (Hemerobiidae) were observed feeding on *P. ulmi*, but they did not appear to be common or capable of maintaining low densities of *P. ulmi*.

17.2.3 Citrus Red Mite

The citrus red mite (***Panonychus citri***) was found more often in the southern part of the San Joaquin Valley, perhaps because it moved into almond orchards from adjacent citrus groves (see Figure S17.5on the CD). Aerial movements of *P. citri* into almonds must occur every year because this species lacks a diapause and cannot overwinter on almonds, which are deciduous. Citrus red mites rarely reached damaging levels, but they may have added to the damage caused by *Tetranychus* species. Like *P. ulmi*, it produces little webbing.

17.2.4 *Tetranychus* Species

Three *Tetranychus* species (*T. urticae*, *T. pacificus*, and *T. turkestani*) could occur in the same orchards in the southern half of the Central Valley, and mixed populations could be found on the same leaf (see Figure S17.6on the CD). In the middle and more northern portions of the Central Valley, *T. urticae* and *T. pacificus* predominated. *Tetranychus* species produce copious webbing and cause similar feeding damage (see Figure S17.6on the CD). These mites overwinter in diapause as adult females that are either red or orange and lack feeding spots. During the winter, they can be found under bark scales and in leaf litter and detritus. Once they move onto the foliage, they feed and develop in colonies on the underside of the leaves, but they will move to the top of leaves if population densities become high. Populations of *Tetranychus* mites are found in the central and lower portions of the trees from March to May after they migrate out of their overwintering sites. Populations may increase in June and become rather evenly distributed over the tree, unless they are washed off the lower parts of the trees by sprinkler irrigation. When *Tetranychus* populations are very high, the branches and even entire trees can become covered with webbing, making it very difficult to control these mites with pesticide applications.

During the hot, dry summer months of June, July, and August, population outbreaks often began in *hot spots*, areas that are poorly irrigated or have compacted soil. They spread out from these hot spots throughout the orchard. Other hot spots could be found on trees located along dirt roads, which are exposed to heat and dust, thus favoring spider mite outbreaks. In some cases, entire orchards became defoliated, and severe damage occurred in less than a week (see Figure S17.7on the CD). Regular and frequent monitoring was essential to determine if acaricide applications were necessary. Unfortunately, because these spider mites disperse aerially, an orchard can be inoculated with large numbers of *Tetranychus* mites from a neighbor's orchard, and severe damage can occur in a short time during the summer. In late August and September, populations of *Tetranychus* species typically decline because females enter diapause and seek out overwintering sites.

17.3 RESEARCH ON CONTROL TACTICS

17.3.1 Biological Control

Natural enemies of spider mites found in unsprayed California almond orchards included the phytoseiid *Metaseiulus occidentalis*, six-spotted thrips (*Scolothrips sexmaculatus*), green lacewings (*Chrysoperla carnea*), ladybeetle (*Stethorus picipes*), *Orius*, *Geocoris*, and cecidomyiid larvae. Brown lacewings (*Hemerobius* sp.) fed on European red mites in unsprayed orchards, as well (Hoy et al. 1978, 1979) (see Figures S17.8to S17.10on the CD). When research for the IMM program began, these natural enemies were not recognized as being sufficiently useful that they could provide control of mites in almonds. Surveys conducted during 1977 and 1978 indicated that *M. occidentalis* was common in most orchards, particularly where *Tetranychus* species were present (Hoy et al. 1978, 1979). A survey was conducted to determine whether populations of this predator could survive the OP pesticides used at that time to control the NOW. Previous work by S.C. Hoyt in Washington apple orchards indicated that *M. occidentalis* had responded to field selection and was able to survive OPs applied to control the codling moth (see Chapter 16). This discovery revolutionized IMM in Washington apple orchards, making it possible to reduce or even eliminate sprays to control spider mites.

17.3.2 Pesticide Selectivity Achieved through Laboratory Selection (Genetic Improvement)

Bioassays of *Metaseiulus occidentalis* populations collected from California almond orchards (and vineyards) showed that, indeed, *M. occidentalis* had responded to field selection with OP insecticides used to control the NOW (see Figure 12.11 in Chapter 12). Many *M. occidentalis* populations were resistant to OPs (Hoy et al. 1978, Hoy 1985b). Unfortunately, populations in different orchards had different levels of OP resistance, probably related to the selection intensity (past treatment history) in that specific orchard (see Figure 12.11 in Chapter 12). Without knowing the history of pesticide use in each orchard, it was difficult to predict whether the predators in a specific almond orchard would survive an application of azinphosmethyl. All *M. occidentalis* colonies from almonds were found to be susceptible to carbaryl, a carbamate insecticide, which was used to control the NOW (Roush and Hoy 1980, 1981a,b). Carbaryl was known to cause spider mite outbreaks, probably due to the toxic effect of this product on *M. occidentalis* and, perhaps, to the stimulatory effect of carbaryl (hormoligosis) on spider mite reproduction. One of the novel aspects of this project was the fact that we conducted laboratory selection on populations of *M. occidentalis* to develop a strain that was highly resistant to both OPs and carbaryl (Roush and Hoy 1980, 1981a,b, Hoy 1985b). We wished to determine whether we could **genetically improve** a predator and demonstrate that it could provide improved control of spider mites in the almond orchards.

17.3.3 Cultural Practices

Despite the presence of *Metaseiulus occidentalis* in almond orchards, *Tetranychus* outbreaks commonly occurred in July and August in most almond orchards during the 1970s (Hoy et al. 1978, 1979). The outbreaks were caused by multiple factors, including water stress due to inadequate irrigation, excessive dust from nearby roads (which is thought to inhibit the effectiveness of predators), high temperatures, and the application of pesticides such as carbaryl or, in some orchards, organophosphates. Spider mite outbreaks were exacerbated by early harvest, a management practice designed to reduce NOW feeding damage (Micke et al. 1966). Early harvest can be induced if irrigation is stopped shortly after almond hull split begins early in July. The dry conditions produced by the lack of irrigation induce rapid drying and splitting of the hull, which allows the nuts to shake out when a mechanical tree shaker is used. Unfortunately, the lack of irrigation in July makes the foliage more suitable for spider mite reproduction. Dehydrated foliage is warmer than well-watered foliage, so spider mite development and reproduction are enhanced due to the higher temperatures.

17.4 COMBINED TACTICS OF THE IMM PROGRAM

The IMM program involved the use of laboratory-selected, pesticide-resistant strains of *Metaseiulus occidentalis* in combination with the use of lower-than-label rates of selective acaricides (propargite and cyhexatin), monitoring of predator-to-prey ratios, dust management, spot treatments of small portions of the orchard early in the season, and adequate irrigation (Hoy 1985a) (Table 17.1).The program was based on the following concepts: (1) spider mites and other mites can be *primary pests* (not secondary or induced pests) in almonds, due to the sensitivity of almonds to feeding by *Tetranychus* mites; (2) chemical control of spider mites may be necessary to maintain spider mite populations below the 120-spider-mite-day level estimated to be just under the level that can cause defoliation of almonds; and (3) pesticides (OPs and carbamates) used to control insect pests, such as the NOW, were disruptive to the natural enemies of mites. Predators would eventually recolonize the pesticide-treated trees, but they would build up in adequate numbers too late to suppress the mites and prevent defoliation.

Metaseiulus occidentalis populations had the ability to track *Tetranychus* populations, responding both numerically and functionally to increases in spider densities, with only a small lag time (see Figure S17.7on the CD). The success obtained in the Washington State apple IMM program (see Chapter 16) biased the viewpoint of some, who argued that *M. occidentalis* should be *sufficient* to suppress spider mite populations. In many California almond orchards, however,natural enemies alone did not reduce pest mites sufficiently to keep the number of mite days accumulated under 120, and acaricides were needed because almonds are more sensitive than apples to mite feeding. Unfortunately, the acaricides themselves perpetuated the spider mite problem by disrupting predator populations in one of two ways: by direct mortality or by indirect mortality.

Direct mortality was due to the mortality of predators from the acaricides applied at the label rates. **Indirect mortality** was less obvious but occurred when this obligate predator was unable to survive due to the short-term loss of its prey, causing *Metaseiulus occidentalis* to disperse out of the crop or die of starvation. Because acaricides never kill all spider mites (due to inadequate coverage),

Table 17.1 Components of the Integrated Mite Management (IMM) Program for California Almonds Developed from 1977 to 1984

1. Almonds do not tolerate spider mites and may defoliate if more than 120 spider-mite days accumulate per leaf. Growers did not want any defoliation, so the program was based on reducing mite days below that number. The actual economic injury level was not determined, due to the time and costs such experiments would require.
2. The key pest in almonds was the navel orangeworm (NOW). Although elimination of 'mummy' nuts containing the NOW during the winter reduced populations during the subsequent growing season and parasitoids had been introduced in a classical biological control program, chemical control was still needed. Applications were typically azinphosmethyl, diazinon, or carbaryl.
3. Studies indicated that *Metaseiulus occidentalis* could provide substantial control of *Tetranychus* mites if not disrupted by pesticides applied to control the NOW; however, *M. occidentalis* varied in its resistance levels to azinphosmethyl and was susceptible to carbaryl.
4. A laboratory selection (genetic improvement) project resulted in a strain of *M. occidentalis* that was resistant to azinphosmethyl, diazinon, and carbaryl. Laboratory trials indicated that the resistances were inherited as dominant genes and that there were no detectable fitness costs.
5. Small-scale field trials indicated that the carbaryl-OP-resistant strain of *M. occidentalis* could establish, multiply, survive field rates of carbaryl or azinphosmethyl, and provide substantial control of spider mites.
6. A mass-rearing method was developed to conduct larger-scale field trials.
7. Monitoring methods were developed and lower-than-label rates of propargite were applied, if required, to ensure that growers cooperating in the large-scale field trials would not suffer defoliation during the period when the resistant predators were establishing in the orchard.
8. The rate of propargite applied was 1/10 to 1/20 of the label rate. This rate reduced spider mite populations by approximately 50% yet allowed these obligate predators to survive. A full rate of propargite suppressed spider mites so much that *M. occidentalis* starved, and a rebound of spider mites occurred because *M. occidentalis* populations lagged behind.
9. Spot treatments were made early in the growing season at low rates to reduce spider mite populations. Additional sprays for mites were not needed in the orchard, as a result, in many cases.
10. With the help of University of California Cooperative Extension staff, larger-scale field trials were conducted to confirm that the released predators could establish, multiply, survive carbaryl, and control spider mites.
11. Experiments on aerial dispersal within and out of almond orchards indicated that the resistant strain would disperse but that, if almond growers wanted the carbaryl-OP-resistant strain, each grower needed to release them.
12. Implementation of the program was greatly assisted by the efforts of influential pest control advisors and Cooperative Extension Agents who promoted the IMM program.

Source: Based on information from Hoy (1982, 1984a, 1985a), Hoy et al. (1978, 1979, 1982, 1984, 1985), and Roush and Hoy (1980, 1981).

spider-mite populations usually rebound. Because predator populations lag behind their prey, crop damage may occur due to the lag in numerical and functional responses by the decimated predators. Thus, *M. occidentalis* often required assistance in almond orchards. This came in the form of using selective pesticides against the primary pest (the NOW in this case), as well as an acaricide that was less toxic to the predators than it was to the spider mites. The IMM program was based on the recognition that phytoseiids (or other natural enemies) may not be able to do the job alone. Mite control was achieved by a combination of tactics (Table 17.1).

17.4.1 Resistant Predator Releases

Carbaryl- and OP-resistant *Metaseiulus occidentalis* were mass reared by commercial companies using starter cultures provided by the University of California at Berkeley (Hoy et al. 1982). Several pest-management consultant companies were instrumental in making releases into large acreages and providing essential advice to almond growers who wanted to adopt the program. Growers were interested in adopting the program because it had the potential to reduce their acaricide costs and provide spider mite control that was as good as, or better than, that achieved with acaricides alone (Table 17.2)(also see Figure S17.7on the CD).

Table 17.2 Requirements for Implementation of the Integrated Mite Management (IMM) Program in California Almonds

1. Outstanding efforts by University of California Cooperative Extension agents, and other UC scientists, in the counties to provide training and assistance to growers wishing to use *Metaseiulus occidentalis*, the monitoring methods, and the reduced rates of acaricides
2. Education of growers by publishing results of all stages of the research conducted to develop the program in publications such as *California Agriculture*, which was read by many growers
3. Support from the Almond Board of California, whose financial support was critical, and which allowed presentations at the annual meetings as a means to educate many almond growers about IMM
4. Commercial producers of *M. occidentalis*, who, when provided starter cultures of the carbaryl-OP-resistant strain and information on rearing methods, produced large numbers of resistant *M. occidentalis* for sale to growers wishing to use them
5. Promotion of the program by key professional crop advisors, which ensured that it was adopted by almond growers
6. Economic benefits of the program (e.g., growers could save $44/acre/year by monitoring and applying lower-than-label rates of propargite to assist *M. occidentalis* in suppressing pest mites, and if releases of the carbaryl-resistant predator strain were required the growers could save $24/acre that year and $44/year subsequently)
7. Control of spider mites by *M. occidentalis* that was as good as, or better than, the traditional acaricide program, and which allowed growers to save money (see prior entry)

Predators were released into orchards in several different ways, and all were effective in obtaining establishment and distribution throughout the orchard. A common pattern involved the release of 100 adult females into every third tree in every third row (one tree in nine) early in the growing season (as soon as a few spider mites were detected). Typically, the resistant *Metaseiulus occidentalis* strain multiplied and spread throughout the orchard within a few weeks. Growers, however, were cautioned that predators might require a year to achieve full establishment; therefore, careful monitoring and subsequent application of a low rate of acaricide, if necessary, were required to prevent damage to their trees from spider mite populations. Growers were told that, if releases were made later in the growing season, then control of spider mites should not be expected until the second season. To ensure that the carbaryl-OP-resistant predators became established, at least one (prior to the release) or two carbaryl treatments were applied during the first growing season to control the NOW and to select for the released resistant *M. occidentalis*.

Monitoring of carbaryl-resistant *Metaseiulus occidentalis* indicated that the resistant strain dispersed as far as 800 meters between 1981 and 1983 (Hoy 1982, Hoy et al. 1984). This is a relatively slow dispersal rate, and we had to advise growers that, if they wanted this pesticide-resistant strain, they should purchase the strain and make their own releases rather than rely on natural dispersal. Movement of carbaryl-susceptible *M. occidentalis* into the orchards appeared to be sufficiently low that dilution of the carbaryl resistance was undetectable over a 2- to 4-year period.

17.4.2 Monitoring Tools

A key tool in the IMM program was the development of monitoring methods for spider mites and *Metaseiulus occidentalis* so growers and pest control advisers could make informed decisions. The monitoring that was required at least every week (Wilson et al. 1984, Zalom et al. 1984a,b) allowed growers to adjust the predator-to-prey ratio *early* in the growing season, preferably before hot weather began in June. Two monitoring methods were developed: (1) a brush-and-count method that is labor intensive but very accurate, and (2) a presence–absence sequential sampling method that was useful from mid-June on and involved determining the relative abundance of spider mites and *M. occidentalis*. The *Hilgardia* papers describing the presence–absence sampling method are available on the accompanying CD. The goal of monitoring was to prevent defoliation, making the assumption that defoliation would reduce tree growth and crop yield the following season. The brush-and-count count sampling method used different treatment guidelines for May, June, July, or August (Hoy 1984b). As an example, if no predators were detected in May, if the total number of spider mite days for the season was greater than 10, and if the mean number of spider mites per leaf had increased each week for 3 weeks, then treatment was recommended immediately, using a lower-than-label rate of propargite or cyhexatin (Hoy 1984b). If the ratio of predators was greater than or equal to 1 predator to 10 spider mites per leaf, then no treatment was required, and biological control might be expected very quickly. In June, the predator-to-prey ratio had to be greater than 1 predator for every 5 spider mites, and the total accumulated spider mite days had to be less than 10 for a "no treatment" recommendation. July and August guidelines were even more stringent than those for May or June, due to the hotter, drier weather and the trees being less able to tolerate feeding damage from mites.

17.4.3 Acaricide Applications

A modified acaricide program was adopted to utilize the pesticide-resistant *Metaseiulus occidentalis*. Label rates of propargite and cyhexatin resulted in excellent control of the spider mites, but the control was apparently too good because it led to a dearth of prey for *M. occidentalis* and predator population crashes (Hoy 1984b, Hoy and Conley 1987). To maintain low prey populations in the orchard throughout the growing season, and from season to season, for *M. occidentalis* to feed on, reduced rates of cyhexatin and propargite were applied, if required. The reduced rates were typically one tenth to one twentieth of the lowest label rate. At these rates, the *M. occidentalis* populations could survive, but the acaricide reduced the spider mite populations by about 50 to 80%. Depending on the density of the spider mites (and the number of mite days accumulated), a single low-rate application of acaricide could eliminate the need for subsequent acaricide sprays, yet maintain low densities of both predator and prey mites. If growers monitored their orchards, especially early in the season, to detect hot spots then acaricides could be applied at the lower-than-label rate to a small area. These spot treatments could be sufficient to suppress spider mites and prevent aerial dispersal throughout the orchard. As a result, some growers were able to reduce their acaricide treatments to these small spot treatments.

17.4.4 Cultural Practices

We encouraged growers to maintain effective irrigation programs to maintain tree vigor and to reduce dust by paving or oiling roads near the orchard.

17.4.5 Costs and Benefits

The benefits of the IMM program in almonds were substantial (Headley and Hoy 1986, 1987). Growers adopted the program because they saved $24 to $44 per acre in acaricide costs per year, even after paying for monitoring by a professional pest-control consultant. The $24 return occurred if predator releases were required, and the $44 rate was achieved when no releases were required. Because a high rate of adoption occurred in the 360,000 acres of California almonds during this interval, an estimated $20 million per year was saved in acaricide costs. Environmental and health benefits were not included in this calculation. This project remains one of a few successful examples of an IPM program that used a *laboratory-selected* predator that had undergone *genetic improvement* by selecting it for resistance to carbaryl. Prior to this, it was believed that laboratory selection would inevitably result in laboratory-adapted, inbred, and unfit natural enemies that would not perform well under field conditions (Hoy 1985a,b).

17.5 UPDATED ALMOND PEST MANAGEMENT PROGRAM

As conditions change and more information becomes available, IPM programs are modified. They often become more information intensive. In 2005, a survey indicated that approximately 97% of almond growers were relying on pest control advisors (PCAs) or consultants (Brodt et al. 2005). Approximately 64% worked with a PCA who was affiliated with an agricultural products supplier, and 31% worked with an independent PCA (not affiliated with pesticide sales). Growers with larger acreages were more likely to use an independent PCA for advice on almond IPM and to use more complex pest-monitoring methods and control practices (Brodt et al. 2005). The University of California Statewide IPM program continually provides updates of guidelines for managing insects and mites in almonds, and the trend has been toward the use of fewer and less-toxic pesticides (Klonsky et al. 1990, Hendricks 1995, Epstein et al. 2000). A survey conducted in 2000 suggested that the use of OP sprays was reduced, as well as the use of dormant sprays. When dormant sprays are used, oil alone is used. In addition, up to 41% are applying *Bacillus thuringiensis* (Barnett et al. 1993) or spinosad to control peach twig borer, and 8% are using pheromone-mating disruption for this pest. New NOW sanitation standards have been recommended (a more stringent threshold of 0.2 mummies per tree) to reduce damage and the risk of aflatoxin contamination of the nuts (Higbee and Siegel 2008). Monitoring for European red mite and brown almond mite eggs (and scales) on dormant spurs now is recommended, and if no mite eggs (or scales) are found in the initial sample of 20 spurs, no treatment is required. If 20% or more of the spurs are infested with brown mite or European red mite, sprays are recommended. Narrow range oil is recommended as a less-toxic treatment that preserves natural enemies.

The almond IPM program thus continues to improve and innovate. In 2007, the Almond Pest Management Alliance was awarded an IPM Innovator Award from the California Department of Pesticide Regulation. The Almond Pest Management Alliance is a cooperative that includes the Almond Board of California, the Almond Hullers and Processors Association, the University of California Statewide IPM Program, the University of California Cooperative Extension, pest control advisors and growers, and Region 9 of the U.S. Environmental Protection Agency. The Almond Pest Management Alliance published the *Seasonal Guide to Environmentally Responsible Pest Management Practices in Almonds*, which provides a decision guide for almond growers to reduce pesticide use without decreased yields or increased reject levels (Pickel et al. 2004).

SELECTED REFERENCES

Andrews, K.L. and M.M. Barnes. 1981. Spider mites on almond in the southern San Joaquin Valley of California. *Environ. Entomol.* 10:6–9.

Andrews, K.L. and L.F. LaPre. 1979. Effects of Pacific spider mite on physiological processes of almond foliage. *J. Econ. Entomol.* 72:651–654.

Barnes, M.M. and K.L. Andrews. 1978. Effects of spider mites on almond tree growth and productivity. *J. Econ. Entomol.* 71:555–558.

Barnett, W.W., J.P. Edstrom, R.L. Coviello, and F.P. Zalom. 1993. Insect pathogen "Bt" controls peach twig borer on fruits and almonds. *Calif. Agric.* 47(5):4–6.

Brodt, S., F. Zalom, R. Krebill-Prather, W. Bentley, C. Pickel, J. Connell, L. Wilhoit, and M. Gibbs. 2005. Almond growers rely on pest control advisers for integrated pest management. *Calif. Agric.* 59(4):242–245.

Epstein, L., S. Bassein, and F.G. Zalom. 2000. Almond and stone fruit growers reduce OP, increase pyrethroid use in dormant sprays. *Calif. Agric.* 54(6):14–19.

Headley, J.C. 1983. Economic analysis of navel orangeworm control in almonds. *Calif. Agric.* 27(5/6):26–29.

Headley, J.C. and M.A. Hoy. 1986. The economics of integrated mite management in almonds. *Calif. Agric.* 40(1):28–30.

Headley, J.C. and M.A. Hoy. 1987. Benefit/cost analysis of an integrated mite management program for almonds. *J. Econ. Entomol.* 80:555–559.

Hendricks, L.C. 1995. Almond growers reduce pesticide use in Merced County field trials. *Calif. Agric.* 49(1):5–10.

Higbee, B.S. and J.P. Siegel 2009. New navel orangeworm sanitation standards could reduce almond damage. *Calif. Agric.* 63(1):24–28.

Hoy, M.A. 1982. Aerial dispersal and field efficacy of a genetically improved strain of the spider mite predator *Metaseiulus occidentalis*. *Entomol. Exp. Appl.* 32:205–212.

Hoy, M.A. 1984a. Genetic improvement of a biological control agent: Multiple pesticide resistance and nondiapause in *Metaseiulus occidentalis* (Nesbitt) (Phytoseiidae). In: D.A. Griffiths and C.E. Bowman (eds.), *Acarology VI*, Vol. 2 (pp. 673–679). Chichester, U.K.: Ellis Horwood.

Hoy, M.A. 1984b. *Managing Mites in Almonds: An Integrated Approach*, UC IPM Publication 1. Davis: University of California.

Hoy, M.A. 1985a. Almonds: Integrated mite management for California almond orchards. In: W. Helle and M.W. Sabelis (eds.), *Spider Mites: Their Biology, Natural Enemies, and Control*, Vol. 1B (pp. 269–310). Amsterdam: Elsevier.

Hoy, M.A. 1985b. Recent advances in genetics and genetic improvement of the Phytoseiidae. *Annu. Rev. Entomol.* 30:345–370.

Hoy, M.A. and J. Conley. 1987. Toxicity of pesticides to western predatory mite. *Calif. Agric.* 41(7):12–14.

Hoy, M.A., N.W. Ross, and D. Rough. 1978. Impact of NOW insecticides on mites in northern California almonds. *Calif. Agric.* 32(5):10–12.

Hoy, M.A., R.T. Roush, K.B. Smith, and L.W. Barclay. 1979. Spider mites and predators in San Joaquin Valley almond orchards. *Calif. Agric.* 33(10):11–13.

Hoy, M.A., D. Castro, and D. Cahn. 1982. Two methods for large scale production of pesticide-resistant strains of the spider mite predator *Metaseiulus occidentalis* (Nesbitt) (Acarina, Phytoseiidae). *Z. Angew. Entomol.* 94:1–9.

Hoy, M.A., H. van de Baan, R. Groot, and R.P. Field. 1984. Aerial movements of mites in almonds: Implications for pest management. *Calif. Agric.* 38(9):21–23.

Klonsky, K., F. Zalom, and B. Barnett. 1990. California's almond IPM program. *Calif. Agric.* 44(5):21–24.

Micke, W.E., D.E. Keser, A.D. Rizzi, and C.V. Carlson. 1966. Early harvesting of almonds. *Calif. Agric.* 20(12):14–16.

Murua, J.R., H. Carman, and J. Alston. 1993. California leads Spain in almond production, exports to the world. *Calif. Agric.* 47(6):11–14.

Pickel, C., W.J. Bentley, J.H. Connell, R. Duncan, and M. Viveros. 2004. *Seasonal Guide to Environmentally Responsible Pest Management Practices in Almonds*, UC ANR Publication No. 21619. Berkeley: Agriculture and Natural Resources, University of California.

Rice, R.E. 1978. Insect and mite pests of almonds. In: W. Micke and D. Kester (eds.), *Almond Orchard Management* (pp. 117–127), UC ANR Publication No. 4092. Berkeley: Agriculture and Natural Resources, University of California.

Rice, R.E. and R.A. Jones. 1978. Mites in almonds and stone fruits. *Calif. Agric.* 32(4):20–21.

Rice, R.E., M.M. Barnes, and C.E. Curtis. 1978. Integrated pest management in almonds. *Calif. Agric.* 32(2):18.

Roush, R.T. and M.A. Hoy. 1980. Selection improves Sevin resistance in spider mite predator. *Calif. Agric.* 34(1):11–14.

Roush, R.T. and M.A. Hoy. 1981a. Genetic improvement of *Metaseiulus occidentalis*: Selection with methomyl, dimethoate, and carbaryl and genetic analysis of carbaryl resistance. *J. Econ. Entomol.* 74:138–141.

Roush, R.T. and M.A. Hoy. 1981b. Laboratory, glasshouse, and field studies of artificially selected carbaryl resistance in *Metaseiulus occidentalis*. *J. Econ. Entomol.* 74:143–147.

Sibbett, G.S. and R.A. Van Steenwyk. 1993. Shredding "mummy" walnuts is key to destroying navel orangeworm in winter. *Calif. Agric.* 47(5):26–28.

Summers, F.M. 1950. Brown almond mites. *Calif. Agric.* 4(7):6.

Summers, F.M. and C.R. Stocking. 1972. Some immediate effects on almonds leaves of feeding by *Bryobia rubrioculus* (Scheuten). *Acarologia* 14:170–178.

UC IPM. 1985. *Integrated Pest Management for Almonds*, UC ANR Publication No. 3308. Oakland: University of California Statewide Integrated Pest Management Program.

USDA. 1999. *Crop Profile for Almonds in California*. Washington, D.C.: U.S. Department of Agriculture (http://www.ipmcenters.org/cropprofiles/docs/caalmonds.pdf).

Welter, S.C., M.M. Barnes, I.P. Ting, and Y.T. Hayashi. 1984. Impact of various levels of late spider mite (Acari: Tetranychidae) feeding damage on almond growth and yield. *Environ. Entomol.* 13:52–55.

Wilson, L.T., M.A. Hoy, F.G. Zalom, and J.M. Smilanick. 1984. Sampling mites in almonds. I. The within-tree distribution and clumping pattern of mites with comments on predator–prey interactions. *Hilgardia* 52(7):1–13.

Zalom, F.G., M.A. Hoy, L.T. Wilson, and W.W. Barnett. 1984a. Sampling mites in almonds. II. Presence–absence sequential sampling for *Tetranychus* mite species. *Hilgardia* 51(7):14–24.

Zalom, F.G., L.T. Wilson, M.A. Hoy, W.W. Barnett, and J.M. Smilanick. 1984b. Sampling *Tetranychus* spider mites in almonds. *Calif. Agric.* 38(5):17–19.

CHAPTER 18

Integrated Mite Management in Citrus in Florida and California

18.1 CITRUS PRODUCTION IN FLORIDA AND CALIFORNIA

Florida and California are major citrus-producing states with different climates, pest mite species, and management programs. A comparison of the citrus systems in these two geographic areas provides a perspective on how climate, crop variety, and crop use affect integrated crop management (IMM) programs.

Florida (the number one total citrus producer in the United States) produced 79% of the 2008–2009 U.S. orange crop and about 20% of the world crop on about 460,000 acres; approximately 95% of the orange crop was processed for juice and the rest for fresh market. Florida produced 69% of the 2008–2009 U.S. grapefruit crop and 54% of the world grapefruit crop on approximately 58,000 acres, 50% of which was for fresh market sales. The production of fruit for juice, rather than fresh market, affects pest management practices, particularly when it comes to managing mites (USDA 2010).

California (number two in citrus production in the United States) produced 80% of the lemons, 28% of the tangerines, 21% of the oranges, and 10% of the grapefruit grown in the United States in 2002. Of the approximately 266,250 acres producing citrus fruit, oranges were grown on about 193,000 acres, lemons on 49,500 acres, grapefruit on 14,000 acres, and other minor citrus varieties on the remainder. Most citrus produced in California is for the fresh market, with 82% of oranges, 70% of tangerines and grapefruit, and 50% of lemons being eaten fresh (USDA 2003).

Biological control has played a crucial role in integrated pest management (IPM) of citrus pests (McCoy 1985, Caltagirone and Doutt 1989, Browning 1994a,b, Browning and McCoy 1994, Browning and Stimac 1994, Childers 1994a, Smith et al. 1997, McCoy et al. 2009). Predatory ants (*Oecophila smaragdina*), for example, have been used to manage citrus pests in China for hundreds of years. Another landmark example is the classical biological control of cottony cushion scale in California's citrus by the Vedalia beetle (*Rodolia cardinalis*) (Caltagirone and Doutt 1989). Importation of the Vedalia beetle into California over 100 years ago was an outstanding success and has often served as a model for what classical biological control could be: inexpensive, permanent, and highly effective. Unless Vedalia beetles are disrupted by the use of toxic pesticides, control of the cottony cushion scale has been stable in California (and Florida) (Grafton-Cardwell and Gu 2003). Vedalia beetles have been transplanted successfully to other citrus-growing regions around the world. This remarkable success has led to a phenomenon I refer to as the 'cottony cushion scale syndrome,' because many researchers and funding agencies have developed the unrealistic expectation that classical biological control always is rapid, inexpensive, perpetual, and fully effective.

18.2 DIVERSITY IN CALIFORNIA CLIMATES AND MITE SPECIES

Citrus is grown in four major areas in California, each with distinctly different climates, ranging from the Desert to the Coastal–Intermediate regions (USDA 2003, UC IPM 2010). Pest management practices and pests vary with citrus production region. Approximately 65% of California citrus is grown in the San Joaquin Valley, where summers are hot and dry, there are no summer rains (trees are irrigated), and winters are cold and wet. A major insect pest in the San Joaquin Valley is the California red scale (*Aonidiella aurantii*), which is frequently treated with pesticides and has developed resistance to many. Some growers release the parasitoid *Aphytis melinus* to augment control of the red scale by the endemic natural enemies, in part because *A. aurantii* is resistant to many pesticides (Grafton-Cardwell and Vehrs 1995, Grafton-Cardwell et al. 2006). Citrus thrips (*Scirtothrips citri*) are significant pests of citrus in the San Joaquin Valley because the thrips feed on epidermal cells of the fruit, leaving scabby scars on the rind, which causes a conspicuous ring. This cosmetic flaw makes the fruit unmarketable as fresh fruit but has no effect on fruit quality (Hare et al. 1990, Grafton-Cardwell et al. 1995). Resistance to pesticides in citrus thrips also has developed (Morse and Brawner 1986, Kahn and Morse 1998), so their control by predators is important.

The predatory mite *Euseius tularensis* (Phytoseiidae) is an effective predator of citrus thrips, as are some generalist predators (including spiders, lacewings, and minute pirate bugs) (McMurtry 1985). Thrips problems are increased in groves with intense pesticide treatments for scales due to disruption of these natural enemies. Other insect pests of citrus in California include the variegated cutworm (*Peridroma saucia*), which may feed on young fruit, especially in the San Joaquin Valley.

18.2.1 Citrus Red Mites

The citrus red mite (***Panonychus citri***) is the most common mite in California citrus. Outbreaks occur if their natural enemies are killed by pesticides (Jones and Parrella 1984, McMurtry 1985, Zalom et al. 1986, Hare et al. 1990, Grafton-Cardwell et al. 1995) (see Figure S18.1A on the CD). This mite has developed resistance to pesticides (Jeppson et al. 1958), so effective biological control is important. The economic injury level for the citrus red mite was revised recently for San Joaquin Valley navel oranges and coastal lemons when it was discovered that these varieties could tolerate higher populations than previously thought. Treatment normally is not required in healthy orchards under IPM. This change in economic injury levels is not unusual; economic injury levels rarely are well defined in perennial crops and often are too stringent. Populations of *P. citri* are under biological control by *Euseius tularensis* (the phytoseiid that also feeds on citrus thrips) (see Figure S18.2 on the CD). *Euseius tularensis* for a long time was considered to be *E. hibisci* but was subsequently found to be a different species (Congdon and McMurtry 1985). *Euseius tularensis* is the most important natural enemy of citrus red mite, in part because it can feed on alternative food sources (pollen, citrus thrips larvae, nectar, and honeydew) (see Figure S18.2 on the CD). Populations of *E. tularensis* vary in their tolerance to miticides and insecticides; some populations are fairly resistant due to past selection in the groves. Buildup of *E. tularensis* populations on citrus red mite prey allows this predator to do a better job of suppressing citrus thrips populations later. Other predators of citrus red mite include the phytoseiid *E. stipulatus*, a lady beetle (*Stethorus picipes*), a predatory dustywing (*Conwentzia barretti*), and the six-spotted thrips (*Scolothrips sexmaculatus*). In addition, a viral disease specific to the citrus red mite is widespread and can become important under warm, moderately dry conditions when mite densities are high (Reed et al. 1975, McCoy et al. 2009, UC IPM 2010). Growers in the San Joaquin Valley depend on the heat of summer and the virus to reduce high citrus red mite populations (see Figure S18.1B on the CD). Monitoring of citrus red mite populations can be achieved by a presence–absence sampling protocol (Jones and Parrella 1984, Zalom et al. 1986).

18.2.2 Citrus Bud Mite

The citrus bud mite (*Eriophyes sheldoni*) is a pest of Coastal–Intermediate region lemons and other citrus in that area, but rarely is it important in other areas (see Figure S18.3 on the CD). The mites feed inside the buds, killing them or causing a rosette-like growth of the foliage and distorted flowers and fruit, which may reduce yield. The life cycle consists of eggs, larva, nymphochrysalis, nymph, imagochrysalis, and adults (Sternlicht and Goldenberg 1971).The mites are arrhenotokous, and males deposit stalked spermatophores. Growers monitor orchards to detect bud mites, checking buds on green angular twigs from mid-spring to autumn. If 40 to 50% of the buds are infested with live mites, economic loss of fruit buds and distorted fruit are likely. Generalist predators feed on bud mites when they are not protected within the buds, and phytoseiids will feed on them when they are in the buds. Chemical control, if required, is applied 2 to 3 months before bloom to protect fruits.

18.2.3 Broad Mite

The broad mite (*Polyphagotarsonemus latus*) is a pest of lemons in the Coastal–Intermediate region of California beginning in March but rarely is an economic pest in the Interior, San Joaquin Valley, or desert citrus-growing regions. Broad mites cause scarred fruits that crack as the fruit grows, leaving a characteristic pattern of scars and new tissue. Monitoring involves visual inspection of leaves for curling and for fruit clusters. There are no cultural or specific biological controls. Chemical controls include wettable sulfur.

18.2.4 Citrus Flat Mite

The citrus flat mite (*Brevipalpus lewisi*) is often an important pest on tangerines in the desert region and a sporadic pest in the San Joaquin Valley and interior regions. Feeding by the citrus flat mite causes scabbing on fruit. The flat mite usually is a secondary invader, feeding on rind tissue damaged by leafhoppers, thrips oviposition, or wind. There are no cultural or specific biological controls for this mite, but the mites can be suppressed by wettable sulfur.

18.2.5 Citrus Rust Mite (Silver Mite)

The citrus rust mite (***Phyllocoptruta oleivora***) is a pest of Coastal–Intermediate region citrus, where humidity is relatively high, at least compared to other citrus-growing regions in California (see Figure S18.4 on the CD). It feeds on exposed surfaces of fruit, producing silvery damage on lemons, rust-brown damage on mature oranges, or black damage on green oranges (see Figure S18.5A,B on the CD). Damage occurs from late spring to late summer. Monitoring for rust mite should be done during early spring through summer. When growers find one or more infested fruit, and if rust mites were a problem the previous year, the orchard is watched closely. Threshold levels depend on the previous year's rust mite damage and current market conditions. Control is achieved by minimizing dust within the orchard (paving roads, using water trucks to wet dirt roads). Trees may be washed with water to remove dust. No effective natural enemies are known, but generalist predators such as *Stethorus*, predaceous dustywings, and six-spotted thrips feed on rust mites. If rust mite populations increase quickly or if scarring occurs, treatment is required. Spot treatments may be sufficient if the infestation is localized. Chemicals used include wettable sulfur.

18.2.6 Yuma Spider Mite

The Yuma spider mite (***Eotetranychus yumensis***) is a pest of grapefruit and lemons in the Desert region, where it can cause defoliation. Other common mites in the Desert region include the Texas citrus mite, citrus red mite, and citrus flat mite. Again, dust reduction can help reduce mite outbreaks, but there are no specific biological control agents for the Yuma spider mite. Damage usually is not sufficiently severe to justify treatment, but sulfur is effective.

18.2.7 Six-Spotted Mite and Two-Spotted Mite

The six-spotted mite (***Eotetranychus sexmaculatus***) and the two-spotted mite (***Tetranychus urticae***) are considered to be of minor importance in California citrus. The six-spotted mite is a minor pest in some Coastal–Intermediate region groves, and feeding on foliage may result in some defoliation. Control is by dust reduction and biological control by the six-spotted thrips, *Stethorus picipes*, *Orius* spp., and the phytoseiid *Euseius tularensis*. When chemical control is required, wettable sulfur can be applied. *Tetranychus urticae* is an occasional pest in the San Joaquin Valley, and its effects are exacerbated by water stress and heat, resulting in defoliation and fruit drop. *Tetranychus urticae* primarily is found on the underside of leaves. High populations in summer and fall may require treatments, but thresholds have not been established. Adequate irrigation will reduce the effects of their feeding, and generalist predators (six-spotted thrips, *Stethorus*, *Orius*, and *E. tularensis*) are important if not disrupted by chemical controls. Chemical control can be obtained by using narrow range oil.

The importance of pests varies by climatic region and citrus variety (Table 18.1).The California red scale, citrus thrips, citricola scale, cottony cushion scale, fork-tailed katydid, citrus cutworm, native gray ant, southern imported fire ant, and Fuller rose beetles are the top pests, in descending order of importance in the San Joaquin Valley.Note that no mites are on this list for the San Joaquin Valley. For the Coastal–Intermediate region, the citrus bud mite ranks as the number one pest on lemons, with Argentine ant, citrus thrips, California red scale, brown garden snail, rust and broad mites, black scale, whiteflies, mealybugs, and Fuller rose beetle being less important. In the Desert region, citrus thrips, California red scale, citrus flat mites, Yuma mites, Texas citrus mites, citrus red mites, woolly whitefly,brown garden snail, citrus peel miner, and Fuller rose beetle are pests in descending order of importance. Note that spider mites are important in these hot, dry desert areas.

Table 18.1 Influence of Climate, Crop Use, and Cultivar on Mite Management in Citrus in Florida and California

Climate	The mite species that are pests in Florida citrus are different from those that are pests in California citrus. In Florida, for example, the citrus rust mite (*Phyllocoptruta oleivora*) is the most important mite pest. In California, the species of mites that are important vary by geographic region, with significant pest species differing along the cooler and wetter Coastal, in the drier Central Valley, or in the Desert regions.
Crop use	In Florida, prior to the invasion of the Asian citrus psyllid and citrus leafminer, the key mite pest was the citrus rust mite. If growers were growing their fruit for fresh market, they had to apply acaricides to prevent any russeting of the fruit. On the other hand, if growers were growing their fruit for processing into juice, then acaricide treatments could be reduced, because the rust mite damage to the exterior of the fruit did not affect the quality of the product. In fact, it is possible that fruit with citrus rust mites are a little sweeter.
Cultivar	Lemons grown along the Coastal region of California are especially vulnerable to damage from the citrus bud mite (*Eriophyes sheldoni*). Cultivars grown for juice in Florida are less likely to require acaricide applications than cultivars grown for fresh market.

In the interior, no mites are among the top nine pest arthropods; thus, spider mites are primarily a problem in the hot, dry Desert valleys. The only eriophyoid that is a serious pest (citrus bud mite) survives inside buds in the more humid Coastal–Intermediate area where it is a pest of lemons.

During the 1990s, the glassy-winged sharpshooter (*Homalodisca vitripennis*) (Cicadellidae) established in California and increased the transmission of the causative agent of Pierce's disease (*Xylella fastidiosa*) in grapes. Unfortunately, the sharpshooter overwinters in citrus groves and then moves into neighboring vineyards, where it can spread Pierce's disease (Grafton-Cardwell et al. 2008). This has resulted in a search for IPM-compatible pesticides to suppress the glassy-winged sharpshooter in citrus that do not disrupt natural enemies of citrus pests. Pest management practices in California citrus groves may have to be modified yet again in the coming years due to the invasion of the Asian citrus psyllid (*Diaphorina citri*), which has just become established in California, although at the time of this writing it is not known whether the causative agent of greening disease is present in California. If greening disease is found in California, then the IPM program will have to be modified because it is difficult to control the spread of greening disease without increased pesticide use.

18.3 MANAGING MITES IN FLORIDA CITRUS GROVES

Considerable effort has been placed on the use of biological control in Florida's citrus groves over the years. Biological control was, for many years, the primary management tactic (McCoy 1985, Browning 1994a,b, Browning and McCoy 1994, Browning and Stimac 1994, Childers 1994a); however, recent invasions of arthropod pests and diseases have created chaos in the citrus IPM program in Florida. Until recently, conservation of natural enemies and classical biological control were the primary IPM tactics used, and substantial or even complete biological control of whiteflies, scales, mealybugs, aphids, and other pests was possible. Two diseases, citrus greening (also known as *huanglongbing*) caused by *Candidatus* Liberibacter asiaticus (Spann et al. 2010) and citrus canker caused by *Xanthomonas axonopodis* (Graham and Dewdney 2009), have disrupted this biological-control-based IPM program.

Insecticides applied to control the Asian citrus psyllid (*Diaphorina citri*) and the citrus leafminer (*Phyllocnistis citrella*) have changed the Florida IPM program to a pesticide-based one, at least for the present (Knapp et al. 1998, Futch and Albrigo 2010, Spann et al. 2010). The change in approach is due to the fact that the psyllid is a vector of the causal agent of citrus greening disease, and the citrus leafminer provides openings in foliage that enhance the ability of the causative agent of canker disease to infect citrus foliage. The dramatically increased use of pesticides to control the psyllid has led to concerns about the development of resistance in the psyllid (Rogers 2010), as well as about secondary pest outbreaks.

Florida's pest management program always has had to evolve over time because new pests have regularly invaded; for example, between 1992 and 2002, three citrus pests—the citrus leafminer, the brown citrus aphid (*Toxoptera citricida*), and the Asian citrus psyllid—invaded Florida (Hoy et al. 1995, Hoy and Nguyen 1997, 2000a,b, Skelley and Hoy 2004, Persad et al. 2007). Because the IPM program in Florida's citrus was so heavily dependent on biological control, it was on a 'biological control treadmill' rather than the more common **pesticide treadmill** (Hoy 2000).

Now citrus growers are so concerned about canker and greening diseases that they are using chemical control as their primary IPM tactic, applying 8 to 28 pesticide applications a year to suppress the Asian citrus psyllid (Rogers 2010, Spann et al. 2010). How long this can continue without resistance to pesticides developing in the psyllid and other insect pests is unknown. The current program to manage greening disease is considered unsustainable due to the high costs of scouting for diseased trees and their removal, as well as the costs of applying pesticides to protect them. Research is underway to develop IPM-compatible methods for controlling canker and greening disease, but it will take time to produce results that can be deployed, because plant pathogens vectored by insects are very difficult to control. As a result, the following comments are based on the status of mites in Florida's citrus prior to 2006. Once again, the citrus program in Florida illustrates the point that IPM programs rarely are static. As new invasive pests or diseases disrupt established programs, new management tools are needed. This takes time and requires a substantial investment in new research.

Although variation in soil and climate exists in Florida's citrus-growing regions, the differences are much fewer than in California. As a result, the insect and mite pests found (with the exception of invasive pests such as citrus root weevils, which may take some years to spread throughout the state) appear more uniform in their distribution. The most important mite pest of citrus in Florida has been the citrus rust mite (*Phyllocoptruta oleivora*) (Muma 1961, Allen et al. 1994, Yang et al. 1995). Also found on citrus in Florida, although of less importance, are another eriophyoid, the pink citrus rust mite (*Aculops pelekassi*); three tetranychids—Texas citrus mite (*Eutetranychus banksi*), citrus red mite (*Panonychus citri*), and six-spotted mite (*Eotetranychus sexmaculatus*); a tarsonemid, the broad mite (*Polyphagotarsonemus latus*); and three tenuipalpids (*Brevipalpus californicus, B. obovatus*, and *B. phoenicis*) (Muma 1961, Childers et al. 1991, 2003a, Childers 1994a,b, 1995, Childers and Fasulo 1995, Timmer and Duncan 1999, Futch et al. 2008).

18.3.1 Citrus Rust Mite

Phyllocoptruta oleivora is found throughout Florida and may be associated with the pink citrus rust mite. These two species develop high populations at different times of the year and are most important as pests of fresh market fruits. Both feed on stems, foliage, and fruits. Mite populations begin to increase in late April to early May on new foliage, reaching a peak in mid-June to mid-July. Citrus rust mite populations usually decline in late August, but increase again in late October or early November (McCoy 1976, Allen et al. 1994). During the summer, citrus rust mites are abundant on fruit and foliage on the outer margins of the tree. Generally, the north bottom section of the tree is preferred. These mites disperse aerially and can move from grove to grove on air currents (Bergh 2001, Bergh et al. 2002).

Injury varies with variety and fruit maturity (Allen 1978, 1979). If epidermal cells are destroyed early in the growing season, smaller fruits may result. Further growth leads to a break-up of dead epidermis, resulting in injury known as *russeting*. Damage to mature fruit (after September) differs from russeting; epidermal cells die and become brownish-black (called *bronzing*). Fall-damaged bronzed fruit can be cleaned (polished) because the natural cuticle and wax layer remain intact. Damage from rust mites can result in reduced fruit grade, reduced fruit size, increased water loss, increased fruit drop, and reduced juice quality. Leaf injury from feeding by rust mites can result in damage similar to russeting, and lower leaf surfaces may show mesophyll collapse.

Tree vigor affects citrus rust mite populations. Dense canopies result in lower temperatures and higher relative humidities, which are less favorable for a rapid increase in citrus rust mite populations; thus, citrus variety, fertilizers, irrigation, and other cultural practices affect this important mite pest.

Rust mite damage (cosmetic blemishes) on the fruit is less important when the fruit is grown for processing as juice. As a result, control efforts could be greatly reduced for fruits grown for juice prior to 2006. In most circumstances, a summer spray was required for greasy spot control, and a miticide such as oil was included at that time to suppress citrus rust mites. A fungal pathogen, *Hirsutella*, can suppress high rust mite populations in the summer if fungicides such as copper are not applied that disrupt growth of *Hirsutella* (see Figure 14.4 in Chapter 14).

Monitoring to resolve whether pesticides are required can be conducted in several ways: (1) determine the percentage of fruit and leaves infested with rust mites, (2) apply a qualitative rating scale, and (3) count adult mites on fruits from randomly selected trees. Mite monitoring should be initiated as soon as populations are detected in the spring and continue every 2 to 3 weeks throughout the season. Monitoring by estimating mite density per square centimeter involves sampling selected trees at random throughout a 10- to 40-acre block of a single variety with uniform horticultural practices. Fruit should be chosen from the four quadrants of the tree and midway in the canopy (between the interior and exterior). The number of rust mites per square centimeter should be recorded and averaged for the 10 acres represented by 20 trees with 4 fruit per tree (or 80 fruit per 10 acres). Six rust mites per square centimeter would indicate that pesticide applications might be required within 10 to 14 days. Ten rust mites per square centimeter would indicate that treatment is required as soon as possible. For fresh market fruit, the action threshold is lower, with an average of two citrus rust mites per square centimeter. Registered chemical controls for citrus rust mite in Florida include sulfur and petroleum oil.

18.3.2 Tetranychidae

Three spider mite species are found in Florida's citrus groves: Texas citrus mite (***Eutetranychus banksi***) (Childers 1995, Hall and Simms 2003), citrus red mite (***Panonychus citri***), and six-spotted mite (***Eotetranychus sexmaculatus***) (Muma 1961, Timmer and Duncan 1999). These mites are most abundant in groves during the relatively dry interval between March and June, because spider mites prefer dryer weather and lower relative humidities (30 to 60%). The Texas citrus mite is more common than the citrus red mite (Hall and Simms 2003) (see Figure S18.6 on the CD), but the six-spotted mite is a sporadic pest. Spider mites feed on foliage, removing cellular contents and causing cell destruction and reduced photosynthesis. High populations of spider mites may feed on fruit. Spider mites can be suppressed by several species of predatory mites, insects, and pathogens (Childers 1994a, McCoy 1996, McCoy et al. 2009). If populations averaging 5 to 10 active stages per leaf develop between September and May, chemical control may be necessary, especially if the trees are water stressed. Petroleum oil provides control, although it has a short residual activity, and care must be taken to reduce the likelihood of phytotoxicity.

18.3.3 Tarsonemidae

The broad mite (***Polyphagotarsonemus latus***) is a problem on citrus grown in greenhouses or shade-house conditions and on lemons and limes in the field. Broad mites often are found in depressions on fruit, where the females lay their distinctive eggs (dimpled, translucent, and white). The active stages are yellowish in color, and adult females have a white stripe on the back. Broad mites feed on fruit and leaves, preferring young fruits. Feeding results in scarring of fruits or curling of leaves. Control can be achieved by applying wettable sulfur sprays. No specific natural enemies have been identified for these mites on citrus in Florida.

18.3.4 Tenuipalpidae

Brevipalpus phoenicis and ***B. obovatus*** were reported to feed on tangerines, causing sufficient injury that defoliation has occurred (Childers 1994b). *Brevipalpus* mites feed on fruit, stems, branches, and foliage of citrus and are most commonly found on the lower surface near the midrib or veins. False spider mites are reddish, slow moving, and small. They have relatively long life cycles compared to other mite families, and their populations usually increase slowly. Their main claim to pest status is due to the fact that these mites are associated with a disease called **leprosis** (also called 'nailhead rust'). Leprosis was a serious disease in Florida's citrus before the late 1920s, but the disease seemed to disappear in the 1920s, probably because sulfur began to be applied for mite control (Childers et al. 2003b, Bastianel et al. 2010). Leprosis can occur on fruit, leaves, shoots, and large limbs. There is concern that this disease can reappear in Florida (Childers et al. 2003b). Details of the biology of the citrus mite species can be found in the references cited below. In addition, many websites contain photographs and descriptions of their biology and damage. Photographs are available for most of the life stages to help confirm the identity of any mites encountered on citrus in California or Florida. Again, if you cannot be sure of your species (you could have discovered a newly introduced pest), submit specimens (adult males and females) to a taxonomic specialist in the groups.

18.4 RESEARCH NEEDS

The citrus pests and their importance vary in both California and Florida, indicating that IMM programs must be tailored to the specific environment in which the crop is grown (Table 18.1). Generally, we know too little about the economic injury levels of mites on the various cultivars of citrus. Many spray programs are based on what a grower will tolerate with regard to visible damage to foliage or fruits. In Florida prior to 2006, a substantial number of sprays for citrus rust mite could be eliminated if growers were realistic in determining whether their crop was destined for juice extraction or for fresh market. Approximately 80% of all fruit was destined for juice, yet many treated their crops as though they had to be blemish free for fresh market sale. Excessive sprays are costly, can disrupt natural enemies, and can be harmful to the environment and to non-target species.

The amount of foliar damage that can be tolerated from mites in citrus is not well resolved for good reason—such experiments are costly and lengthy. Yield reductions due to foliar damage are difficult to quantify because citrus trees contain substantial amounts of stored nutrients. Thus, yield studies must be conducted for several years to detect changes in tree growth rate or yield. Varietal, climatic, and cultural differences influence such studies, making it necessary to have large numbers of replicates to obtain good data. Historically, growers and pest managers have erred on the side of treating pests more frequently than can be scientifically justified. Resistance to pesticides in many pests has been a consequence of overtreating, as well as disruption of natural enemies by the pesticide treatments.

Pest managers of mites in citrus may need to conduct experimental trials, at least on a small scale, to determine whether effective mite control can be achieved if pesticide applications are reduced or modified to allow more natural enemies to survive. Wettable sulfur applications or oils (highly refined to reduce phytotoxicity) historically have been effective in suppressing pest mites if coverage is adequate and applications are timed to deal with newly hatched mites. Resistance to oils has never been demonstrated in mites, and resistance to sulfur has been documented only in spider mites found on grapes in California. Phytotoxicity issues should be considered carefully when applying oils. Finally, improved management tactics for greening and canker are needed that will be compatible with the extensive suite of natural enemies of the other pests of citrus in Florida and California. For additional information on citrus mites see Gerson (2003), Vacante (2010a,b), and Smith et al. (1997).

SELECTED REFERENCES

Allen, J.C. 1978. The effect of citrus rust mite damage on citrus fruit drop. *J. Econ. Entomol.* 71:746–750.

Allen, J.C. 1979. Effect of citrus rust mite damage on citrus fruit growth. *J. Econ. Entomol.* 72:195–201.

Allen, J.C., Y. Yang, J.L. Knapp, and P.A. Stansly. 1994. The citrus rust mite story: A modeling approach to a fruit–mite–pathogen system. In: D. Rosen, F.D. Bennett, and J.L. Capinera (eds.), *Pest Management in the Subtropics: Biological Control—A Florida Perspective* (pp. 617–639). Andover, U.K.: Intercept.

Bastianel, M, V.M. Novelli, E.W. Kitajima, K.S. Kubo, R.B. Bassanezi, M.A. Machado, and J. Freitas-Astua. 2010. Citrus leprosis: Centennial of an unusual mite–virus pathosystem. *Plant Dis.* 94(3):284–292.

Bergh, J.C. 2001. Ecology and aerobiology of dispersing citrus rust mites (Acari: Eriophyidae) in central Florida. *Environ. Entomol.* 30:318–326.

Bergh, J.C., C.W. McCoy, and B.S. Otieno. 2002. Diel periodicity of emergence of adult citrus rust mites in central Florida. *Exp. Appl. Acarol.* 26:169–185.

Browning, H.W. 1994a. Classical biological control of citrus scale insects. In: D. Rosen, F.D. Bennett, and J.L. Capinera (eds.), *Pest Management in the Subtropics: Biological Control—A Florida Perspective* (pp. 49–78). Andover, U.K.: Intercept.

Browning, H.W. 1994b. Early classical biological control on citrus. In: D. Rosen, F.D. Bennett, and J.L. Capinera (eds.), *Pest Management in the Subtropics: Biological Control—A Florida Perspective* (pp. 27–46). Andover, U.K.: Intercept.

Browning, H.W. and C.W. McCoy. 1994. Friendly fungi and the citrus tree snail. In: D. Rosen, F.D. Bennett, and J.L. Capinera (eds.), *Pest Management in the Subtropics: Biological Control—A Florida Perspective* (pp. 11–25). Andover, U.K.: Intercept.

Browning, H.W. and J.L. Stimac. 1994. Classical biological control of whiteflies on citrus. In: D. Rosen, F.D. Bennett, and J.L. Capinera (eds.), *Pest Management in the Subtropics: Biological Control—A Florida Perspective* (pp. 79–100). Andover, U.K.: Intercept.

Caltagirone, L.E. and R.L. Doutt. 1989. The history of the vedalia beetle importation to California and its impact on the development of biological control. *Annu. Rev. Entomol.* 34:1–16.

Childers, C.C. 1994a. Biological control of phytophagous mites on Florida citrus utilizing predatory arthropods. In: D. Rosen, F.D. Bennett, and J.L. Capinera (eds.), *Pest Management in the Subtropics: Biological Control—A Florida Perspective* (pp. 255–288). Andover, U.K.: Intercept.

Childers, C.C. 1994b. Feeding injury to 'Robinson' tangerine leaves by *Brevipalpus* mites (Acari: Tenuipalpidae) in Florida and evaluation of chemical control on citrus. *Florida Entomol.* 77:265–271.

Childers, C.C. 1995. *Texas Citrus Mite*, Publication No. ENY-818. Gainesville: University of Florida IFAS Extension (http://edis.ifas.ufl.edu/ch022).

Childers, C.C. and T.R. Fasulo. 1995. *Citrus Red Mite*, Publication No. ENY-817. Gainesville: University of Florida IFAS Extension (http://edis.ifas.ufl.edu/ch021).

Childers, C.C., M.M. Abou-Setta, and M.S. Nawar. 1991. Biology of *Eutetranychus banksi*: Life tables on 'Marsh' grapefruit at different temperatures (Acari: Tetranychidae). *Int. J. Acarol.* 17:29–35.

Childers, C.C., J.V. French, and J.C.V. Rodrigues. 2003a. *Brevipalpus californicus*, *B. obovatus*, *B. phoenicis*, and *B. lewisi* (Acari: Tenuipalpidae): A review of their biology, feeding injury and economic importance. *Exp. Appl. Acarol.* 30:5–28.

Childers, C.C., J.C.V. Rodrigues, K.S. Derrick, D.S. Achor, J.V. French, W.C. Welbourn, R. Ochoa, and E.W. Kitajima. 2003b. Citrus leprosis and its status in Florida and Texas: Past and present. *Exp. Appl. Acarol.* 30:181–202.

Congdon, B.D. and J.A. McMurtry. 1985. Biosystematics of *Euseius* on California citrus and avocado with a description of a new species (Acari: Phytoseiidae). *Int. J. Acarol.* 11:23–30.

Futch, S. and G. Albrigo. 2010. Management factors for HLB: Sprout control and killing trees in place. *Citrus Indust.* 91(3):18–20.

Futch, S.H., C.C. Childers, and C.W. McCoy. 2008. *A Guide to Citrus Mite Identification*, Publication No. HS-806. Gainesville: Institute of Food and Agricultural Sciences, University of Florida (http://edis.ifas.ufl.edu/ch179).

Gerson, U. 2003.Acarine pests of citrus: Overview and non-chemical control. *Syst. Appl. Acarol.* 8:3–12.

Grafton-Cardwell, B., A. Eller, and N. O'Connell. 1995. Integrated citrus thrips control reduces secondary pests. *Calif. Agric.* 49(2):23–28.

Grafton-Cardwell, E.E. and P. Gu. 2003. Conserving vedalia beetle, *Rodolia cardinalis* (Mulsant) (Coleoptera: Coccinellidae), in citrus: A continuing challenge as new insecticides gain registration. *J. Econ. Entomol.* 96:1388–1398.

Grafton-Cardwell, E.E. and S.L.C. Vehrs. 1995. Monitoring for organophosphate- and carbamate-resistant armored scale (Homoptera: Diaspididae) in San Joaquin Valley Citrus. *J. Econ. Entomol.* 88:495–504.

Grafton-Cardwell, E.E., J.E. Lee, J.R. Stewart, and K.D. Olsen. 2006. Role of two insect growth regulators in integrated pest management of citrus scales. *J. Econ. Entomol.* 99:733–744.

Grafton-Cardwell, E.E., J.E. Lee, S.M. Robillard, and J.M. Gorden. 2008. Role of imidacloprid in integrated pest management of California citrus. *J. Econ. Entomol.* 101:451–460.

Graham, J. and M. Dewdney. 2009. Management recommendations for citrus canker in 2009. *Citrus Indust.* 90(2):10–13.

Hall, D.G. and M.K. Simms. 2003. Damage by infestations of Texas citrus mite (Acari: Tetranychidae) and its effect on the life of "Valencia" leaves in an irrigated citrus grove. *Florida Entomol.* 86:15–28.

Hare, J.D., J.E. Pherson, T. Clemens, J.L. Menge, C.W. Coggins, T.W. Embleton, and J.L. Meyer. 1990. Effects of managing citrus red mite (Acari: Tetranychidae) and cultural practices on total yield, fruit size and crop value of "Navel" orange. *J. Econ. Entomol.* 83:976–984.

Hoy, M.A. 2000. The David Rosen lecture: Biological control in citrus. *Crop Prot.* 10(8–10):657–664.

Hoy, M.A. and R. Nguyen. 1997. Classical biological control of the citrus leafminer *Phyllocnistis citrella* Stainton (Lepidoptera: Gracillariidae): Theory, practice, art, and science. *Trop. Lepid.* 8(Suppl. 1):1–19.

Hoy, M.A. and R. Nguyen. 2000a. Classical biological control of Asian citrus psylla: Update on *Tamarixia radiata* releases and first releases of *Diaphorencyrtus aligarhensis*. *Citrus Indust.* 81(12):48–50.

Hoy, M.A. and R. Nguyen. 2000b. Classical biological control of brown citrus aphid: Release of *Lipolexis scutellaris*. *Citrus Indust.* 81(10):24–26.

Hoy, M.A., R. Nguyen, D. Hall, M. Pomerinke, J. Pena, H. Browning, and P. Stansly. 1995. Establishment of citrus leafminer parasitoid *Ageniaspis citricola* in Florida. *Citrus Indust.* 76(12):12–17.

Jeppson, L.R., M.J. Jesser, and J.O. Complin. 1958. Resistance of citrus red mite to organic phosphates in California. *J. Econ. Entomol.* 51:232–233.

Jones, V.P. and M.P. Parrella. 1984. Presence–absence sampling of citrus red mite on lemons. *Calif. Agric.* 40(3):31–32.

Kahn, I. and J.G. Morse. 1998. Citrus thrips (Thysanoptera: Thripidae) resistance monitoring in California. *J. Econ. Entomol.* 91:361–366.

Knapp, J.L., S. Halbert, R. Lee, M. Hoy, R. Clark, and M. Kesinger. 1998. The Asian citrus psyllid and citrus greening disease. *Citrus Indust.* 79(10):28–29.

McCoy, C.W. 1976. Leaf injury and defoliation caused by the citrus rust mite, *Phyllocoptruta oleivora*. *Florida Entomol.* 59:403–410.

McCoy, C.W. 1985. Citrus: Current status of biological control in Florida. In: M.A. Hoy and D.C. Herzog (eds.), *Biological Control in Agricultural IPM Systems* (pp. 481–499). Orlando, FL: Academic Press.

McCoy, C.W. 1996. Pathogens of eriophyoids. In: E.E. Lindquist, M.W. Sabelis, and J. Bruin (eds.), *Eriophyoid Mites: Their Biology, Natural Enemies, and Control* (pp. 481–490). Amsterdam: Elsevier.

McCoy, C.W., R.A. Samson, D.G. Boucias, L.S. Osborne, J.E. Peña, and L.J. Buss. 2009. *Pathogens Infecting Insects and Mites of Citrus*. Winter Park, FL: Friends of Microbes.

McMurtry, J.A. 1985. Citrus. In: W. Helle and M.W. Sabelis (eds.), *Spider Mites: Their Biology, Natural Enemies, and Control*, Vol. 1B (pp. 339–347). Amsterdam: Elsevier.

Morse, J.G. and O.L. Brawner. 1986. Toxicity of pesticides to *Scirtothrips citri* (Thysanoptera: Thripidae) and implications to resistance management. *J. Econ. Entomol.* 79:565–570.

Muma, M.B. 1961. *Mites Associated with Citrus in Florida*, University of Florida Agricultural Experiment Station Bull. 640. Gainesville: University of Florida.

Persad, A.B., M.A. Hoy, and R. Nguyen. 2007. Establishment of *Lipolexis oregmae* in a classical biological control program for the brown citrus aphid in Florida. *Florida Entomol.* 90:204–213.

Reed, D.K., H. Tashiro, and J.B. Beavers. 1975. Determination of mode of transmission of the citrus red mite virus. *J. Invert. Pathol.* 26:239–246.

Rogers, M.E. 2010. Five tips for successful insecticide resistance management. *Citrus Indust.* 91(5):16–17.

Skelley, L.H. and M.A. Hoy. 2004. A synchronous rearing method for the Asian citrus psyllid and its parasitoids in quarantine. *Biol. Control* 29:14–23.

Smith, D., G.A.C. Beattie, and R. Broadley. 1997. *Citrus Pests and Their Natural Enemies: Integrated Pest Management in Australia*. Brisbane: Department of Primary Industries, State of Queensland.

Spann, T.M., R.A. Atwood, M.N. Dewdney, R.C. Ebel, R. Ehsani, G. England, S. Futch, T. Gaver, T. Hurner, C. Oswalt, M.E. Rogers, F.M. Roka, M.A. Ritenour, and M. Zekri. 2010. IFAS guidance for huanglongbing (greening) management. *Citrus Indust.* 91(4):6–13.

Sternlicht, M. and S. Goldenberg. 1971. Fertilisation, sex ratio and postembryonic stages of the citrus bud mite *Aceria sheldoni* (Ewing) (Acarina, Eriophyidae). *Bull. Entomol. Res.* 60:391–397.

Timmer, L.W. and L.W. Duncan (eds.) 1999. *Citrus Health Management*. St. Paul, MN: American Phytopathological Society Press.

UC IPM. 2010. *Integrated Pest Management Guideline for Citrus*, UC ANR Publication No. 3441. Oakland: University of California Statewide Integrated Pest Management Program.

USDA. 2003. *Crop Profile for Citrus in California*. Washington, D.C.: National Institute of Food and Agriculture, U.S. Department of Agriculture (http://www.ipmcenters.org/cropprofiles/docs/cacitrus2.pdf).

USDA. 2006. *Crop Profile for Citrus (Major) Oranges/Grapefruit in Florida*. 2006. Washington, D.C.: National Institute of Food and Agriculture, U.S. Department of Agriculture (http://www.ipmcenters.org/CropProfiles/docs/FLcitrus(major).pdf).

Vacante, V. 2010a. *Citrus Mites: Identification, Bionomy, and Control*. Wallingford, U.K.: CAB International.

Vacante, V. 2010b. Review of phytophagous mites collected on citrus in the world. *Acarologia* 50:221–241.

Yang, Y., J.C. Allen, J.L. Knapp, and P.A. Stansly. 1995. Relationship between population density of citrus rust mite (Acari: Eriophyidae) and damage to "Hamlin" orange fruit. *Environ. Entomol.* 24:1024–1031.

Zalom, F.G., L.T. Wilson, C.E. Kennett, N.V. O'Connell, D.L. Flaherty, and J.G. Morse. 1986. Presence–absence sampling of citrus red mite. *Calif. Agric.* 40(3):15–16.

CHAPTER 19

Managing Mites on Ornamental Plants

19.1 TYPES OF ORNAMENTAL PLANTS

Pest mites occur on ornamental plants during their production in greenhouses and nurseries, within homes and atriums in commercial spaces, and in the general landscape in lawns and parks. Ornamental plants include many species and cultivars: bulbs, woody ornamental trees and shrubs, herbaceous annuals, and perennials such as orchids. Growing conditions vary dramatically. Ornamental plants are grown in the ground, in pots and hanging baskets, and in large planters. As a result, managing mites in ornamentals is a complex topic, and integrated mite management (IMM) programs for mites will vary with crop and location. Furthermore, mite management is dependent to a large degree on the other pest problems (insects, diseases) in the system (see Figures S19.1to S19.3 on the CD). The main acarine pests of ornamental plants are spider mites (Tetranychidae), eriophyoid mites (Eriophyoidea), and broad mites (Tarsonemidae). False spider mites (Tenuipalpidae) are occasional pests, and bulb mites (Acaridae) may be pests in certain crops. The two-spotted spider mite (*Tetranychus urticae*) is a premier pest of foliage of ornamental plants around the world, except in very humid tropical conditions (Table 19.1).

19.2 TACTICS FOR MANAGING PESTS OF ORNAMENTALS

Because ornamental plants are sold on the basis of their beauty, a high premium is placed on producing plants that lack obvious pests or damage from pests, so the economic injury level typically is very low (van de Vrie 1985, van Lenteren and Woets 1988). As a result, chemical control has been a dominant management tactic until relatively recently; however, resistance to pesticides has developed in many of the key mites and insects, often within a year or two after their introduction (van de Vrie 1985). Due to concerns about resistance, the limited availability and high costs of effective pesticides, and worker safety, pest managers are attempting to apply IMM tactics to manage mite pests on ornamentals (Gilkeson 1984, Hussey and Scopes 1985, Applied Bio-nomics 1993, Casey 1997, Dreistadt 2001, Helyer et al. 2003, Zhang 2003, Gerson and Weintraub 2007, Greer and Diver 2010). IMM of ornamentals will include a variety of approaches (Table 19.2).

19.2.1 Cultural Controls

Cultural methods require extensive horticultural and pest-biology information, as well as diligence in their implementation. Because plants produced in greenhouses or nurseries must be managed intensively, *prevention* is an important component in managing arthropod pests. It is crucial that new plantings be free of arthropod pests or diseases. Inspect roots and foliage of all planting stock before purchase or delivery; new shipments of plants should be quarantined to be sure the

Table 19.1 Significant Pest Mites in Commercial Ornamental Greenhouse Plant Production

Species	Crops	Importance
	Tetranychidae (spider mites)	
Tetranychus urticae (two-spotted spider mite)	Most	Most important
T. cinnabarinus (carmine spider mite)	Tomato and carnation	Important
T. kanzawai (Kanzawa spider mite)	Many ornamentals	Occasional
Bryobia kissophilia	Ivy	Rare
Eotetranychus lewisi (poinsettia spider mite)	Poinsettia	Common
E. sexmaculatus (six-spotted mite)	Azaleas	Occasional
Panonychus citri (citrus red mite)	Ornamentals	Occasional
	Tenuipalpidae (false spider mites)	
Brevipalpus obovatus (privet mite)	50 genera of ornamentals	Minor
B. phoenicis (red and black flat mite)	50 genera of ornamentals	Minor
B. russulus	Cacti and succulents	Minor
Tenuipalpus pacificus	Orchids	Moderate
	Tarsonemidae	
Polyphagotarsonemus latus (broad mite)	Many	Very important around the world
Phytonemus pallidus (cyclamen mite)	Cyclamen, begonia, fuchsia, geranium	Important in North America, Europe, Asia, and Australia
	Eriophyoidea	
Aceria aloinis (aloe wart mite)	Aloe species	Important where aloe is grown
A. barberton (Gerbera erineum mite)	Gerbera	Important in South Africa
A. dianthi	Carnations	Finland
A. genistae	Broom	Europe
A. georghioui	Carnations	Occasional in Cyprus and California
A. lantanae (Lantana gall mite)	*Lantana camara*	Important
A. paradiathi	Dianthus	Occasional
A. tulipae	Tulips	Important in the Netherlands and Japan
Cosetacus camelliae	*Camellia japonica*	California
Epitrimerus alinae (chrysanthemum leaf rust mite)	Chrysanthemums	Occasional
Eriophyes lowi (lilac bud mite)	*Syringa vulgaris*	Europe
Eriophyes spiraeae (bridal wreath gall mite)	*Spirea densiflora* (bridal wreath spirea)	North America and Europe
Paraphytoptus chrysanthemi (chrysanthemum rust mite)	Chrysanthemums	Occasional
	Acaridae	
Rhizoglyphus robini	Bulbs and corms	Important
R. echinopus	Bulbs and corms	Important
Tyrophagus putrescentiae	Ornamentals	Occasional
T. longior	Ornamentals	Occasional

Source: Based on information from Buxton (1989), Gerson and Weintraub (2007), Jeppson et al. (1975), Meyer (1996), and Zhang (2003).

Table 19.2 Managing Pests in Greenhouses Using Quarantines, Cultural Practices, Monitoring, and Augmentative Biological Control

1. Remove all crop debris and disinfect soil each year to start the new growing season with fewer pests. Allowing greenhouses to freeze or heat to very high temperatures between uses can reduce the carryover of pests.
2. Make sure any plants brought into the greenhouse at the start of the season are free of arthropod pests or diseases.
3. Choose cultivars less prone to pest problems, if available. Apply the optimal amount of fertilizer and water to promote healthy growth.
4. Ensure that the temperatures and relative humidity are optimal for crop growth, but do not encourage plant pathogens or arthropod pests.
5. Make sure greenhouse workers do not move pests from one greenhouse to the next on clothing or equipment; disinfest equipment and clothing.
6. Reduce the likelihood of pests entering the greenhouse by maintaining clean, weed-free areas around greenhouses.
7. Identify all pests for each crop and learn their life cycle.
8. Reduce pests with strong sprays of water on foliage and prune off infested foliage or branches.
9. When considering making releases, identify a company that can provide the appropriate natural enemies for the pests anticipated to occur. Contact the company ahead of the growing season to be sure they can fill orders when needed.
10. Monitor the crop at least each week using a hand lens, as well as noting while watering or pruning whether problems are developing. Carefully examine plants using a specific system, such as monitoring 10 to 20 leaves per row from every third row. Record the pests and their abundance. Identify hot spots where pest problems are consistent and monitor these carefully. Use indicator plants, if possible.
11. When pest or disease infestations are limited to a few plants or leaves, remove plants or foliage and destroy.
12. Establish a release rate and schedule for natural enemies based on the recommendations of the supplier. Know how many pests are present to order the correct number of natural enemies. Resolve whether an inoculative or inundative release will be made, and order natural enemies accordingly. Keep records of all costs associated with the releases.
13. Monitor results of releases of natural enemies.
14. When pesticides are necessary to prevent economic damage, use ones that do not disrupt the natural enemies and do not cause phytotoxic effects. Make sure all greenhouse workers know what and when to spray.

Source: Based on information from Gilkeson (1984), van de Vrie (1985), and Zhang (2003).

planting stock is free of pests. Segregate stock propagation plants from young plants to prevent spread of pests. Grow bedding plants from seed. Screen vents and fan intakes to reduce the invasion of larger arthropod pests into greenhouses (although mesh sufficiently small to eliminate mites can create problems with air flow). Minimize the movement of plants, equipment, and people between greenhouses to reduce the transport of mites, and other pests, from site to site (Steiner and Elliott 1987).

Growing cultivars that are less susceptible to damage from arthropod pests or diseases is another important component of an integrated pest management (IPM) program. Making sure that the crop is grown in the appropriate soil, with adequate fertilizer, light, and water to ensure a healthy plant, enhances the likelihood that the crop will be commercially acceptable.

In conservatories and other large indoor spaces, start with the correct drainage, soil preparation, and fertility. Choose plants suited to the light levels and other environmental conditions available, and follow with management tactics that keep the plants in optimal health. These include providing good air circulation, reducing temperatures in sunny conditions, and taking measures to increase relative humidity in dry conditions, as well as pruning, thinning, and staking. Choose plants that are poor hosts for common greenhouse and nursery pests. Sometimes cultivars of the same species differ dramatically in their propensity to become infested. In landscape or interiorscape plantings, a rule of thumb is to ensure the diversity of plants by choosing no more than 20% of the plants from any one family and no more than 10% from any one genus (the 20-10 planting rule). Cultivars that are more resistant to common pests should be chosen (Dreistadt 1994).

19.2.2 Monitoring Is Essential

Monitoring for pests and diseases is essential, and the correct identification of the pest is important so the appropriate management option can be employed (Zhang 2003). It is not possible to monitor too often or too much. Most managers monitor too little. It is necessary to know the life cycle of the pest to monitor it and to devise ways to reduce the likelihood of new infestations. Although many pests produce characteristic damage, use a hand lens or dissecting microscope to examine the infested material to correctly identify the pest (and to be sure that the pest is still there). Reference books, websites, and a variety of other resources can assist in the diagnosis of mite pests. If you believe you have a novel pest (which could be a new invasive pest), send specimens to a taxonomic specialist for identification. Collect samples using methods described in Chapter 4. When monitoring, look at the top and bottom of leaves and growing tips. Note whether curled, twisted, rolled, or discolored leaves and buds are apparent. Look for webbing and yellowish stippling on leaves (an indicator of spider mites). Where many types of plants are grown, monitor intensively those plants that are more susceptible to pests (**indicator plants**). As long as indicator plants are free of pests, the remainder of the plants should be uninfested. If the indicator plants are infested, monitor other plants. Palms, Easter lilies, *Dieffenbachia*, and citrus are good indicator plants for spider mites. Keep written records of all counts, as well as maps of damage for future reference. Monitoring should be conducted when pests are active. Obviously, cooler weather results in slower growth of pest populations, so monitoring schedules can be adjusted during cooler weather.

19.2.3 How Many Mites Are Too Many?

Economic injury levels vary with the location and timing of the problem, as well as the plant type. Ways to define injury include recording the percentage of leaves damaged on a particular plant, the percentage of plants affected at a specific location, the number of pests counted, the number of pests counted in relation to the number of natural enemies found, and the number of complaints received from the public about a problem. Plants that are sold in nurseries or grown in interiorscapes are highly visible to the public, and typically less damage is tolerated. Acceptable injury levels in nurseries and greenhouses depend on whether the pest is likely to spread and what effect it has on the growth rate or long-term health of the plant. Plants often can be cleaned up with a chemical spray just prior to sale. Plants in the lawn and park environment are often able to tolerate more pests, and such pests may be less noticeable. Generalist natural enemies may assist in suppressing landscape pests.

19.2.4 Control Tactics

Action levels depend on the control tactic planned. If the control acts quickly, then higher pest population levels can be tolerated. If the control is slower acting (e.g., systemic pesticide, growth regulator, biological control agent), then fewer pests can be tolerated. In the case of augmentative releases of natural enemies, *earlier is nearly always better* because predators typically lag behind their prey populations. (See Chapter 5 for a discussion of augmentative releases of natural enemies.) Release rates will vary with temperature, cultivar, and pest density. If few pests are present, a common release ratio is one predatory mite (phytoseiid) for each ten spider mites (all stages). If substantial damage has already occurred, then even more predators would have to be released to suppress pest populations (perhaps one predator for every two spider mites). Follow the guidelines (if available) provided with the purchased natural enemies; otherwise, assume that experiments must be conducted to resolve release rates and timing for a specific situation. If little damage can be tolerated, consider using a pesticide that has low toxicity to natural enemies (such as soap, oil, or sulfur).

Predatory insects may not be effective in offices or malls where plants are spaced far apart, because the adult insects may fly to the windows and appear to be pests (Steiner and Elliot 1987). Predatory mites are more effective in these circumstances because they cannot fly, but they must be placed directly on the infested plants because they do not disperse rapidly, which increases labor costs. Because predatory mites are small, people do not notice them. Physical methods for controlling indoor and nursery pests include strong sprays of water to control some insects and mites, removing infested leaves by hand, or pruning out infested branches.

Relatively few pesticides are registered for use indoors (e.g., malls, offices). The pesticides available are less toxic and less persistent. Less-toxic materials include growth regulators, soaps, natural pyrethrins (very short residual), and horticultural oils. Evaluate both soaps and horticultural oils to be sure they do not cause phytotoxicity prior to general use, and always maintain plants in a well-watered condition. Regular applications (calendar sprays) are not recommended because aphids, thrips, and mites in greenhouses develop resistance rapidly to most products (except, so far, no resistance is known to horticultural oils, and sulfur resistance is very are). Sometimes sprays are used to clean up pest outbreaks prior to starting a biological control program or before shipping bedding plants or nursery stock for sale. After treatments with chemicals or natural enemies, monitor the results using appropriate methods to determine whether the treatment has worked. If plants are to be sold, residues from pesticides are often visible on the foliage and may reduce the acceptability of the ornamental plants being sold (van de Vrie 1985).

19.3 MITES ON ORNAMENTAL PLANTS

19.3.1 Tetranychidae

The most common spider mite in ornamental plants in greenhouses or in interior plantscapes is the two-spotted spider mite (*Tetranychus urticae*) (de Moraes and Tamai 1999) (see Figure S19.4on the CD). It especially is a problem in hot, dry environments, so increasing relative humidity and maintaining adequate irrigation of plants can slow their developmentrate. *Tetranychus cinnabarinus* is a closely related species, but adult females are reddish in coloration (confusion remains as to whether they are separate species). The European red mite (*Panonychus ulmi*) can be found on nursery roses, and the citrus red mite (*Panonychus citri*) can occur on citrus, palms, and other subtropical plants.

The phytoseiid *Phytoseiulus persimilis* can be most effective when released at the first sign of *Tetranychus urticae* populations. Increasing the relative humidity to favor the survival of *P. persimilis* can improve control, although care should be taken so the higher relative humidity does not foster fungal pathogens of the crop. *Phytoseiulus persimilis* has a rapid developmental rate (1 week at 25°C), and a high fecundity (up to 80 progeny). It is fast moving and feeds on *Tetranychus* mites, especially on low-growing plants and shrubs. It is unable to survive once its prey is eliminated, and it will starve or migrate out of the release site. It is mass reared by a number of commercial producers and often is released by sprinkling the predators mixed with a carrier of bran or grits over the plants.

Other species of predatory mites sold commercially include *Metaseiulus occidentalis*, which survives well in lower relative humidities but has a slower reproductive rate and lower prey consumption rate than *Phytoseiulus persimilis.* It must be released early in the spider mite infestation (Field and Hoy 1986). *Metaseiulus occidentalis*, and other phytoseiid species, can persist longer than *P. persimilis* when prey populations are low, so they may require fewer repeat releases. *Amblyseius fallacis* is effective on mites occupying woody ornamental plants. The lady beetle *Stethorus punctillum* has a higher prey consumption rate than phytoseiids, so it can clean up higher densities rapidly. *Stethorus* will disperse from the plants when prey densities decline. See Chapter 6 for additional information on the spider mites, and Chapters 12 and 13 for more information on predatory insects and mites.

19.3.2 Tenuipalpidae

False spider mites (Tenuipalpidae) can be pests on ornamental plants, especially when grown in greenhouses where temperatures and relative humidities are high. The phalaenopsis mite (***Tenuipalpus pacificus***) and the oncidium mite (***Brevipalpus oncidii***) are fairly specific to their orchid hosts (see Figure S19.5 on the CD). The red and black mite (***Brevipalpus phoenicis***) is found on palms, privet, citrus, and many other plants. The omnivorous mite (*B. californicus*) and the privet mite (***B. inornatus***) feed on a variety of woody ornamentals. False spider mites puncture the epidermis of the host plant and suck out the liquid contents, causing a spot that may later turn brown. Damage from tenuipalpids is similar to that of tetranychids, but the onset of symptoms is slower because the development and reproduction rates are lower in tenuipalpids. These mites can be controlled by applications of sulfur. Oil combined with an adjuvant (Silwet® L-77) can improve coverage and control (Cating et al. 2010). See Chapter 9 for more information about this family.

19.3.3 Tarsonemidae

The Tarsonemidae is a large family of small mites. Their biology was discussed in Chapter 7. There are two main pests of ornamental plants in greenhouses (Zhang 2003). ***Polyphagotarsonemus latus*** (broad mite) is most important because it is a pest of many crops and ornamentals around the world (see Table 19.1). ***Phytonemus*** (or ***Steneotarsonemus***) ***pallidus*** (cyclamen mite or strawberry mite) is a pest of both strawberries and ornamentals in greenhouses around the world (see Figure S19.6 on the CD). These mites can reproduce throughout the year and disperse by walking or by phoresy on insects (Fan and Petitt 1998). Control of broad mites (*P. latus*) can be achieved using releases of phytoseiids such as *Amblyseius barkeri*, *A. californicus*, and *A. cucumeris*. Other phytoseiids have been used locally, such as *A. fallacis* and *Metaseiulus occidentalis* (Huffaker and Kennett 1953).

Hemitarsonemus tepidariorum (fern mite) is a pest of new growth of ferns in dark, warm, and humid greenhouses, attacking *Asplenium*, *Polystichum*, and *Pteris* (Cameron 1925, Pritchard 1951, Zhang 2003). Feeding causes brownish areas to appear, and the tender new growth causes the fronds to become wrinkled and misshapen. Pritchard (1951) suggested that this mite can be controlled with sulfur, and Cameron (1925) indicated that control could be achieved by sanitation, inspection, elimination of infested plants, enhancing ventilation, providing adequate drainage, and controlling temperature and relative humidity. No phytoseiids are known to be effective.

Steneotarsonemus laticeps (bulb scale mite) is a pest of *Amaryllis*, *Narcissus*, *Hippeastrum*, and other species of the Amaryllidaceae (Murdoch 1974, Zhang 2003). This species causes spots on the vegetative growth, red scars on *Amaryllis* bulbs, and weak and malformed flowers. Sanitation and quarantines can be effective methods for control, and hot-water treatment can eliminate mites from bulbs (after 4 hours at 43.3°C) (Jeppson et al. 2003, Zhang 2003). Studies by Messelink and van Holstein-Saj (2006, 2007) determined that the phytoseiid *Amblyseius* (= *Neoseiulus*) *barkeri* is an effective predator of the bulb scale mite because it is present in soil, as well as on bulbs and leaves. *Amblyseius barkeri* has been mass reared for biological control of thrips in greenhouses and thus is commercially available. Other predators present in a survey of *Amaryllis* bulbs in commercial nurseries included *Hypoaspis aculeifer* and *H. angusta*.

Tarsonemus confusus, although known as a fungivore, is a minor pest of ornamentals such as African violets, azalea, *Cissus*, *Cyclamen*, *Gloxinia*, ivy, and *Pilea* in greenhouses (Zhang 2003). *Tarsonemus confusus* currently is not present in the United States, but it is a potential invader because it has been collected on ornamental plants at ports of entry (Childers and Rodrigues 2005). This mite is a pest of fungal cultures and has broad food habits (Reiss and Caroline 1953).

19.3.4 Eriophyoidea

The biology of the Eriophyoidea is discussed in Chapter 8. Most species of eriophyoids are relatively host specific, causing bud deformities, erinea, or witches' broom (see Figure S19.7on the CD). As a result, the geographic distribution of these mites may be limited to the original distribution of their ornamental host plant; however, their distribution can change because ornamental plants are moved around the world by tourists and by commerce. Tiny eriophyoids are difficult to detect when there are low populations on plants (that lack symptoms), even when plants go through a quarantine inspection. Invasions of eriophyoids into new geographic areas are to be expected, so the distributions noted here may change (Navia et al. 2010).

Twenty-four species (Eriophyidae) attack ornamental plants. ***Epitrimerus alinae*** (chrysanthemum rust mite) is an occasional pest of greenhouse-grown *Chrysanthemum* in Europe (Meyer 1996, Zhang 2003). Control of these mites is feasible with applications of oil or sulfur. Removal of infested stems and foliage may reduce its spread. In addition, some predatory mites (Tydeidae, Phytoseiidae) are known to feed on them. Chemical control requires precise and effective applications of acaricides. ***Cosetacus cameliae*** is a pest of *Camellia japonica* in California, causing russeting and premature flower drop (Meyer 1996). ***Eriophyes lowi*** (lilac bud mite) is a pest in Europe, causing witches' broom. ***Eriophyes spiraeae*** (bridal-wreath gall mite) is found in North America and Europe, where gall-like buds instead of flowers are produced.

Aceria aloinis (aloe wart mite) is a pest of *Aloe* in California, Florida, and southern Africa (Meyer 1996). It is difficult to control this mite because it occurs in galls or outgrowths. Quarantines and cultural practices (removing affected plants or plant parts) can reduce its spread. ***Aceria barbertoni*** (*Gerbera* erineum mite) is present in South Africa and causes a woolly erineum on the lower leaf surfaces (Meyer 1996). ***Aceria genistae*** (broom gall mite) is a pest of broom (species of Leguminosae) in Europe, causing deformities of the shoots. ***Aceria lantanae*** (*Lantana* gall mite) is found in the Caribbean, Florida, and Central and South America. Galls are produced from a mass of very small leaves and distorted flowerbuds. This mite has been evaluated as a possible biological control agent in areas where *Lantana* is considered a weed. ***Aceria proteae*** (*Protea* witches' broom mite) is found in South Africa, where it damages flowers (Meyer 1996).

Three species of *Aceria* are pests of carnations (Meyer 1996, Zhang 2003). ***Aceria dianthi*** is a leaf vagrant on carnations in Finland, causing stunting of plants. ***Aceria paradianthi*** is known from *Dianthus* species in Europe, Argentina, and the United States, where it produces distorted and stunted plants. ***Aceria georghioui*** has been found in California and Cyprus on carnations, causing infested plants to be distorted. ***Aceria (=Eriophyes) tulipae*** attacks bulbs (tulips and *Allium*, as well as garlic and onion) in the Netherlands and Japan (Conijn et al. 1996). The taxonomy of this mite is confusing in the literature. Formerly, *A. tulipae* was thought to attack plants in the Gramineae (maize, wheat) as well as bulbs, but now the mites attacking the Gramineae are considered to be a different species. *Aceria tulipae* can be a serious pest when bulbs are stored under temperatures that allow rapid development of mite populations. Bulb mites rarely are a problem during the growing season if bulbs are planted that are free of mites. Symptoms of damage include the development of superficial red to purple or cream to yellow spots on the normally white outer-bulb scales of stored bulbs. These spots enlarge and may eventually cover the whole bulb, and the bulb scales will become desiccated. Mites invade the space between the bulb scales and multiply, damaging the inner bulb scales when densities reach many thousands. Plants grown from slightly or moderately infested bulbs develop slowly and produce a stunted plant.

Aceria tulipae does not survive for long without food and water, so it is unlikely to survive in clean, empty storage rooms. Sanitation of the storage rooms can ensure a lack of mites to infest newly stored bulbs. Bulbs brought into the storage room from the field typically have low mite densities. Spread from bulb to bulb is slow in the field but rapid when bulbs are stored in large volumes under suitable temperatures. Control can be achieved by storing bulbs for 4 weeks at 20°C, followed by up to 4 weeks at 17°C and 9 to 18 weeks at 5 or 9°C (Conijn et al. 1996). Another cultural practice is to treat stored bulbs with hot water, but the amount of time and the temperature used vary with the type of bulb, and damage to the bulb can occur. Garlic bulbs, for example, are treated with hot water (55 to 60°C) for 10 to 15 minutes, or 2.25 hours at 39°C, or 1.5 hours at 41°C, or 1 hour at 43°C. Treating tulips for 4 hours at 45°C reduces mite densities but does not completely eliminate them.

Biological control of bulb mites can be achieved by the release of the predator *Hypoaspis aculeifer* (Lesna et al. 2000). Companies producing these predators recommend that a hot-water treatment of the bulbs be done prior to planting to aid the effectiveness of the predators. *Hypoaspis aculeifer* also can be released during the cultivation of bulbs in the greenhouse, where they will feed on sciarid fly larvae, root aphids, and thrips pupae in the soil. These predatory mites are commercially available from several companies.

19.3.5 Acaridae *Rhizoglyphus robini* and *Rhizoglyphus echinopus*

Rhizoglyphus robini and *Rhizoglyphus echinopus* are important pests of bulbs and corms of ornamental plants, including *Narcissus*, *Eucharis*, lilies, orchids, gladioli, hyacinth, tulip, dahlia, and *Freesia*, during storage, in greenhouses, and in fields around the world (Zhang 2003). Both are common in decaying organic matter and soil. Several predatory mites are biological control agents, including *Hypoaspis aculeifer* (Lesna et al. 1995, 2000). This mite must be released at a ratio of at least one predator to every two or five prey mites to obtain adequate control. Heat treatments of bulbs and corms can eliminate the mites. These mites are discussed in Chapter 11.

Species of *Tyrophagus* are inhabitants of stored foods and nests, and they are considered to be fungivores and saprophages. Some species will damage bulbs or other crops that are grown in greenhouses. For example, ***Tyrophagus putrescentiae*** is referred to as the mold mite in stored products and homes, but *T. putrescentiae* will attack ornamentals in greenhouses in Europe and is a pest of corms of *Freesia* and crocus bulbs (Zhang 2003). Using soil or other growing media that has been sterilized, using fungicides to reduce fungi, and releasing predatory mites (*Hypoaspis*) will reduce mite populations in the soil. If *T. putrescentiae* are damaging foliage or flowers, they may be controlled by the release of *Amblyseius* (= *Neoseiulus*) *cucumeris* or *A.* (= *N.*) *barkeri*.

Tyrophagus similis can be a pest of greenhouse plants including cucumber, French beans, *Phlox*, spinach, and *Narcissus* bulbs (Zhang 2003). It can be controlled by soil sterilization in greenhouses (Misayo et al. 2005). *Tyrophagus similis* also feeds on nematodes (Walter et al. 1986), so its presence may not always be detrimental. ***Tyrophagus longior*** is a pest of stored products and an occasional pest of greenhouse cucumbers in Europe (Zhang 2003), and it also damages *Verbena*, *Delphinium*, and *Cymbidium* orchids in greenhouses because it vectors fungal spores that prevent the orchid flowers from opening. ***Tyrophagus neiswanderi*** is found in stored products and nests (Sanchez-Ramos et al. 2007), but in European greenhouses it can be found on cucumber foliage; chrysanthemum cuttings and flowers; *Gerbera* and cyclamen flowers; bulbs of *Narcissus*, tulip, and *Hippeastrum*; and *Freesia* corms. It also attacks *Cymbidium* and *Phalaenopsis* orchids (Zhang 2003). Workman and Martin (2002) found that *T. neiswanderi* (called the pollen-cap mite by *Cymbidium* orchid growers in New Zealand) moves into the flowers, where it feeds on pollen. Releases of *Hypoaspis* predatory mites into the orchid media can suppress *T. neiswanderi* populations, thereby preventing damage to the orchid flowers.

19.4 PREDATORY MITE RELEASE METHODS IN GREENHOUSES

Predatory mites are released in greenhouses by several methods, each of which has advantages and disadvantages. The predators often are released directly on the infested foliage by sprinkling predators mixed with a carrier such as vermiculite, bran, or grit from bottles (Hamlen 1978, Hamlen and Lindquist 1981, Hussey and Scopes 1985, Osborne 1987, Gough 1991, Pratt and Croft 2000, IOBC–OILB 2010). This method is labor intensive and expensive, and achieving an even distribution can be difficult. Dispersal of the predators to adjacent plants or plant parts can be slow, allowing damage to the plants or plant parts that fail to receive predatory mites. Alternative approaches to reduce labor costs include the development of blowers to spray predators over the crop, although spraying may result in considerable predator mortality (Opit et al. 2005).

Another release method involves the use of **banker plants**. Banker plants are plants that help in the development and dispersal of predators for the control of plant pests (Hendrickson 1980, Pratt and Croft 2000, Matteoni 2003, Osborne and Barrett 2005, Osborne 2010). The plant provides food for the predators, perhaps by producing pollen or nectar or by harboring prey mites. Banker plants typically are added to the crop prior to the development of a pest problem. The natural enemies are able to multiply on the banker plant, dispersing from it onto the crop to provide long-term control of the pest. Osborne (2010) has developed a list of issues to consider when using banker plants. An advantage of banker plants is that natural enemies are produced continuously, and they can disperse from the banker plants to the crop to suppress pests, thereby reducing costs to distribute the predators. A disadvantage to banker plants could be the spread of pests (perhaps the prey provided for the predators to feed on) from the banker plants to the crop if a judicious choice of prey species is not made.

Slow-release bags or sachets provide yet another release method for some phytoseiid species. *Amblyseius cucumeris* and *A. swirskii* can be released preventatively in crops, and the sachets can produce predatory mites over a period of 4 to 6 weeks. The bags or sachets are hung up in the crop, and the phytoseiids within the sachets feed on grain mites (*Acarus siro*) that are feeding on bran. Not all phytoseiid species will feed and develop well on factitious prey or pollens (Castagnoli and Simoni 1990, Duso and Camporese 1991, van Rijn and Tanigoshi 1999, Sengonca and Drescher 2001, Vantornhout et al. 2004). The species of predatory mites that often are released in greenhouses are listed in Table 19.3.

19.5 THE FUTURE OF PEST MANAGEMENT IN GREENHOUSE ORNAMENTALS

In 1992, Parrella et al. discussed the limitations to the enhanced use of augmentative biological control, including the high cost of natural enemies, problems associated with the availability and quality of natural enemies, and the lack of rigorous research projects documenting success with specific release rates and economic analyses. Many of these concerns remain today, although additional research has provided new tools for IMM in greenhouse ornamentals.

Improved monitoring tools are now available (Parrella et al. 1989, Karlik et al. 1995, Sanderson and Zhang 1995, Zhang and Sanderson 1995, Opit et al. 2003, Alatawi et al. 2005). Economic analyses of chemical control and biological control options have been performed, and a general model has been produced that could be applied to a variety of ornamental crops (Schumacher et al. 2006). Schumacher et al. (2006) reported that about 50% of floricultural producers in the United States use or have used biological controls in their production systems. Their general bioeconomic

Table 19.3 Some Predatory Mites Commercially Produced for Greenhouse Pest Management on Ornamentals

Species[a]	Target Prey (Alternative Foods)
Phytoseiidae	
Phytoseiulus persimilis	Webspinning *Tetranychus* mites (none)
Amblyseius (= *Neoseiulus*) *barkeri*	Broad mite, *Polyphagotarsonemus latus* (pollen, grain mites)
Amblyseius (= *Neoseiulus* or *Typhlodromus*) *californicus*	Spider mites, cyclamen mite (pollen)
Amblyseius (= *Neoseiulus*) *cucumeris*	Thrips, broad mites, cyclamen mites (pollen, bran mites)
Amblyseius fallacis	Cyclamen mite, spider mites, rust mites (pollen)
Amblyseius swirskii	Whiteflies, thrips, broad mites, spider mite eggs and nymphs (scale insects, pollen, honeydew)
Metaseiulus (*Typhlodromus* or *Galendromus*) *occidentalis*	Webspinning *Tetranychus* mites (none)
Laelapidae (or Hypoaspidae)	
Hypoaspis (*Geolaelaps*) *aculeifer* (multiple biotypes?)	Nematodes, acarid mites, sciarid fly larvae, bulb mites, and thrips
Hypoaspis (*Stratiolaelaps*) *scimitus*	Fungus gnats, thrips, shore flies

[a] The names of these predators vary in the literature.

model of floriculture production looked at ways to maximize profits. Under various model assumptions of chemical control and with or without predator releases, different profits were obtained. Interestingly, the authors discovered that "there is potential to reduce the frequency of pesticide applications in greenhouse floriculture production by educating consumers on the benefits of predator arthropods for biological control of pests." They concluded that "further economic research on the social value of reducing pesticide applications is needed. The potential social benefits may warrant policy that provides economic incentives to growers to increase the use of biological controls in the future" (Schumacher et al. 2006).

Communication between researchers involved in developing IPM tools for ornamental plant production in greenhouses is enhanced by publications from the International Organization for Biological Control (IOBC), which holds conferences and publishes bulletins, as well as a newsletter called *STING*. Several commercial producers of natural enemies provide extensive information on their websites regarding augmentative biological control. The issue of whether releases of multiple predatory species results in intraguild predation or improved pest control in greenhouse ornamentals remains a question (Rott and Ponsonby 2000, Schausberger and Walzer 2001, Venzon et al. 2001).

European scientists have led the way since the 1960s in developing practical biological-control-based management programs for greenhouse vegetables and ornamentals (Hussey and Scopes 1985, van de Vrie 1985, van Lenteren and Woets 1988, van Lenteren 2000, 2003, 2006, Pilkington et al. 2010). van Lenteren (2000) summarized the progress in managing pests in European greenhouse vegetables. Today, about "100 species of beneficial organisms are commercially available for control of all important insect and mite pests. The change to IPM was not based on idealism about a cleaner or healthier environment, but was rooted on clear advantages for the grower." Furthermore, efforts to control diseases using biological control have begun and efforts to control pests on ornamentals are increasing. van Lenteren (2000) predicted that "greenhouse crops will be produced without the need to use conventional pesticides in the very near future." In fact, pest management in European greenhouses is on a **biological control treadmill**, in which most pests are under biological control or are managed with tactics that are compatible with biological control. If new pests invade, research is conducted immediately to ensure that the current biological-control-based program is not disrupted with pesticides.

What a difference a continent makes. North America lags behind Europe in implementing biological-control-based pest management in greenhouses. When Warner and Getz (2008) surveyed industry leaders, research scientists, retail distributors, and customers in North America between 2004 and 2006, they discovered that 22 North American insectaries produce only 38 natural enemy species, which constitute less than 10% of the biologically based pest control market. No new insectaries had been established since 1996, several had declared bankruptcy, and few new species had been brought into production. The market appeared to be static, with low demand and low prices obtained for the species. The commercial natural enemy industry in North America is "generally conducted at a small scale by operations that are generally poorly capitalized. Augmentative biological control is highly dependent upon international and intercontinental movement for the natural enemies commercially produced. The existing permitting system in the U.S. appears inappropriate for scientifically reared and monitored commercial insectaries" (Warner and Getz 2008).

Why is there such a difference between the North American and European greenhouse industry? Possibly, it has to do with the regulation of pesticides. North American experts identified the cancellation of pesticide registration as potentially helping the industry. Warner and Getz (2008) indicated that small business loan programs could assist producers of natural enemies in expanding and improving quality of their natural enemies. Mechanisms are needed to facilitate the international movement of commercially produced natural enemies. Research is needed that is targeted to help the industry with market development, new product lines, field demonstrations, and quality-control initiatives. "Public initiatives to address these challenges will be essential to fulfilling the oft-stated potential of augmentative biological control" in North America (Warner and Getz 2008).

SELECTED REFERENCES

Alatawi, D.J., G.P. Opit, D.C. Margolies, and J.R. Nechols. 2005. Within-plant distribution of twospotted spider mites (Acari: Tetranychidae) on impatiens: Development of a presence–absence sampling plan. *J. Econ. Entomol.* 98:1040–1047.

Applied Bio-nomics, Ltd. 1993. *Biological Technical Manual.* Victoria, B.C.: Applied Bio-nomics, Ltd. (http://www.appliedbio-nomics.com/technical-manual.html).

Buxton, J.H. 1989. *Tyrophagus longior* (Gervais) (Acarina: Acaridae) as a pest of ornamentals grown under protection. *Plant Pathol.* 38:447–448.

Cameron, W.P.L. 1925. The fern mite (*Tarsonemus lepidariorum* Warburton). *Ann. Appl. Biol.* 12:93–112.

Casey, C. (ed.) 1997. *Integrated Pest Management for Bedding Plants*, IPM No. 407. Ithaca, NY: Cornell University Cooperative Extension.

Castagnoli, M. and S. Simoni. 1990. Biological observations and life table parameters of *Amblyseius cucumeris* (Oud.) (Acarina: Phytoseiidae) reared on different diets. *Redia* 73:569–583.

Cating, R.A., M.A. Hoy, and A.J. Palmateer. 2010. Silwet L-77 improves the efficacy of horticultural oils for control of Boisduval scale *Diaspis boisduvalii* (Hemiptera: Diaspididae) and the flat mite *Tenuipalpus pacificus* (Arachnida: Acari: Tenuipalpidae) on orchids. *Florida Entomol.* 93:100–106.

Childers, C.C. and J.C.V. Rodrigues. 2005. Potential pest mite species collected on ornamental plants from Central America at port of entry to the United States. *Florida Entomol.* 88:408–414.

Conijn, C.G.M., J. van Aartrijk, and I. Lesna. 1996. Flower bulbs. In: E.E. Lindquist, M.W. Sabelis, and J. Bruin (eds.), *Eriophyoid Mites: Their Biology, Natural Enemies, and Control* (pp. 651–659). Amsterdam: Elsevier.

de Moraes, G.J. and M.A. Tamai. 1999. Biological control of *Tetranychus* spp. on ornamental plants. In: G. Fischer and A. Angarita (eds.), *Proceedings of the International Symposium on Cut Flowers in the Tropics* (pp. 247–252), Acta Horticulture No. 482. Leuven, Belgium: ISHS.

Dreistadt, S.H. 1994. *Pests of Landscape Trees and Shrubs: An Integrated Pest Management Guide*, UC ANR Publication No. 2259. Oakland: University of California Statewide Integrated Pest Management Program.

Dreistadt, S.H. 2001. *Integrated Pest Management for Floriculture and Nurseries*, UC ANR Publication No. 3405. Oakland: University of California Statewide Integrated Pest Management Program.

Duso, C. and P. Camporese. 1991. Developmental times and oviposition rates of predatory mites *Typhlodromus pyri* and *Amblyseius andersoni* (Acari: Phytoseiidae) reared on different foods. *Exp. Appl. Acarol.* 13:117–128.

Fan, Y. and F.L. Petitt. 1998. Dispersal of the broad mite, *Polyphagotarsonemus latus* (Acari: Tarsonemidae), on *Bemisia argentifolli* (Homoptera: Aleyrodidae). *Exp. Appl. Acarol.* 22:411–415.

Field, R.P. and M.A. Hoy. 1986. Evaluation of genetically improved strains of *Metaseiulus occidentalis* (Nesbitt) (Acarina: Phytoseiidae) for integrated control of spider mites on roses in greenhouses. *Hilgardia* 54(2):1–31.

Gerson, U. and P.G. Weintraub. 2007. Mites for the control of pests in protected cultivation. *Pest Manag. Sci.* 63:658–676.

Gilkeson, L.A. 1984. *Biological Control Methods for Pests in Commercial Greenhouses*, EAP Publication No. 52. Sainte-Anne-de-Bellevue, Quebec: Ecological Agriculture Projects (http://eap.mcgill.ca/publications/eap52.htm).

Gough, N. 1991. Long-term stability in the interaction between *Tetranychus urticae* and *Phytoseiulus persimilis* producing successful integrated control on roses in southeast Queensland. *Exp. Appl. Acarol.* 12:83–101.

Greer, L. and S. Diver. 2010. *Integrated Pest Management for Greenhouse Crops*. Butte, MT: ATTRA, National Sustainable Agriculture Information Service, National Center for Appropriate Technology (http://attra.ncat.org/attra-pub/gh-ipm.html).

Hamlen, R.A. 1978. Biological control of spider mites on greenhouse ornamentals using predaceous mites. *Proc. Florida State Hort. Soc.* 91:247–249.

Hamlen, R.A. and R.K. Lindquist. 1981. Comparison of two *Phytoseiulus* species as predators of two-spotted spider mites on greenhouse ornamentals. *Environ. Entomol.* 10:524–527.

Helyer, N., K. Brown, and N.D. Cattlin. 2003. A *Color Handbook of Biological Control in Plant Protection*. Portland, OR: Timber Press.

Hendrickson, R.M. 1980. Continuous production of predacious mites in the greenhouse. *N.Y. Entomol. Soc.* 88:252–256.

Huffaker, C.B. and C.E. Kennett. 1953. Developments toward biological control of cyclamen mite on strawberries in California. *J. Econ. Entomol.* 46:802–812.

Hussey, N.W. and N. Scopes (eds.) 1985. *Biological Pest Control: The Glasshouse Experience*. Ithaca, NY: Cornell University Press.

IOBC–OILB. 2010. *STING 33: Newsletter on Biological Control in Greenhouses*. Jokioinen, Finland: MTT Agrifood Research Finland, Plant Production (https://portal.mtt.fi/portal/page/portal/iobc/Publications).

Jeppson, L.R., H.H. Keifer, and E.W. Baker. 1975. *Mites Injurious to Economic Plants.* Berkeley: University of California Press.

Karlik, J.F., P.B. Goodel, and G.W. Osteen. 1995. Sampling and treatment thresholds for spider mite management in field-grown rose plants. *HortScience* 30:1268–1270.

Lesna, I., M.W. Sabelis, H.R. Bolland, and C.G.M. Conijn. 1995. Candidate natural enemies for control of *Rhizoglyphus robini* Claparede (Acari: Astigmata) in lily bulbs: Exploration in the field and pre-selection in the laboratory. *Exp. Appl. Acarol.* 19:655–669.

Lesna, I., C.G.M. Conijn, M.W. Sabelis, and N.M. van Straalen. 2000. Biological control of the bulb mite, *Rhizoglyphus robini*, by the predatory mite, *Hypoaspis aculeifer*, on lilies: Predator–prey dynamics in the soil, under greenhouse and field conditions. *Biol. Contr. Sci. Technol.* 10:179–193.

Matteoni, J.A. 2003. Economics of banker plant systems in Canadian greenhouse crops. In: *Proceedings of the 1st International Symposium on Biological Control of Arthropods* (pp. 154–155), FHTET-03-05. Washington, D.C.: U.S. Department of Agriculture Forest Service (http://www.bugwood.org/arthropod/day2.html).

Messelink, G.J. and R. van Holstein-Saj. 2006. Potential for biological control of the bulb scale mite (Acari: Tarsonemidae) by predatory mites in amaryllis. *Proc. Neth. Entomol. Soc. Meet.* 17:113–118.

Messelink, G.J. and R. van Holstein-Saj. 2007. Biological control of the bulb scale mite *Steneotarsonemus laticeps* (Acari: Tarsonemidae) with *Neoseiulus barkeri* (Acari: Phytoseiidae in amaryllis. *IOBC/WPRS Bull.* 30(5):81–85.

Meyer, M.K.P.S. 1996. Ornamental flowering plants. In: E.E. Lindquist, M.W. Sabelis, and J. Bruin (eds.), *Eriophyoid Mites: Their Biology, Natural Enemies, and Control* (pp. 641–650). Amsterdam: Elsevier.

Misayo, M, N. Tomohiko, O. Daigo, and F. Toshio. 2005. Control of acarid mites *Tyrophagus similis* Volgin by soil sterilization. *Annu. Rep. Kansai Plant Prot. Soc.* 47:1–8.

Murdoch, G. 1974. Bulb scale mite (*Steneotarsonemus laticeps*) on *Narcissus* in the United Kingdom. *Acta Hortic.* 47:157–163.

Navia, D., R. Ochoa, C. Welbourn, and F. Ferragut. 2010. Adventive eriophyoid mites: A global review of their impact, pathways, prevention and challenges. *Exp. Appl. Acarol.* 51:225–255.

Opit, G.P., D.C. Margolies, and J.R. Nechols. 2003. Within-plant distribution of twospotted spider mite, *Tetranychus urticae* Koch (Acari: Tetranychidae), on ivy geranium: Development of a presence–absence sampling plan. *J. Econ. Entomol.* 96:482–488.

Opit, G.P., J.R. Nechols, D.C. Margolies, and K.A. Williams. 2005. Survival, horizontal distribution and economics of releasing predatory mites (Acari: Phytoseiidae) using mechanical blowers. *Biol. Control* 33:344–351.

Osborne, L.S. 1987. Biological control of *Tetranychus urticae* Koch on ornamental foliage plants in Florida. *Bull. SROP* 10(2):144–148.

Osborne, L.S. and J.E. Barrett. 2005. Using banker plants. *Ornamental Outlook* Sept.:26–27.

Osborne, L.A. 2010. *Banker Plants*. Gainesville: Institute of Food and Agricultural Sciences, University of Florida (http://mrec.ifas.ufl.edu/lso/banker/banker.htm).

Parrella, M.P., V.P. Jones, M.S. Malais, and K.M. Heinz. 1989. Advances in sampling in ornamentals. *Florida Entomol.* 72:394–403.

Parrella, M.P., K.M. Heinz, and L. Nunney. 1992. Biological control through augmentative releases of natural enemies: A strategy whose time has come. *Am. Entomol.* 38(3):172–179.

Pilkington, L.J., G. Messelink, J.C. van Lenteren, and K. Le Mottee. 2010. "Protected biological control": Biological pest management in the greenhouse industry. *Biol. Control* 52:216–220.

Pratt, P.D. and B.A. Croft. 2000. Banker plants: Evaluation of release strategies for predatory mites. *J. Environ. Hort.* 18:211–217.

Pritchard, E.A. 1951. The fern mite. *Calif. Agric.* July:10.

Reiss, F. and L. Caroline. 1953. *Tarsonemus confusus*: Contaminant of fungus cultures. *Arch. Derm. Syphilol.* 68:728–730.

Rott, A.S. and D.J. Ponsonby. 2000. Improving the control of *Tetranychus urticae* on edible glasshouse crops using a specialist coccinellid (*Stethorus punctillum* Weise) and a generalist mite (*Amblyseius californicus* McGregor) as biocontrol agents. *Biocontr. Sci. Technol.* 10:487–498.

Sanches-Ramos, I., F. Alvarez-Alfageme, and P. Castanera. 2007. Reproduction, longevity and life table parameters of *Tyrophagus neiswanderi* (Acari: Acaridae) at constant temperatures. *Exp. Appl. Acarol.* 43:213–226.

Sanderson, J.P. and Z.-Q. Zhang. 1995. Dispersion, sampling, and potential for integrated control of twospotted spider mite (Acari: Tetranychidae) on greenhouse roses. *J. Econ. Entomol.* 88:343–351.

Schausberger, P. and A. Walzer. 2001. Combined versus single species release of predaceous mites: Predator–predator interactions and pest suppression. *Biol. Control* 20:269–278.

Schumacher, S.K., T.L. Marsh, and K.A. Williams. 2006. Optimal pest control in greenhouse production of ornamental crops. *Agric. Econ.* 34:39–50.

Sengonca, C. and K. Drescher. 2001. Laboratory studies on the suitability of *Thrips tabaci* Lindeman (Thysanoptera: Thripidae) as prey for the development, longevity, reproduction and predation of four predatory mite species of the genus *Amblyseius* (Acari: Phytoseiidae). *Z. Pflanzenk. Pflanzenschutz* 108:66–76.

Steiner, M.Y. and D.P. Elliot. 1987. *Biological Pest Management for Interior Plantscapes*, 2nd ed. Edmonton: Alberta Public Affairs Bureau.

van Lenteren, J.C. 2000. A greenhouse without pesticides: Fact or fantasy? *Crop Prot.* 19:375–384.

van Lenteren, J.C. (ed.) 2003. *Quality Control and Production of Biological Control Agents: Theory and Testing Procedures*. Wallingford, U.K.: CAB International.

van Lenteren, J.C. 2006. How not to evaluate augmentative biological control. *Biol. Control* 39:115–118.

van Lenteren, J.C. and J. Woets. 1988. Biological and integrated pest control in greenhouses. *Annu. Rev. Entomol.* 33:239–269.

van de Vrie, M. 1985. Control of Tetranychidae in crops: Greenhouse ornamentals. In: W. Helle and M.W. Sabelis (eds.), *Spider Mites: Their Biology, Natural Enemies, and Control* (pp. 273–283). Amsterdam: Elsevier.

van Rijn, P.C.J. and L.T. Tanigoshi. 1999. Pollen as food for the predatory mites *Iphiseius degenerans* and *Neoseiulus cucumeris* (Acari: Phytoseiidae): Dietary range and life history. *Exp. Appl. Acarol.* 23:785–802.

Vantornhout, I., H.L. Minnaert, L. Tirry, and P. De Clercq. 2004. Effect of pollen, natural prey and factitious prey on the development of *Iphiseius degenerans*. *BioControl* 49:627–644.

Venzon, M., A. Jassen, and M.W. Sabelis. 2001. Prey preference, intraguild predation and population dynamics of an arthropod food web on plants. *Exp. Appl. Acarol.* 25:785–808.

Walter, D.E., R.A. Hudgens, and D.W. Freckman. 1986. Consumption of nematodes by fungivorous mites, *Tyrophagus* spp. (Acarina: Astigmata: Acaridae). *Oecologia* 70:357–361.

Warner, K.D. and C. Getz. 2008. A socio-economic analysis of the North American commercial natural enemy industry and implications for augmentative biological control. *Biol. Control* 45:1–10.

Workman, P.J. and N.A. Martin. 2002. Effect of pesticides on cymbidium orchid pollen-cap mite and its predator *Hypoaspis* sp. *N.Z. Plant Prot.* 55:380–384.

Zhang, Z.-Q. 2003. *Mites of Greenhouses: Identification, Biology, and Control*. Wallingford, U.K.: CAB International.

Zhang, Z.-Q. and J.P. Sanderson. 1995. Twospotted spider mite (Acari: Tetranychidae) and *Phytoseiulus persimilis* (Acari: Phytoseiidae) on greenhouse roses: Spatial distribution and predator efficacy. *J. Econ. Entomol.* 88:352–357.

PART V

Soil Mites and Agriculture

The soil is a complex and diverse ecosystem that contains many species of arthropods (including springtails, mites, termites, symphylans, centipedes, spiders, pseudoscorpions, sowbugs, and beetles). In addition, soil contains algae, bacteria, fungi, protozoa, earthworms, small vertebrates, and nematodes (Figure V.1)(also see Figure VS.1 on the CD). A shovel full of soil from a forest can contain more species than live in the above-ground portion of a rainforest. A handful of soil may contain billions of bacteria and more than 16 km of fungal hyphae, and the abundance of oribatid mites may reach 400,000 per square meter (Krivolutsky and Druk 1986).

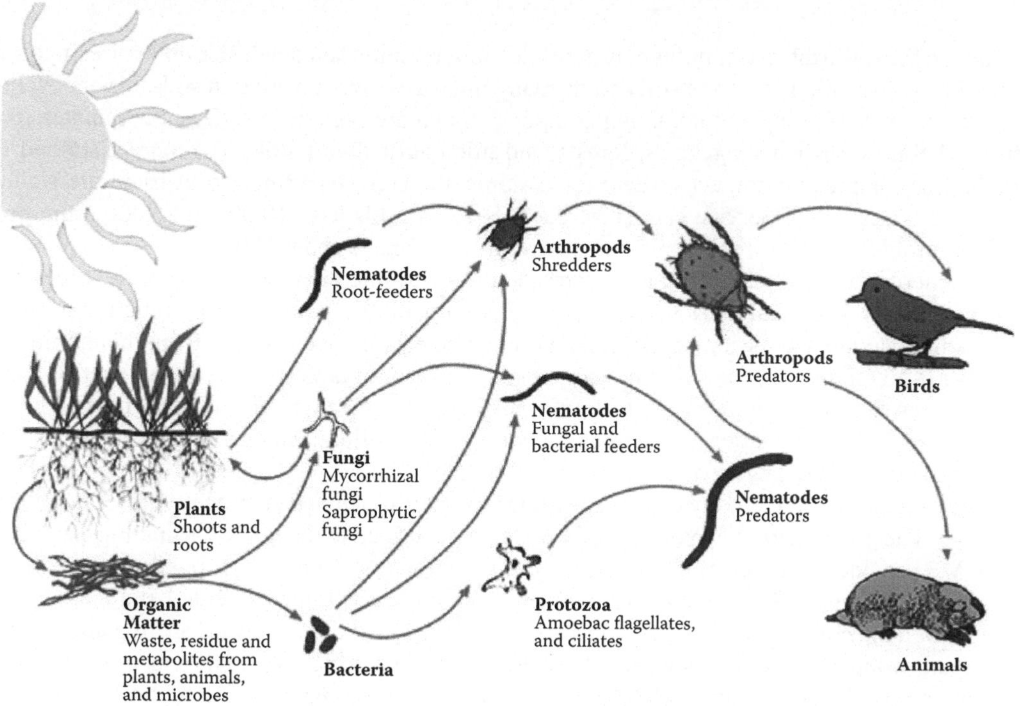

Figure V.1 Soil is a complex ecosystem with complex interactions between the different organisms living there. Mites may be important shredders and predators in the system, as well as serve as prey. Some feed on nematodes, fungi, and other organic matter. Mites distribute fungi and bacteria, aerate the soil, and are involved in nutrient cycling. (From Tugel, A. et al., Eds., *Soil Biology Primer*, Revised ed., Soil and Water Conservation Society, Ankeny, IA, 2000. With permission.)

Figure V.2 This sample includes diverse occupants of leaf or soil litter after being extracted in a Berlese funnel. Small arthropods, including mites, nematodes, collembola, ants, and beetles, are present. The mites may be predators, fungivores, detritus feeders, or saprophages. (Photograph by Valerie Behan-Pelletier, Canadian National Collection of Insects and Arachnids, Ottawa.)

Among the soil arthropods, mites play multiple roles. Despite their small size, mites are important in aeration of the soil, mixing the soil and its many microbes, and assisting in soil-nutrient cycling. Soil is a complex of living and nonliving components that are present in different combinations in different habitats, such as rangelands, forests, and tilled agricultural soils. The more disturbed the soil, the less complex the soil ecosystem; for example, the U.S. Department of Agriculture Natural Resources Conservation Service estimates that U.S. forest soils have 10,000 to 25,000 arthropods per square foot (0.093 square meters), but agricultural soils have about 100.

Soil microarthropods shred organic material, stimulate microbial activity, mix microbes with their food, mineralize plant nutrients, enhance soil aggregation, aerate the soil through their movements, stimulate the succession of species, and, in some cases, control pests. Even though the role of mites in soil mixing may be small in comparison with that of larger invertebrates, such as earthworms, mites exercise an important function in mineral turnover and vegetation succession and as decomposers of organic matter. In combination with the microflora, which they disperse, soil mites help in decomposing organic matter.

The abundance and diversity of mites decline with soil depth and with the types of cultivation employed. The great majority of soil mites appear to be confined to the top 7 to 8 cm. Figure V.2 and Figure V.3 provide examples of the diversity of microarthropods and mites found in the soil (also see Figures VS.2 and VS.3 on the CD); however, soil mites have limited mobility and must survive extremes of temperature and moisture. Most soil mites are found under vegetation, in plant residues, in decaying wood, or under rocks where the relative humidity is high. A single square meter of soil can contain up to 200,000 arthropods, the majority of which may be mites. Tilling to a greater depth and frequency reduces soil organisms in agricultural soils, while no-till and strip tillage will maintain more soil organisms and diversity. Soil compaction and pesticides also reduce soil organisms in agricultural soils. The types of crops used as cover crops or in crop rotations affect the community of organisms in the soil. Unfortunately, these relationships are not well understood at this time. Among the mites, the **Oribatida** (= **Cryptostigmata**) are especially common in soils, and they

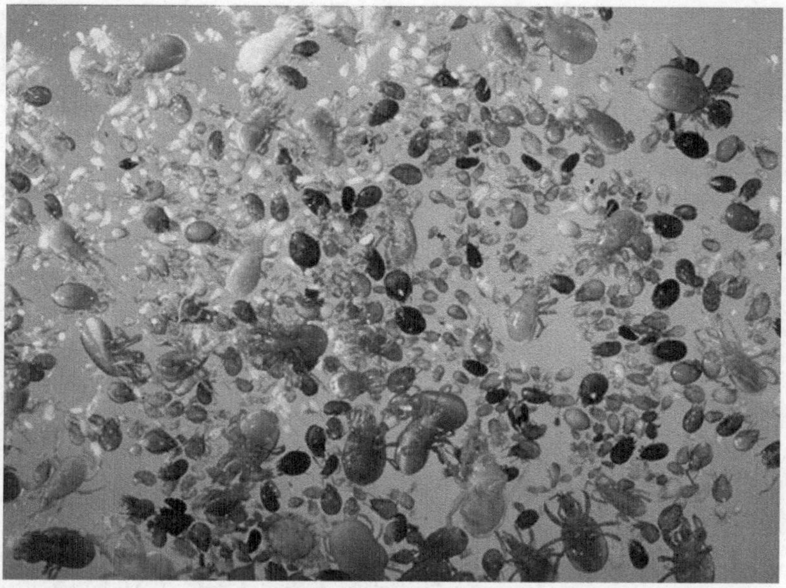

Figure V.3 Assorted mites extracted using a Berlese funnel include species in the Oribatida, Gamasida, Acaridida, and Actinedida. The diversity is most extensive in less-disturbed habitats. Common agricultural processes, such as plowing, weeding, and compaction, reduce the diversity and number of soil mites. (Photograph by Valerie Behan-Pelletier, Canadian National Collection of Insects and Arachnids, Ottawa.)

are often referred to as *soil* or *beetle mites* because many are heavily sclerotized and dark in color (Balogh 1972, Balogh and Mahunka 1983, Marshall et al. 1987). Mites in the Actinedida, Gamasida, and Acaridida are also common, although they are not as numerous. Many of the Gamasida are predators of other small invertebrates, including mites, collembola, and nematodes.

An estimated 9000 species of oribatids represent 172 families of Oribatida (Norton and Behan-Pelletier 2009). A generalized biology is as follows: Oval eggs are laid singly or in small clumps in old exuviae, in the axils of moss bracts, under detritus, and in soil pores. Often the female retains a number of eggs until all are mature and she deposits them all at once. The length of time to reach adulthood depends on temperature; smaller species have a shorter developmenttime and life span than larger species. Rate of development also is influenced by whether the diet is adequate and whether the mites are crowded. In general, small species may have several generations a year, while larger species may have one per year. A few oribatid species require 2 to 3 years to complete their life cycle. Some oribatid life spans are as long as 4 or 5 years. Norton and Behan-Pelletier (2009) suggested that the long life span many oribatids exhibit has resulted in selection for strong defensive modifications, such as protective setae, waxy exudates, defensive glands, cuticular hardening, and body forms that provide protection from predation. Many have thickened and darkened exoskeletons and can retract their legs into cavities of their body for protection (Figure V.4)(also see Figure VS.4 on the CD).

Mating takes place when oribatid males deposit spermatophores that are taken up by the female (indirect sperm transfer). Spermatophores are stalked and often deposited in groups on the substrate, where many probably are eaten. Copulation has never been observed in oribatids. Females typically deposit 1 to 12 eggs at a time. Some oribatids are thelytokous, having only females.

Oribatids choose optimal temperatures (or at least avoid unfavorable temperatures). Humidity requirements are species specific; some are able to survive for days at low relative humidity while others require a higher relative humidity to survive. Salinity and soil pH are important; soils rich in organic matter have a higher pH.

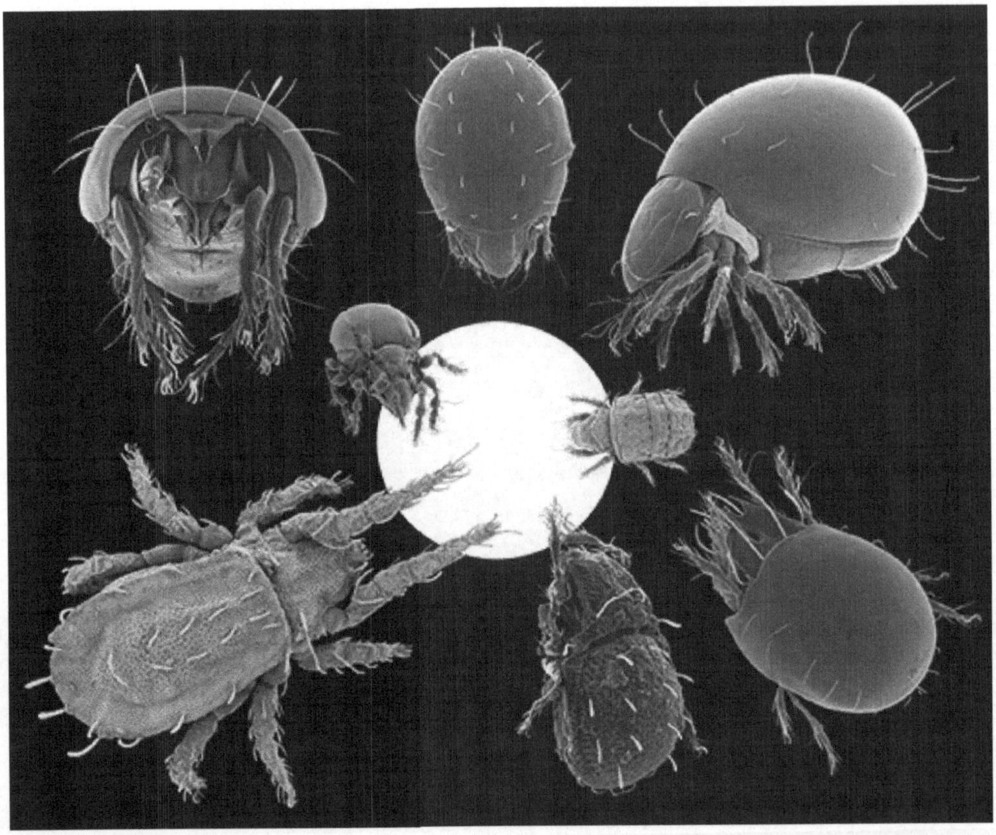

Figure V.4 Diversity of form and structure of mites found in the soil. Note that some mites have sclerotized structures that allow the legs to be withdrawn for protection. Many species have dark and thickened exoskeletons. The white circle in the center represents the size of a period at the end of this sentence and allows comparisons of the sizes of these mites. (Illustration by David E. Walter, Royal Alberta Museum, Canada.)

Oribatids have been classified as having different types of feeding behaviors. The largest species typically feed on leaf litter or are unspecialized. The smaller species feed on fungi, bacteria, and other microbes. Some are macrophytophages (feeding on higher plant material) and have a role in litter decomposition. Some are microphytophages feeding on microflora, some are panphytophages (unspecialized plant feeders), some are zoophages feeding on living organisms, some are necrophages feeding on carrion, and some are coprophages feeding on fecal material. The panphytophages feed on pollen, small pieces of algae, lichens, highly decomposed humus, and fungal hyphae (Behan-Pelletier et al. 1983, Kaneko 1988). Some oribatids can be maintained on cricket powder in the laboratory, and nematodes may be eaten by some species. Laboratory rearing of some species can be accomplished with dried mushrooms, chopped lichens, decomposing leaves, or an artificial diet of dextrose and casein or brewer's yeast. Some oribatids are considered pests. Oribatid species may serve as intermediate hosts for cestode parasites (Denegri 1993). At least 47 species of oribatid mites in 32 genera transmit 12 species of tapeworms (cestodes) (Haq 2001). Affected animals include horses, sheep, humans, macaque monkeys, marmots, voles, and cows. Apparently, the mammal ingests the mites when feeding, which then transmit the parasites.

Only a few soil mite species have a direct role in altering soil morphology. Most exercise indirect effects through stimulation of microbial activity and spore distribution. Pesticides can have a negative effect on nutrient cycling by permitting energy to remain bound up in undecomposed plant debris. Pesticides can positively affect nutrient cycling by killing predators of saprophagous arthropods, thus releasing detritus-feeding populations. Oribatids are thought to be especially important decomposers of plant debris. On mulched plots, they make up 60% of the total arthropods, compared to 8.5% on mowed or fallow plots. Distribution of microarthropods is irregular under annual crops, in part due to cultivation. Fewer mites are found in grasslands than in forest communities. Microarthropod distribution is influenced by water, pore space, oxygen saturation deficits, temperature variations, ground cover, fungi, flooding, cropping, cultivation, organic matter, disturbance, soil compaction, soil type and texture, predation, and feeding habits. It has been observed that oribatids are slow to colonize degraded land, requiring 4 to 7 years.

Oribatid species present in a particular area respond to agricultural practices, such as the use of pesticides, fertilizers, or plowing (Behan-Pelletier 1999). Research has been conducted to determine whether the oribatid species present may be used to determine whether agricultural soil quality is being degraded or enhanced. Agricultural soils could be improved if management practices would reduce or eliminate cultivation and the use of heavy machinery and general biocides. Incorporating residues of crops and other organic material would improve the release of nutrients and water availability. Research should be conducted to learn how best to apply organic materials and associated microorganisms, synchronize management practices with crop and soil biota phenology, and improve our knowledge of soil species with regard to ecosystem processes. Much remains to be learned about the role that mites play in soil biology and ecology, in part because diverse habitats remain unsampled and because specialized sampling methods are required to extract the mites from the substrate.

New species still are being discovered in unusual soil habitats; for example, although most mites are within the top 15 cm of soil, some mites may be found at depths of two meters or more. One such mite (*Gordialycus*) is truly bizarre, appearing more like a nematode than a mite (Prostigmata: Nematalycidae) (Schubart 1973). This mite is long and worm-like and has two pair of legs at the anterior of the body. The third and fourth pair of legs are greatly reduced and located distally from the first two pair. The body is a very elongated behind the posterior legs (Figure V.5). Norton et al. (2008) found that *Gordialycus* mites associated with fine sand, where they move by using rows of plate-like cuticular structures that "interlace and operate together with an underlying muscle layer." Unfortunately, little is known about the biology of these mites.

Soil mites can inhabit extreme environments; for example, *Alaskozetes antarcticus* (Oribatida: Podacaridae) is found in the Antarctic, where it lives in intertidal debris, lichens, penguin dung, and algae (Block and Convey 1995). It is nearly 1 mm in length and is the largest free-living

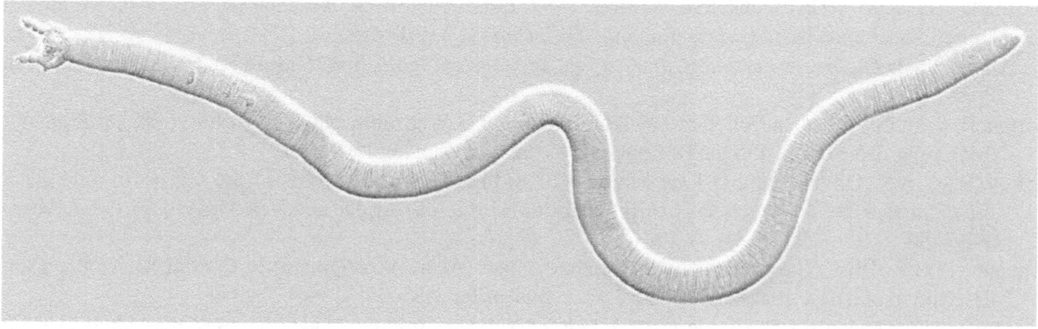

Figure V.5 Scanning electron micrograph of *Gordialycus* (Nematalycidae). This unusual elongated soil mite is found in sandy soils and it is one of the most unusual mites known. Its first two pairs of legs are anterior, near the mouthparts. The third and fourth pairs of legs are greatly reduced and distal. Behind the fourth pair of legs is a very elongated opisthosoma. This genus contains the most elongated mites known. This mite moves among sand particles by using the transverse rows of plate-like cuticular structures that interlace and operate together with an underlying muscle layer. (Photograph provided by Roy A. Norton, State University of New York, Syracuse.)

indigenous arthropod in the Antarctic. This mite can occur in large numbers, although a life cycle can take 5 years or more. It is able to survive winters using glycerol as an antifreeze. For a more detailed discussion of the ecology of this and the other mites found in Antarctica, see Walter and Proctor (1999).

Soil fertility and nutrient balances in the soil promote healthy crops that are less susceptible to damage from arthropod pests and diseases. Maintaining a healthy cropping system requires an understanding of the effects that cover cropping, crop rotations, composts, and plant residues play in soil health, as well as understanding the role of the flora and fauna in the soil. For additional information on soil biology and ecology in all its complexity, see Nardi (2003), Coleman et al. (2004), and Gobat et al. (2004). See also the *Soil Biology Primer* (Tugel et al. 2000).

SELECTED REFERENCES

Balogh, J. 1972. *The Oribatid Genera of the World.* Budapest: Akadémiai Kiadó.

Balogh, J. and S. Mahunka. 1983. T*he Soil Mites of the World. Primitive Oribatids of the Palaearctic Region.* Amsterdam: Elsevier.

Behan-Pelletier, V.M. 1999. Oribatid mite biodiversity in agroecosystems: role for bioindication. *Agric. Ecosys. Environ.* 74:411–423.

Behan-Pelletier, V.M. and S.B. Hill. 1983. Feeding habits of sixteen species of Oribatei (Acari) from an acid peat bog, Glenamoy, Ireland. *Revue d'Écologie et de Biologie du Sol* 20(2):221–267.

Block, W. and P. Convey. 1995. The biology, life cycle and ecophysiology of the Antarctic mite *Alaskozetes antarcticus*. *J. Zool.* 236:431–449.

Coleman, D.C., D.A. Crossley, Jr., and P.F.Hendrix. 2004. *Fundamentals of Soil Ecology*, 2nd ed. San Diego, CA: Academic Press.

Denegri, G.M. 1993. Review of oribatid mites as intermediate hosts of tapeworms of the Anoplocephalidae. *Exp. Appl. Acarol.* 17:567–580.

Gobat, J.M., M. Aragno, and W. Matthey. 2004. *The Living Soil: Fundamentals of Soil Science*. Enfield, NH: Science Publishers.

Haq, M.A. 2001. Potential oribatid mite vectors of cestode parasites in Kerala, India. In: R.B. Halliday, D.E. Walter, H.C. Proctor, R.A. Norton, and M.J. Colloff (eds.), *Acarology, Proceedings of the 10th International Congress* (pp. 569–575). Melbourne, Australia: CSIRO Publishers.

Kaneko, N. 1988. Feeding habits and cheliceral size of oribatid mites in cool temperate forest soils in Japan. *Revue d'Écologie et de Biologie du Sol* 25(3):353–363.

Krivolutsky, D.A. and A.Y. Druk. 1986. Fossil oribatid mites. *Annu. Rev. Entomol.* 31:533–545.

Marshall, V.G., R.M. Reeves, and R.A. Norton. 1987. Catalogue of the Oribatida (Acari) of the continental United States and Canada. *Mem. Entomol. Soc. Canada* 139:1–418.

Nardi, J.B. 2003. *The World Beneath Our Feet: A Guide to Life in the Soil.* Oxford, U.K.: Oxford University Press.

Norton, R.A. and V.M. Behan-Pelletier. 2009. Oribatida. In: G.W. Krantz and D.E. Walters (eds.), *A Manual of Acarology*, 3rd ed. (pp. 430–564). Lubbock: Texas Tech University Press.

Norton, R.A., A.R. Oliveira, and G.J. de Moraes. 2008. First Brazilian records of the acariform mite genera *Adelphacarus* and *Gordialycus* (Acari: Acariformes: Adelphacaridae and Nematalycidae). *Int. J. Acarol.* 34:91–94.

Schubart, H.O.R. 1973. The occurrence of Nematalycidae (Acari, Prostigmata) in Central Amazonia with a description of a new genus and species. *Acta Amazonica* 3:53–57.

Tugel, A., A. Lewandowski, D. Happe-vonArb (eds.) 2000. *Soil Biology Primer*, Revised ed. Ankeny, IA: Soil and Water Conservation Society (http://soils.usda.gov/sqi/concepts/soil_biology/biology.html).

Walter, D. and H. Proctor. 1999. *Mites: Ecology, Evolution, and Behaviour.* Wallingford, U.K.: CAB International.

PART VI

Pest Mites of Honey Bees

European honey bees (or honeybees) (*Apis mellifera*) are a significant component of agriculture in many countries. According to Southwick and Southwick (1992), honey bee pollination of 62 agricultural crops provides economic benefits valued at \$1.6 to \$8.3 billion in the United States. The lower value assumes that wild pollinators could partially replace honey bees as pollinators of crops; however, honey bees are responsible for at least 80% of the pollination of more than 100 different crops (Southwick and Southwick 1992). Bees also produce honey and wax, although the value of these products is much less than the value of their pollination services (Winston 1987, Doebler 2000). More recently, vanEngelsdorp and Meixner (2010) estimated that, globally, honey production during 2006 was valued at \$1.25 billion; they observed that 52 of the 115 leading global food commodities depend on honey bee pollination for either fruit or seed set, and 35% of the human diet is thought to benefit from pollination. They further concluded that globally the value of insect pollination is estimated at \$212 billion, "which represents about 9.5% of the total value of agricultural production" (vanEngelsdorp and Meixner 2010).

Many mite species associate with honey bees and other bees in the Apoidea. Over 20,000 bee species can be found in 11 families, with most bees constructing nests and provisioning cells with pollen and nectar (Winston 1987). As expected, with such an abundant food resource, a wide variety of mites is associated with these bee nests, with some acting as saprophages (especially fungivores), as predators of other organisms in bee nests, and some as true parasites of bees. Two true parasites recently have become very significant pests of *Apis mellifera* around the world.

Sometime around 1950, *Varroa jacobsoni* (Gamasida or Mesostigmata: Varroidae), an Asian honey bee mite originally found on *Apis florea*, was accidentally moved to Europe from Asia where it attacked *A. mellifera*. In the 1970s, *Varroa* was transferred to North Africa and in 1971 to South America (Sanford 2003). *Varroa* invaded the United States in 1987 and is now the most serious mite pest of *A. mellifera* around the world (see Chapter 20). The second parasitic mite species, *Acarapis woodi*, a European tarsonemid mite that infests the tracheae of bees, was first found in South America in the 1950s, in Mexico in 1980, and in the United States in 1984. It also is a serious pest of *A. mellifera* (see Chapter 21). Other mites are found on bees in other parts of the world, including another varroid *(Euvarroa sinhai)* that parasitizes *Apis florea*. *Tropilaelaps clareae* (Gamasida: Laelapidae) occurs on bees in Asia and can kill a colony of European honey bees more quickly than *Varroa*. *Tropilaelaps clareae* can live on all *Apis* species and is best suited to the tropics. If these mites invade the Western Hemisphere, additional bee management problems would occur.

In the fall of 2006, beekeepers in a number of countries (Europe, China, and the United States) began to be concerned about European honey bee colonies disappearing in large numbers without any clear reason (Neumann and Carreck 2010, vanEngelsdorp et al. 2010). This 'disappearing disease' or 'colony collapse disorder' caused 30 to 90% of hives to die out during the winter of 2006 in the United States. The relationship between colony collapse disorder and the bee mites *Varroa* and *Acarapis* remains unclear at the time of this writing. Some beekeepers and scientists think that the bees are disappearing due to multiple stressors, possibly including the negative effects of *Varroa* and *Acarapis* in combination with bacterial, fungal, and viral diseases that afflict honey bees. Many believe that pesticides are affecting bees negatively or that a new and unknown pathogen could be leading to these losses (Cox-Foster et al. 2007, Higes et al. 2008, vanEngelsdorp et al. 2010). Poor nutrition (quality and quantity of food), colony overcrowding, lack of adequate nutrition going into the winter period, and the stresses of moving bee hives long distances to provide pollination services could play a role in colony collapse disorder (LeConte et al. 2010). The biology of the honey bee is complex and has been studied extensively (Winston 1987). The bee has been bred for a variety of attributes as a semidomesticated insect for many years by beekeepers, and it is clear that genetic variability can be found that can be used to improve pest and disease resistance further. It is likely that understanding the genome of the honey bee will open up other opportunities for genetic improvement of the honey bee (Honeybee Genome Sequencing Consortium 2006).

The literature suggests that similar large losses of honey bee colonies have occurred previously (in the 1880s, 1920s, and 1960s), but it is unclear if those losses were caused by the same factors (Underwood and vanEngelsdorp 2007). Funding for research into the cause or causes of the disappearing bees has been raised, and research is underway to answer the many questions arising in both Europe and the United States (Pettis and Delaplane 2010). For more information on the research designed to understand and manage colony collapse disorder and links to other relevant sites, go to http://www.ars.usda.gov/is/br/ccd/ccd_actionplan.pdf. To learn more about the European efforts, see www.coloss.org.

SELECTED REFERENCES

Cox-Foster, D.L., S. Conian, E.C. Holmes, G. Palacios, J.D. Evans, N.A. Moran, P.L. Quan, T. Briese, M. Hornig, D.M. Geiser, V. Martinson, D. vanEngelsdorp, A.L. Kalkstein, A. Drysdale, J. Hul, J. Zhai, L. Cui, S.K. Hutchinson, J.F. Simons, M. Egoim, J.S. Pettis, and W.L. Lipkin. 2007. A metagenomic survey of microbes in honey bee colony collapse disorder. *Science* 318:283–287.

Doebler, S.A. 2000. The rise and fall of the honeybee. *BioScience* 50:738–742.

Higes, M., R. Martin-Hernandez, C. Botias, E.G. Ballon, A.V. Gonzalez-Porto, L. del NozalBarrios, J.L. Bernal, J.J. Jimenez, P.G. Palencia, and A. Meana. 2008. How natural infection by *Nosema ceranae* causes honeybee colony collapse. *Environ. Microbiol.* 10:1659–1668.

Honeybee Genome Sequencing Consortium. 2006. Insights into social insects from the genome of the honeybee *Apis mellifera*. *Nature* 443:931–949.

LeConte, Y., M. Ellis, and W. Ritter. 2010. *Varroa* mites and honey bee health: Can *Varroa* explain part of the colony losses? *Apidologie* 41(3):353–363.

Neumann, P. and N.L. Carreck. 2010. Honey bee colony losses. *J. Apic. Res.* 49:1–6.

Pettis, J.S. and K.S. Delaplane. 2010. Coordinated responses to honey bee decline in the USA. *Apidologie* 41:256–263.

Sanford, M.T. 2003. *Diseases and Pests of the Honey Bee*, UF IFAS Publication No. CIR766. Gainesville: Department of Entomology and Nematology, University of Florida Institute of Food and Agricultural Sciences (http://edis.ifas.ufl.edu/AA090).

Southwick, E.E. and L. Southwick, Jr. 1992. Estimating the economic value of honey bees (Hymenoptera: Apidae) as agricultural pollinators in the United States. *J. Econ. Entomol.* 85:621–633.

Underwood, R. and D. vanEngelsdorp. 2007. Colony collapse disorder: Have we seen this before? *Bee Culture* 35:13–18.

vanEngelsdorp, D. and M.D. Meixner. 2010. A historical review of managed honey bee populations in Europe and the United States and the factors that may affect them. *J. Invert. Pathol.* 103:S80–S95.

vanEngelsdorp, D., J. Hayes, Jr., R.M. Underwood, and J.S. Pettis. 2010. A survey of honey bee colony losses, fall 2008 to spring 2009. *J. Apic. Res.* 49:7–14.

Winston, M.L. 1987. *The Biology of the Honey Bee*. Cambridge, MA: Harvard University Press.

CHAPTER 20

Varroa jacobsoni

20.1 BIOLOGY AND TAXONOMY

Because the *Varroa* mite (Gamasida or Mesostigmata: Varroidae) has spread around the world, beekeepers have had to alter their management practices in dramatic ways (Sammarato et al. 2000, vanEngelsdorp and Meixner 2010). Left untreated, *Varroa* can kill most bee colonies within one or two years. *Varroa* mites have driven feral honey bee populations in the United States extinct, thereby reducing the pollination benefits of these bees for home gardeners and on unmanaged lands (Sanford et al. 1998, Sanford 2003). *Varroa* entered the United States in 1986 and spread to every mainland state within 9 years, and it has now reached Hawaii (DeJong et al. 1982, Eickwort 1994, Sammataro et al. 2000, MAAREC 2005).

Where did *Varroa* come from? About 30 years ago, there were three known species of *Varroa* parasitizing honey bees (*V. jacobsoni*, *V. underwoodi*, and *V. rindereri*) (DeJong et al. 1982). *Varroa jacobsoni* was first described as an ectoparasite of the Asian honey bee (*Apis cerana*) in Asia, but it later switched to the European honey bee (*A. mellifera*) and was thought to have become a pest of *A. mellifera* around the world. Taxonomic problems complicated the picture (again), and Anderson and Trueman (2000) found that *V. jacobsoni* actually is a complex of at least two species. It is now recognized that the species attacking *A. mellifera* is *Varroa destructor*. Furthermore, it appears that there are biotypes of *V. destructor* with different levels of virulence (Anderson and Fuchs 1998). Molecular tests suggest that a South Korean biotype of *V. destructor* and a Japan/Thailand biotype occur (Anderson and Trueman 2000). The Japan/Thailand biotype was found on populations of *A. mellifera* in Japan, Thailand, and the Americas. The Korean biotype appears more pathogenic to *A. mellifera* than the Japan/Thailand biotype. Both biotypes appear to be present in the United States, but the Korean biotype appears to be the only one in Europe.

Much of the older literature on *Varroa* does not discriminate between biotypes or species of *Varroa* (Oldroyd 1999). *Varroa destructor* currently is considered the major pest of European honey bees in most parts of the world (deGuzman and Rinderer 1999, Sammataro et al. 2000). *Varroa destructor* apparently requires high temperatures and high relative humidity to do well and reproduces on both capped drones (males) and worker brood (sterile female workers) of *Apis mellifera* (Le Conte et al. 1989, Garrido and Rosenkranz 2003). Adult females of *Varroa* are parasites of adult bees; they are heavily sclerotized and reddish-brown in color (Figure 20.1). Despite their relatively large size, *Varroa* females can be difficult to detect on adult bees.

Adult *Varroa* are large for mites, averaging 1 to 1.8 mm in length and 1.5 to 2 mm in width (Needham et al. 1988). They are dorsoventrally flattened and curved to fit into the crevices of the abdomen of the adult bee so they are less likely to be removed during grooming by the bee (Figure 20.2). Adult males are lighter in color and smaller, averaging 0.75 to 1 mm in length and 0.7 to 0.9 mm in width; they are not found on adult bees.

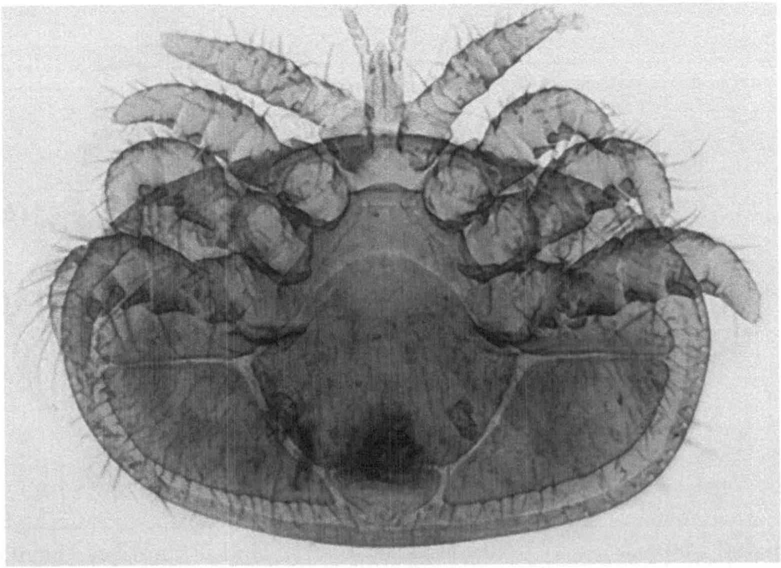

Figure 20.1 Slide-mounted and partially cleared specimen of *Varroa* female from honey bees in Florida, ventral view. (Photograph by Patricia Toth, Department of Entomology and Nematology, University of Florida, Gainesville.)

Varroa females suck hemolymph from holes in the intersegmental membranes of their hosts (DeJong et al. 1982, Rosenkrantz et al. 2010). Females commonly are found on the abdomen, under the abdominal sclerites, or between the thorax and abdomen. They hold tightly onto the bee with their legs but can move rapidly over the surface of the bee. The reproductive cycle of *Varroa* requires that the female move from adult bees to the cells of bee larvae to reproduce. *Varroa* females enter the brood cell containing a larval bee shortly before the cell is sealed, apparently due to arrestment of the mite by chemicals produced by 8-day-old larvae (Rickli et al. 1994).

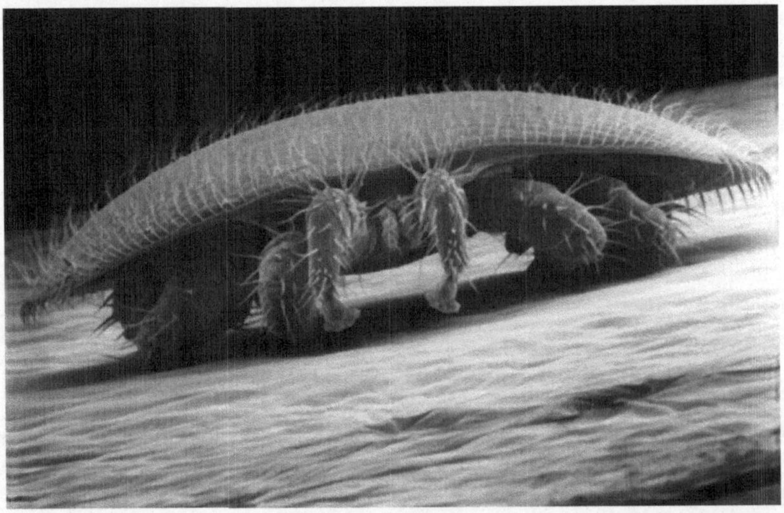

Figure 20.2 Scanning electron micrograph of *Varroa* female showing the mouthparts and the flattened body shape, which is slightly curved. (Photograph by Harvey Cromroy, Department of Entomology and Nematology, University of Florida, Gainesville.)

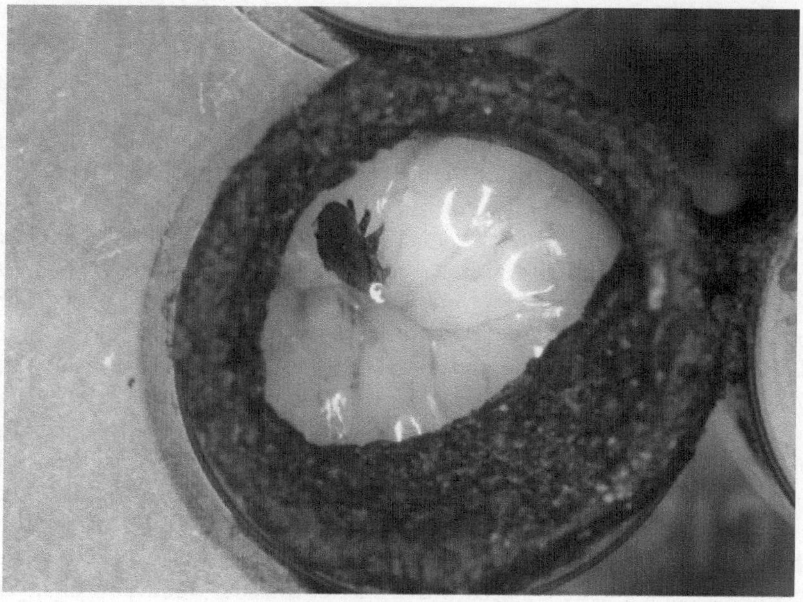

Figure 20.3 An adult *Varroa* female feeding on a bee pupa within the honey bee brood cell after the cap has been partially removed. The mite is a dark reddish-brown. (Photograph by Jim Castner, Department of Entomology and Nematology, University of Florida, Gainesville.)

The *Varroa* female may feed on the bee larva but quickly crawls underneath the larva and immerses herself in the brood food provided by the worker bees for the bee larva. The *Varroa* female remains in the brood food, ventral side toward the opening of the cell, until the bee larva eats the food, which cleans and frees the *Varroa* female. As many as 21 adult *Varroa* females can be found within a single brood cell. After the *Varroa* females are freed from the brood food, most die, but those that survive begin feeding on the hemolymph of the larval and later the pupal bee (Figure 20.3) (also see Figure S20.1 on the CD). The mites swell noticeably so there are spaces between the ventral plates. *Varroa* females then deposit eggs one at a time on the brood cell walls. The larval stage of *Varroa* develops within the first 24 hours but remains inside the egg, molting to a protonymph that hatches about 47 hours after the egg is laid. The protonymph feeds on hemolymph of the bee pupa for several days, then molts to a deutonymph. The deutonymph continues to feed for several more days before molting to the adult. Males are small and pale; their chelicerae are highly modified for sperm transfer with a spermatodactyl, and they do not feed. Mating takes place within the cell. Development for males requires 6 to 7 days; females, 8 to 9 days. Males and the many mites that do not reach maturity remain in the cell and die, while the newly mated females leave the cell with the emerging bee.

Virgin *Varroa* females apparently cannot produce viable male progeny, as would be expected if this species were arrhenotokous (haploid males and diploid females). Apparently, females must be mated to produce any progeny and, furthermore, must mate immediately after molting to adulthood to produce progeny. This suggests that *Varroa* has a parahaploid genetic system, as is found in phytoseiids (Mesostigmata or Gamasida), in which all eggs are fertilized and those embryos destined to be males lose half their chromosomes during embryonic development.

An average of 1.4 *Varroa* adult females is produced in worker brood cells and 2.4 *Varroa* adult females from drone (male) brood. *Varroa* females parasitize adult bees after they emerge from the brood cells (see Figure S20.2 on the CD). They are able to discriminate between different aged bees and prefer younger bees (Eickwort 1994, Sammataro et al. 2000, Sanford 2003). Transfer of *Varroa* to new colonies is not well understood but probably is due to dispersal of foraging bees. *Varroa* mites overwinter in hives and can survive for 2 months on adult bees in summer and at least 5 months in winter (Amdam et al. 2004). An infested colony may have thousands of *Varroa* (see Figure S20.3 on the CD). An individual worker adult may have up to 5 *Varroa* and a drone up to about 12.

Varroa is a more serious pest of *Apis mellifera* than of *A. cerana* and quickly develops to destructive levels in hives of *A. mellifera*. Reduced honey production and weak or killed colonies are the result. Immature bees with up to six mites per cell will be deformed or killed. If densities are lower than six mites per cell, a few bees will have damaged wings, even though 50% or more of the bees are infested. *Apis cerana* worker bees apparently recognize *Varroa* as a parasite and rapidly remove the mites from their body and out of the brood. This grooming behavior slows mite population growth (deGuzman et al. 2001).

Apis mellifera that are infested by *Varroa* are more prone to diseases (Bowen-Walker et al. 1999, Nordstrom 2003, Chen et al. 2004, De Miranda et al. 2010). *Varroa* can transfer acute paralysis virus from unhealthy bees to healthy bees. Alternatively, *Varroa* may stress the bees so much that the endogenous virus is activated. Bacteria also can be transferred from bee to bee by the mites through the open wounds caused by the mite during feeding. De Rycke et al. (2002) reported that *Varroa* mites can transmit American foulbrood (a bacterium) from infected to healthy bee colonies. Mites tear the integuments of the bees with their chelicerae when they take in hemolymph. The loss of hemolymph leads to weakened bees with a shorter life span, reduced weight, and deformed abdomens. Drone bees also are damaged and reduced in size and have smaller seminal vesicles and fewer spermatozoa when infested with more than three mites. The duration of the mating flight by infested drones is reduced by half. Colony collapse occurs due to reduced vigor of the bees and to secondary infections with pathogens.

The complete DNA sequence of the mitochondrion of *Varroa destructor* has been obtained, which should allow detailed taxonomic and phylogenetic studies to be conducted on these and related mites (Evans and Lopez 2002). Steiner et al. (1982) found that males and females of *Varroa* have 7 and 14 chromosomes, respectively. The genome of this mite is being sequenced, and the results may provide new insight into its biology.

20.2 MONITORING FOR *VARROA*

The symptoms of colony infestations vary with mite density (Sammataro et al. 2000). Low-level infestations are difficult to detect, but larger infestations result in brood cells that are empty and malformed workers and drones with deformed wings and small abdomens. Such bees may not be able to fly. Infested brood may be seen at the hive entrance if the workers uncap and remove them. Mites also are discovered by pulling the caps off brood cells. *Varroa* appear as brown or whitish spots on the white pupae. Guanine, the fecal material of the mite, can be seen as white spots on the walls of brood frames in highly infested colonies. Several assay methods have been developed, including the ether roll, sugar shake, and sticky board methods.

20.2.1 Ether Roll

Collect about 300 to 500 adult bees in a clear glass jar and spray them briefly with engine starter (ether) fluid. Shake the jar to dislodge the mites and then roll it so the mites adhere to the sides of the jar. Add alcohol or soapy water to the jar and agitate the contents again to displace the remaining mites. Pour the liquid through coarse mesh to strain out the bees. Filter the mites through a fine filter and count them.

20.2.2 Sugar Shake Method

Shake bees from a brood frame into a plastic tub, concentrate the bees, scoop half a cup of bees, and dump them into a wide-mouth quart jar; cover with a screen lid. Add two tablespoons of powdered sugar through the screen lid and ensure that all bees are well covered. Keep the jar upright while shaking the jar vigorously from side to side, coating the bees. Bees should look completely white. Turn the jar upside down and shake vigorously over a baking tray or other light colored surface, moving the jar around so sugar and mites are spread evenly. Count the mites, then dump the tray or wipe the surface clean and repeat the shaking process. (Usually, mites move their legs after falling onto the tray, which makes it easier to see them.) Repeat shaking and counting until no more mites or sugar falls out of the jar. Record the number of mites removed, and pour the bees back into the hive.

20.2.3 Sticky Board

Place white paper or a plastic sheet covered with petroleum jelly (or cooking oil) on the bottom board of a colony and smoke the hive with pipe tobacco in a smoker. After closing the hive for 10 to 20 minutes, remove the board and count the mites. Alternatively, a sticky board can also be left in place for 1 to 3 days, which is less labor intensive (see Figure S20.3 on the CD).

20.2.4 How Many *Varroa* Are Too Many?

Economic injury levels and damage have not been reliably calculated, but drops of more than 100 mites per day onto white sticky boards were estimated to justify treatment in Georgia (Sammataro et al. 2000). Others believe this number is too high. Tarpy et al. (2007) observed that *Varroa* can kill colonies within one or two years after initial infestation. Finding mites during a honey flow precludes chemical treatment because the pesticides will show up in the honey.

20.3 CONTROL OF *VARROA*

Options include the development of resistant bee strains through selective breeding, chemical control, and cultural controls (Table 20.1).

20.3.1 Chemical Control

The treatment must not kill the bees and, ideally, must kill mites sealed within the brood cells (Lindberg et al. 2000). Because chemical fumigants cannot penetrate the brood cells, multiple treatments are required because complete control cannot be obtained with a single application. A variety of chemicals have been tried, but none is fully satisfactory, and the mites have developed resistance to fluvalinate and coumaphos, at least in some regions (Martin et al. 2002). Furthermore, pesticide residues must not be present in honey sold for human consumption.

Chemical control requires monitoring all colonies and treating on a regular basis, perhaps several times a year. Monitoring and treatment can add significantly to the cost of honey production and pollination services. Apistan® plastic strips (containing fluvalinate) have been used since 1988 in the United States. It is a contact pesticide, so mites must come in contact with the strips. Colonies typically are treated in late summer or early fall to improve winter survival. Treatment before the honey flow in early spring or the summer dearth, when spring honey is removed and fall honey

Table 20.1 Possible Control Tactics for Managing *Varroa* in Honey Bee Colonies

1. Pesticide resistance is a problem in many populations of *Varroa*. Resistance to Apistan® (fluvalinate) or CheckMite™ (coumaphos) occurs, so chemical control is not always feasible. Pesticides can appear in wax and honey products unless applied at the proper time (which can allow *Varroa* populations to increase to high levels prior to treatment).
2. Formic acid kills *Varroa* but is dangerous to use, and temperatures must be appropriate or the treatment can be toxic to bees (and humans). Other organic acids, such as oxalic and lactic, applied in sugar syrup trickled on bees, require broodless bees and may cause bee mortality.
3. Smoking can dislodge mites, and the mites can be trapped with a sticky board at the bottom of the hive in small apiaries.
4. Combs of drone brood can be used to attract and trap *Varroa* mites, and the drone brood can then be destroyed.
5. Heating hives to about 44°C will kill *Varroa* mites, but the sealed bee brood will survive.
6. Nonchemical control methods include trapping mites by removing the capped drone brood, the use of screened floors, and placement of sticky traps on the bottom board.
7. Less-toxic materials include powdered sugar applied to bees to induce grooming behavior to remove *Varroa* from adult bees.
8. One long-term solution is to breed bees that are resistant to *Varroa*. Several breeding programs have produced bees that are improved in their hygienic or their grooming behavior. The University of Minnesota's Hygienic stock is commercially available in the United States and can control American foulbrood and chalkbrood diseases. Colonies developed by the U.S. Department of Agriculture (USDA) are commercially available; the VSH (*Varroa*-sensitive hygiene) line developed by the USDA is available through Glenn Apiaries (Rinderer et al. 2010), and the Russian hybrid bees developed by the USDA have been provided to the Russian Honeybee Breeder's Association to continue selection. These mites can be used, without pesticides or with reduced pesticides, to control *Varroa*. Similar breeding programs are under way in Europe.
9. Additional breeding work could provide lines of bees adapted to different climates and crops that are resistant to *Varroa* using a variety of genetic mechanisms.

Source: Based on information from Rinderer (2010), Rosenkranz et al. (2010), Sammataro et al. (2000), Underwood and Currie (2005).

supers are not yet present in the colony, may also be important. Coumaphos (Bayer Bee Strips® or CheckMite™), an organophosphate, is embedded in plastic strips. Both fluvalinate and coumaphos should be applied only when there is no honey flow. Resistance to both products has been found in mite populations in different geographic areas, so the development of other control methods is a high priority.

Formic acid (Mite-Away II™) has been used to control *Varroa*, although it can be toxic to bees (and beekeepers) if used improperly (Satta et al. 2005, Underwood and Currie 2005, vanEngelsdorp et al. 2008). Fumigation with formic acid can kill mites in capped worker brood and adult bees, without harming queens or uncapped brood (van Engelsdorp et al. 2008). Formic acid also provides control for the tracheal mite (*Acarapis woodi*). Essential oils also are being used to control *Varroa* (Calderone 1999, Melathopoulos et al. 2000, Strange and Sheppard 2001, Rice et al. 2002).

20.3.2 Cultural Controls

Cultural control methods are labor intensive and thus often unused in large apiaries, but they may provide good control if applied consistently (Sammataro et al. 2000). These approaches include **smoking and dropping mites**; partial control in lightly infested apiaries can be obtained with tobacco smoke or other materials that cause mite knockdown. Mites should fall on a sticky board (see Figure S20.3 on CD). Mites can be trapped on drone brood. Because *Varroa* prefers drones, combs of the drone brood can be used to attract, trap, and remove mites. Worker brood can be removed. Heat kills *Varroa* mites at or around 111°F, while sealed brood survives.

20.3.3 Biological Controls

Biological control has been investigated by several researchers, but has not yet become a useful integrated mite management (IMM) tool. Chandler et al. (2001) provided an overview of natural enemies of *Varroa*. Fungi have been evaluated as microbial pesticides of *Varroa* (Peng et al. 2002, Meikle et al. 2007), and methods for delivering fungi have been developed (Kanga et al. 2010).

20.3.4 Host Resistance

Some beekeepers let all susceptible colonies die and then rear queens from the survivors to start new colonies, thus conducting genetic selection programs. The mechanisms of the resistance to *Varroa* are due to multiple factors, including hygienic behavior (Rothenbuhler 1964), grooming, and modified brood attractiveness (Nazzi and Milani 1996, Spivak 1996, Martin et al. 1997, Harris and Harbo 1999, Spivak and Reuter 2001a,b, Ibrahim et al. 2007, Harris 2008, Spivak et al. 2009). The U.S. Department of Agriculture (USDA) imported Russian honey bees into the United States and evaluated them for genetic resistance to *Varroa* (Rinderer et al. 1999, 2001, 2010, Danka and Beaman 2007, deGuzman et al. 2007, Ward et al. 2008). These resistant bees have been compared to Italian honey bees for flight activity, pollination, and resistance to *Varroa*, with promising results.

20.3.4.1 Hygienic Behavior

In the 1960s, Walter Rothenbuhler coined the term 'hygienic behavior' to explain why some honey bee workers remove dead brood from their colonies. Rothenbuhler (1964) proposed that two genes were involved: one gene (*u*) controlled uncapping of brood cells containing dead pupae, and a second gene (*r*) controlled the removal of these dead pupae. If a worker has two copies of *u* (*uu*), she will uncap (but *Uu* or *UU* females will not do so). Only if she has two copies of *rr* will she remove the dead pupae (*Rr* or *RR* females will not do so). For the complete behavior to occur, hygienic worker (female) bees must have two copies of both genes (*uu*, *rr*). Subsequent work by Lapidge et al. (2002) suggested that the genetic basis of hygienic behavior might involve additional genes. Such hygienic behavior reduces mite populations. In the United States, Spivak and colleagues selected the University of Minnesota's Hygienic stock and released it to commercial queen producers for use by beekeepers (Spivak et al. 2009), and the USDA has developed two breeding programs. One program involves a line that both suppresses mite reproduction and increases hygiene (Harris and Harbo 1999, Harris 2008). The second program involves the production of a hybrid line produced by crossing Russian honey bees resistant to *Varroa* (de Guzman et al. 2007, Tarpy et al. 2007, Rinderer et al. 2010). The hybrid Russian line has been provided to the Russian Honeybee Breeder's Association for additional selection and deployment in the United States.

20.3.4.2 Grooming Behavior

Honey bee adults clean themselves and their hive mates, which may injure or kill *Varroa* or may induce the mites to move to other hosts or to be lost. Colonies of bees vary in the amount of their grooming behavior, and this is a heritable trait. Mites groomed off bees can be trapped on bottom board traps (Rinderer et al. 2010).

20.3.4.3 Brood Attractiveness

The larvae of most European bees are highly attractive to *Varroa*; however, strains of bees are known that are less attractive to the mites. *Varroa* may not reproduce well on some honey bee lines because female mites do not lay eggs, *Varroa* females are delayed in depositing eggs, or *Varroa* females die before oviposition (Martin et al. 1997). Mites with low or no fertility had few or no sperm in their seminal receptacle. The selection of bee queens that suppress mite reproduction has been successful. Rinderer et al. (2010), however, questioned whether sufficient knowledge is available to use brood attractiveness successfully in breeding programs and discussed whether chemical cues can be identified from food or combs, or from castes that could enhance the ability to select.

20.4 INTEGRATED *VARROA* MANAGEMENT

Although resistant lines have been identified or selected for, resistance to *Varroa* is not complete, and other control methods are required (Rosenkranz et al. 2010). Furthermore, it is always possible that *Varroa* populations can develop resistance to the resistant bee strains. Control should be undertaken after monitoring colonies for *Varroa*. Lee et al. (2010) provide practical sampling plans for both colonies and apiaries based on studies of 31 commercial apiaries; for example, beekeepers can estimate the density of mites in a single colony by dislodging mites from approximately 300 adult bees taken from one brood box frame in the colony. Sampling whole apiaries would require sampling eight colonies in the same manner as for a single colony.

Control tactics deployed need to be appropriate to the local climatic and bee-keeping conditions. Factors to consider include the pest density in the hive and surrounding areas, the need to avoid chemical treatments during nectar flows, and the need to reduce stress on bees to reduce the likelihood they will succumb to colony collapse disorder, the cause of which has not yet been resolved. Miticides that are applied to control *Varroa* should be the least toxic available to reduce stress on the bees. Timing of treatments is crucial; control should occur prior to the generation that will overwinter, because only strong, healthy bees can successfully overwinter, especially in cold climates. Because *Varroa* can develop resistance to pesticides, multiple control tactics should be used to delay the development of resistance.

As noted by Rosenkranz et al. (2010), "There is neither a *Varroa* treatment available which fulfills all the criteria 'safe, effective and easy to apply' nor a honey bee which is sustainably tolerant to varroosis under temperate climatic conditions." The authors concluded that additional research is required and that long-term solutions will require long-term study and international cooperation.

SELECTED REFERENCES

Amdam, C.V., K. Hartfelder, K. Norberg, A. Hagen, and S.W. Omholt. 2004. Altered physiology in worker honey bees (Hymenoptera: Apidae) infested with the mite *Varroa destructor* (Acari: Varroidae): A factor in colony loss during overwintering? *J. Econ. Entomol.* 97:741–747.

Anderson, D.L. and S. Fuchs. 1998. Two genetically distinct populations of *Varroa jacobsoni* with contrasting reproductive abilities on *Apis mellifera*. *J. Apic. Res.* 37:69–78.

Anderson, D.L. and J.W.H. Trueman. 2000. *Varroa jacobsoni* (Acari: Varroidae) is more than one species. *Exp. Appl. Acarol.* 24:165–189.

Bowen-Walker, P.L., S.J. Martin, and A. Gunn. 1999. The transmission of deformed wing virus between honeybees (*Apis mellifera*) by the ectoparasitic mite *Varroa jacobsoni* Oud. *J. Invert. Pathol.* 73:101–106.

Calderone, N.W. 1999. Evaluation of formic acid and a thymol-based blend of natural products for the fall control of *Varroa jacobsoni* (Acari: Varroidae) in colonies of *Apis mellifera* (Hymenoptera: Apidae). *J. Econ. Entomol.* 92:253–260.

Chandler, D., K.D. Sunderland, B.V. Ball, and G. Davidson. 2001. Prospective biological control agents of *Varroa destructor* n.sp., an important pest of the European honeybee, *Apis mellifera*. *Biocontrol Sci. Technol.* 11:429–448.

Chen, Y.-P., J.S. Petis, J.D. Evans, M. Kramer, and M.F. Feldlaufer. 2004. Transmission of Kashmir bee virus by the ectoparasitic mite *Varroa destructor*. *Apidologie* 35:411–448.

Danka, R.G. and L.D. Beaman. 2007. Flight activity of USDA-ARS Russian honeybees (Hymenoptera: Apidae) during pollination of lowbush blueberries in Maine. *J. Econ. Entomol.* 100:267–272.

deGuzman, L.I. and T.E. Rinderer. 1999. Identification and comparison of *Varroa* species infesting honey bees. *Apidologie* 30:89–95.

deGuzman, L.I., T.E. Rinderer, J.A. Stelzer, L. Beaman, G.T. Delatte, and C. Harper. 2001. Hygienic behavior by honey bees from far-eastern Russia. *Am. Bee J.* 142:58–60.

deGuzman, L.I., T.E. Rinderer, and A.M. Frake. 2007. Growth of *Varroa destructor* (Acari: Varroidae) populations in Russian honey bee (Hymenoptera: Apidae) colonies. *Ann. Entomol. Soc. Am.* 100:187–195.

DeJong, D., R.A. Morse, and G.C. Eickwort. 1982. Mite pests of honey bees. *Annu. Rev. Entomol.* 27:229–252.

De Miranda, J., G. Cordoni, and G. Budge. 2010. The acute bee paralysis virus–Kashmir bee virus–Israeli acute paralysis virus complex. *J. Invert. Pathol.* 103:S30–S47.

De Rycke, P.H., J.J. Joubert, S.H. Hosseinian, and F.J. Jacobs. 2002. The possible role of *Varroa destructor* in the spreading of American foulbrood among apiaries. *Exp. Appl. Acarol.* 27:313–318.

Eickwort, G.C. 1994. Evolution and life-history patterns of mites associated with bees. In: M.A. Houck (ed.), *Mites: Ecological and Evolutionary Analyses of Life-History Patterns* (pp. 281–251). New York: Chapman & Hall.

Evans, J.D. and D.L. Lopez. 2002. Complete mitochondrial DNA sequence of the important honey bee pest, *Varroa destructor* (Acari: Varroidae). *Exp. Appl. Acarol.* 27:69–78.

Garrido, C. and P. Rosenkranz. 2003. The reproductive program of female *Varroa destructor* mites is triggered by its host, *Apis mellifera*. *Exp. Appl. Acarol.* 31:269–273.

Harris, J.W. 2008. Effect of brood type on *Varroa*-sensitive hygiene by worker honey bees (Hymenoptera: Apidae). *Ann. Entomol. Soc. Am.* 101:1137–1144.

Harris, J.W. and J.R. Harbo. 1999. Low sperm counts and reduced fecundity of mites in colonies of honey bees (Hymenoptera: Apidae) resistant to *Varroa jacobsoni* (Mesostigmata: Varroidae). *J. Econ. Entomol.* 92:83–90.

Ibrahim, A., G.S. Reuter, and M. Spivak. 2007. Field trial of honey bee colonies bred for mechanisms of resistance against *Varroa destructor*. *Apidologie* 38:67–76.

Kanga, L.H.B., J. Adamczyk, J. Patt, C. Gracia, and J. Cascino. 2010. Development of a user-friendly delivery method for the fungus *Metarhizium anisopliae* to control the ectoparasitic mite *Varroa destructor* in honey bee, *Apis mellifera*, colonies. *Exp. Appl. Acarol.* 52(4):327–342.

Lapidge, K.L., B.P. Oldroyd, and M. Spivak. 2002. Seven suggestive quantitative trait loci influence hygienic behavior of honey bees. *Naturwissenschaften* 89:565–568.

Lee, K.V., R.D. Moon, E.C. Burkness, W.D. Hutchison, and M. Spivak. 2010. Practical sampling plans for *Varroa destructor* (Acari: Varroidae) in *Apis mellifera* (Hymenoptera: Apidae) colonies and apiaries. *J. Econ. Entomol.* 103:1039–1050.

Le Conte, Y., G. Arnold, J. Trouiller, C. Masson, B. Chappe, and G. Ourisson. 1989. Attraction of the parasitic mite *Varroa* to the drone larvae of honey bees by simple aliphatic esters. *Science* 245:658–659.

Lindberg, C.M., A.P. Melathopoulos, and M.L. Winston. 2000. Laboratory evaluation of miticides to control *Varroa jacobsoni* (Acari: Varroidae), a honey bee (Hymenoptera: Apidae) parasite. *J. Econ. Entomol.* 93:189–198.

MAAREC. 2005. *Varroa Mites*, Publication No. 4.7. State College, PA: The Pennsylvania State University, Mid-Atlantic Apiculture & Extension Consortium (MAAREC.cas.psu.edu).

Martin, S., K. Holland, and M. Murray. 1997. Non-reproduction in the honeybee mite *Varroa jacobsoni*. *Exp. Appl. Acarol.* 21:539–549.

Martin, S.J., P.J. Elzen, and W.R. Rubink. 2002. Effect of acaricide resistance on reproductive ability of the honey bee mite *Varroa destructor*. *Exp. Appl. Acarol.* 27:195–207.

Meikle, W.G., G. Mercadier, N. Holst, C. Nansen, and V. Girod. 2007. Duration and spread of an entomopathogenic fungus, *Beauveria bassiana* (Deuteromycota: Hypomycetes), used to treat *Varroa* mites (Acari: Varroidae) in honeybee (Hymenoptera: Apidae) hives. *J. Econ. Entomol.* 100:1–10.

Melathopoulos, A.P., M.L. Winston, R. Whittington, H. Higo, and M. LeDoux. 2000. Field evaluation of neem and canola oil for the selective control of the honey bee (Hymenoptera: Apidae) mite parasites, *Varroa jacobsoni* (Acari: Varroidae) and *Acarapis woodi* (Acari: Tarsonemidae). *J. Econ. Entomol.* 93:559–567.

Nazzi, F. and N. Milani. 1996. The presence of inhibitors of the reproduction of *Varroa jacobsoni* Oud. (Gamasida: Varroidae) in infested cells. *Exp. Appl. Acarol.* 20:617–623.

Needham, G.R., R.E. Page, M. Delfinado-Baker, and C.E. Bowman (eds.) 1988. *Africanized Honey Bees and Bee Mites.* London: Prentice Hall.

Nordstrom, S. 2003. Distribution of deformed wing virus within honey bee (*Apis mellifera*) brood cells infested with the ectoparasitic mite *Varroa destructor. Exp. Appl. Acarol.* 29:293–302.

Oldroyd, B.P. 1999. Coevolution while you wait: *Varroa jacobsoni*, a new parasite of western honeybees. *Trends Ecol. Evol.* 14:312–315.

Peng, C.Y.S., X. Zhou, and H.K. Kaya. 2002. Virulence and site of infection of the fungus, *Hirsutella thompsonii*, to the honey bee ectoparasitic mite, *Varroa destructor. J. Invert. Pathol.* 81:185–195.

Rice, N.D., M.L. Winston, R. Whittington, and H.A. Higo. 2002. Comparison of release mechanisms for botanical oils to control *Varroa destructor* (Acari: Varroidae) and *Acarapis woodi* (Acari: Tarsonemidae) in colonies of honey bees (Hymenoptera: Apidae). *J. Econ. Entomol.* 95:221–226.

Rickli, M., P.A. Diehl, and P.M. Guerin. 1994. Cuticle alkanes of honeybee larvae mediate arrestment of bee parasite *Varroa jacobsoni. J. Chem. Ecol.* 20:2437–2453.

Rinderer, T.E., G.T. Delatte, L.I. de Guzman, J. Williams, J.A. Steizer, and V. Kuznetsov. 1999. Evaluations of the *Varroa* resistance of honey bee imported from far-eastern Russia. *Am. Bee J.* 139:287–290.

Rinderer, T.E., L.I. de Guzman, G.T. Delatte, J.A. Stelzer, V.A. Lancaster, V. Kuznetsov, L. Beaman, R. Watts, and J.W. Harris. 2001. Resistance to the parasitic mite *Varroa destructor* in honey bees from far-eastern Russia. *Apidologie* 32:381–394.

Rinderer, T.E., J.W. Harris, G.J. Hunt, and L.I. deGuzman. 2010. Breeding for resistance to *Varroa destructor* in North America. *Apidologie* 41(3):409–424.

Rosenkranz, P., P. Aumeier, and B. Ziegelmann. 2010. Biology and control of *Varroa destructor. J. Invert. Pathol.* 103:S96–S119.

Rothenbuhler, W. 1964. Behavior genetics of nest cleaning in honey bees. I. Response of four inbred lines to disease-killed brood. *Anim. Behav.* 12:578–583.

Sammataro, D., U. Gerson, and G. Needham. 2000. Parasitic mites of honey bees: Life history, implications, and impact. *Annu. Rev. Entomol.* 345:519–548.

Sanford, M.T. 2003. *Diseases and Pests of the Honey Bee*, UF IFAS Publication No. CIR766. Gainesville: Department of Entomology and Nematology, University of Florida Institute of Food and Agricultural Sciences (http://edis.ifas.ufl.edu/AA090).

Sanford, M.T., H.A. Denmark, H.L. Cromroy, and L. Cutts. 1998. *Featured Creatures: Varroa Mite*. Gainesville: Department of Entomology and Nematology, University of Florida Institute of Food and Agricultural Sciences (http://entnemdept.ifas.ufl.edu/creatures/misc/bees/varroa_mite.htm).

Satta, A., I. Floris, M. Eguaras, P. Cabras, V.L. Garau, and M. Melis. 2005. Formic acid-based treatments in control of *Varroa destructor* in a Mediterranean area. *J. Econ. Entomol.* 98:267–273.

Spivak, M. 1996. Honey bee hygienic behavior and defense against *Varroa jacobsoni. Apidologie* 27:245–260.

Spivak, M. and G.S. Reuter. 2001a. Resistance to American foulbrood disease by honey bee colonies *Apis mellifera* bred for hygienic behavior. *Apidologie* 32:555–565.

Spivak, M. and G.S. Reuter. 2001b. *Varroa destructor* infestation in untreated honey bee (Hymenoptera: Apidae) colonies selected for hygienic behavior. *J. Econ. Entomol.* 94:326–331.

Spivak, M., G.S. Reuter, K. Lee, and B. Ranum. 2009. The future of the MN hygienic stock of bees is in good hands! *Am. Bee J.* 149:965–967.

Steiner, J., S. das Gracas Pompolo, C.S. Takahashi, and L.S. Goncalves. 1982. Cytogenetics of the acarid *Varroa jacobsoni. Rev. Brasil. Genet.* 4:841–844.

Strange, J.P. and W.S. Sheppard. 2001. Optimum timing of miticide applications for control of *Varroa destructor* (Acari: Varroidae) in *Apis mellifera* (Hymenoptera: Apidae) in Washington State, USA. *J. Econ. Entomol.* 94:1324–1331.

Tarpy, D.R., J. Summers, and J.J. Keller. 2007. Comparison of parasitic mites in Russian-hybrid and Italian honey bee (Hymenoptera: Apidae) colonies across three different locations in North Carolina. *J. Econ. Entomol.* 100:258–266.

Underwood, R.M. and R.W. Currie. 2005. Effect of concentration and exposure time on treatment efficacy against *Varroa* mites (Acari: Varroidae) during indoor winter fumigation of honey bees (Hymenoptera: Apidae) with formic acid. *J. Econ. Entomol.* 98:1802–1809.

vanEngelsdorp, D. and M.D. Meixner. 2010. A historical review of managed honey bee populations in Europe and the United States and the factors that may affect them. *J. Invert. Pathol.* 103:S80–S95.
vanEngelsdorp, D., R.M. Underwood, and D.L. Cox-Foster. 2008. Short-term fumigation of honey bee (Hymenoptera: Apidae) colonies with formic and acetic acids for the control of *Varroa destructor* (Acari: Varroidae). *J. Econ. Entomol.* 101:256–264.
Ward, K., R. Danka, and R. Ward. 2008. Comparative performance of two mite-resistant stocks of honey bees (Hymenoptera: Apidae) in Alabama beekeeping operations. *J. Econ. Entomol.* 101:654–659.

CHAPTER 21

Tracheal Mite (*Acarapis woodi*)

21.1 BIOLOGY

The tracheal mite, *Acarapis woodi* (Actinedida or Prostigmata: Tarsonemidae), is the *only* internal mite attacking honey bees. The tracheal mite was first found in the tracheae of *A. mellifera* in 1919 after many bees and colonies died in Europe from what was called the Isle of Wight disease (De Jong et al. 1982, Needham et al. 1988, Eickwort 1994, Sammataro et al. 2000). This mite now is found throughout Europe, Asia, Africa, South America, and North America and can be found affecting *A. mellifera*, *A. cerana*, and *A. dorsata*. Despite a quarantine to prevent its entry into the United States, this mite was discovered in this country by 1984. In addition to *A. woodi*, two other *Acarapis* species (*A. dorsalis* and *A. externus*) feed on honey bee hemolymph, but these are external parasites. These species also have moved out of Europe and, in 1960, were discovered in North America, Australia, New Zealand, and New Guinea.

Acarapis woodi are so small that they can infest the tracheal system of adult bees, where they feed on hemolymph by piercing the tracheae with their sharp stylets (Figure 21.1 and Figure 21.2) (also see Figure S21.1 on the CD). Females are about 120 to 189 μm (0.12 to 0.19 mm) long, and males are about 96 to 102 μm. The tracheal mite is associated with colony deaths, but there is some controversy regarding whether the deaths were caused by the mite or by bacterial or viral pathogens such as chronic bee paralysis virus (McMullan and Brown 2009). It appears that infestation with *A. woodi* impairs the ability of the bees to fly and weakens them so they die of starvation or exposure. Individual bees with high levels of *Acarapis* mites have lower oxygen consumption rates, and they are unable to properly maintain colony temperatures in cold weather (Otis and Scott-Dupree 1992). It is unclear as to how detrimental *A. woodi* is, because some colonies parasitized with mites may show no obvious symptoms despite infestations that have persisted for years. When the colony is stressed, however, especially in cold weather, *A. woodi*, perhaps in combination with other maladies, can lead to the loss of the colony (Tarpy et al. 2007). For example, prior to the invasion of this mite, an average of 11% of bee colonies died during the winter in northern states in the United States. After *A. woodi* invaded Pennsylvania, these losses increased to 31%. Losses of infested colonies over the winter can be as high as 90% (Frazier et al. 1994).

All stages of the mite are found in the tracheae of the bee, and several generations may develop in a single bee. Eventually, newly fertilized adult *Acarapis* females leave the tracheae to colonize a new bee, apparently attracted to the hydrocarbons found in the cuticle of bees less than 4 days old. When a suitable host is found, the female enters a trachea and lays approximately 20 to 25 eggs over 25 to 30 days. Some daughters disperse when the host bee is about 13 days old, with dispersal of daughters peaking between 15 and 25 days (Pettis and Wilson 1996). Mites searching for a bee's spiracle are vulnerable to desiccation or starvation and die after a few hours if they have not found a suitable spiracle to enter (Sammataro and Needham 1996).

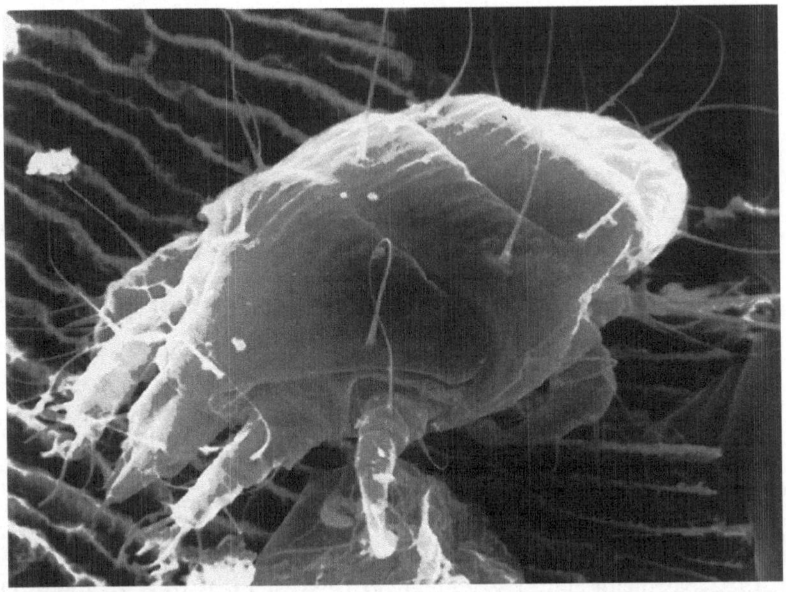

Figure 21.1 Scanning electron micrograph of an adult female tracheal mite (*Acarapis woodi*) within the trachea of a honey bee. (Photograph courtesy of U.S. Department of Agriculture, Agricultural Research Service.)

The life cycle ranges from 12 to 21 days, and a hive has multiple generations during the year. The sex ratio appears to be skewed to females (3:1 to 4:1). This mite is unusual in that it has no protonymphal or deutonymphal stages; the life history is egg → larva → pharate adult → adult male and female. The egg is large, and the larva actively feeds. Although the larva has six legs, only one pair is well developed. The others are rudimentary. The larva does not molt, but becomes inactive and transforms into an adult that is visible within the larval exoskeleton (pharate adult). Males guard pharate females and, after the female emerges, they mate. Sib matings are common in this species,

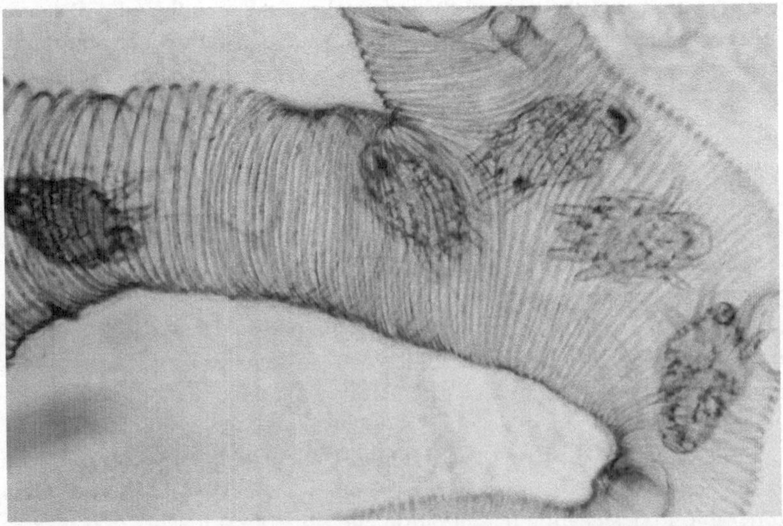

Figure 21.2 Tracheal mites (*Acarapis woodi*) within the thoracic trachea of a honey bee. (Photograph by Gloria de Guzman, U.S. Department of Agriculture, Agricultural Research Service.)

Table 21.1 Tactics for Control of the Tracheal Mite *Acarapis woodi*

1.	Start hives with queens free of tracheal mites.
2.	Practice good hive management to reduce the incidence of, and effects of, *Acarapis*. Reduce stresses by maintaining adequate numbers of bees in the hives so they can maintain temperatures during the winter. Wrap hives during the winter in cold environments. Maintain hives where pollen and nectar sources are adequate to maintain strong colonies.
3.	Quarantines are important. Don't collect swarms that could be infected. Reduce movement of colonies, and avoid crowded areas. Don't introduce colonies from infected areas or that have not been tested for *A. woodi*.
4.	Monitoring methods have been developed to detect *Acarapis*, involving dissection and staining; however, if bees are crawling in front of the hive, display a lack of vigor, and exhibit the K-wing behavior, it is likely they are heavily infected.
5.	Chemical controls must be effective against the tracheal mite but harmless to bees and must not accumulate in hive products (wax, honey). The chemical must reach the mites within the bee tracheae by being inhaled and be lethal only to the mite. Chemicals can be applied only when honey supers can be removed.
6.	Registered products in the United States include pure menthol crystals or formic acid. Menthol crystals can be applied to hives during moderate temperatures, but application during high temperatures causes the bees to be repelled, and during cold weather the crystals do not release their vapors. Formic acid fumigation can be effective, as well, but is hazardous to handle.
7.	Rear resistant bees. Resistant bee strains have been developed through field and laboratory selection. Resistance is achieved by the bees grooming and preventing females of *A. woodi* from entering tracheae. When the mites have entered the tracheae, there is no resistance mechanism to eliminate them.

Source: Based on information from Lorenzen and Gary (1986) and Sammataro et al. (2000).

enhancing the probability that selection for acaricide resistance can occur rapidly. Infested tracheae typically are discolored about ten days after initial infestation from the damage caused by feeding mites (Pettis and Wilson 1996). Adult fertilized daughters apparently transfer from the bee in which they develop to another in the hive, and dispersal from hive to hive is apparently aided by humans or by infested bees drifting between hives.

Most studies on this mite have focused on how to control it (Bailey and Perry 2002). The mite is associated with decreased honey production, decreased brood production, and increased winter mortality of colonies (Sammataro et al. 2000). Mite infestations of package bee colonies are a serious problem; however, some strains of bees are resistant to *Acarapis woodi*, and it appears that well-maintained healthy colonies are less likely to have heavy infestations.

21.2 INTEGRATED CONTROL OF *ACARAPIS WOODI*

Integrated mite management (IMM) options include host resistance, cultural controls, and chemical controls, with monitoring for infestation being a key tool (Table 21.1).

21.2.1 Monitoring

Detecting tracheal infestations is not simple or easy (Lorenzen and Gary 1986). Monitoring involves dissection and examination of the prothoracic tracheae of individual bees, which makes it slow and expensive to monitor a large number of hives for infection. Beekeepers may notice, however, that bee life spans are shortened, and severely infested bees are unable to fly. Large numbers of bees may be seen crawling on the ground near the hive, and infested bees may be seen in clusters near the hive. Many infested adults extend their wings outward, which is described as *K-wing*.

Infested colonies may be slow to build up populations in the spring and lack vigor. In the fall, there may be a high rate of die-off before the onset of winter. Although these behavioral traits are suggestive, they are not conclusive.

Multiple monitoring methods have been developed, and each has its pros and cons (Colin et al. 1979, Camazine 1985, Peng and Nasr 1985, Lorenzen and Gary 1986, Tomasko et al. 1993). To confirm the presence of *Acarapis woodi*, a dissecting microscope is necessary, as well as special clearing techniques. Because drones are larger and easier to dissect, they should be examined, if possible (Sammataro 2006). Fresh bees are easier to dissect than those preserved in alcohol, but alcohol-preserved bees (50% ethanol or propanol) can be examined.

The method described by Lorenzen and Gary (1986) provides clear diagrams of the steps involved and should be consulted. First, place the bee ventral side up in a wax-filled dissecting dish, inserting a number-two insect pin vertically through the middle of the thorax and then tilting the head up slightly. With a fine forceps, grasp the coxae of the first pair of legs and pull to remove the legs and head from the body. This exposes the first pair of thoracic tracheae. If the infestation is heavy, the tracheae may be darkened due to scarring from feeding by the mites, and many mites may be visible. If so, no further dissection is needed. If no mites or darkening can be seen, insert another insect pin through the thorax and make two cuts into the thorax with a razor blade. These two cuts create two flaps of exoskeleton that can be pulled aside, using forceps, to examine each trachea. The tracheae should not be removed, because most female *Acarapis woodi* oviposit just inside the spiracular opening and can be seen with the dissecting microscope. Higher magnification (with a compound microscope) will reveal the mites. A light infestation may involve one trachea, while heavy infestations involve both.

Frazier et al. (2000) developed a sequential sampling plan that provides a decision table. The tables allow the beekeeper to choose specific injury levels and to examine the sequence of mite-infested and uninfested bees using microscopic examinations.

21.2.2 Host Resistance

Several lines of bees resistant to tracheal mites have been developed and are commercially reared and available to beekeepers (Danka and Villa 1996, 1998, 2000, Nasr et al. 2001, De Guzman et al. 2002, 2005, Villa and Rinderer 2008). The development of Buckfast resistant lines began shortly after the mites were discovered in England (Adam 1968); however, it is also likely that every population of bees exhibits genetic variability for tracheal mite resistance, and it is possible to select rapidly for resistance (Gary and Page 1987, Nasr et al. 2001). In the case of selected highly resistant Russian bees crossed with highly susceptible colonies from the United States, the resistance in the Russian bees is "likely polygenic, but there may be a number of genes with major dominance interacting with minor genes" (Villa and Rinderer 2008). Resistance seems to be achieved by increased grooming behavior of the bees (Pettis and Pankiw 1998, Sammataro et al. 2000, DeGuzman et al. 2002, Villa 2006, Villa and Rinderer 2008). Villa and Rinderer (2008) concluded that, "Selected resistant Russian bees are a valuable tool to deal with the persistent problems with tracheal mites experienced by beekeepers, particularly in colder climates."

21.2.3 Cultural Controls

Colonies should be initiated from queens known to be free of *Acarapis woodi* and, ideally, using strains that are resistant. Do not acquire hives that come from infected areas or that have not been tested for *A. woodi*. Bees that are stressed are more prone to problems from *A. woodi*. Stresses can be due to diseases, too many colonies being located together and resulting in interference and robbing, inadequate nectar and pollen, and severe weather. McMullan and Brown (2009) found that tracheal-mite-infested colonies can die out in winter or early spring due to an inability to thermoregulate, and they indicated that wrapping hives during the winter could reduce colony deaths. Key factors that influence the severity of infestation include hive insulation, winter climate, and bee strain. Another factor could be the amount of stored honey (chemical energy) available in the winter to provide energy for the bees, allowing them to maintain adequate temperatures within the hive. Reduced adult bee densities compared to brood density influence mortality during winter because there are fewer adults to maintain an adequate temperature. Maintaining colonies in good condition may allow bees to survive even though they have a moderate infestation of tracheal mites. Infestations with *Varroa* mites or pathogens debilitate colonies, making them more vulnerable to infestations with *A. woodi*. In other words, maintaining healthy colonies through proper management of all pests is critically important.

21.2.4 Chemical Control

Any chemical control must be of low toxicity to the bees yet must penetrate the tracheae where the mites are located. Tracheal mites can be managed without synthetic acaricides. Products that can be used to control *Acarapis woodi* include menthol, formic acid, or vegetable oil (Melathopoulos et al. 2000, Rice et al. 2002, Underwood and Currie 2009). Oils sprayed on bees protect them from infestation by tracheal mites, perhaps by making it difficult for female mites to successfully invade the tracheae of a new host. Another control method involves feeding bee colonies grease patties (two parts white sugar to one part vegetable shortening) during the fall and winter (Sammataro and Needham 1996). The vegetable shortening does not kill female tracheal mites, but the oil may conceal the odor of young bees, so the females continue to wander and fail to find an acceptable host. Because young bees emerge continuously, the patty must be present for an extended period in fall and spring when mite densities increase. Pure menthol crystals, extracted from the plant *Mentha arvensis*, can control *A. woodi* (Sammataro et al. 2000). During cold weather, the menthol is ineffective because an insufficient amount of vapor is released. At high temperatures, the vapors can repel bees from the hive.

SELECTED REFERENCES

Adam, Brother. 1968. "Isle of Wight" or acarine disease; its historical and practical aspects. *Bee World* 49:6–18.

Bailey, L. and J.N. Perry. 2002. The natural control of the tracheal mite of honey bees. *Exp. Appl. Acarol.* 25:745–749.

Camazine, S. 1985. Tracheal floatation: rapid method for the detection of honey bee acarine disease. *Am. Bee J.* 125:104–105.

Colin, M.E., J.P. Faucon, A. Giaufret, and C. Sarrazin. 1979. A new technique for the diagnosis of acarine infestation in honeybees. *J. Apicult. Res.* 13:222–224.

Danka, R.G. and J.D. Villa. 1996. Influence of resistant honey bee hosts on the life history of the parasite *Acarapis woodi*. *Exp. Appl. Acarol.* 20:313–322.

Danka, R.G. and J.E. Villa. 1998. Evidence of autogrooming as a mechanism of honey bee resistance to tracheal mite infestation. *J. Apic. Res.* 37:39–46.

Danka, R.G. and J.D. Villa. 2000. Inheritance of resistance to *Acarapis woodi* (Acari: Tarsonemidae) in first-generation crosses of honey bees (Hymenoptera: Apidae). *J. Econ. Entomol.* 93:1602–1605.

De Guzman, L.I., T.E. Rinderer, G.T. Delatte, J.A. Stelzer, G. Beaman, and V. Kuznetsov. 2002. Resistance to *Acarapis woodi* by honey bees from far-eastern Russia. *Apidologie* 33:411–415.

De Guzman, L.I., T.E. Rinderer, M. Bigalk, H. Tubbs, and S.J. Bernard. 2005. Russian honey bee (Hymenoptera: Apidae) colonies: *Acarapis woodi* (Acari: Tarsonemidae) infestations and overwintering survival. *J. Econ. Entomol.* 98:1766–1801.

De Jong, D., R.A. Morse, and G.C. Eickwort. 1982. Mite pests of honey bees. *Annu. Rev. Entomol.* 27:229–252.

Eickwort, G.C. 1994. Evolution and life-history patterns of mites associated with bees. In: M.A. Houck (ed.), *Mites: Ecological and Evolutionary Analyses of Life-History Patterns* (pp. 281–251). New York: Chapman & Hall.

Frazier, M.T., J. Finley, C.H. Collison, and E. Rajotte. 1994. The incidence and impact of honey bee tracheal mites and nosema disease on colony mortality in Pennsylvania. *Bee Sci.* 3:94–100.

Frazier, M.T., J. Finley, W. Harkness, and E.G. Rajotte. 2000. A sequential sampling scheme for detecting infestation levels of tracheal mites (Heterostigmata: Tarsonemidae) in honey bee (Hymenoptera: Apidae) colonies. *J. Econ. Entomol.* 93:551–558.

Gary, N.E. and R.E. Page, Jr. 1987. Phenotypic variation in susceptibility of honey bees, *Apis mellifera*, to infestation by tracheal mites, *Acarapis woodi*. *Exp. Appl. Acarol.* 3:291–305.

Lorenzen, K. and N.E. Gary. 1986. Modified dissection technique for diagnosis of tracheal mites (Acari: Tarsonemidae) in honey bees (Hymenoptera: Apidae). *J. Econ. Entomol.* 9:401–403.

McMullan, J.B. and M.J.F. Brown. 2009. A qualitative model of mortality in honey bee (*Apis mellifera*) colonies infested with tracheal mites (*Acarapis woodi*). *Exp. Appl. Acarol.* 47:225–234.

Melathopoulos, A.P., M.L. Winston, R. Whittington, H. Higo, and M. Le Doux. 2000. Field evaluation of neem and canola oil for the selective control of the honey bee (Hymenoptera: Apidae) mite parasites, *Varroa jacobsoni* (Acari: Varroidae) and *Acarapis woodi* (Acari: Tarsonemidae). *J. Econ. Entomol.* 93:559–567.

Nasr, M.E., G.W. Otis, and C.D. Scott-Dupree. 2001. Resistance to *Acarapis woodi* by honey bees (Hymenoptera: Apidae): divergent selection and evaluation of selection progress. *J. Econ. Entomol.* 94:332–338.

Needham, G.R., R.E. Page, Jr., M. Delfinado-Baker, and C.E. Bowman (eds.) 1988. *Africanized Honey Bees and Bee Mites*. London: Prentice Hall.

Otis, G.W. and C.D. Scott-Dupree. 1992. Effects of *Acarapis woodi* on overwintered colonies of honey bees (Hymenoptera: Apidae) in New York. *J. Econ. Entomol.* 85:40–46.

Peng, Y.S. and M.E. Nasr. 1985. Diagnosis of honey bee tracheal mites (*Acarapis woodi*) by simple staining techniques. *J. Invert. Pathol.* 46:325–331.

Pettis, J.S. and W.T. Wilson. 1996. Life history of the honey bee tracheal mite (Acari: Tarsonemidae). *Ann. Entomol. Soc. Am.* 89:368–374.

Pettis, J.S. and T. Pankiw. 1998. Grooming behavior by *Apis mellifera* L. in the presence of *Acarapis woodi* (Rennie) (Acari: Tarsonemidae). *Apidologie* 29:241–253.

Rice, N.D., M.L. Winston, R. Whittington, and H.A. Higo. 2002. Comparison of release mechanisms for botanical oils to control *Varroa destructor* (Acari: Varroidae) and *Acarapis woodi* (Acari: Tarsonemidae) in colonies of honey bees (Hymenoptera: Apidae). *J. Econ. Entomol.* 95:221–226.

Sammataro, D. 2006. An easy dissection technique for finding the tracheal mite, *Acarapis woodi* (Rennie) (Acari: Tarsonemidae), in honey bees, with video link. *Int. J. Acarol.* 32:1–5.

Sammataro, D. and G.R. Needham. 1996. Host-seeking behaviour of tracheal mites (Acari: Tarsonemidae) on honey bees (Hymenoptera: Apidae). *Exp. Appl. Acarol.* 20:121–136.

Sammataro, D., U. Gerson, and G. Needham. 2000. Parasitic mites of honey bees: life history, implications, and impact. *Annu. Rev. Entomol.* 345:519–548.

Tomasko, M., J. Finley, W. Harkness, and E. Rajotte. 1993. A sequential sampling scheme for detecting the presence of tracheal mite (*Acarapis woodi*) infestations in honey bee (*Apis mellifera* L.) colonies. *Penn State Agric. Exp. Stn. Bull.* 871.

Tarpy, D.R., J. Summers, and J.J. Keller. 2007. Comparison of parasitic mites in three different locations in North Carolina. *J. Econ. Entomol.* 100:258–266.

Underwood, R.M. and R.W. Currie. 2009. Indoor winter fumigation with formic acid for control of *Acarapis woodi* (Acari: Tarsonemidae) and *Nosema*. *J. Econ. Entomol.* 102:1729–1736.

Villa, J.D. 2006. Autogrooming and bee age influence the migration of tracheal mites to Russian and susceptible worker bees (*Apis mellifera* L.). *J. Apic. Res.* 45:28–31.

Villa, J.D. and T.E. Rinderer. 2008. Inheritance of resistance to *Acarapis woodi* (Acari: Tarsonemidae) in crosses between selected resistant Russian and selected susceptible U.S. honey bees (Hymenoptera: Apidae). *J. Econ. Entomol.* 101:1756–1759.

PART VII

Parasitic Mites of Mammals and Birds

Mites associated with domesticated animals function as scavengers in nests or as obligatory parasites feeding on living tissues of their host. Many parasitic mites are pests because they are vectors of protozoan, rickettsial, bacterial, and viral disease agents. Most parasitic mite species exhibit some degree of structural and biological adaptation to their lifestyle, ranging from modifications of their chelicerae for piercing the skin of their host to a reduction in sclerotinization to allow for expansion of the body during feeding. Some have highly modified legs for clasping hairs or feathers. Others have simplified their life cycle, eliminating one stage (usually the deutonymphal). Some species retain their eggs until they hatch (**ovoviviparity**). Chapter 22 provides an overview of the key aspects of tick biology that make them direct pests, as well as vectors of human and animal disease agents. Chapter 23 discusses some of the families of parasitic mites that may attack poultry, livestock, companion animals, and humans.

What is not discussed in these chapters is **delusory parasitosis**, which is a belief that there are bugs, worms, or mites that are "biting, crawling over or burrowing into, under, or out of your skin" (Kimsey and Mussen 2003). People afflicted with delusory parasitosis often are convinced that these organisms are present in their homes or furniture. Afflicted people may go from doctor to doctor in an attempt to get a diagnosis of the problem. Eventually, entomologists or acarologists may be consulted when the medical doctors are unable to find any living organisms associated with the symptoms. If the person is, in fact, infected with a mite or other parasite, dermatologists may recognize the problem, although the problem may be caused by stress, reactions to drugs, allergies, dry skin, irritation by fiberglass filaments, vitamin or other dietary deficiencies, diabetes mellitus, or hypothyroidism (Hinkle 2000). If a dermatologist does find arthropods, they should be sent to an entomologist or acarologist for identification. Because people believe they are infested, they may attempt to control the problem themselves by using harsh soaps or toxic chemicals, and these can cause itching, inflammation, and other skin problems. For additional information on this problem, see Kimsey and Mussen (2003) and Hinkle (2000).

REFERENCES

Hinkle, N.C. 2000. Delusory parasitosis. *Am. Entomol.* 46:17–25 (http://www.ent.uga.edu/pubs/delusory.pdf).

Kimsey, R.B. and E.C. Mussen. 2003. *Delusory Parasitosis: What's Attacking Me?* Publication No. 7443. Davis: Cooperative Extension, University of California (http://www.ipm.ucdavis.edu/PMG/PESTNOTES/pn7443.html).

CHAPTER 22

Ticks (Argasidae and Ixodidae)

22.1 TICKS AS PESTS

Ticks (Metastigmata or Ixodida in the Parasitiformes) are obligate ectoparasites and the largest of the Acari (Evans et al. 1961, Sonenshine 1991, 1993, Mullen and Durden 2002, Nicholson et al. 2009). Ticks attack all vertebrates, except fish, and are important vectors of microorganisms causing diseases in humans and domestic animals (Hoogstraal 1966, 1981, Bram 1978, Nutting 1984, Spielman et al. 1985, Lane et al. 1991, Stafford 2007). Worldwide, ticks are the most important vectors of disease agents of veterinary importance, and second only to mosquitoes as vectors of disease agents of humans (Nicholson et al. 2009). Because of their economic importance as direct pests and as vectors of disease agents, as well as their larger size (2 to 30 mm in length), ticks are better studied than any other group of acarines (Klompen et al. 1996).

The two largest tick families are the **Ixodidae** (hard ticks) and the **Argasidae** (soft ticks). The family Nuttalliellidae contains only one species and occurs in Africa, where it is rare. Hard ticks (Ixodidae) have a single dorsal shield that, in unfed females, covers the front half of the dorsum. In ixodid males, the scutum covers the entire dorsum. Ixodid ticks are divided into the **Prostriata** (which contains one genus, *Ixodes*, with 241 species) and the **Metastriata** (11 genera with 442 species) (Nicholson et al. 2009). Ticks in the Prostriata have an anal groove that extends *anterior* of the anus and encloses the anus. Ticks in the Metastriata have an anal groove *posterior* to the anus. Soft ticks (Argasidae) lack heavily sclerotized plates and have a tough, leathery integument that can become greatly distended when they engorge with blood.

The life cycle of ticks varies. Some have a single host, some have two hosts, and some have three; however, the life cycle typically involves egg → larva → nymph(s) → adult male and female. See Figure 22.1 for the life stages of the American dog tick (*Dermacentor variabilis*) (Ixodidae).

Ticks have a gnathosoma (also called the **capitulum**) with an elongate **hypostome** armed with recurved teeth (**retrorse teeth**) for piercing and holding onto their hosts (Figure 22.2) (also see Figure S22.1 on the CD). The chelicerae serve as cutting organs, which permits insertion of the hypostome into the skin. The hypostome is an anchoring organ that makes it difficult to remove a tick from the skin without leaving the mouthparts attached. If the mouthparts are left in the wound, the wound can become infected, producing an inflamed sore, or ulcer.

The best tick removal method is to *pull slowly and steadily by grasping the head of the tick very close to its insertion into the skin* (Stafford 2007). Pulling must be sufficiently gentle so the mouthparts do not break off. After removal, use an antiseptic on the site to prevent infection. Use gloves, or even a paper towel, to protect your skin from potential contamination with any pathogens the tick may carry. Wash your hands after handling a tick because tick feces and secretions can be infective. If you are concerned about the possibility of infection with a tick-transmitted disease agent (such as Lyme disease), save the tick in a closed container in the freezer so it can be identified and tested (Table 22.1).

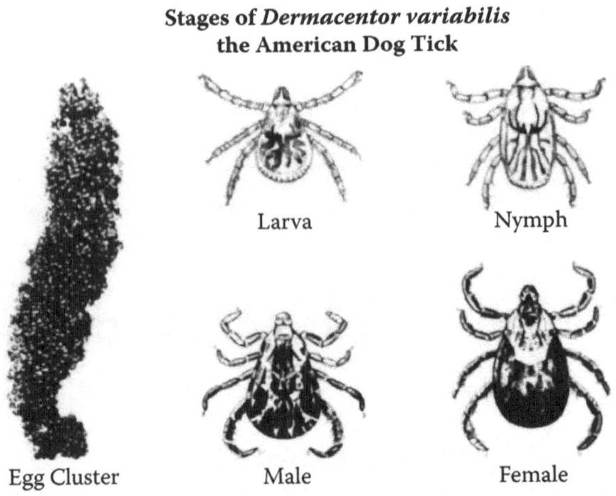

FIGURE 22.1 The life stages of the American dog tick (*Dermacentor variabilis*) (Ixodidae) include egg, larva, nymphs, and adult male and female. Males and females are dimorphic. (Redrawn from the Centers for Disease Control and Prevention, Atlanta, GA; www.cdc.gov/ncidod/dvbid/lyme/resources/handbook.pdf.)

There are about 866 species of ticks worldwide, with about 683 in the Ixodidae and about 183 in the Argasidae (Nicholson et al. 2009). Of the 75 tick species in the United States, 55 are ixodids and 20 are argasids. It is thought that the ancestors of ticks evolved as obligate ectoparasites of reptiles, although it is possible that they evolved earlier with amphibians as hosts (Klompen et al. 1996). The Ixodidae is considered the more primitive family, because they are less host specific and attack a larger number of hosts, including mammals, birds, and reptiles. By contrast, the Argasidae appears to have evolved with animals occupying dens, nests, or burrows. Argasids, therefore, became relatively host and habitat specific, as well as resistant to desiccation and starvation.

Figure 22.2 Highly magnified retrorse teeth on the hypostome of a tick. They make it difficult to remove a tick once it has inserted its mouthparts into the host. (Photograph by Janice Carr, Centers for Disease Control and Prevention, Atlanta, GA.)

Table 22.1 How to Remove an Attached Tick

1. Grasp the tick firmly and very close to its insertion into the skin.
2. Pull the tick slowly and gently so the mouthparts do not break off. If done slowly enough, the tick may withdraw its mouthparts.
3. Use gloves or a paper towel to protect your hands from contamination with any pathogens the tick might carry.
4. Use an antiseptic on the bite after the tick is removed to prevent infection.
5. Wash your hands after handling a tick because tick feces and secretions could be infective.
6. If you are concerned about whether the tick is infected with a disease, save the tick in a closed container in the freezer or in 70% alcohol for possible identification and testing.
7. Do *not* burn the tick or use chemicals in an attempt to get the tick to remove its mouthparts; this will likely kill the tick and leave the mouthparts embedded.

Source: Based on information from Chan and Kaufman (2008) and Stafford (2007).

Ticks are significant pests because feeding causes general irritation and itching and can cause secondary infections (Table 22.2). Tick saliva is complex and contains many components. Slow-feeding tick species may deposit a cement to anchor the mouthparts in the host. The saliva of ticks also includes enzymes and enzyme inhibitors, histamine agonists and antagonists, anticoagulant factors, antiplatelet factors, prostaglandins, and paralysis toxins (Sauer et al. 1995). These cause pain and allergic reactions. In most cases, the bite of a tick is not felt at first due to the salivary secretions injected. Removal of blood from a host by large infestations (exsanguination) can lead to anemia and death (see Figure S22.3 on the CD). Most importantly, ticks can transmit microorganisms that cause disease in humans, domestic animals, and wildlife.

Tick paralysis is a special case of envenomization. Not all tick species can cause paralysis, but approximately 20 species are known to do so. Tick paralysis is a paralysis of the muscles that is induced by tick bites on humans, cattle, dogs, and other animals (Sauer et al. 1995, Lysyk and Majak 2003, Edlow 2010, Lysyk 2010). It apparently is a severe sensitization reaction to salivary secretions or toxins in tick saliva. Tick paralysis is characterized in humans by an ascending motor paralysis and a temperature of 104°F (40°C). Paralysis begins in the legs, gradually ascends, and affects the arms, thorax, and throat. At this stage, it is difficult to swallow. If the heart or respiratory centers are affected, death will occur. Young children are most prone to tick paralysis, especially when the tick is on the neck or head. Complete recovery can occur within a few hours or days after the tick (or ticks) are removed. As few as 30 adult female *Dermacentor andersoni* can cause paralysis in cattle, and paralysis begins within 5 to 8 hours of attachment. Other ticks that can cause paralysis include *Ixodes holocyclus*, *I. brunneus*, *I. rubicundus*, *Argas walkerae*, *Rhiphicephalus evertsi*, and *D. variabilis* (Nicholson et al. 2009).

Table 22.2 Why Ticks Are Significant Pests

1. Ticks cause general irritation and itching, and they can cause secondary infections of the wounds.
2. Ticks result in weight loss, reduced milk production, or reduced egg production in cattle and poultry.
3. Ticks inoculate toxic fluids in their saliva, which contains anticoagulants and proteolytic enzymes that can cause tick paralysis in humans, cattle, dogs, and other animals.
4. Ticks cause pain and allergic reactions from their feeding.
5. Ticks cause exsanguination, leading to anemia or death when they are present in large numbers.
6. Ticks transmit disease-causing agents (viruses, bacteria, rickettsia, spirochetes, protozoans) to humans, domestic animals, and wildlife.
7. Ticks can invade the ear canal (otoacariasis).

Source: Based on information from Nicholson et al. (2009) and Nutting (1984).

Table 22.3 Tickborne Diseases in Humans and Animals

Disease	Disease-Causing Agent	Tick Vectors
African swine fever (pigs, wild boars)	Iridovirus	*Ornithodoros moubata*
Avian spirochetosis (turkeys, chickens, birds)	*Borrelia anserina*	*Argas persicus*
Cattle tick fever	*Babesia bigemina* protozoan	*Rhipicephalus* (*B.*) *annulatus*, *R.* (*B.*) *microplus*
East coast fever (cattle, buffalo)	*Theileria parva*	Rhipicephalus appendiculatus
Ehrlichiosis (dogs, ruminants)	*Ehrlichia canis*, *E. ewingii*, *E. phagocytophila*	*Ixodes ricinus*, *Rhipicephalus sanguineus*, *Amblyomma americanum*, others
Heartwater (ruminants)	*Cowdria ruminantium*	*Amblyomma hebraeum*, *A. variegatum*
Louping ill (sheep)	Flavivirus	*Ixodes ricinus*
Lyme disease (humans)	*Borrelia burgdorferi* and other *Borrelia* spp.	*Ixodes scapularis*, *I. ricinus*, *I. pacificus*, *I. persulcatus*
Q fever (humans)	*Coxiella burnetti*	Many tick species
Rocky Mountain spotted fever (humans)	*Rickettsia rickettsii*	*Dermacentor variabilis*, *D. andersoni*, *Amblyomma americanum*
Tick-bite allergies (humans, cattle, sheep, goats, dogs, chickens)	Proteins injected	*Ixodes pacificus*, *Argas reflexus*, *Ornithodoros coriaceus*
Tick paralysis (humans, cattle, sheep, goats, dogs, chickens)	No pathogen, toxic saliva	*Dermacentor variabilis*, *D. andersoni*, *Ixodes holocyclus*
Tularemia (sheep, horses, rabbits, humans)	*Francisella tularensis*	*Dermacentor andersoni*

Source: Adapted from Mullen, G. and Durden, L., Eds., *Medical and Veterinary Entomology*, Academic Press, New York, 2002.

On the east coast of Australia, *Ixodes holocyclus*, a three-host tick, causes a particularly virulent form of tick paralysis in companion animals. The toxin is similar in action to botulinum. Tick paralysis induced by *I. holocyclus* causes the death of 10,000 calves and cattle each year (Doube and Kemp 1975, Grattan-Smith et al. 1997). Australian veterinarians must treat tick paralysis in domestic animals and farm animals with drugs. Prevention is important and involves examining pets daily during the tick season and removing them, although it is difficult to do this with farm animals. If an animal becomes paralyzed, veterinary services must be provided or the animal could die. Doube and Kemp (1975) found that 10 ticks could cause paralysis of young calves, and 20 ticks were required to induce paralysis in older calves. Nine of ten calves that were paralyzed by *I. holocyclus* died. Strategies to decrease the risk of animals and humans being affected by tick paralysis include removing vegetation where their natural hosts live and keeping the grass short. An antitoxin can decrease paralysis cases in production animals, dogs, and cats, but severe allergic reactions can occur to the antitoxin. Attempts are being made to develop a vaccine against *I. holocylus*.

In addition to causing tick paralysis, ticks can transmit viral, bacterial, spirochete, rickettsial, and protozoan disease agents (Table 22.3). For additional information on rickettsial diseases transmitted to humans by ticks in the United States, see the Centers for Disease Control and Prevention website (www.cdc.gov/ticks/index.html).

Ticks are highly effective vectors of disease agents because they have few natural enemies. They feed intermittently (Argasidae) or on a sequence of hosts (most Ixodidae), and they are long lived (several years to more than 5) so the pathogens have time to multiply within the host. Transmission of pathogens occurs by **transstadial transmission** (from one instar to the next instar) or **transovarial transmission** (from mother to progeny through the egg) (Burgdorfer and Varma 1967), and they have a high reproductive potential (Table 22.4).

Table 22.4 Biological Attributes of Ticks That Make Them Pests of Veterinary and Human Importance

1. Ticks have a high reproductive potential (argasids can deposit 500 to 1000 eggs, and one ixodid was recorded depositing 18,000 eggs).
2. Ticks can survive long periods without food. Argasids live 1 to 2 years (and up to 5).
3. Many ixodid species are able to attack a large number of host species. These can be reservoirs for disease-causing agents.
4. Ticks can transmit viral, bacterial, spirochete, rickettsial, and protozoan disease-causing agents transstadially and transovarially.
5. Argasids live in nests, burrows, or cracks and crevices between intermittent feeding on their hosts, so they have a high rate of success in finding hosts.
6. Ticks have few natural enemies.
7. Ticks develop resistance to pesticides.

Source: Based on information from Burgdorfer and Varma (1967), Nicholson et al. (2009), Samish and Rehacek (1999), and Tuininga et al. (2009).

22.2 BIOLOGY OF THE IXODIDAE

Ticks have complex life cycles that influence their ability to transmit disease and to be managed. Eggs typically are deposited in a single batch of thousands. There is only one nymphal stage, and larvae, nymphs, and adults feed once in each stage. Males and females are dimorphic; males often die after mating, and females die after ovipositing. They usually live outdoors and attach to passing hosts. They generally need several days to complete their feeding. Three-host ticks are the most likely to be vectors of disease agents.

22.2.1 Three-Host Tick Life Cycle

The majority of ixodids are three-host ticks. Adult females mate on the host, engorge with blood, and drop to the ground to deposit eggs, typically under stones and lumps of soil or in the crevices of walls or cracks in wood. The eggs are small and deposited in large masses. Oviposition may require several days to several weeks. The largest number of eggs recorded (18,000) was deposited by a Gulf Coast tick female, *Amblyomma maculatum*, an important pest in the Western Hemisphere (Teel et al. 2010). After laying eggs, the female dies; males die after mating. Males feed on blood but do not become engorged. The enormous number of eggs deposited by females suggest that the likelihood of a larva finding a suitable host is slight, and the probability that it will reach adulthood is even less (Figure 22.3). The larva is called a seed tick. Tick larvae must find a host on which to feed. Some actively seek out a host, attracted by scent, carbon dioxide, or temperature. Others seek a host by climbing up low-growing vegetation and waiting for a small vertebrate to pass by. If a suitable host comes along, the tick becomes very active, reaching out with its front legs (**questing**) and grasping the fur. After a blood meal, the engorged larva drops to the ground and molts into an eight-legged nymph, if it is not killed by desiccation or predation. The nymph must find another host and feed, after which it again drops to the ground and molts into an adult, if the environment is suitable. Adult males and females repeat the process. Typically, larvae and nymphs of three-host ticks feed on small mammals and birds, and the adults feed on larger mammals. The first phase of feeding, lasting several days, is relatively slow, but rapid engorgement occurs during the final 12 to 24 hours before the ticks detach. During the slow phase, the females increase in weight about tenfold. During the rapid phase of feeding, an additional tenfold increase in weight occurs. The hazards facing three-host ticks are severe, but their biology increases the likelihood that some ticks will complete their life cycle. Multiple-host ticks have a high reproductive potential, the ability to survive long periods without food, and a general lack of specificity regarding hosts upon which they can feed.

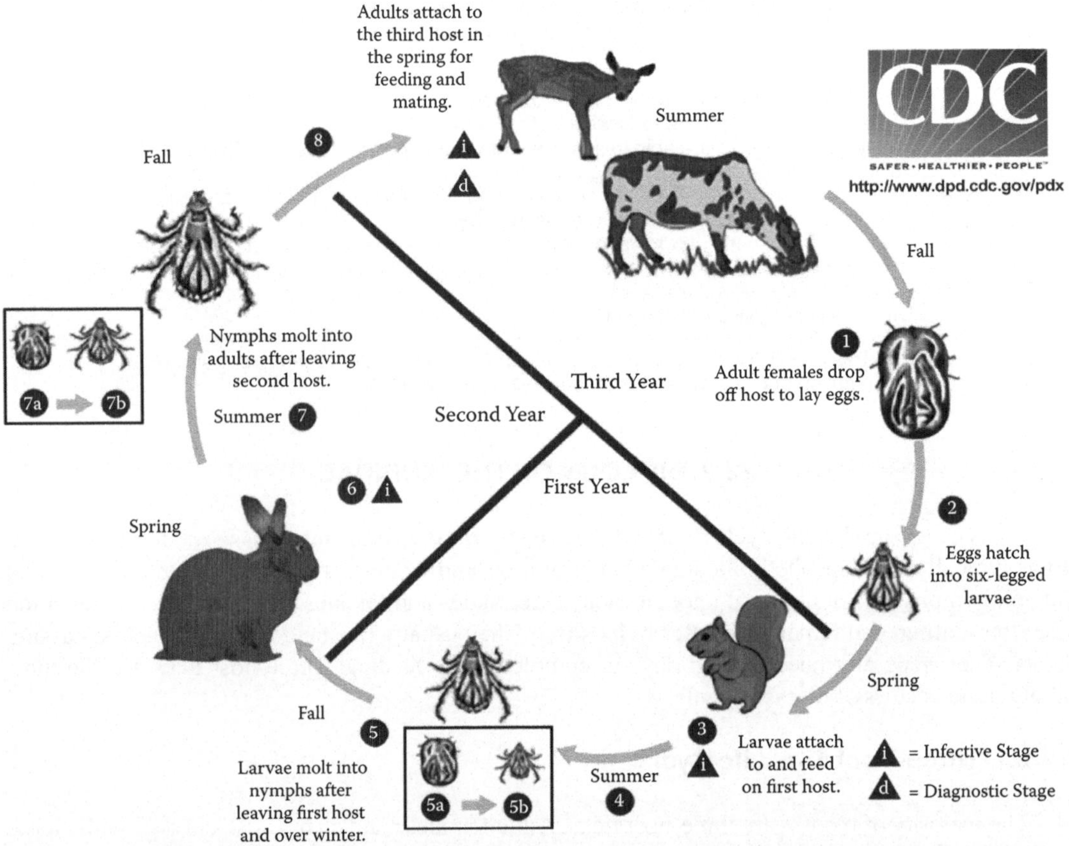

Figure 22.3 An example of a typical life cycle of a three-host ixodid tick (Ixodidae). Most ixodids have a 3-year life cycle. Adult females (1) drop off a host to deposit eggs. The eggs hatch into larvae (2), which overwinter and feed on a small mammal in the spring (3). After feeding, the larva drops off the host and molts to a nymph (4, 5). After overwintering, nymphs attach to a new, often larger, host and feed (6). After feeding, they drop off (7) and molt to adults. After overwintering (8), the adults attach to a third, and larger, host where they feed and mate. After mating, females begin the cycle again (1). The three hosts may be the same species, and humans can serve as first, second, and third hosts. (Redrawn from the Centers for Disease Control and Prevention, Atlanta, GA; http://www.dpd.cdc.gov/dpdx/html/Frames/S-Z/Ticks/body_Ticks_three_host_cycle.htm.)

22.2.2 Two-Host Ticks

In some ixodid species, the larva remains on the host and molts into a nymph. The nymph then feeds on the original host and drops to the ground to molt into an adult. Only the adult must seek a new host. This lifestyle is less hazardous than that of three-host ticks.

22.2.3 One-Host Ticks

In these ixodid species, the larvae, nymphs, and adults remain on the same host. The female leaves only to lay her eggs on the ground. An example of a one-host tick is the cattle tick, ***Rhipicephalus (Boophilus) annulatus***.

22.3 GENERA OF THE IXODIDAE

22.3.1 *Ixodes*

The *Ixodes* genus in the Prostriata contains approximately 241 species and approximately 14 subgenera and occurs around the world (Nicholson et al. 2009). The genus *Ixodes* includes species that are important vectors of disease agents for both humans and animals; for example, the black-legged tick (***Ixodes scapularis***) transmits Lyme disease caused by *Borrelia burgdorferi* (a spirochete) and human babesiosis (a protozoan) to humans. *Ixodes scapularis* also transmits tick-borne fever (*Anaplasma*, a rickettsia) to domestic and wild ruminants, horses, dogs, and humans and borrelioses (*Borrelia*) to dogs, cats, and cattle. In earlier literature, this tick was called *Ixodes dammini* (Patnaude and Mather 2008) (see Figures S22.1, S22.2, and S22.3 on the CD).

Lyme disease was first identified in 1975 as a disease and is named for Lyme, Connecticut, where it was first described (Vail et al. 1994). A bull's-eye type of swelling and redness may occur where the tick has bitten and transmitted the Lyme disease agent (Figure 22.4). Lyme disease currently is the most commonly reported vector-borne disease of humans in the eastern half of the United States (Chapman et al. 2006). The tick is found from Florida to central Texas, north to Maine, and westward to Minnesota and Iowa. This three-host tick transmits the spirochete *Borrelia burgdorferi*, the causative agent of Lyme disease, to deer and other large mammals. Typically, each active stage of *Ixodes scapularis* feeds upon a different host. Larvae feed on small mammals, such as mice and birds. Nymphs feed on a variety of slightly larger hosts. Adults feed on large mammals, especially white-tailed deer. Females deposit 1000 to 3000 eggs after feeding. Larvae that are free of spirochete must acquire the pathogen by feeding on a **reservoir host** and retain the spirochete through its molt to the nymphal stage (transstadial transmission). For more information on managing ticks to reduce infection with Lyme disease, see Spielman et al. (1985), Lane et al. (1991), Stafford (2007), and Patnaude and Mather (2008).

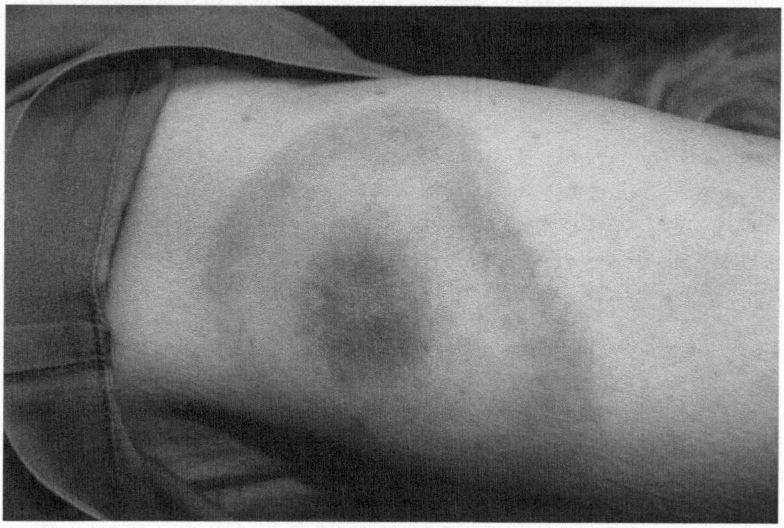

Figure 22.4 Typical bull's-eye swelling and redness may occur where *Ixodes scapularis* has transmitted *Borrelia burgdorferi*, the causative agent of Lyme disease. (Photograph courtesy of the Centers for Disease Control and Prevention, Atlanta, GA.)

Ixodes ricinus (the castor bean tick or sheep tick) can transmit Lyme disease, human babesiosis, and tick-borne encephalitis to humans (Nicholson et al. 2009). It also transmits louping ill (a flavivirus) to sheep; tick-borne fever (*Anaplasma*) to domestic and wild ruminants, horses, dogs, and humans; canine ehrlichiosis to dogs; and borrelioses (*Borrelia*) to dogs, cats, and cattle (see Figure S22.4 on the CD). ***Ixodes pacificus*** (the western black-legged tick) transmits Lyme disease and anaplasmosis to humans. It also transmits tick-borne fever to domestic and wild ruminants, horses, dogs, and humans, as well as borrelioses to dogs, cats, and cattle.

22.3.2 *Dermacentor*

The *Dermacentor* genus has approximately 33 species and includes the American dog tick (*Dermacentor variabilis*), Rocky Mountain wood tick (*D. andersoni*), Pacific Coast tick (*D. occidentalis*), and winter tick (*D. albipictus*), which is a pest of cattle, horses, deer, and moose. Tick-caused anemia during the winter months can cause the death of these animals. Loss of blood results in weight loss and loss of milk production in cattle. Tick bites are primary sites of attack by screwworm flies, where screwworms remain a problem.

Dermacentor variabilis occurs in the United States, east of the Rocky Mountains, and is commonly found on dogs (Chan and Kaufman 2008) (see Figure S22.5 on the CD). It is a three-host tick, feeding on smaller mammals as a larva and a nymph and on larger mammals as an adult (Figure 22.1 and Figure 22.3). Nymphs can survive 6 months without a blood meal, and adults can survive 2 years without feeding. A tick quests for a host by climbing on a blade of grass or other vegetation; it clings to it with its third pair of legs and waves its anterior legs as a potential host approaches. The tick then grabs onto the host as it brushes against the vegetation. Adults of *Dermacentor variabilis* will attack cattle, horses, and humans in addition to dogs. This tick can vector the pathogens causing Rocky Mountain spotted fever and tularemia. *Dermacentor variabilis* also can cause tick paralysis on dogs, especially if it attaches to the neck or head of the dog and is allowed to feed for 5 to 6 days. Upon removal of the tick, the dog usually recovers rapidly. This tick also can cause tick paralysis in children. *Dermacentor variabilis* occurs in wooded, shrubby, or weedy areas. Regular grooming, washing, and examination of dogs are recommended to control the American dog tick. Veterinarians can prescribe topical treatments to prevent tick infestations.

22.3.3 *Rhipicephalus*

The *Rhipicephalus* genus contains 80 species and includes the brown dog tick (***Rhipicephalus sanguineus***) (see Figure S22.6 on the CD) and the brown ear tick (***R. appendiculatus***) (Nicholson et al. 2009). This genus typically parasitizes mammals. *Rhipicephalus sanguineus* is distributed around the world and is unusual in that it can complete its entire life cycle indoors; in fact, *R. sanguineus* can build up to high numbers in the home, with females depositing up to 5000 eggs (Lord 2008). It is a three-host tick that requires three blood meals. Preventing engorgement of the ticks on dogs is critical to managing this pest in the home. Treatments with pesticides prescribed by a veterinarian can reduce tick infestations, but multiple treatments may be required. High levels of infestation can cause skin irritation, and *R. sanguineus* can vector canine ehrlichiosis and canine babesia in the United States. In Europe, Asia, and Africa, this tick is a vector of *Rickettsia conorri* (the causative agent of Mediterranean spotted fever, boutenneuse fever, or tick typhus) (Lord 2008).

The subgenus *Boophilus*, originally considered as a genus, includes the cattle tick, ***Rhipicephalus* (*B.*) *annulatus***, and the tropical fever tick or southern cattle tick, ***Rhipicephalus* (*B.*) *microplus***. Much of the earlier literature refers to the vectors of cattle fever as *Boophilus annulatus* or *B. microplus*. *Rhipicephalus annulatus* is a vector of bovine babesiosis, a disease of cattle and water buffalo, as is *R. microplus*. ***Rhipicephalus appendiculatus*** (brown ear tick) can transmit east coast fever (*Theileria parva*) to cattle and Cape buffalo.

Rhipicephalus (*B.*) *microplus* is considered the most important tick parasite of cattle in the world (Angus 1996). Colonists of the Western Hemisphere brought *R. annulatus* and *R. microplus* with them on tick-infested cattle. By the late 19th century, bovine babesiosis (also known as Texas cattle fever) was a major deterrent to the development of a strong cattle industry in the southern United States (Bram et al. 2002). Direct and indirect losses in 1906 were estimated to be $130,500,000, which would be the equivalent of at least a billion dollars today. As a result, the United States initiated a national *Boophilus* tick eradication program in 14 southern states and a portion of southern California. By 1943, the national eradication campaign was completed, although cattle ticks remain along the Rio Grande River between Texas and Mexico.

The cattle fever eradication program conducted by the Animal and Plant Health Inspection Service (APHIS) of the U.S. Department of Agriculture (USDA) is still active in Texas, because both *Rhipicephalus* (*B.*) *annulatus* and *R. microplus* are present in a permanent quarantine area for about 800 km along the Rio Grande River. Both tick species are abundant in Mexico, and tick-infested cattle, horses, and wildlife can cross the Rio Grande River during the dry season into the quarantine zone, with the potential to reintroduce the ticks. Unfortunately, these ticks have become resistant to multiple pesticides, which makes it difficult to eliminate ticks on livestock that enter this country from Mexico using standard tick dips (see Figure S22.7 on the CD). Furthermore, wildlife, such as white-tailed deer, also carry the ticks into the United States (Pound et al. 2010). The economic value of the *Boophilus* eradication program accrues from potential losses of cattle from babesiosis; the mortality of cattle can be as high as 90% from this disease. There is concern that reintroduction and the spread of these ticks (and the babesiosis pathogens they transmit) could be very difficult to eradicate today due to resistance to pesticides, restrictions by state and federal regulatory agencies, and public concerns about the use of pesticides. Furthermore, dense populations of white-tailed deer would make it difficult to use the vacant pasture option previously used to help eradicate the ticks. Thus, concern about bovine babesiosis in cattle in the United States is increasing (Bram et al. 2002).

Australia also has had an ongoing problem with the cattle tick *Rhipicephalus* (*B.*) *microplus* (Angus 1996). Research in Australia during the past 100 years has led to an effective integrated pest management (IPM) program that is based, in part, on the development of an effective vaccine against the tick. In addition, cross-breeding of cattle for tick resistance has yielded positive results. The Australian IPM program involves strategic dipping of cattle with pesticides, pasture rotation (to eliminate the ability of ticks to find hosts), and developing tick-resistant cattle. Angus (1996) noted, "Although the new tick vaccine offers longer-term control of the parasite, it too needs to be used as a component of an integrated pest-management strategy" and concluded that "no one method alone is sufficient over a longish period."

22.3.4 *Hyalomma*

The *Hyalomma* genus contains 21 species, many of which live in dry environments in the Eastern Hemisphere, where they parasitize wild mammals, livestock, birds, and reptiles. ***Hyalomma marginatum*** (considered by some to be a species complex) is a vector of the Crimean–Congo hemorrhagic fever virus to humans, and *H. detritum* is a vector of bovine tropical theileriosis (Nicholson et al. 2009).

22.3.5 *Amblyomma*

The *Amblyomma* genus contains 129 species that attack nearly all terrestrial vertebrates around the world (Nicholson et al. 2009). It includes the Gulf Coast tick (***A. maculatum***), the lone star tick (***A. americanum***) (see Figure S22.8 on the CD), and the bont tick (***A. hebraeum***). In Africa, the bont tick and the tropical bont tick (***A. variegatum***) attack livestock. These ticks are vectors

Table 22.5 Comparison of the Biology of the Ixodidae and Argasidae

Ixodidae	Argasidae
Approximately 650 species worldwide	Approximately 150 species worldwide
More primitive family that is less host specific and attacks larger number of hosts	Less primitive family that is found in dens, burrows, and nests
Single dorsal shield that covers half of the dorsum of females and the entire dorsum of males	Lack heavily sclerotized plates; have leathery, expandable integument
One-, two-, or three-host life cycles; most (600 species) have three hosts (the most important vectors of animal and human disease)	Typically feed on a single host species, although they may feed on multiple individuals
Larvae, nymphs, and adults feed once each stage	Intermittent feeders for short periods in nests, burrows, and dens; can survive long periods without food
Deposit hundreds to thousands of eggs, up to 18,000 for one species	Deposit 500 to 1000 eggs

Source: Adapted from Nicholson, W.L. et al., in *Medical and Veterinary Entomology*, 2nd ed., Mullen G.R. and Durden, L.A., Eds., Elsevier, Amsterdam, 2009, pp. 493–542.

of the rickettsial pathogen *Ehrlichia ruminantium* that causes heartwater fever. They also vector *Rickettsia africae*, the cause of African tick-bite fever in humans. Because these ticks also occur on islands in the Caribbean, there is concern that they could invade the United States.

22.4 BIOLOGY OF THE ARGASIDAE

The Argasididae (soft ticks) are intermittent feeders. They feed for a short period of time and then find a hidden part of the nest, burrow, or cracks and crevices in which to rest. They sometimes are considered plural-host ticks, because many different hosts *of the same species* may be involved in their life cycle. Argasids feed much like bedbugs; they hide during the day and come out to feed at night (Table 22.5 and Figure 22.5). Argasids may have many nymphal instars. After each feeding, the nymph molts into a somewhat larger instar. The adults feed intermittently, and, in contrast to the ixodids, females lay a small batch of eggs (20 to 50) after each blood meal, but a single female can deposit 500 to 1000 eggs in her lifetime. Argasids are hardy and can survive long periods without food. Some are very long lived and can live 1 to 2 years without feeding. There are four genera in the Argasidae: *Argas*, *Ornithodoros*, *Carios*, and *Otobius*. Ticks in the ***Argas*** genus (57 species) include the fowl tick (*A. persicus*) and the pigeon tick (*A. reflexus*). This genus is distributed around the world. The 87 species of the ***Carios*** genus include parasites of bats and birds. The 38 species in the genus ***Ornithodoros*** are parasites of reptiles, birds, and mammals around the world; *O. hermsi* is a vector of relapsing fever spirochetes to humans and other animals. ***Otobius*** ticks are found in North America, Africa, and Asia; the genus contains two species, *O. megnini* (spinose ear tick) and *O. lagophilus*. *Otobius megnini* infests cattle, horses, and most domestic ruminants. It also attacks wild ruminants (deer, antelope, mountain sheep) and humans. This tick feeds in the ears, causing injury to the ear canal and secondary infections (**otoacariasis**).

22.5 PEST MANAGEMENT OF TICKS

Ticks are considered to cause the greatest economic losses to livestock production in the world of any arthropod pest. Ticks are difficult to manage because they have few natural enemies (Samish and Rehacek 1999, Nicholson et al. 2009, Tuininga et al. 2009). Only a few birds feed on them, and insect predators of adults are rare, although ants may kill larvae in the soil. Desiccation of larvae can cause high rates of mortality, so cultural control can be important. Parasitoids are found in the hymenopteran family Encyrtidae, but they are not considered significant natural enemies (Hu et al. 1998, Collatz et al. 2009). Research is being conducted to determine whether fungal pathogens can be used to control ticks (Bharadwaj and Stafford 2010). The reproductive potential of ticks is high, which makes control difficult. Ticks also develop resistance to pesticides rapidly. The accompanying CD includes is a PDF of the *Tick Management Handbook* (with the permission of Kirby C. Stafford). It is an excellent source of detailed information about tick management for homeowners, pest control operators, and public health officials. See also Table 22.6.

22.5.1 Managing Tick Infestations of Humans

Avoid known tick-infested areas, especially in spring and summer (Stafford 2007). Information on the local distribution and abundance of ticks in the United States can be obtained from the Centers for Disease Control and Prevention (Chapman et al. 2006), local health departments, and

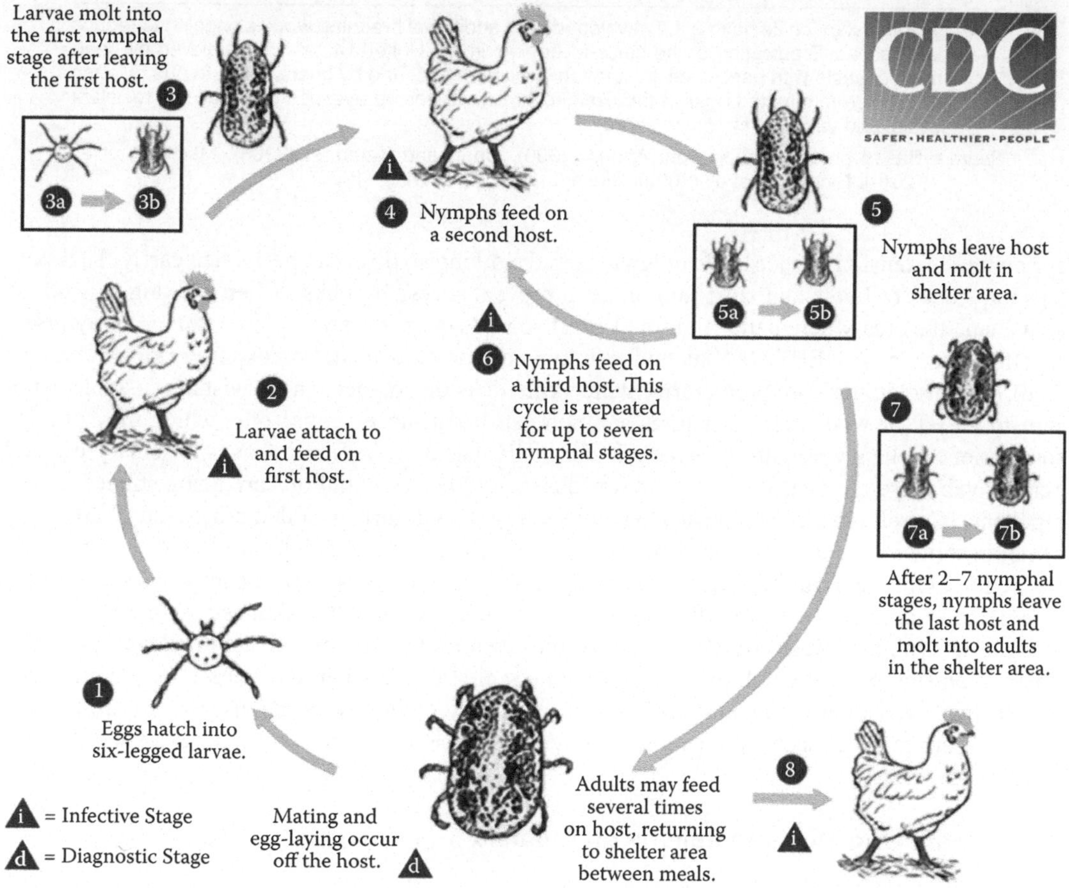

Figure 22.5 Typical life cycle of an argasid soft tick (Argasidae). Argasids take smaller, but more numerous, blood meals than the ixodids. Eggs are deposited by an argasid female in a sheltered area, and the eggs hatch into a six-legged larva (1). The larva attaches to and feeds on a host (2). The larva molts to the first nymphal stage after leaving the first host (3). Nymphs feed on a second host (4) and molt off the host (5) in a sheltered area. After two to seven nymphal stages (6), nymphs molt to the adult stage (7). Adults feed several times on the host, returning to the sheltered area between meals (8). These ticks often are found in nests quite close to their hosts so they may feed repeatedly on the same host species (or even the same individuals). Females may lay egg batches after each meal. (Redrawn from the Centers for Disease Control and Prevention, Atlanta, GA; http://www.dpd.cdc.gov/dpdx/html/Frames/S-Z/Ticks/body_Ticks_argasid_cycle.htm.)

Table 22.6 Methods for Controlling Ticks Attacking Humans and Animals

1. Wear light-colored clothing so ticks can be seen and removed before they attach.
2. Wear a hat and long-sleeved shirt for added protection.
3. Tuck pant legs into socks or boots and shirt into pants to reduce tick access.
4. Apply repellent containing DEET on clothes and exposed skin, or treat clothing with permethrin. Alternatively, use essential oils as a brief repellent.
5. Walk in the center of trails to avoid questing ticks on brush and taller grasses.
6. Remove clothing promptly after returning and wash and dry at a high temperature to kill ticks.
7. Inspect your body carefully, and remove any ticks, being careful to avoid leaving mouthparts imbedded and crushing the tick.
8. Save the tick in a sealed container in the freezer for identification if disease transmission is suspected.
9. Use deer fencing to reduce access of deer to yards, remove leaf litter, and mow grass around homes to reduce relative humidity and enhance desiccation of immature ticks.
10. Use tick dips or sprays to achieve chemical control of ticks on livestock if resistance has not developed; pets can be protected by tick repellents or acaricides prescribed by a veterinarian.
11. Vaccinate whenever possible for tickborne diseases.

Note: Tick-resistant cattle have been developed, and additional breeding work is ongoing to reduce infestations. Eradication of the cattle-fever ticks in the United States was achieved by treatment of cattle with pesticides, by rotation of pasture use, and by quarantines. In Australia, an IPM program directed against the cattle-fever tick utilizes sprays, pasture rotations, resistant cattle, and vaccines in combination.

Source: Based on information from Angus (1996), Chan and Kaufman (2008), Machado et al. (2010), Nicholson et al. (2009), Stafford (2007), and Wikel (1996).

agricultural extension services. Wear light-colored clothing so ticks can be spotted easily. Tuck pant legs into socks or boots and shirt into pants to prevent access by ticks. When walking in wooded areas, tape the area where pants and socks meet so ticks cannot crawl under clothing. Spray insect repellent containing DEET (*N,N*-diethyl-*m*-toluamide) on clothes and on exposed skin (except the face), or treat clothing with permethrin, which kills ticks on contact. Those wishing to avoid exposure to DEET or who prefer a natural solution can try using essential oils, which may provide some brief repellency. Possibilities include citronella, eucalyptus, lemon leaves, rosemary oil, and pennyroyal; these are considered safe in low dosages. Other essential oils are being studied as tick repellents (Carroll et al. 2010). Wear a hat and a long-sleeved shirt for added protection. Walk in the center of trails to avoid grass and overhanging brush where questing ticks are found. After returning, remove clothing promptly and wash and dry it at a high temperature. Inspect your body closely. If you find a tick, remove it carefully by grasping the tick as close to the skin as possible and pulling straight out with a slow, steady movement. Avoid crushing the tick, which could release infectious microorganisms. Save the tick by placing it into a sealed container in the freezer for identification and analysis later in case you suspect Lyme disease or another tick-borne disease. Environmental modifications to residences (application of insecticides, use of deer fencing, and removal of leaf litter) also may help.

22.5.2 Managing Tick Infestations on Animals

Several tactics are available, including tick dips (see Figure S22.7 on the CD) and eradication programs. Chemical control of ticks on livestock reduces the tick load and may also reduce the transmission cycle of tick-borne diseases. Typically, cattle are treated periodically with acaricides, which may be necessary every 6 months. Efforts are being made to develop tick-resistant breeds of livestock (Machado et al. 2010). Such breeds may support tick infestations at much lower levels than do susceptible breeds (Wikel 1996, Machado et al. 2010). Vaccination against tick-borne diseases is being achieved for tick fever in Australia and the control of East Coast fever in Kenya. In addition, vaccines have been produced to control heartwater fever and theileriosis.

SELECTED REFERENCES

Angus, B.M. 1996. The history of cattle tick *Boophilus microplus* in Australia and achievements in its control. *Int. J. Parasitol.* 26:1141–1355.

Bharadwaj, A. and K.C. Stafford III. 2010. Evaluation of *Metarhizium anisopliae* strain F52 (Hypocreales: Clavicipitacea) for control of *Ixodes scapularis* (Acari: Ixodidae). *J. Med. Entomol.* 47(5):862–867.

Bram, R.A. (ed.) 1978. *Surveillance and Collection of Arthropods of Veterinary Importance*, Agric. Handbook No. 518. Washington, D.C.: U.S. Department of Agriculture.

Bram, R.A., J.E. George, R.E. Reichard, and W.J. Tabachnick. 2002. Threat of foreign arthropod-borne pathogens to livestock in the United States. *J. Med. Entomol.* 39:405–416.

Burgdorfer, W. and M.G.R. Varma. 1967. Trans-stadial and transovarial development of disease agents in arthropods. *Annu. Rev. Entomol.* 12:347–376.

Carroll, J.F., G. Paluch, J. Coats, and M. Kramer. 2010. Elemol and amyris oil repel the ticks *Ixodes scapularis* and *Amblyomma americanum* (Acari: Ixodidae) in laboratory bioassays. *Exp. Appl. Acarol.* 51:383–392.

Chan, W.H. and P.E. Kaufman. 2008. *Featured Creatures: American Dog Tick*. Gainesville: Department of Entomology and Nematology, University of Florida Institute of Food and Agricultural Sciences (http://entomology.ifas.ufl.edu/creatures/urban/medical/american_dog_tick.htm).

Chapman, A.S. and the Tickborne Rickettsial Diseases Working Group. 2006. Diagnosis and management of tickborne rickettsial diseases: Rocky Mountain spotted fever, ehrlichioses, and anaplasmosis—United States. *Morbid. Mortal. Weekly Rep.* 55(RR04):1–27.

Collatz, J., A. Fuhrmann, P. Selzer, R.M. Oehme, K. Hartelt, P. Kimmig, T. Meiners, U. Mackenstedt, and J.L.M. Steidle. 2009. Being a parasitoid of parasites: Host finding in the tick wasp *Ixodiphagus hookeri* by odours from mammals. *Entomol. Exp. Appl.* 134:131–137.

Doube, B.M. and D.H. Kemp. 1975. Paralysis of cattle by *Ixodes holocyclus* Neumann. *Aust. J. Agric. Res.* 26:615–640.

Edlow, J.A. 2010. Tick paralysis. *Curr. Treat. Options Neurol.* 12:167–177.

Evans, G.O., J.G. Sheals, and D. MacFarlane. 1961. *The Terrestrial Acari of the British Isles: An Introduction to their Morphology, Biology, and Classification.* London: British Museum of Natural History.

Grattan-Smith, P.J., J.G. Morris, H.M. Johnston, C. Yiannikas, R. Malik, R. Russell, and R.A. Ouvrier. 1997. Clinical and neurophysiological features of tick paralysis. *Brain* 120:1975–1987.

Hoogstraal, H. 1966. Ticks in relation to human diseases caused by viruses. *Annu. Rev. Entomol.* 11:261–308.

Hoogstraal, H. 1981. Changing patterns of tickborne diseases in modern society. *Annu. Rev. Entomol.* 26:75–99.

Hu, R., K.E. Hyland, and J.H. Oliver. 1998. A review on the use of *Ixodiphagus* wasps (Hymenoptera: Encyrtidae) as natural enemies for the control of ticks (Acari: Ixodidae). *Syst. Appl. Acarol.* 3:19–28.

Klompen, J.S.H., W.C. Black, J.E. Keirans, and J.H. Oliver. 1996. Evolution of ticks. *Annu. Rev. Entomol.* 41:151–161.

Lane, R.S., J. Piesman, and W. Burgdorfer. 1991. Lyme borreliosis: Relation of its causative agent to its vectors and hosts in North America and Europe. *Annu. Rev. Entomol.* 36:587–609.

Lord, C.C. 2008. *Featured Creatures: Brown Dog Tick.* Gainesville: Department of Entomology and Nematology, University of Florida Institute of Food and Agricultural Sciences (http://creatures.ifas.ufl.edu/urban/medical/brown_dog_tick.htm).

Lysyk, T.J. 2010. Tick paralysis caused by *Dermacentor andersoni* (Acari: Ixodidae) is a heritable trait. *J. Med. Entomol.* 47:210–214.

Lysyk, T.J. and W. Majak. 2003. Increasing the paralyzing ability of a laboratory colony of *Dermacentor andersoni* Stiles. *J. Med. Entomol.* 40:85–194.

Machado, M.A., A.L.S. Azevedo, R.L. Teodoro, M. deFatima, A. Pires, M.G.C.D. Peixoto, C. deFritas, M.C.A. Prata, J. Furlong, M.V.G.B. daSilva, S.E.F. Guimaraes, L.C.A. Regitano, L.L. Coutinho, G. Gasparin, and R.S. Verneque. 2010. Genome wide scan for quantitative trait loci affecting tick resistance in cattle (*Bos taurus* x *Bos indicus*). *BMC Genomics* 11:280.

Mullen, G. and L. Durden (eds.) 2002. *Medical and Veterinary Entomology*. New York: Academic Press.

Nicholson, W.L., D.E. Sonenshine, R.S. Lane, and G. Uilenberg. 2009. Ticks (Ixodida). In: G.R. Mullen and L.A. Durden (eds.), *Medical and Veterinary Entomology*, 2nd ed. (pp. 493–542). Amsterdam: Elsevier.

Nutting, W.B. (ed.) 1984. *Mammalian Diseases and Arachnids*. Vol. I. *Pathogen Biology and Clinical Management*. Vol. II. *Medico-Veterinary, Laboratory and Wildlife Diseases and Control*. Boca Raton, FL: CRC Press.

Patnaude, M.R. and T.N. Mather. 2008. *Featured Creatures: Blacklegged or Deer Tick*. Gainesville: Department of Entomology and Nematology, University of Florida Institute of Food and Agricultural Sciences (http://creatures.ifas.ufl.edu/urban/medical/deer_tick.htm).

Pound, J.M., J.E. George, D.M. Kammlah, K.H. Lohmeyer, and R.B. Davey. 2010. Evidence for role of white–tailed deer (Artiodactyla: Cervidae) in epizootiology of cattle ticks and southern cattle ticks (Acari: Ixodidae) in reinfestations along the Texas/Mexico border in south Texas: A review and update. *J. Econ. Entomol.* 103:211–218.

Samish, M. and J. Rehacek. 1999. Pathogens and predators of ticks and their potential in biological control. *Annu. Rev. Entomol.* 44:159–182.

Sauer, J.R., J.L. McSwain, A.S. Bowman, and R.C. Essenberg. 1995. Tick salivary gland physiology. *Annu. Rev. Entomol.* 40:245–267.

Sonenshine, D.E. 1991. *Biology of Ticks*, Vol. 1. New York: Oxford University Press.

Sonenshine, D.E. 1993. *Biology of Ticks*, Vol. 2. New York: Oxford University Press.

Spielman, A., M.L. Wilson, J.F. Levine, and J. Piesman. 1985. Ecology of *Ixodes dammini*-borne human babesiosis and Lyme disease. *Annu. Rev. Entomol.* 30:429–460.

Stafford, K.B. 2007. *Tick Management Handbook: An Integrated Guide for Homeowners, Pest Control Operators, and Public Health Officials for the Prevention of Tick-Associated Disease*, Revised ed. New Haven: Connecticut Agricultural Experiment Station (www.caes.state.ct.us; also available on the accompanying CD).

Teel, P.D., H.R. Ketchum, D.E. Mock, R.E. Wright, and O.F. Strey. 2010. The Gulf Coast tick: A review of the life history, ecology, distribution, and emergence as an arthropod of medical and veterinary importance. *J. Med. Entomol.* 47:707–722.

Tuininga, A.R., J.L. Miller, S.U. Morath, T.J. Daniels, R.C. Falco, M. Marchese, S. Sahabi, D. Rosa, and K.C. Stafford. 2009. Isolation of entomopathogenic fungi from soils and *Ixodes scapularis* (Acari: Ixodidae) ticks: Prevalence and methods. *J. Med. Entomol.* 46:587–565.

Vail, S., G. Smith, and C. Lord. 1994. Population biology of *Ixodes scapularis*, the vector of Lyme disease in the eastern and north central United States. In: M.E. Scott and G. Smith (eds.), *Forensic and Infectious Diseases* (pp. 263–277). Orlando, FL: Academic Press.

Wikel, S.K. 1996. Host immunity to ticks. *Annu. Rev. Entomol.* 41:1–22.

CHAPTER 23

Pest Mites of Farm and Companion Animals

Approximately 250 species of mites are parasitic on humans and domestic animals (Mullen and O'Connor, 2002). These mites can be minor pests, causing temporary skin irritations from their bites, or they may cause persistent dermatitis because they invade the skin or hair follicles. Some may induce allergic responses; transmit pathogens; transmit tapeworms; invade the respiratory passages, ears, and internal organs; or cause **delusory acariasis** (when individuals believe they are being attacked by mites but, in fact, they are not). This chapter briefly discusses parasitic mites in the Gamasida, Actinedida, and Acaridida that are pests of poultry, cattle, sheep, pigs, horses, companion animals, and humans. For additional discussions of mites and ticks of veterinary importance, see Baker et al. (1956), Evans et al. (1961), Williams et al. (1985), Baker (1999), Mullen and O'Connor (2002), Eldridge and Edman (2004), and Williams (2009).

23.1 POULTRY RED MITE OR ROOST MITE, *DERMANYSSUS GALLINAE* (DERMANYSSIDAE)

The poultry red mite or roost mite (***Dermanyssus gallinae***) is a blood-sucking ectoparasite of chickens, wild birds (pigeons, sparrows), and cage birds (Chauve 1998, Nordenfors et al. 1999, McCrea et al. 2005). It can attack humans, as well (Figure 23.1). Severe infestations can lead to anemia, reduced egg production and growth, or even death of the birds. This mite is considered one of the most important **direct pests** of domestic birds (Nordenfors et al. 1999) and is especially important as a pest in the Palearctic Region and in the United States. A direct pest is one that causes damage or harm by itself, not through the transmission of a pathogenic agent.

The roost or poultry red mites live in bird nests. They move onto the birds to feed on their blood at night and return to darkened areas during the day. When bird nests are abandoned, the mites may move into homes and attack people, producing painful bites and dermatitis. Various viruses have been isolated from *Dermanyssus gallinae* found in bird nests, suggesting that the mite may transmit fowl pox virus, fowl spirochetosis, Newcastle virus, pullorum disease, fowl typhoid, and fowl cholera (Chauve 1998); however, evidence supporting the ability of *D. gallinae* to transmit these viruses to humans, or birds, is conflicting (Mullen and O'Connor 2002). Eggs produced by infested poultry can be downgraded because the eggs are stained by blood from squashed mites. Birds also may suffer from anemia and reduced egg production. The mites will attack humans working on poultry farms.

The adults are red when engorged with blood. If they contain partially digested blood, they are black, gray, or white. Females are about 1 mm long. Mites rarely are found on the skin of the host during the day, because most of the life cycle is spent in hiding places in the nest or roost. All active stages, except the larvae, feed on blood (Nordenfors et al. 1999). The life cycle requires from 7 to 13 days. Each female deposits four or five batches of about 8 eggs in crevices or debris, for a

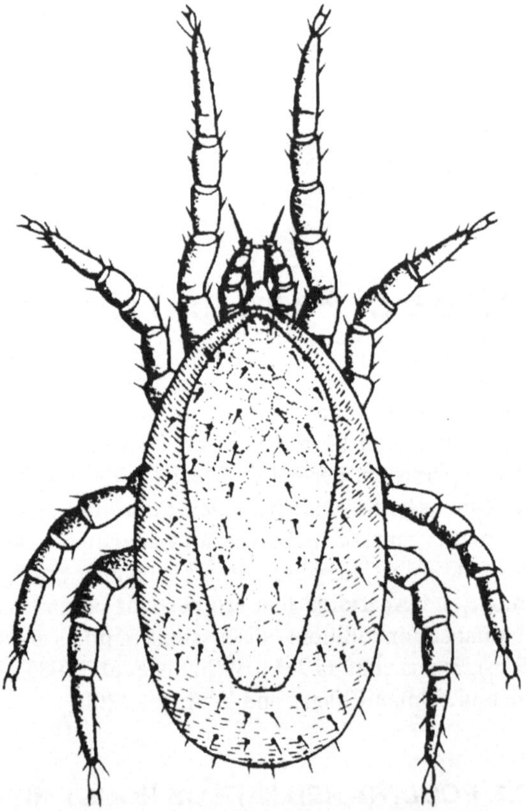

Figure 23.1 Female of *Dermanyssus gallinae*, the poultry red mite or roost mite. This mite is an important pest of domestic birds and lives in nests, moving onto the birds at night to feed on their blood. (Adapted from the U.S. Centers for Disease Control and Prevention, Atlanta, GA.)

total of about 40 eggs; each batch is preceded by a blood meal. Eggs may hatch within 47 to 72 hours. The larvae do not feed and molt to the protonymphal stage, which feeds, then molts to the deutonymphal stage. Deutonymphs take a blood meal, then molt to adults. The mites can survive without food for up to 9 months if the temperature is between 5 and 25°C and the relative humidity is high (Nordenfors et al. 1999). This ability to survive under such diverse environmental conditions makes it difficult to eliminate an infestation; however, reducing the relative humidity and freezing or heating (>45°C) can eliminate mites on empty egg flats, nest boxes, and perches and in empty poultry houses (Nordenfors et al. 1999). Monitoring for mites may be inaccurate, because the mites are found on the bird at night, when few pest managers are willing to work. Look for mites in their hiding places during the day or on the bird at night.

Chemical control historically has been used to control *Dermanyssus gallinae* but is becoming less desirable due to problems with registration of new products, loss of registration of pesticides, and the development of resistance to pesticides by the mites. Furthermore, in some locations, chemical control can only be used in empty poultry houses to avoid chemical residues in eggs and meat. Because this mite has become resistant to pesticides (Chauve 1998), alternative controls have been attempted, including the use of entomopathogenic fungi, silica dust or other desiccants, cultural controls (improved chicken house cleaning, heating henhouses to eliminate infestations prior to introduction of poultry), and the use of biological control agents (predators) (Mul and Koenraadt 2009).

Table 23.1 Control Methods for the Poultry Red Mite (*Dermanyssus gallinae*)

Preventing the establishment of *D. gallinae* is the key to managing the poultry red mite. Steps to reduce or eliminate infestations include:

1. Eliminate cracks in poultry houses to eliminate resting sites.
2. Remove wild bird nests from homes and near poultry houses.
3. Heat the henhouse to temperatures above 55°C before restocking.
4. Regularly wash poultry houses.
5. Treat walls and floors with silica dust prior to introducing new birds.
6. Monitor to ensure persons and materials entering the farm are free of mites.
7. Prevent entry of other pests that are potential carriers of mites.
8. Treat edges of feeding troughs with silica dust.
9. Treat the manure conveyor belt.
10. Monitor egg trays and be sure they are clean.
11. Obtain birds that are free of *D. gallinae* (inspect to be sure) and transport them in mite-free crates.
12. Monitor regularly for the presence of *D. gallinae* in rearing facilities.
13. Maintain as long an interval as possible between depopulating and repopulating poultry houses.

Source: Based on information from Chauve (1998) and Mul and Koenraadt (2009).

Inert dusts, including silica and diatomaceous earth, are thought to kill mites by desiccation and are being applied to control *Dermanyssus gallinae* in Europe (Kilpinen and Steenberg 2009). Comparisons of various products, including kaolin, diatomaceous earth, and silica formulations, indicate that relative humidity during treatment is important. At 85% relative humidity (RH), efficacy was significantly reduced compared to when products were tested at 75% RH. The results also demonstrated that exposure for 24 hours to surfaces treated with low concentrations is sufficient to kill the mites, but thorough treatment of all surfaces in a poultry house is essential (Kilpinen and Steenberg 2009). Despite the fact that an adult female poultry mite ingests about 0.2 mg of blood per meal, the "most efficient inert dusts … caused almost complete loss of water within 24 h at 75% RH (and 25°C)" (Kilpinen and Steenberg 2009).

Preventing the establishment of *Dermanyssus gallinae* is the key to managing this pest (Chauve 1998, Mul and Koenraadt 2009). Mul and Koenraadt (2009) have provided an overview of the potential sources of mite infestation, risk of spread, and cultural methods for reducing the risk of introducing and spreading *D. gallinae* (Table 23.1). Monitoring free-range poultry for *D. gallinae* can be complicated (Zenner et al. 2009). Inspecting traps and bird droppings in nest boxes or on perches provides evidence of infestation, although the efficacy varies. Monitoring also allows the pest manager to determine whether control measures are effective.

Currently, biological control of *Dermanyssus gallinae* is not employed as a pest management tactic; however, several species of predatory mites may become important in the future in augmentative releases. The predatory mites *Hypoaspis aculeifer* and *Androlaelaps casalis* were able to feed on *D. gallinae* and did not feed on the blood of the other bird species tested (although that may not be true for all bird species) (Lesna et al. 2009). Additional research may allow these, or other, predators to be developed as natural enemies of this poultry pest. Managing *D. gallinae* is very difficult once an infestation has occurred, because these mites can survive up to 8 months without a blood meal. Control tactics must take this behavior into account. Research also is being conducted to develop a vaccine for poultry that reduces or eliminates infestations with *D. gallinae* (Wright et al. 2009).

Distinguishing the northern fowl mite *Ornithonyssus sylviarum* from *Dermanyssus gallinae* is important. Although both are blood feeders and cause extensive damage, managing the two pests requires different tactics. Because *O. sylviarum* spends its entire life on its host, the birds themselves must be treated when they become infested. Weisbroth (1960) noted that adult females of the two species could be discriminated easily. If a female mite is slide mounted in temporary mounts in water, ventral side up under a cover slip, and examined under 100× magnification, the anal plate of a *O. sylviarum* female will be teardrop shaped (Figure 23.2), and that of *D. gallinae* will be truncate in shape.

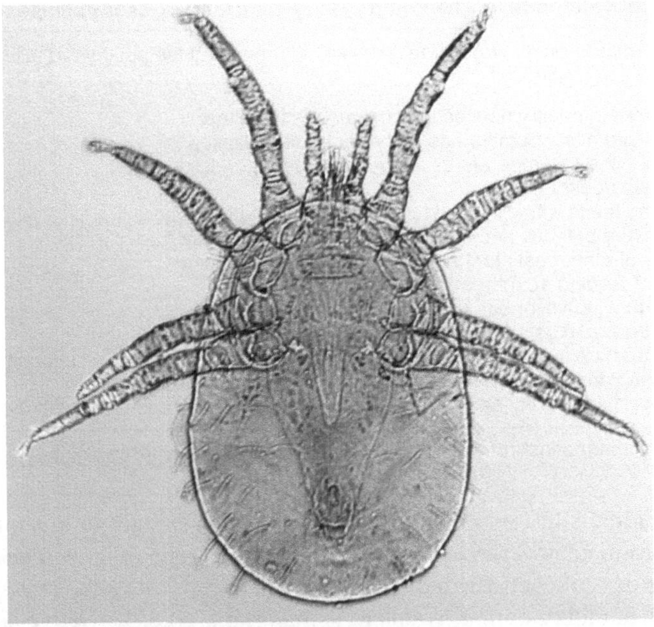

Figure 23.2 Slide-mounted adult female of the northern fowl mite (*Ornithonyssus sylviarum*) (Macronyssidae). This mite is the most serious pest of chickens, especially in temperate regions. It spends its entire life on its host. (Photograph by J.F. Butler, Department of Entomology and Nematology, University of Florida, Gainesville.)

23.2 EUROPEAN OR NORTHERN FOWL MITE, *ORNITHONYSSUS SYLVIARUM* (MACRONYSSIDAE)

The European, or northern, fowl mite (*Ornithonyssus sylviarum*) attacks poultry and wild birds in temperate regions around the world (Hogsette et al. 1991, Knee and Proctor 2007, Kaufman et al. 2009). It is considered the most serious pest of chickens. Unlike *Dermanyssus gallinae*, *O. sylviarum* spends its entire life on its host and feeds during the day and night. The entire life cycle takes 5 to 7 days. This pest causes irritation, blood loss, and anemia in poultry; when it bites humans, it can produce intense itching. It is a nasty pest of egg handlers, readily biting humans and producing a range of symptoms from irritation to severe allergic reactions. Although these mites may be found all over the body, infestations with *O. sylviarum* are concentrated around the vent of the chicken (Figure 23.3). (The vent is the external opening to the cloaca, a chamber where the digestive, reproductive, and excretory tracts come together.) Birds infested with mites eat less, lose weight, and have a pale pink comb. Crusting of feathers around the vent is due to the collection of mite excrement. Mite levels and effects on egg production seem related to the strain of birds and the diet on which they are maintained. Young birds are more susceptible to high infestations. Large mite populations can cause severe anemia (exemplified by a pale comb). Death can result from blood loss if 200,000 (or more) mites are on a bird. Mites thrive when temperatures are cooler and with low relative humidity. It is not clear whether these mites serve as vectors of disease agents (Mullen and O'Connor 2002, Knee and Proctor 2007).

An *Ornithonyssus sylviarum* female deposits two to five eggs over her life, either on the host or in the nest; the life stages include egg, larva, feeding protonymph, nonfeeding deutonymph, and feeding adult males and females. The protonymph feeds at least twice on the host and molts. Deutonymphs do not feed and become adults within 3 or 4 days. A total generation can develop

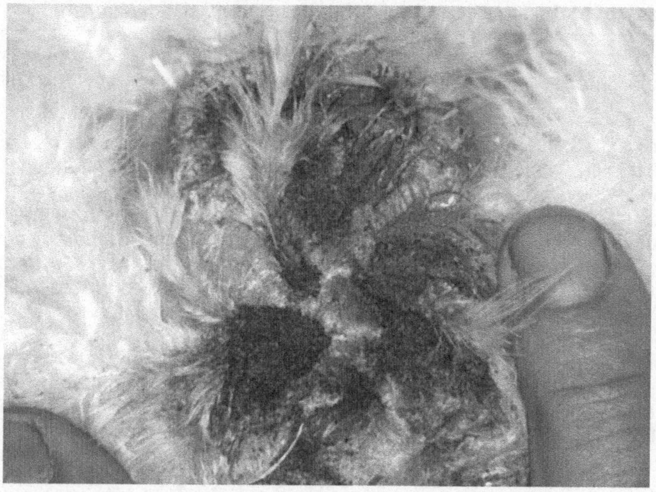

Figure 23.3 Northern fowl mites (*Ornithonyssus sylviarum*) cluster around the vent of a chicken. Their feces can cause crusts to form on the feathers. (Photo by J.F. Butler, Department of Entomology and Nematology, University of Florida, Gainesville.)

from egg to adult in 5 to 7 days, assuming that the protonymph is able to find a host quickly. A large population can develop rapidly due to the short generation time. Under ideal conditions, and with no mortality, a population of 10 mites could turn into 1 billion mites in 10 weeks!

Removing all poultry from an infested house and returning them within 2 weeks may not be sufficient to break the life cycle; in fact, a "vacancy period of >5 wk in summer or >8 wk in winter in southern California should allow for mite mortality high enough to eradicate residual populations" (Chen and Mullens 2008). Although this mite spends most of its time on its host, mites can move off the bird and conceal themselves in litter, manure, and cracks in the structure where pesticides do not completely eliminate them. Mites can be transmitted from bird to bird and can be introduced into a clean coop by bringing in infested birds or by mites moving off of wild birds. Mites can be transported from farm to farm on egg flats (Kells and Surgeoner 1996, 1997).

Control in the past was based on chemical control, but preventing the introduction of mites into poultry houses is the only dependable means of keeping the birds free of mites (Table 23.2). Chemical control of mites is difficult because the birds' water-repellent feathers protect the mites from sprays. When spraying poultry, it is important to deliver a good volume of insecticide under high pressure underneath the bird, especially in the vent area.

Monitoring for this mite typically has involved examining birds, focusing on the vent area, although complaints by farm workers of bites also are an indicator of infestations. Harris et al. (2000) devised a presence–absence sequential sampling method for these mites on caged layer hens, using an infestation rate of 25% or more as economically important, and found that examining as few as five hens could be sufficient to reach a treatment decision. Mullens et al. (2000), however, observed that the mites can be found on eggs shortly after the eggs are laid, which might serve as a more rapid and easier method for detecting high-level infestations.

23.3 TROPICAL FOWL MITE, *ORNITHONYSSUS BURSA* (MACRONYSSIDAE)

The tropical fowl mite (*Ornithonyssus bursa*) is found in the southern United States and in other subtropical and tropical areas around the world (Denmark and Cromroy 1987). It is found on birds and mammals and can be a serious pest of domestic fowl and wild birds. This mite may be found

Table 23.2 Control of the Northern Fowl Mite (*Ornithonyssus sylviarum*)

Prevention is the best method for managing *O. sylviarum*. Chemical control may be needed if infestations occur. Because mites can become resistant to pesticides, preventing infestations is a high priority:

1. Monitor birds prior to introducing them into production facilities to confirm they are free of mites.
2. Disinfest all egg flats, carts, cartons, crates, and racks before they enter the facility.
3. Prevent mite-infested wild birds or rodents from entering the poultry houses.
4. Have personnel change clothing and other gear before moving between poultry houses or farms to prevent the accidental transport of mites from infected birds to new poultry houses.
5. Monitor eggs and adult birds weekly or bimonthly to determine infestation levels and whether the economic injury level (EIL) has been reached. One method involves monitoring newly deposited eggs, another requires examining hens.
6. Mites have become resistant to multiple pesticides in the United States, so use cultural practices when possible.
7. If applying pesticides, deliver adequate amounts to the skin in the vent area.
8. After birds are removed from a poultry house, spray cages and other equipment to control mites, or expose the residual mites to high heat or freezing for several weeks.
9. Do not introduce mite-free poultry into empty houses until more than a month has passed to reduce the likelihood that mites from the preceding flock have survived off their hosts and could infest the new poultry.
10. Improve true resistance to mites through breeding; some strains of birds appear to be more resistant to mites establishing on them.
11. Reduce stresses on the birds to reduce the impact of mites on their health and productivity.

Source: Based on information from Axtell and Arends (1990), Hinkle and Hickle (1999), Kells and Surgeoner (1997), Mullens et al. (2001, 2004), and Yazwinski et al. (2005).

in nests of birds around buildings and can invade homes and poultry houses, affecting both birds and their human caretakers. The life cycle of this mite is similar to that of the northern fowl mite. On chickens, the mites are found in the fluffy feathers and are numerous around the vent. Control around homes and poultry houses involves removing bird nests and washing the area with a strong spray of water or steam. This mite can only live about 10 days away from bird hosts, so its effect on humans is short term.

23.4 CHIGGERS (ACTINEDIDA: TROMBICULIDAE)

The larvae of the family Trombiculidae typically parasitize vertebrates and are called **chiggers** (Sasa 1961, Zhang 1998). Humans are accidental hosts of chiggers. Generally, chiggers are nuisances, causing dermatitis; however, some species also can transmit rickettsial disease agents that cause scrub typhus (several types) in much of Asia (Sasa 1961). The geographical distribution of scrub typhus is a roughly triangular area between Japan, west India, and Australia (Sasa 1961, Varma 1969). In the Western Hemisphere, chiggers are nuisances to humans, but none is known to transmit rickettsia.

Trombiculid adults and nymphs are bright red and may be seen crawling over the soil surface. They look like velvety red spiders. Adults and nymphs are predators, but the larvae are parasitic on rodents, birds, poultry, rabbits, livestock, snakes, toads, and humans. Larval chiggers are very small (Figure 23.4) and typically are found in shady areas where they are less likely to become desiccated. When a potential host moves by, multiple chiggers may move toward it. The larval chigger can move quickly over the ground and crawl onto the feet or legs of its host, attracted by the carbon dioxide the host produces. Once on a human host, the chigger moves about until it reaches a place where the skin is thinner, which often is around the ankles, under socks, or behind knees (see Figure S23.1 on the CD). People sitting on chigger-infested ground may become infested with chiggers around the waistline or under elastic bands of their underwear (Figure 23.5).

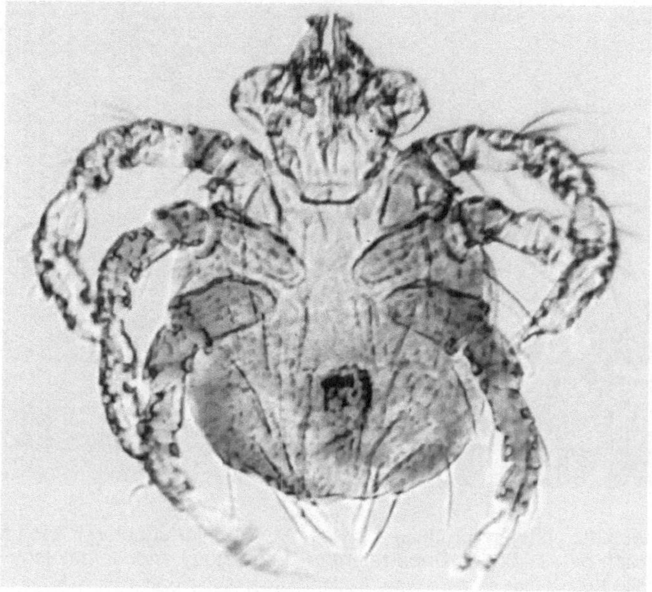

Figure 23.4 Slide-mounted chigger (larval Trombiculidae). The larva is parasitic, but nymphs and adults are predators. In parts of Asia, chiggers can transmit a rickettsial disease (Tsutsugamushi fever) to humans, although it usually is a parasite of rodents. In most parts of the world, chiggers are only a nuisance, causing dermatitis. (Photograph by J.F. Butler, Department of Entomology and Nematology, University of Florida, Gainesville.)

Chiggers do not burrow into the skin. They pierce the skin (often near a hair follicle) and inject a fluid that prevents the blood from clotting. The fluid causes tissues to become inflamed, producing a characteristic red welt with a white, hard center. The reddened, swollen tissues camouflage the tiny red chigger in the center of the itchy bump. Fluid secreted by the chigger liquefies the tissues that are then ingested. It is the injected fluids that cause the itching and the reddened bump. The secreted fluids literally dissolve the skin cells, and the skin responds by producing a tube-like structure called a

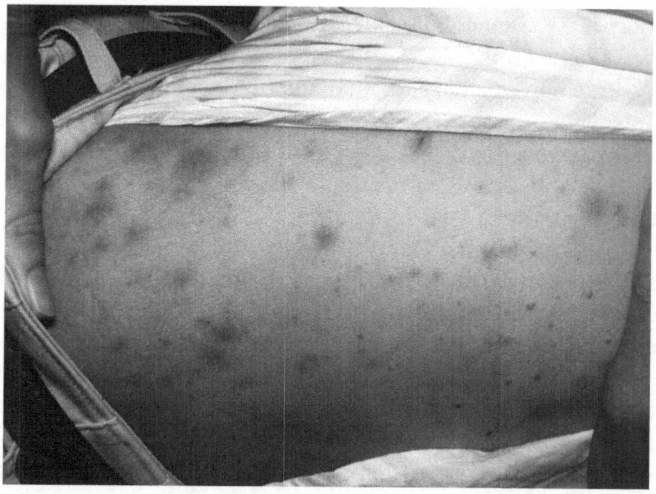

Figure 23.5 Chiggers, larval trombiculids, cause itchy, red skin and welts by their feeding. (Photograph by J.F. Butler, Department of Entomology and Nematology, University of Florida, Gainesville.)

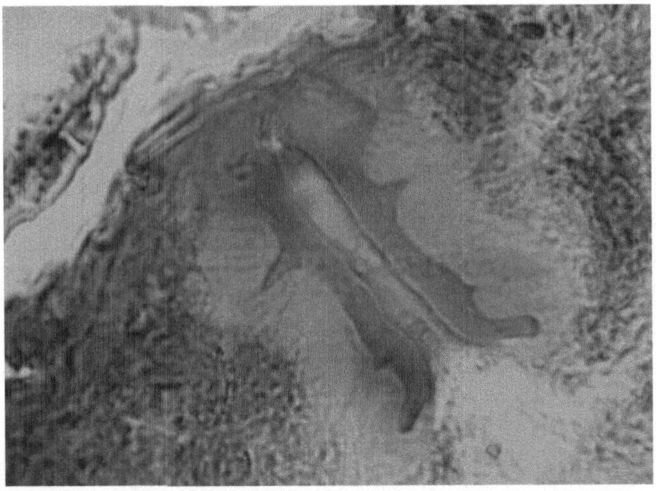

Figure 23.6 During feeding, chiggers produce a stylostome (tubular structure), seen here in a skin section. (Photograph by J.F. Butler, Department of Entomology and Nematology, University of Florida, Gainesville.)

stylostome, which walls off the injected saliva and also functions like a feeding tube for the chigger (Figure 23.6) (also see Figure S23.2 on the CD) (Hase et al. 1978, Shatrov 2009). The longer the chigger feeds, the larger the stylostome becomes and the larger the reddened bump that results. The intense itching is due to an allergic reaction to the saliva of the chigger and to the stylostome itself, which must be dissolved and carried away by the immune system. The chigger can feed for several hours to several days before finally detaching and dropping into the soil to enter a quiescent stage before it molts to the nymphal stage.

The life cycle is egg → parasitic larva (or chigger) → nymphochrysalis (resting stage) → active predatory nymph → imagochrysalis (resting stage) → adult predatory males and females. Male chiggers deposit spermatophores that are taken up by females (indirect sperm transfer). Eggs tend to be deposited in groups, which can explain the clustered distribution of chigger larvae.

Chiggers in temperate climates most often are found in overgrown, brushy or grassy areas, especially where small rodents are common. Females deposit eggs on the ground in groups of up to 400, picking damp but well-drained sites. As a result, their distribution can be quite patchy. Chiggers are found in yards, parks, camps, picnic sites, and other recreational areas. They may be especially common along streams, under trees, in orchards, and in berry thickets. Chiggers can be trapped or monitored by placing plastic discs or dishes in areas of heavy infestation (Krantz and Walter 2009). The chiggers will move on and under them within a few minutes.

Chiggers have one generation a year in temperate climates; the larval stage is most common during July, August, or early September. In tropical areas, two or more generations per year are possible, and chiggers are more active during the wetter seasons. To find a host, larval chiggers move to the tip of grasses and fallen leaves to grab onto a passing host. Rodents are common hosts, but chiggers can attack larger animals and humans. Chiggers move to a feeding spot (ears of rodents, around the eyes of birds, or where clothing is tight on humans) and attach themselves to the skin. Once they have fed, they fall to the ground and molt to the nymphal stage. Lipovsky (1954) found that nymphs and adults of two chigger species, *Trombicula alfreddugesi* and *T. splendens*, could be reared on quiescent stages of other small arthropods, but primarily used eggs of arthropods as their food.

23.4.1 Control

Remove debris (wood, brush piles) from sites. Mow regularly and remove brush to create a less favorable habitat for chiggers and their wild hosts. Short grass allows sunlight to penetrate and dry out the area. Treating a well-mowed lawn with pesticides to control chiggers is of little benefit, because chiggers avoid direct sunlight and normally will not infest areas that are mowed frequently to keep grass short. Wearing shirts with long sleeves and long pants made of tightly woven fabric will reduce the chances of being attacked, as will tucking pants into boots to avoid exposing skin. Avoid walking through unmowed fields and brushy or other overgrown areas, especially during chigger season (normally July, August, and early September in temperate climates). Walk in the center of mowed trails. When hiking or camping in chigger-infested areas, wear long pants tucked into boots or socks to keep chiggers out.

Apply a repellent such as DEET (*N,N*-diethyl-*m*-toluamide) to shoes, cuffs, socks, and pant legs. Be aware, though, that DEET has been associated with adverse effects in humans because it is absorbed through the skin, and it must be used according to the label. The label includes the warnings that DEET should not be used under clothing or on damaged skin and that it should be washed off after it is no longer needed (Sudakin and Trevathan 2003). Eamsobhana et al. (2009) tested several essential oils as repellents and found that 4 of the 13 tested were effective. It is advisable to shower or bathe with abundant soap *immediately* after coming indoors; this will remove many chiggers that have not attached because they do not settle down to bite for several hours. Wash clothing in hot water to remove chiggers. (See Figures S21.1 and S23.2 on the accompanying CD for examples of the reddened bites and the stylostome.)

During World War II, chiggers transmitted the causal agent of scrub typhus, ***Rickettsia tsutsugamushi***, to soldiers. The disease reduced the effectiveness of the U.S. armed forces in Asia. At least 5441 cases (with 283 deaths) occurred from the disease among the soldiers. Scrub typhus remains endemic in many parts of Southeast Asia (Williams 1944, Mohr 1947, Dohany 1978, Phasomkusolsil et al. 2009). At least seven species of trombiculid mites in the subgenus *Leptotrombidium* vector *Rickettsia tsutsugamushi* (Hase et al. 1978). Because only the larva feeds on vertebrate hosts, such as rats, from which it can acquire the rickettsia, the rickettsia is passed transstadially and transovarially to the next generation of trombiculid larvae (Phasomkusolsil et al. 2009).

When it has been injected into the human host, the rickettsia undergoes an incubation period of 6 to 21 days, and the disease manifests itself as fever with gastrointestinal, respiratory, or central nervous system symptoms. Illness can be mild or severe, and death can occur in 1 to 30% of untreated cases (Durand et al. 2004). If infected persons are treated promptly with an antibiotic, deaths are rare and recovery is rapid.

Consistent control of rat populations, reservoirs of the rickettsia, is important in managing the disease. A **reservoir host** is any person, animal, plant, soil, or substance in which an infectious agent normally lives and multiplies. The reservoir host typically harbors the infectious agent without injury to itself and serves as a source from which other individuals can be infected. The infectious agent primarily depends on the reservoir host for its survival. A drawback to this approach is that if all rats are removed from an area then relatively more humans may be infected shortly afterward because the infected chiggers have no other hosts upon which to feed.

No vaccine is available to protect against scrub typhus, so prevention is crucial. Two molecular tests involving the polymerase chain reaction have been developed to diagnose the rickettsial pathogen in humans (Jiang et al. 2004). The poster shown in Figure 23.7 advised soldiers during World War II on how to avoid the chiggers that transmit scrub typhus in Asia, and these guidelines are still helpful. Scrub typhus does not occur in the United States.

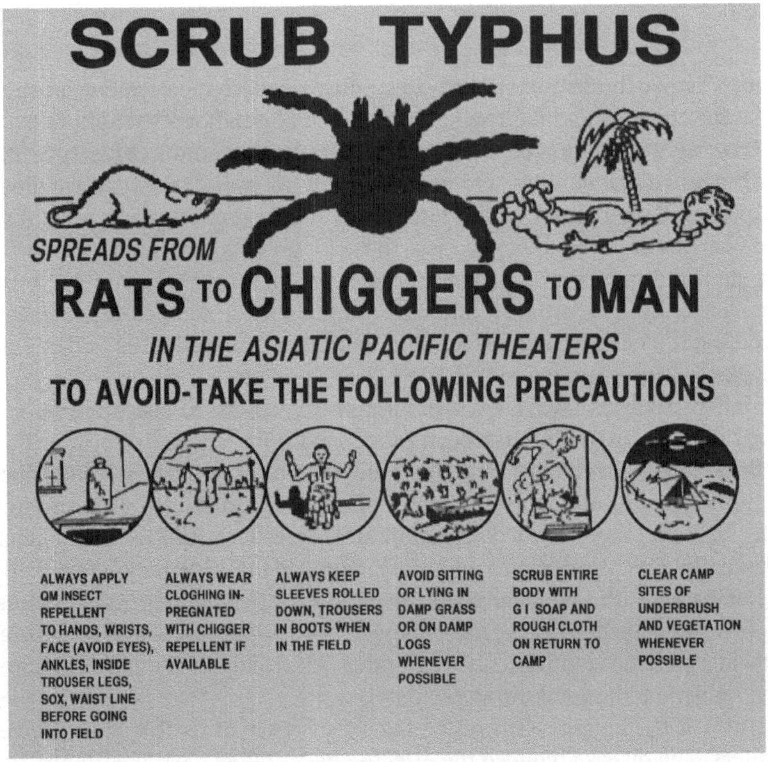

Figure 23.7 An U.S. Armed Forces poster was used during World War II in Asia to train troops to prevent scrub typhus (caused by *Rickettsia tsutsugamushi*), which is transmitted by chiggers (larval Trombiculidae) during their feeding. Rats are the reservoir hosts for the rickettsia.

23.4.2 The Spanish Moss Myth

Many people who live in the southern United States, where Spanish moss, *Tillandsia usneoides* (Bromeliacea), can be found hanging from trees, believe that it is the source of chiggers (or "red bugs"). Small children often are told not to handle Spanish moss for this reason. When Whitaker and Ruckdeschel (2010) reviewed the literature and conducted an analysis of the arthropod inhabitants of both Spanish moss and ball moss (*Tillandsia recurvata*), they discovered that, although many arthropods can be found in Spanish moss, they found no chiggers. The most common organisms found were mites (but not chiggers) and ticks, followed by Psocoptera, Collembola, Araneae, Coccidae, Thysanoptera, Formicidae, Isopoda, Diplopoda, Coleoptera, and Lepidoptera larvae. Although chiggers parasitize lizards and skinks, including those that climb trees, it is unlikely that they would persist in the moss (Whitaker and Ruckdeschel 2010). The authors concluded that, "The adult mites would have to live in the moss and deposit their eggs there for the larvae (chiggers) to be present and attach to passing vertebrates. And, if a chigger were attached to a lizard that climbed a tree and the chigger fell, it would molt into the next non-biting stage of the life cycle." So, it appears that this myth is not based on accurate biology.

23.4.3 Chiggers as Direct Pests

A few species of chiggers may be pests of domestic cats, causing dermatitis, while dogs appear less often to be attacked (Mullen and O'Connor 2002). Grazing livestock, including pigs, sheep, goats, and cattle, can be attacked by chiggers, resulting in skin irritation and, occasionally,

Table 23.3 Control and Management of Chiggers

Chiggers are accidental pests of humans, having diverse other vertebrates as their primary hosts. In most areas, chiggers are only nuisances, causing dermatitis and intense itching, which can result in secondary bacterial infections. In parts of southeast Asia, chiggers can transmit a serious rickettsial disease (scrub typhus, or tsutsugamushi fever) to humans. Cultural controls include removing debris (wood, brush piles, and other debris) from sites to reduce habitat for rodents. Other measures include:

1. Mow regularly and remove brush to create a less favorable (drier) habitat for chiggers and their wild hosts. Short grass allows sunlight to penetrate and dries out the area.
2. Do not assume that treating a well-mowed lawn with pesticides to control chiggers will be of benefit, as chiggers avoid direct sunlight and normally will not infest lawns that are well maintained.
3. Wear shirts with long sleeves and long pants made of tightly woven fabric to reduce the chances of attack; also, tuck pants into boots to avoid exposing skin.
4. Avoid walking through uncut fields, brush, or other overgrown areas, especially during chigger season (July, August, and early September in temperate climates); walk in the center of mowed trails.
5. Apply a repellent, such as DEET, to shoes, cuffs, socks, and pant legs.
6. Do not sit or lie on bare ground or on the grass; use a ground cover, preferably one that has been treated with a repellent.
7. Shower or bathe immediately after coming indoors during chigger season. Because chiggers do not settle down to feed for several hours, bathing soon after exposure removes chiggers that have not yet attached.
8. Wash clothing in very hot water to remove chiggers.

Source: Based on information from Dohany (1978) and Mohr (1947).

secondary infections. Chiggers also may attack chickens. Although chiggers are parasites of lizards and are, therefore, detrimental to them, more than 150 species of lizards in five families (Iguanidae, Chamaeleonide, Gekkonidae, Lacertidae, Scincidae) have **mite pockets** in the neck, axilla, groin, and postfemoral region that are inhabited by chiggers (Arnold 1986, Benton 1987). These special mite pockets provide a protected, warm, humid site for the chigger. The mite pockets have a specialized structure, with scales smaller than normal, and a good blood supply that allows the chiggers to feed more easily. Lizards that have mite pockets are more likely to be infested with chiggers than closely related species lacking the pockets, but the pockets are not a direct response to chigger infestations. Pockets occur in young, uninfested lizards. One hypothesis is that the pockets are "a form of damage limitation for the lizards" resulting in contained damage from which the lizard can recover rapidly (Arnold 1986). Clearly, chiggers and lizards with mite pockets have evolved an interesting relationship (see Table 23.3).

23.5 FOLLICLE MITES, *DEMODEX* (ACTINEDIDA: DEMODICIDAE)

At least 86 species of mites in the genus *Demodex* (Demodicidae) burrow into skin follicles and feed on subcutaneous secretions, particularly sebum (Desch 2009). The *Demodex* life cycle includes a larva and two nymphal stages. All stages occur within the skin follicles. Adult mites are very small (0.1 to 0.4 mm long) and are highly modified, having short legs and no body setae. Their needle-like chelicerae are used to pierce skin cells on which they feed. They can be seen only under high magnification (Figure 23.8). Species in this genus are strongly host specific, and it is likely that most vertebrates harbor one or more species of *Demodex*, with new species continuing to be found as additional hosts are examined.

Demodex mites are highly modified for living within the skin follicles of mammals (Desch and Nutting 1977). Their exoskeleton is thin, colorless, and transparent. The legs of the immatures appear to be reduced to a single segment and function only as holdfast organs during feeding. Unlike most acarines, the legs of adult *Demodex* mites have four segments, only three of which are movable. Body muscles of *Demodex* appear to be used for adjusting the hydrostatic pressure of the body. *Demodex* mites have salivary glands, but they lack a hindgut, and feeding is primarily a chemical process. The small volume of the gut lumen and the relatively large salivary glands

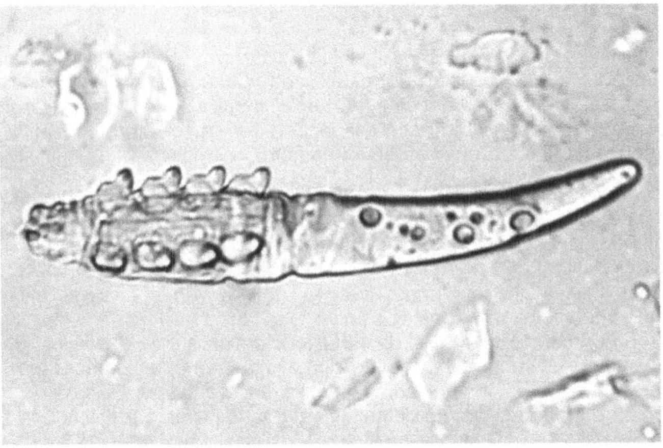

Figure 23.8 *Demodex follicularum* (Demodicidae), highly magnified, showing the reduced legs and body modified for inhabiting the pores of skin and sebaceous glands of humans. Most humans have follicle mites, and the mites normally cause no problems. (Photograph by Lyle Buss, Department of Entomology and Nematology, University of Florida, Gainesville.)

suggest that preoral digestion occurs and that liquefied food is ingested. The mites have a closed hindgut, and "waste products accumulated during the life of the mite which cannot be eliminated by diffusion through the cuticle are, apparently, stored as crystalline material in the gut cells" (Desch and Nutting 1977). These mites lack a tracheal system, so gas exchange probably occurs by diffusion through the cuticle. The life cycle of *Demodex* mites includes egg, larva, protonymph, and adult male and female. Males have an aedeagus. In some species, it is known that males are haploid and females are diploid (arrhenotoky).

Nearly all humans have two species of *Demodex* inhabiting their skin (Nutting 1976, Desch 2009). Both *D. brevis* and *D. folliculorum* are found in the skin of the forehead, nose, and other regions of the human face. *Demodex folliculorum* is found in follicles above the sebaceous glands, and *D. brevis* is found deeper in sebaceous glands (Krantz and Walter 2009). These mites do not usually cause human disease unless the individual has a suppressed immune system. In rare cases, secondary bacterial infections can occur. Mothers are thought to transfer *Demodex* mites to their babies. Apparently, 90 to 100% of all humans have follicle mites (Mullen and O'Connor 2002, Desch 2009).

Desch (2009) reported that, to date, 86 species of *Demodex* have been found in 11 of the 18 orders of placental mammals and in 3 of the 7 marsupial orders. It is likely that more species of *Demodex* will be identified. Occasionally, humans can become infested by *Demodex* species obtained from other hosts, such as pet dogs. Control of the invasive mites from other host species can be obtained by washing the skin with a mild alkaline or sulfur soap, followed by a mild sulfur lotion. Typically, infections can be resolved in a few weeks (Mullen and O'Connor 2002).

Demodex canis appears to be present on most dogs without causing disease (Figure 23.9); however, if the dog is young and has an immature or impaired immune system, *D. canis* can cause follicular or red mange, which is a very serious disease. The actual disease is caused by bacteria (*Staphylococcus*), but it is facilitated by the damage to the follicles caused by the mites. Affected dogs have a characteristic and disagreeable odor.

Demodex caprae attacks goats, especially young animals, pregnant females, and milk goats. These mites are transmitted to the young immediately after birth. *Demodex caprae* produces papules (small rounded bumps in the skin with a defined border), which eventually rupture, followed by hyperkeratinization (excessive development of keratin in the hairs lining the inside of a hair follicle), which can lead to a blocked follicle (Lebel and Nutting 1971). The mites are extruded from the

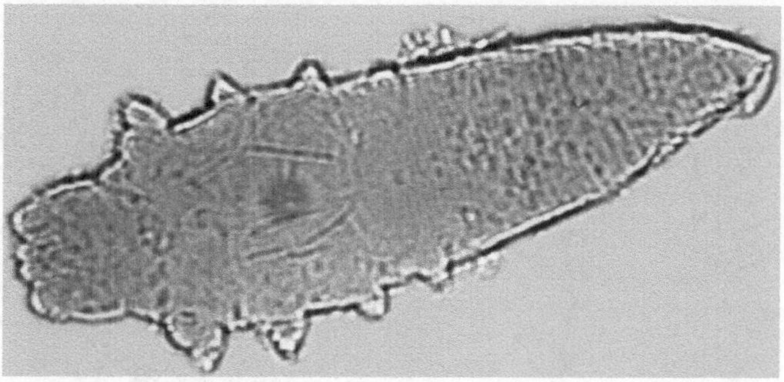

Figure 23.9 *Demodex canis* is a pest of dogs. Feeding damage from the mites allows bacterial infections with *Staphylococcus* to develop into red mange. (Photograph by J.F. Butler, Department of Entomology and Nematology, University of Florida, Gainesville.)

papule when it ruptures. Economic losses from this mite can occur. In several countries in Africa, *D. caprae* can be the most important pest of goats. A single goat may have as many as 100 papules, with a total population of approximately 5 million mites.

Demodex equi attacks horses, and ***Demodex cati*** and ***Demodex gatoi*** cause disease in cats with impaired immune systems. Cattle and sheep are attacked by one or more species of *Demodex*. ***Demodex bovis*** often is found in low numbers on cattle but is rarely a serious problem. Mites apparently are transmitted from mothers to their calves (Fisher et al. 1980); however, if mite populations increase dramatically, hair loss can occur over large areas and the hides can be damaged. Damage may be observed first on the heads and front legs of cattle. Deep skin scrapings of the damaged areas are done to diagnose *Demodex* mites. The scrapings must be examined under high magnification to see these tiny mites. Another species, ***Demodex tauri***, is found in hair follicles and sebaceous glands of the eyelids of cattle. Veterinarians should diagnose and control *Demodex* in livestock or pets. Concerns about *Demodex* species in humans should be discussed with a dermatologist.

23.6 STRAW- OR HAY-ITCH MITES, *PYEMOTES* (ACTINEDIDA: PYEMOTIDAE)

Pyemotes tritici and, occasionally, *P. ventricosus*, are ectoparasites of grain-infesting insects (Hymenoptera, Coleoptera, and Lepidoptera). These mites can be found in grain, dried beans and peas, and wheat straw, as well as hay and other dried grasses, where they feed on insect larvae and pupae. *Pyemotes* mites will bite all parts of the bodies of farm workers handling grain and hay, and can cause a severe skin eruption, resembling hives, chicken pox, or scabies (Swan 1934). These mites also are called the **hay-itch mite** or **straw-itch mite** (Figure 23.10). Humans are not a normal host for the mite, and the mites do not stay long.

Pyemotes tritici produces a mixture of venoms that immobilizes its insect host (Figure 23.11) (Tomalski et al. 1989). The potency of the venom is very high (Tomalski and Miller 1991). Each mite has approximately 34 pg of the toxin, which is about 0.1% of the mite's weight. A very tiny amount is injected into its prey, yet the toxin is able to paralyze insects 150,000 times the size of the mite. The venom is extremely toxic to insects but not acutely toxic to mammals (Tomalski et al. 1989). The more mites present on an insect, the more quickly it becomes paralyzed. When the toxin gene from the mite was cloned and inserted into a nuclear polyhedrosis virus, the genetically modified virus was able to paralyze Lepidoptera larvae and kill them more rapidly (Tomalski and Miller 1991).

The biology of *Pyemotes tritici* is one of the most unusual among the Acari, as both males and females are produced **viviparously.** They are sexually mature when they emerge from their mother (Swan 1934). Mites are most abundant during the hot summer weather, and unfed females

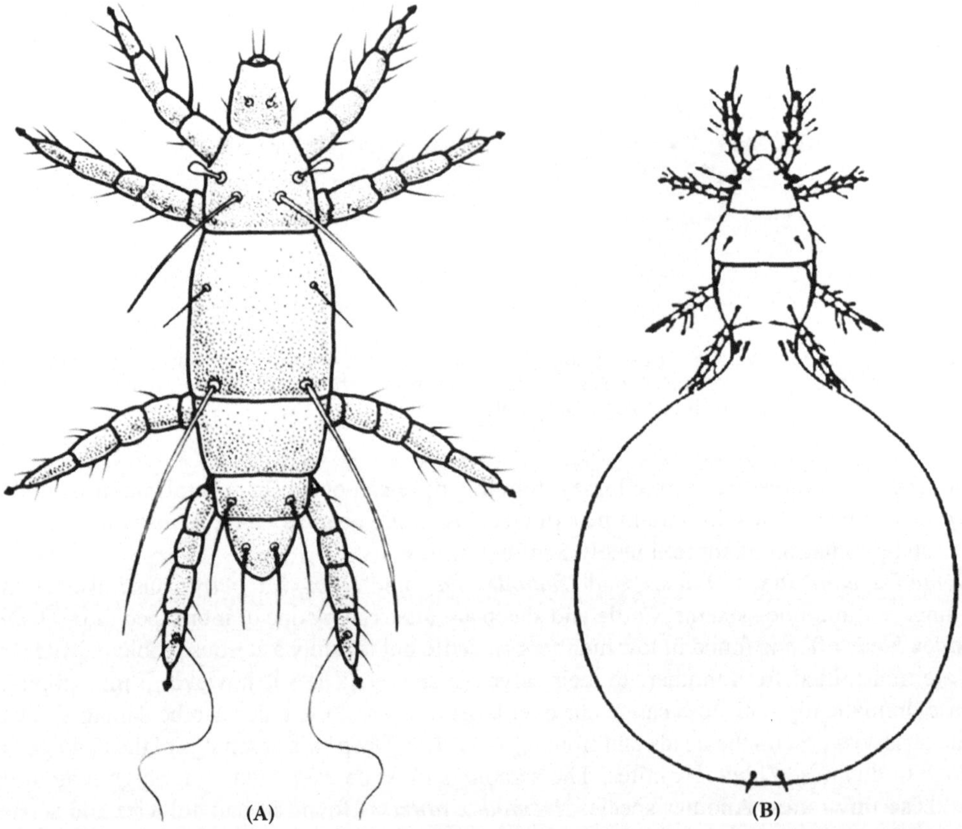

Figure 23.10 (A) Adult female of *Pyemotes ventricosus* (Pyemotidae) prior to feeding and becoming gravid; also known as the hay-itch or straw-itch mite. (B) Adult female of *P. ventricosus* with greatly distended opisthosoma containing many progeny (physogastry). (Adapted from Swan, D.C., *J. Agric. S. Aust.*, 37, 1289–1299, 1934.)

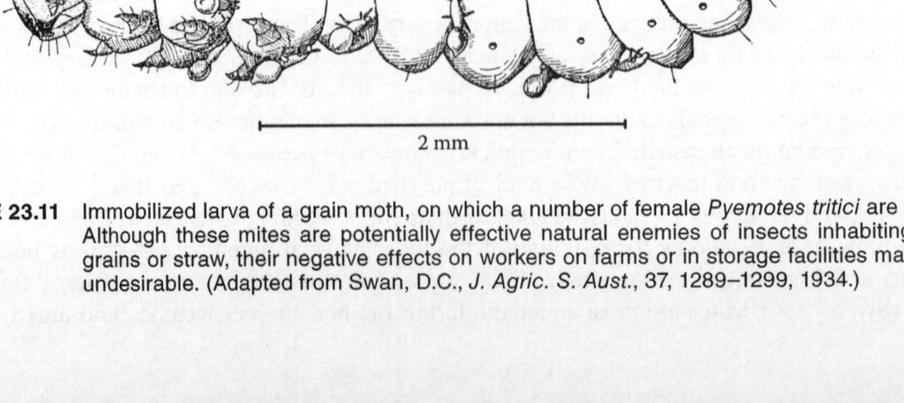

FIGURE 23.11 Immobilized larva of a grain moth, on which a number of female *Pyemotes tritici* are feeding. Although these mites are potentially effective natural enemies of insects inhabiting stored grains or straw, their negative effects on workers on farms or in storage facilities make them undesirable. (Adapted from Swan, D.C., *J. Agric. S. Aust.*, 37, 1289–1299, 1934.)

die rapidly if no hosts are available, especially if the temperature is high and the relative humidity is low. Initially, the female takes short meals and moves about on its insect host. After feeding, however, her body becomes very swollen (**physogastry**), and the female can no longer move. Eggs inside the female develop to the adult stage (see Figure 23.10). A single female may produce 200 to 300 progeny if the food supply is abundant, but if multiple females attack a single insect host they may produce fewer progeny (see Figure 23.11). A meal is a prerequisite for the development of progeny. The male spends his life wandering over the distended body of the physogastric female (Figure 23.12). The sex ratio is skewed, with only about 4% of the progeny males.

The heteromorphic males are born first and feed very little. When the newly developed daughter is about to exit her mother's body, the male (probably her brother) moves to the genital opening. With his powerful hind pair of legs (see Figure 23.12), he seizes the young female and drags her through the opening. They then copulate and the daughters begin the cycle again, with the daughters searching for a new host. Males do not assist at the birth of males, and the young females can emerge without help. If females do not mate, the females produce all male progeny (arrhenotoky). If multiple females attack an insect host, the physogastric females resemble a bunch of grapes (Figure 23.13).

Although these mites are natural enemies of insects in stored hay, straw, or grains, their negative effects on humans on the farm or in food-storage facilities make them pests. Control of *Pyemotes* mites involves elimination of the mite's host insects. Sanitation of grain storage areas is important. Stored products can be fumigated to disinfest them. Storing hay crops in different regions of the field from year to year may reduce the likelihood of infestation by *Pyemotes* mites.

Related species of *Pyemotes*, such as *P. herfsi*, have been associated with human bites in wooded areas (Broce et al. 2006). *Pyemotes herfsi* feeds on midge larvae (*Contarinia* sp., Cecidomyiidae) that live in leaf galls on oak trees or on larvae of the pine tip moth (*Rhyacionia buoliana*), as well as a variety of other insects. The biology of *P. herfsi* is similar to that of *P. tritici*, and the mites apparently are dispersed by wind.

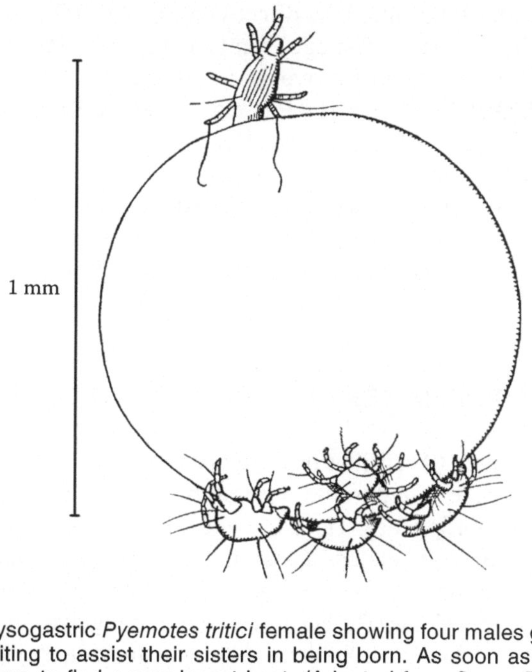

Figure 23.12 A gravid, physogastric *Pyemotes tritici* female showing four males grouped around her genital opening, waiting to assist their sisters in being born. As soon as the females emerge, they mate and leave to find a new insect host. (Adapted from Swan, D.C., *J. Agric. S. Aust.*, 37, 1289–1299, 1934.)

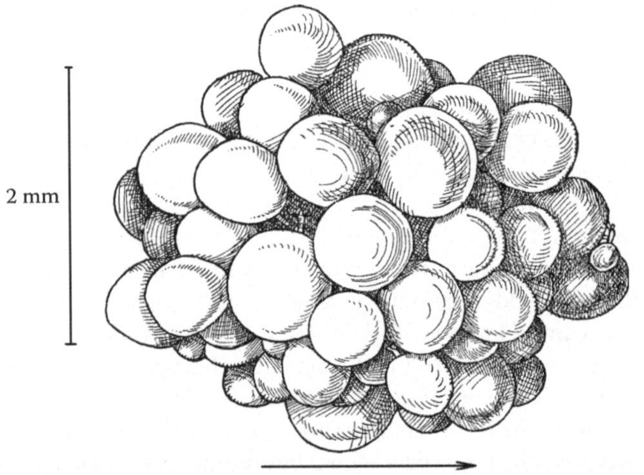

Figure 23.13 After a group of *Pyemotes tritici* females have fed on a grain moth larva, removing much of the contents from the larva and killing it, the physogastric females resemble a bunch of grapes. (Adapted from Swan, D.C., *J. Agric. S. Aust.*, 37, 1289–1299, 1934.)

23.7 FUR MITES, *CHEYLETIELLA* (ACTINEDIDA: CHEYLETIELLIDAE)

Cheyletiella blakei, *C. parasitivorax*, and *C. yasguri* are parasites of cats, rabbits, and dogs, respectively, and are known as fur mites. These external parasites do not burrow into the skin. They feed on lymph by piercing the epidermis with their chelicerae. These mites generally cause a mild dermatitis but occasionally can cause severe dandruff-like dermatitis. They also will attack humans. The mites are easily transmitted by direct contact. All life stages of fur mites occur on the hosts, with the eggs glued to hairs. The eggs can be eaten when the animals groom themselves and passed in the feces. Fur mites can survive up to 10 days off the host and can be picked up from animal bedding, blankets, and carpets. Fur mites often are phoretic on cat and dog fleas. ***Cheyletiella blakei*** is found on the faces of cats. Heavy infestations can result in crusty patches, loss of hair, and intense itching and scratching. ***Cheyletiella yasguri*** is found on dogs and is common in Europe and North America. It causes only occasional problems, perhaps because flea and tick control measures also control this mite. Young dogs are more severely affected than older dogs. ***Cheyletiella parasitivorax*** is a common parasite of domestic rabbits around the world, but is rarely found on wild rabbits.

23.8 ACARIDID MITES AS PARASITES OR SCAVENGERS

A number of acaridid mites (Acaridida or Astigmata) live on vertebrates, either as true parasites or as scavengers. Some are found in the respiratory tracts of their hosts, and others are skin parasites (Sarcoptidae and Psoroptidae). The scavengers feed on hair and feather fragments, skin detritus, and secretions of dermal glands. These mites may have reduced legs and sucker-like structures on the pretarsi. The first two pair of legs of the **Listrophoridae** (fur mites) are modified for clasping hairs. The **Analgesidae** (feather mites) have triangular projections on two or three pair of legs to aid in their attachment to feathers. The **Sarcoptidae** and **Psoroptidae** never have a hypopial deutonymphal stage.

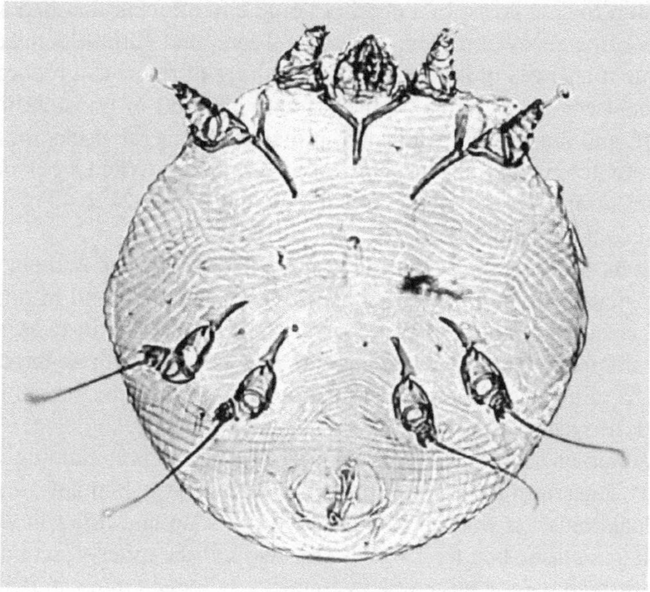

Figure 23.14 *Sarcoptes scabei* (Acaridida: Sarcoptidae) causes human scabies. The females burrow through the skin, causing itching. (Photograph courtesy of the Centers for Disease Control and Prevention, Atlanta, GA.)

23.8.1 Mange Mites (*Sarcoptes*)

Sarcoptes mites (Acaridida: Sarcoptidae), also known as scabies, mange, or itch mites, infest a variety of mammals including humans (Figure 23.14). Each host, including humans, has its own strain or variety. The sarcoptic mites of animals can infest the skin of humans and cause itching and rash but are considered self-limiting, although the infestations can last several months and produce intense itching and lesions around each mite burrow. The life cycle consists of egg, larva, protonymph, and tritonymph (Arlian 1989). *Saroptes scabiei* is known to attack mammals in 17 families and 7 orders, with the mites from different hosts showing few morphological differences. As a result, it is difficult to resolve whether the different hosts have different mite species or varieties; however, the transfer of scabies mites from dogs to mice, rats, guinea pigs, pigs, cattle, goats, and sheep was unsuccessful (Arlian 1989). The reasons for the apparent host specificity of *S. scabiei* are unknown.

The *Sarcoptes* female burrows into the outer layer of the skin (epidermis), where she feeds on tissue fluids. She lays eggs that she cements to the floor of the burrow. The females deposit up to 3 eggs per day for 8 weeks, producing about 150 eggs. These eggs hatch in 3 to 4 days, and the newly hatched larvae move from the burrows onto the surface of the skin and molt to the first nymphal stage. While on the skin, these larval and nymphal mites can be dislodged and fall off the host. The mites will then move toward host odors and infest a new host if environmental conditions are adequate (moderate temperatures and high relative humidity that reduces the likelihood of desiccation). Arlian (1989) reported that scabies mites from a dog could survive 1 to 9 days at 15 to 25°C at 25 to 97% relative humidity. The ability to penetrate the skin of a new host can be reduced if the mite is held under unfavorable conditions.

In humans, live scabies mites often are found in the webs of the fingers and around the wrists, arms, or elbows. Other infestation sites include the soles and insteps of the feet. Scabies mites from humans can be recovered from dust vacuumed from floors, chairs, and mattresses and can be the source for transmission to new hosts. In a nursing home environment that had adequate sanitation, live mites were less often recovered from bedding, floors, and furniture than in private homes, suggesting that frequent changes of bed linens and washing of floors can reduce mite transmission (Arlian 1989). Personal contact is probably the primary method of transmission of scabies mites among patients, staff, and family. Once infested, it may take several weeks for mite populations to increase and for severe itching and lesions to develop in humans. When a person has been infected and recovered, reinfestation can result in itching and lesions within 24 to 48 hours, perhaps because the person was sensitized by the previous infestation.

Scabies infestations affect people from all socioeconomic levels, without regard to age, sex, race, or personal hygiene (Arlian 1989). Clusters of cases can be found in nursing homes, child-care centers, and other institutions. Scabies are transferred by direct skin-to-skin contact, including sexual contact. Transfer from clothing or bed covers can occur only if these have been contaminated immediately before handling. Scabies can be transmitted as long as the person is infected, until the mites and eggs are destroyed.

The treatment for human scabies involves the use of skin lotions containing pesticides, available through a physician's prescription. In addition, all sheets, pillows, blankets, and clothing worn by the infested individuals must be washed in hot water and soap and dried in a dryer. Drycleaning clothing or sealing it in a plastic bag for 2 weeks can also kill the mites. Vacuuming rugs and furniture with which an infested person has come in contact can reduce mites in the home.

Diagnosis of scabies can be made by the identification of mite burrows; however, the burrow is difficult to see without magnification, and burrows usually are seen only when the burrow becomes inflamed. One method for visualizing the burrows involves placing a drop of ink on the suspected burrow. After a few minutes, the ink will seep into the burrow and become visible.

Symptoms of *Sarcoptes scabei* in lightly infected hosts may be minor, but large infestations may result in sarcoptic mange in domestic animals that can be severe. Extensive crust formations may be produced, from which blood and serum exude (Patrick 1999). The skin may become folded and leathery. The mites consume host lymph and lysed tissue, resulting in anemia and weight loss, perhaps due to toxic secretions from the mite. Secondary bacterial infections may occur. If not treated, **sarcoptic mange** can even lead to the death of the host.

In dogs, sarcoptic lesions occur on the head or around the ears, elbows, or axillary and inguinal regions. Lesions develop, and the skin becomes crusty. The skin thickens, secondary bacterial infections develop, allergic reactions flare up, and self-mutilation may occur due to the itching.

In pigs, lesions initially occur on the head, especially the ears, nose, and eyes, and later on the neck, shoulders, and back. The pigs scratch and rub intensely. Inflammation, scurfy skin, and secondary bacterial infections occur, resulting in reduced growth rate and downgraded carcasses. The mite spreads from pig to pig by close skin contact or contact with contaminated surfaces. Scrapings taken from suspicious lesions on the skin and inside the ears with a teaspoon can be spread onto a piece of black paper and left for 10 minutes. Mange mites are rounded in shape and about 0.5 mm in length; they may be just visible to the naked eye. To positively identify the mites, the scrapings should be sent to a diagnostic laboratory. An enzyme-linked immunosorbent assay (ELISA) blood test is available. Check with a veterinarian regarding the appropriate treatment. Cattle with sarcoptic mite infestations are required by law in the United States to be quarantined and treated by federal or state veterinarians (Patrick 1999).

23.8.2 Scaly-Leg or Beak Mites, *Knemidocoptes* (Acaridida: Knemidocoptidae)

Scaly-leg or beak mites (***Knemidocoptes mutans***) are parasites in chickens, turkeys, and pheasants (Butcher and Beck 1996). Other species of *Knemidocoptes* are parasites of wild birds. ***Knemidocoptes laevis*** is the depluming mite of chickens, geese, pheasants, and pigeons throughout

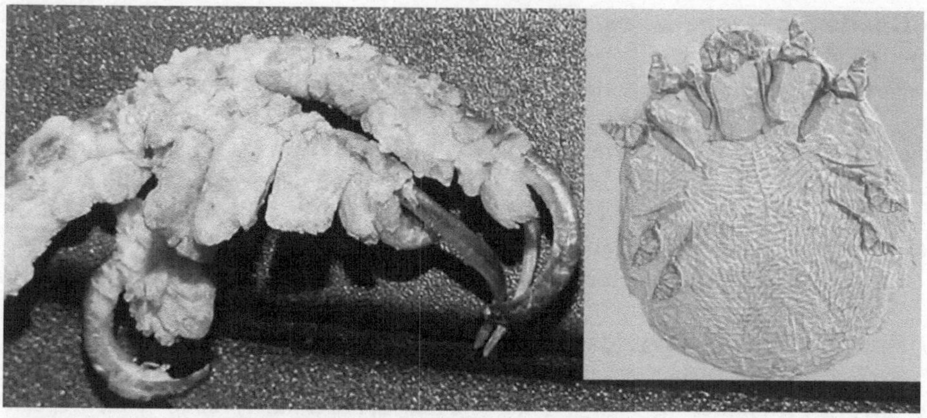

Figure 23.15 A *Knemidocoptes* species causing enlargement, displacement, and loosening of the scales on the foot of a wild bird (left). The *Knemidocoptes* mite found under these skin scales is on the right. (Photographs by Heather Proctor, University of Alberta, Edmonton, Canada.)

North America. Adults are about 0.4 to 0.5 mm in length. These mites burrow in the epidermis of the feet and legs, which causes displacement and loosening of the scales on the feet and legs, as well as inflammation with an exudate that hardens (Figure 23.15). Hyperkeratosis may develop such that legs become thickened and deformed, and lameness may occur. To monitor for these mites, remove a loose skin scale from the leg and examine the underside for mites. ***Knemidocoptes gallinae*** is found in the epidermis of chickens, along the base of the feathers on the back, top of the wing, breast, thighs, and vent. Infestation results in inflammation so feathers break off easily or are pulled out by the birds. Pull a few feathers at the edge of a lesion and examine them for mites. Consult a veterinarian for treatment, and segregate infested birds.

23.8.3 Psoroptic Mange Mites, *Psoroptes* (Acaridida: Psoroptidae)

Psoroptes equi, ***P. ovis***, and ***P. bovis*** cause **psoroptic mange** of sheep, cattle, wild ruminants, goats, and rabbits. Adults are about 0.25 to 0.4 mm long. Mites spend their entire life cycle on the host, where the larvae, nymphs, and adults feed. The life cycle takes about 12 days. The mites are spread when new hosts contact scabs left on fences, handling chutes, and trucks where an infected animal has rubbed or by direct contact. The highly contagious psoroptic mange is caused when mites pierce the skin with their chelicerae at the base of hairs or wool, suck lymph, and cause inflammation. Feeding causes small vesicles filled with lymph to appear, which gradually coalesce.

When bacterial infections develop, the vesicles become pustules. *Psoroptes* mites migrate from these areas to adjacent healthy skin, spreading the disease. The infestation causes intense irritation, and large areas may be denuded of hair or wool. Psoroptic mange can spread to the entire body. Infection results in reduced weight gain and damage to hides. Severely infested animals may even die. If psoroptic mange is detected, infested cattle are required by law in the United States to be quarantined and treated. In horses, lesions begin at the base of the long hairs of the mane, tail, and forelock. In rabbits, this mite causes formation of a crust in the ear and a brown discharge in the external ear canal. Diagnose by finding the mites in skin scrapings from the edges of active lesions. *Psoroptes ovis* has been eradicated from sheep in the United States, and psoroptic mange (*P. equi*) on horses is rare in the United States.

23.8.4 Chorioptic Mange Mites, *Chorioptes* (Acaridida: Chorioptidae)

Chorioptes bovis causes a scab-like disease (**chorioptic mange**, foot mange mite, itchy leg mite) in horses, cows, sheep, goats, rabbits, and llamas. Chorioptic mange is not as severe as psoroptic mange and usually is restricted to the legs. *Chorioptes bovis* feeds on dead epidermal tissues, not on living tissues. The scab-like lesions produced usually are not severe and do not spread rapidly or extensively; however, infested animals may stomp, scratch, rub, or kick, especially at night. Symptoms are more common in fall and winter. The mite can be identified in skin scrapings. In some states, *Chorioptes* mites must be reported and treated; in some cases, quarantines are required.

23.8.5 Feather Mites (Analgoidea, Pterolichoidea, and Freyanoidea)

The three superfamilies of feather mites contain approximately 2000 named species. The Analgoidea, Pterolichoidea, and Freyanoidea have colonized all parts of the integument of birds, with some feeding inside or on feathers. Others are on the skin as parasites, while others consume only feather oils and cause no damage (Proctor 2003). Feather mites are of relatively little economic importance, although heavy infestations cause feather picking in poultry. Two genera, *Syringophilus* and *Dermoglyphus*, can be found inside the quills of poultry and cage birds. All life stages are found inside the quill. Adults move to new feathers and spread the infestation.

23.9 ENDOPARASITES OF LIVESTOCK

Mites have invaded the air sacs and other parts of the respiratory tract, as well as the peritoneal and thoracic cavities; for example, the fowl cyst mite, ***Laminosioptes cysticola*** (Laminosioptidae), is a widespread parasite of domestic poultry and pigeons. The mites invade subcutaneous tissue and cause the formation of small nodules that become calcified when the mites die. Large populations can cause death. Ear mites, ***Otodectes cynotis*** (Psoroptidae), can be found in dogs, cats, and ferrets. Infested dogs that have mites within the ear canal may rub their ears and shake their head. The infestations are usually found in both ears and can be treated by a veterinarian.

SELECTED REFERENCES

Arlian, L.G. 1989. Biology, host relations, and epidemiology of *Sarcoptes scabiei*. *Annu. Rev. Entomol.* 34:139–161.

Arnold, E.N. 1986. Mite pockets of lizards, a possible means of reducing damage by ectoparasites. *Biol. J. Linnean Soc.* 29(1):1–21.

Axtell, R.C. and J.J. Arends. 1990. Ecology and management of arthropod pests of poultry. *Annu. Rev. Entomol.* 35:101–126.

Baker, A.S. 1999. *The Mites and Ticks of Domestic Animals: An Identification Guide and Information Source*. London: The Stationery Office.

Baker, E.W., T.M. Evans, D.J. Gould, W.B. Hull, and H.L. Keegan. 1956. *A Manual of Parasitic Mites of Medical or Economic Importance*. New York: National Pest Control Association.

Benton, M.J. 1987. The mite pockets of lizards. *Nature* 325:391–392.

Brennan, J.M. and M.L. Goff. 1977. Keys to the genera of chiggers of the Western Hemisphere (Acarina: Trombiculidae). *J. Parasitol.* 63:554–566.

Broce, A.B., L. Zurek, J.A. Kalisch, R. Brown, D.L. Keith, D. Gordon, J. Goedeke, C. Welbourn, J. Moser, R. Ochoa, E. Azziz-Baumgartner, F. Yip, and J. Weber. 2006. *Pyemotes herfsi* (Acari: Pyemotidae), a mite new to North America as the cause of bite outbreaks. *J. Med. Entomol.* 43:610–613.

Butcher, G.D. and C. Beck. 1996. *Knemidocoptic Mange in Pet Birds: Scaly Face and Scaly Leg Disease*, Fact Sheet VM 59. Gainesville: Cooperative Extension Service, University of Florida Institute of Food and Agricultural Sciences (http://www.cheboygancoop.com/animalscience/pets/vm02200.pdf).

Chauve, C. 1998. The poultry red mite *Dermanyssus gallinae* (De Geer, 1778): Current situation and future prospects for control. *Vet. Parasitol.* 79:239–245.

Chen, B.L. and B.A. Mullens. 2008. Temperature and humidity effects on off-host survival of the northern fowl mite (Acari: Macronyssidae) and the chicken body louse (Phthiraptera: Menoponidae). *J. Econ. Entomol.* 101:637–646.

Denmark, H.A. and H.L. Cromroy. 1987. *Featured Creatures: Tropical Fowl Mite*. Gainesville: Department of Entomology and Nematology, University of Florida Institute of Food and Agricultural Sciences (http://entnemdept.ifas.ufl.edu/creatures/livestock/tropical_fowl_mite.htm).

Desch, C.E. 2009. Human hair follicle mites and forensic acarology. *Exp. Appl. Acarol.* 49:143–146.

Desch, C.E. and W.B. Nutting. 1977. Morphology and functional anatomy of *Demodex folliculorum* (Simon) of man. *Acarologia* 19:422–462.

Dohany, A.L. 1978. Vector transmission of scrub typhus and control of vector mites. *Malaysian J. Pathol.* 1:15–16.

Durand, A.M., S. Kuartei, I. Togamae, M. Sengebau, L. Demma, W. Nicholson, and M. O'Leary. 2004. Scrub typhus in the Republic of Palau, Micronesia. *Emerging Infect. Dis.* 10:1838–1840.

Eamsobhana, P., A. Yoolek, W. Kongkaew, K. Lerdthusnee, N. Khlaimanee, A. Parsartvit, N. Malainual, and H.S. Yong. 2009. Laboratory evaluation of aromatic essential oils from thirteen plant species as candidate repellents against *Leptotrombidium* chiggers (Acari: Trombiculidae), the vector of scrub typhus. *Exp. Appl. Acarol.* 47:257–262.

Eldridge, B.F. and J.D. Edman (eds.) 2004. *Medical Entomology: A Textbook on Public Health and Veterinary Problems Caused by Arthropods*. Dordrecht: Kluwer Academic.

Evans, G.O., J.G. Sheals, and D. MacFarlane. 1961. *The Terrestrial Acari of the British Isles: An Introduction to their Morphology, Biology, and Classification.* London: British Museum of Natural History.

Fisher, W.F., R.W. Miller, and A.L. Everett. 1980. Natural transmission of *Demodex bovis* Stiles to dairy calves. *Vet. Parasitol.* 7:233–241.

Harris, M.A., M.J. Brewer, and J.A. Meyer. 2000. Presence–absence sequential sampling plan for Northern fowl mite, *Ornithonyssus sylviarum* (Acari: Macronyssidae), on caged-layer hens. *J. Econ. Entomol.* 93:544–549.

Hase, T., L.W. Roberts, P.K. Hildebrandt, and D.C. Cavanough. 1978. Stylostome formation by *Leptotrombidium* mites (Acari: Trombiculidae). *J. Parasitol.* 64:712–718.

Hinkle, N.C. and L.A. Hickle. 1999. California caged layer pest management evaluation. *J. Appl. Poultry Res.* 8:327–338.

Hogsette, J.A., J.F. Butler, W.V.Miller, and R.D. Hall. 1991. *Annotated Bibliography of the Northern Fowl Mite,* Ornithonyssus sylvarum *(Canestrini and Fanzago) (Acari: Macronyssidae)*, Misc. Publication No. 76. Lanham, MD: Entomological Society of America.

Jiang, J., T.C. Chan, J.J. Temenak, G.A. Dasch, W.M. Ching, and A.L. Richards. 2004. Development of a quantitative real-time polymerase chain reaction assay specific for *Orientia tsutsugamushi*. *Am. J. Trop. Med. Hyg.* 70:351–356.

Kaufman, P.E., P.G. Koehler, and J.F. Butler. 2009. *External Parasites of Poultry*, UF IFAS Publication No. ENY-290. Gainesville: Cooperative Extension Service, University of Florida Institute of Food and Agricultural Sciences (http://edis.ifas.ufl.edu/ig140).

Kells, S.A. and G.A. Surgeoner. 1996. Dispersion of northern fowl mites, *Ornithonyssus sylviarum*, between poultry facilities via infested eggs from layer and breeder flocks. *J. Agric. Entomol.* 13:265–274.

Kells, S.A. and G.A. Surgeoner. 1997. Sources of northern fowl mite (*Ornithonyssus sylviarum*) infestation in Ontario egg production facilities. *Appl. Poultry Sci.* 6:221–228.

Kilpinen, O. and T. Steenberg. 2009. Inert dusts and their effects on the poultry red mite (*Dermanyssus gallinae*). *Exp. Appl. Acarol.* 48:51–62.

Knee, W. and H. Proctor. 2007. Host records for *Ornithonyssus sylviarum* (Mesostigmata: Macronyssidae) from birds of North America (Canada, United States and Mexico). *J. Med. Entomol.* 44:709–713.

Krantz, G.W. and D.E. Walter (eds.) 2009. *A Manual of Acarology*, 3rd ed. Lubbock: Texas Tech University Press.

Lebel, R.R. and W.B. Nutting. 1971. Population dynamics of a parasitic mite *Demodex caprae* (Trombidiformes: Demodicidae) In: *Proceedings of the Third International Congress of Acarology, Prague* (pp. 517–521). The Hague: Dr. W. Junk Publishers.

Lesna, I., P. Wolfs, F. Faraji, L. Roy, J. Komdeur, and M.W. Sabelis. 2009. Candidate predators for biological control of the poultry red mite *Dermanyssus gallinae*. *Exp. Appl. Acarol.* 48:63–70.

Lipovsky, L.J. 1954. Studies of the food habits of postlarval stages of chiggers (Acarina, Trombiculidae). *Univ. Kansas Sci. Bull.* 36:943–958.

McCrea, B., J.S. Jeffrey, R.A. Ernst, and A.C. Gerry. 2005. *Common Lice and Mites of Poultry: Identification and Treatment*, UC ANR Publication No. 8162. Oakland: University of California, Division of Agriculture and Natural Resources (http://ucanr.org/freepubs/docs/8162.pdf).

Mohr, C.O. 1947. Notes on chiggers, rats and habitats on New Guinea and Luzon. *Ecology* 28:194–199.

Mul, M.F. and C.J.M. Koenraadt. 2009. Preventing introduction and spread of *Dermanyssus gallinae* in poultry facilities using the HACCP method. *Exp. Appl. Acarol.* 48:167–181.

Mullen, G.R. and B.M. O'Connor. 2002. Mites (Acari). In: G. Mullen and L. Durden (eds.), *Medical and Veterinary Entomology* (pp. 449–516). Boston: Academic Press.

Mullens, B.A., N.C. Hinkle, and C.E. Szijj. 2000. Monitoring northern fowl mites (Acari: Macronyssidae) in caged laying hens: Feasibility of an egg-based sampling system. *J. Econ. Entomol.* 93:1045–1054.

Mullens, B.A., N.C. Hinkle, L.J. Robinson, and C.E. Szijj. 2001. Dispersal of northern fowl mites, *Ornithonyssus sylviarum*, among hens in an experimental poultry house. *J. Appl. Poultry Res.* 10:60–64.

Mullens, B.A., D.R. Kuney, N.C. Hinkle, and C.E. Szijj. 2004. Producer attitudes and control practices for northern fowl mites in southern California. *J. Appl. Poultry Res.* 13:488–492.

Nordenfors, H., J. Hoglund, and A. Uggla. 1999. Effects of temperature and humidity on oviposition, molting and longevity of *Dermanyssus gallinae* (Acari: Dermanyssidae). *J. Med. Entomol.* 36:68–72.

Nutting, W.B. 1976. Hair follicle mites (Acari: Demodicidae) of man. *Intern. J. Dermatol.* 15:79–98.

Patrick, D.C. 1999. *Beef Cattle Handbook: Cattle Scabies*, BCH-3820. Madison: University of Wisconsin Cooperative Extension, in cooperation with the Extension Beef Cattle Resource Committee and its member states (http://www.iowabeefcenter.org/Beef%20Cattle%20Handbook/Scabies.pdf).

Phasomkusolsil, S., P. Tanskul, S. Rtatham, P. Watcharapichat, D. Phulsuksombati, S.P. Frances, K. Lerdthusnee, and K.J. Linthicum. 2009. Transstadial and transovarial transmission of *Orientia tsutsugamushi* in *Leptotrombidium imphalum* and *Leptotrombidium chiangraiensis* (Acari: Trombiculidae). *J. Med. Entomol.* 46:1442–1445.

Proctor, H.C. 2003. Feather mites (Acari: Astigmata): Ecology, behavior and evolution. *Annu. Rev. Entomol.* 48:185–209.

Sasa, M. 1961. Biology of chiggers. *Annu. Rev. Entomol.* 6:221–244.

Shatrov, A.B. 2009. Stylostome formation in trombiculid mites (Acariformes: Trombiculidae). *Exp. Appl. Acarol.* 49:261–280.

Sudakin, D.L. and W.R. Trevathan. 2003. DEET: A review and update of safety and risk in the general population. *J. Toxicol. Clin. Toxicol.* 41:831–829.

Swan, D.C. 1934. The hay itch mite, *Pediculoides ventricosis* (Newport) (Acarina, Pediculoididae) in South Australia. *J. Agric. S. Aust.* 37:1289–1299.

Tomalski, M.D. and L.K. Miller. 1991. Insect paralysis by baculovirus-mediated expression of a mite neurotoxin gene. *Nature* 352:82–85.

Tomalski, M.D., R. Kutney, W.A. Bruce, M.R. Brown, M.S. Blum, and J. Travis. 1989. Purification and characterization of insect toxins derived from the mite, *Pyemotes tritici*. *Toxicon* 27:1151–1167.

Varma, R.N. 1969. Prevalence of *Leptotrombidium deliense*, the scrub typhus vector, in the Eastern Himalayas. *Nature* 222:984–985.

Weisbroth, S.H. 1960. The differentiation of *Dermanyssus gallinae* from *Ornithonyssus sylviarum*. *Avian Dis.* 4:133–137.

Whitaker, J.O. and C. Ruckdeschel. 2010. Spanish moss, the unfinished chigger story. *Southeast. Nat.* 9:85–94.

Williams, R.W. 1944. A check list of the mite vectors and animal reservoirs of tsutsugamushi disease. *Am. J. Trop. Med.* 24:355–357.

Williams, R.E. 2009. *Veterinary Entomology: Livestock and Companion Animals*. Boca Raton, FL: CRC Press.

Williams, R.E., R.D. Hall, A.B. Broce, and P.J. Scholl (eds.) 1985. *Livestock Entomology*. New York: John Wiley & Sons.

Wright, H.W., K. Bartley, A.J. Nisbit, R.M. McDevigtt, N.H.C. Sparks, S. Brocklehurst, and J.F. Huntley. 2009. The testing of antibodies raised against poultry red mite antigens in an *in vitro* feeding assay; preliminary screen for vaccine candidates. *Exp. Appl. Acarol.* 48:81–91.

Yazwinski, T.A., C.A. Tucker, J. Robins, J. Powell, M. Phillips, Z. Johnson, D. Clark, and R. Wolfenden. 2005. Effectiveness of various acaricides in the treatment of naturally occurring *Ornithonyssus sylviarum* (northern fowl mite) infestations of chickens. *J. Appl. Poultry Res.* 14:265–268.

Zenner, L., G. Bon, C. Chauve, C. Nemoz, and S. Lubac. 2009. Monitoring *Dermanyssus gallinae* in free-range poultry farms. *Exp. Appl. Acarol.* 48:157–166.

Zhang, Z.-Q. 1998. Biology and ecology of trombidiid mites (Acari: Trombidioidea). *Exp. Appl. Acarol.* 22:129–155.

PART VIII

Pest Mites of Stored Products and Households

Many types of mites (and insects) share our homes and our stored foods. Some arthropods are consumers, feeding directly on the stored foods, whereas others feed on the detritus, fungi, yeasts, and other organisms growing on and around the stored foods. Some mites prey on or parasitize the pest mites and other small arthropods. The stored-product mites probably originally were associated with the nests of vertebrates (birds, mammals) and have adapted to the rich food supplies that we provide in our homes and food-storage facilities. The mites that are found in foods are discussed in Chapter 24, and the mites known as dust mites, which inhabit our beds, carpets, and other household items, are discussed in Chapter 25. The dust mites have an impact on human health due to their ability to cause allergic responses.

CHAPTER 24

Post-Harvest Pest Mites

24.1 ACARINE PESTS OF STORED FOODS

Pests of stored foods include mites, insects (especially beetles and moths), vertebrates such as rats and mice, and fungi (Koehler 1991). Stored-product insects alone can cause losses of up to 20% or more of foods produced (Heaps 2006, Phillips and Throne 2010). Hagstrum and Subramanyam (2009) noted that more than 1660 insect species in 120 families have been reported associated with stored products, although almost 470 of these insects are natural enemies of the pests.

Mites can be especially serious pests of stored food products in environments with a high relative humidity. A variety of acarid mites (Acaridida or Astigmata) in the families Acaridae, Glycyphagidae, Chortoglyphidae, Carpoglyphidae, Histiostomidae, and Pyroglyphidae can be found in agricultural commodities such as grain, dried fruits, nuts, cheeses, spices, cereals, flower bulbs, and pet foods in storage (Krantz 1971, Hughes 1976). In addition, mite families in the Prostigmata (or Actinedida) or Mesostigmata (or Gamasida) can be found as predators and fungal or detritus feeders (Hughes 1976). Thus, stored products are small ecosystems with diverse arthropods serving as consumers of grains and fungi, as parasites, and as predators. Some of these arthropod species are cosmopolitan, probably because they have been transported around the world in food stores, and some are more limited in their distribution.

24.1.1 The Acaridae and Glycyphagidae

Two families, the Acaridae and Glycyphagidae, are especially common in stored foods (Hughes 1976). The life cycles of the Acaridae and the Glycyphagidae include egg, larva, protonymph, tritonymph, and adult males and females. A hypopus (deutonymphal stage) is produced if conditions are unfavorable for continued development (Figure 24.1 and Figure 24.2). Hypopi may be active or immobile and may be phoretic on insects that are present in the stored grains. Mites in stored products often are similar in species composition to the mites found in the nests of birds and small mammals, where the mites feed on fungi, remains of food, excrement, and other detritus (Evans et al. 1961). One common stored-food mite, *Glycyphagus domesticus*, is commonly found in nests, bat roosts, beehives, and human houses. Some mites are able to feed directly on the stored foods, but others may be feeding on the molds that attack these foods (Hughes 1976). Stored grains provide a suitable habitat for many types of fungi, and the mites may not only feed on the fungi or their spores but also transport the spores to new areas. *Tyrophagus putrescentiae*, a common stored-grain mite, can carry viable spores in its digestive tract as well as on the outside of its body, which aids in the distribution of these fungi. A number of grain mites, such as *T. putrescentiae*, *Acarus siro*, and *A. gracilis*, feed and reproduce on a wide variety of fungi, and it is thought that the presence of these fungi can enhance their reproduction (Hughes 1976).

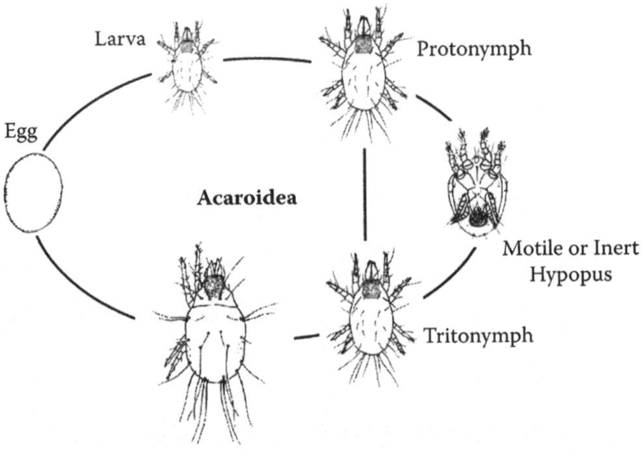

Figure 24.1 Life cycle of the Acaroidea. Under favorable conditions, the protonymph molts to a tritonymph; under unfavorable conditions, a hypopus (motile or inert deutonymph) is produced. Hypopi lack mouthparts and have a thickened exoskeleton. They can survive low relative humidities and are able to survive several months without feeding. Some will become phoretic on insects. When the hypopi arrive in a new environment, they molt and will begin to feed. (Adapted from Colloff, M.J., *Dust Mites*, CSIRO Publishing, Collingwood, Australia, and Springer, Dordrecht, 2009. With permission from CSIRO, Australia.)

24.1.2 Mites Other than the Acaridida

Many types of mites are found in stored food products (Hughes 1976). Occasionally, cryptostigmatid (Oribatida) mites are found in houses and stored foods as strays. The Prostigmata (or Actinedida) contains a number of mites that are found in stored foods and in homes, including species in the Cheyletidae, Bdellidae, Cunaxidae, Tydeidae, Tarsonemidae, Pyemotidae, and Tetranychidae. The Cheyletidae are predatory mites and may be found associated with acarid mites in granaries, warehouses, and barns, as well as in leaf litter and soil. The Bdellidae (snout mites) are free-living mites with an elongated gnathosoma; they are usually red or black. Bdellids are occasionally found in debris in flour and grain bags, where they probably feed on other mites. Tydeid mites may feed on other mites, and some species may be found in grain residues and barn

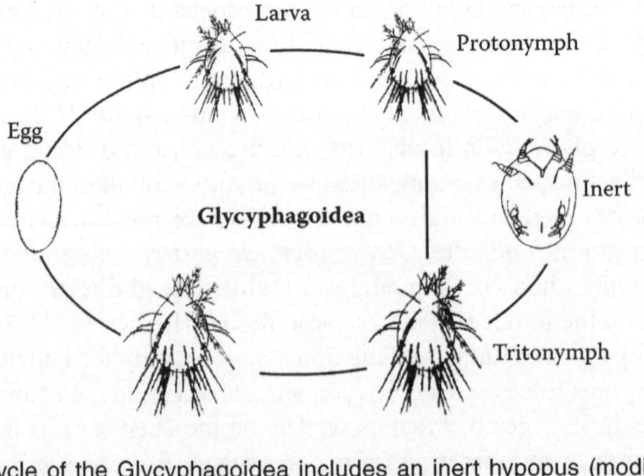

Figure 24.2 The life cycle of the Glycyphagoidea includes an inert hypopus (modified deutonymph) when conditions are unfavorable. Note the lack of mouthparts and the greatly reduced legs. Under favorable conditions, the protonymph molts to a tritonymph, and no deutonymphal (hypopial) stage occurs. (Adapted from Colloff, M.J., *Dust Mites*, CSIRO Publishing, Collingwood, Australia, and Springer, Dordrecht, 2009. With permission from CSIRO, Australia.)

debris. The Tetranychidae are plant feeders, and *Bryobia praetiosa*, the clover mite, may be found in grasses and other plants inside homes (see Chapter 6). Tarsonemid mites include species that feed on plants, fungi, and insects (see Chapter 7). One tarsonemid, *Tarsonemus granarius*, feeds on fungi in stored grains. Pyemotid mites found in stored grains include *Pyemotes ventricosus* (also known as the hay-itch or straw-itch mite), which is a parasite on stored-product insects and feeds on their hemolymph (see Chapter 23 for more information on the biology of *Pyemotes*). This mite is known as the hay-itch mite because it can cause dermatitis in people handling infested grains from its bites, although it cannot remain alive on humans for more than 24 hours. Some gamasid mites are predators of small arthropods, but some feed on fungi, some are coprophagous, and some are scavengers. Gamasid species found in stored foods may be carried there by rodents or by insects. Gamasids such as *Hypoaspis aculeifer* can feed on *Tyrophagus putrescentiae* or *Glycyphagus domesticus*. Another gamasid, *Blattisocius tarsalis*, is a predator of eggs and moth larvae (*Anagasta*, *Plodia*, and *Sitotroga*) that feed on stored grains. The females of *B. tarsalis* attach themselves to adult moths and may be transported by them to new habitats. Related species of *Blattisocius* will feed on the eggs of grain beetles and the eggs of grain mites. Some *Blattisocius* species appear to focus on acarid mites and their eggs as prey.

24.1.3 Sources of Infestation

Stored products (foodstuffs) can be infested at every point from their origin in the field to their final use, including storage bins or granaries, mills, warehouses where foods are held for use or distribution, food-processing plants, food-serving establishments, grocery stores, pantries, and cupboards. The most commonly infested foods are grains, spices, and nuts. Less often, dried fruits, candy, dried dog food, dried flowers, and cosmetics or drugs are infested. Old, unused, or difficult-to-reach products are frequently infested in the home. Pantry and stored-product arthropod pests include both insects and mites, and much of the research on pest management has focused on insects (Hagstrum and Subramanyam 2009). Most often, these pests are introduced into homes via infested food from the grocery store.

24.2 CONTROL OF MITES IN STORED GRAINS AND OTHER FOODS

Acarus gracilis and *A. siro* (Acaridae) are common mites found in stored-food products. The Acaridae includes a large group of species that are saprophagous, graminivorous, fungivorous, or phytophagous and able to live in wet to fairly dry habitats. Other common species include *Glycyphagus domesticus* and *Tyrophagus putrescentiae*, which feed mainly on microorganisms growing on the stored foods and probably cause little direct damage, although they have considerable nuisance value (see Figures S24.1 and S24.2 on the CD). Some species in the genus *Acarus* can feed directly on stored grain. *Acarus siro* is the most important pest of flour and grain; it is primarily found in stored grains that have absorbed moisture. If the grain can be kept dry, it is possible to suppress acarid mite populations (Table 24.1).

Detailed and accurate information on the biology of stored-product pests must be made available if integrated pest management practices (IPM) are to be effective in stored foods (Hagstrum and Subramanyam 2009). Because the different foods and environmental conditions vary greatly, it is difficult to devise an IPM program that is suitable for every situation; however, various tactics are available to manage pests on farms (in storage bins), in processing plants, in grocery stores, and in homes (Schöller 1998, Talbot and Koehler 2002, Vincent et al. 2003, Conyers and Bell 2007, Hagstrum and Subramanyam 2008, 2009, Eaton and Kells 2009). Key issues in the development of integrated mite management (IMM) programs are the ability to monitor, identify the pest species, understand pest biology and behavior, and determine the most appropriate management tactic.

Table 24.1 Methods for Controlling Stored-Product Mites (and Insects)

Chemical Controls

Regulations are reducing the number of chemicals available to manage stored-product pests. Fumigation of storage facilities with methyl bromide is becoming limited due to its damage to the ozone layer in the atmosphere. Residual pesticides (organophosphorus products) have been used to treat the bins or walls of storage facilities, but the development of resistance in the pests has led to research on alternative control measures. Inert dusts, such as silica aerogels and diatomaceous earth, have been used traditionally to protect stored grains and as treatments for empty storage facilities. They have low mammalian toxicity and are persistent and stable. Botanicals, such as azadirachtin and benzyl benzoate, have been used in many parts of the world because of their repellency, antifeedant, growth-inhibition, and antiovipositional effects, as well as their effects on development. Studies on the use of insect growth regulators, juvenile hormone analogs, ecdysone (molting hormone) agonists, and chitin synthesis inhibitors have been conducted.

Monitoring

All new shipments of food should be monitored on a regular basis, and only pest-free shipments should be accepted. A stock rotation program ensures that older foods are used first. Trapping and monitoring allow source populations to be detected and, possibly, eliminated in storage facilities and grocery stores, or suppressed using a variety of tactics.

Cultural Controls

Cultural controls involve the following:

1. Exclude pests to avoid pest problems; monitor and reject infested products or store the foods in pest-resistant, clean containers.
2. Keep the immediate area around warehouses or other storage facilities free of litter and possible food sources for pests. Remove trash and clean trash containers regularly.
3. Keep the roofs and walls of storage facilities in good repair and free from bird nests or roosts. Eliminate rodent nests, which could harbor arthropod pests; monitor for rodents and control them using approved methods.
4. Keep the floors of storage facilities in good repair (free of cracks that would allow entry of pests). Keep doors and windows closed to prevent pest entry. Heating, ventilation, and air conditioning equipment should be filtered or screened and accessible for cleaning. Store all foods away from walls and off floors. Aisles should be sufficiently wide so damage to packaging does not occur.
5. Clean up any spillage immediately, and repair or eliminate damaged containers that have been exposed to possible infestations.
6. Realize that locating and eliminating the source populations can be less expensive and more effective than many tactics, especially when dealing with pantry pests.
7. Use extreme heat or cold to suppress populations in certain circumstances. Most arthropods develop best between 25 and 32°C, and many cannot survive below 13°C or above 35°C.
8. Maintain relative humidity below 70% to reduce the favorability of the crop for fungi and mites; aeration involves pulling air into a storage bin to change the temperature or moisture content of stored foods.
9. Package foods in insect- or mite-proof materials.
10. Reduce pest populations or prevent problems by observing proper sanitation practices in grain bins, storage facilities, shipping containers, grocery stores, and households. Thoroughly clean harvesting equipment and transport trucks.
11. Use low-oxygen and high-carbon dioxide atmospheres in airtight containers to eliminate some stored-product pests.
12. Disinfest packaged pet foods and spices in an autoclave in less than 4 hours using high-pressure and modified atmospheres.

Biological Controls

Biological control of stored-product insects and mites has been studied in the laboratory and in small model systems. These natural enemies tend to be relatively specific to an insect or mite species; however, generalist predators may be present that assist in suppressing pests, especially if chemical controls have not been used.

Source: Based on information from Collins (2006), Fields and White (2002), Mason and Strait (1998), Phillips and Throne (2010), Rust and Reierson (1999), and Vincent et al. (2003).

It is urgent that IPM programs for stored foods be developed because the use of methyl bromide treatments as a space fumigant in flour and feed mills or for fruit and cereals is being phased out because it depletes ozone in the atmosphere (Fields and White 2002). Resistance to pesticides has developed in some stored-product pests, and consumer demands for alternatives to chemicals have driven demand for alternative control tactics.

Alternatives to chemical fumigation include exclusion, trapping, chemical control, the use of heat or cold, aeration, sanitation, inert dusts, or biological controls (Mason and Strait 1999, Rust and Reierson 1999, Collins 2006, Iatrou et al. 2010, Phillips and Throne 2010). The options deployed will depend on the food product and its environment (temperate or humid subtropics or tropics), as well as the size of the storage unit, the length of storage, and many other factors (Table 24.1). Having a thorough knowledge of the biology and behavior of the pests involved is crucial. Although this chapter has focused on mites, stored foods can be infested with insects, fungi, and vertebrate pests such as rats and mice, which requires that control tactics be compatible for all pests.

In pantries, homeowners can deploy cultural practices to control mites (and insects): Keep foods in sealed containers. Place flour and other products that might be stored for a long time into the freezer for several days to kill any pests prior to placing the products on shelves; freezing will kill most pests, but excess moisture could form, which can be a source of mold, so be sure to use the food as soon as possible. Inspect foods periodically and discard infested products. Clean shelves of spilled foods so infestations cannot be maintained. If an infestation is found, determine the source of infestation and examine and discard infested food.

24.3 STORED BULB MITES AND THEIR CONTROL

Several acarid species are economic pests of stored bulbs, including mites in the genus *Rhizoglyphus* (Diaz et al. 2000, Zhang 2003). The two most common pests are *R. robini* and *R. echinopus* (Acaridae). The taxonomy of bulb mites is confusing. There appear to be 13 *Rhizoglyphus* species that are pests of crops around the world, but it is not clear whether some of these species' names are synonyms. *Rhizoglyphus robini* and *R. echinopus* appear to have been spread around the world on infested crops. *Rhizoglyphus* species are similar in appearance, with shiny, oval, smooth, colorless bodies and reddish brown legs and mouthparts. They are relatively large mites, up to 1 mm long, and can be seen with the naked eye. The normal life cycle is egg, larva, protonymph, heteromorphic deutonymph (or hypopus), tritonymph, and adult. The hypopi are nonfeeding and are adapted to surviving long periods without food, as they have a reduced gnathosoma, no mouth or chelicerae, a nonfunctional gut, heavy sclerotinization, and a sucker plate used for attaching to a host, such as beetles, dipterans, or fleas (Diaz et al. 2000). Hypopi may be phoretic on insects or other organisms and can withstand low relative humidity or other adverse conditions. The production of hypopi "is induced by low food quality and quantity, high concentrations of waste products, and extremes in temperature and humidity" (Diaz et al. 2000).

Eggs of *Rhizoglyphus robini* are oval, translucent, white, and about half the size of the adult female. Males and females of *R. robini* mate several times a day. Males may remain in copula for 20 minutes to 6 hours. During this time, a male remains attached to the female by means of anal and tarsal suckers and prevents copulation by other males. Males are polymorphic, and there are fighting morphs, which kill other males. Females of *R. echinopus* can deposit up to 460 eggs. The life cycle takes 17 to 27 days at 23 to 26°C.

Rhizoglyphus mites have a cosmopolitan distribution and attack many bulbs, including dahlia, narcissus, liatris, lily, and amaryllis, as well as carrots, onions, garlic, and potato tubers (Diaz et al. 2000). Bulb mites also attack roots of vegetables, vines, wheat, oats, and other crops. They can be found on fallen fruit, humus, mushrooms, and grains with high moisture content. *Rhizoglyphus* mites do best at high relative humidity and in damp places and are associated with the spread of fungal diseases such as *Fusarium*, *Stromatinia*, and *Pseudomonas*. Damaged bulbs exhibit brown rotting tissue around the neck of the bulb. Thousands of mites and eggs may be found in the damaged tissues. Leaves and flowers that develop may be delayed and abnormal. Severely infested plants may be stunted and have few or no roots. In onions, the mites gather on the roots and cause the plants to collapse and the bulbs fail to enlarge. Bulb mites are not easy to control in the field because the mites hide between bulb scales, inside roots, and within sprouts.

Integrated management programs require methods for identification, monitoring, and control. Methods for monitoring bulb mite populations have not been developed (Diaz et al. 2000). High populations of mites can develop and some type of damage or reduced growth rate can be observed before these mites are identified by a visual inspection of bulbs or plants. Garlic-baited traps have been developed to sample mite phenology and abundance in onion fields (Diaz et al. 2000).

Fumigation with methyl bromide, cyanide, carbon disulfide, and metam sodium has been used in the past to control these mites in storage and prior to planting. Insecticides have been used, but resistance to organophosphates has been found in the bulb mites (Diaz et al. 2000). Alternatively, storing bulbs at −2°C and treating the bulbs prior to planting with a hot-water treatment (2 or 3 hours at 39°C or 41°C) provides good control. Control is more difficult if the bulbs have been planted, because the mites are under the outermost scales of the bulbs. In Israel, solar heating the soil before planting onions or garlic disinfests soil infested with *R. robini*. Solar heating is achieved by covering infested soil with transparent polyethylene sheets (0.04 mm thick). Biological control of *R. robini* by releases of the soil-dwelling predatory mite *Hypoaspis aculeifer* shows promise (Lesna et al. 1996). *Hypoaspis* is well adapted to the top layer of the soil and can feed on a variety of small arthropods (sciarid fly larvae, thrips pupae), as well as the bulb mites. Good control can be achieved if bulbs are treated with hot water to reduce mite population levels, and predators are then introduced. This predator is available commercially, and information is available on how to deploy it.

SELECTED REFERENCES

Collins, D.A. 2006. A review of alternatives to organophosphorus compounds for the control of storage mites. *J. Stored Prod. Res.* 42:395–426.

Colloff, M.J. 2009. *Dust Mites*. Collingwood, Australia: CSIRO Publishing; Dordrecht: Springer.

Conyers, S.T. and C.H. Bell. 2007. Low oxygen requirements for population control of two mite species of stored grain. In: E.J. Donahaye, S. Navarro, C. Bell, D. Jayas, R. Noyes, and T.W. Phillips (eds.), *Proceedings of the Seventh International Conference on Controlled Atmosphere and Fumigation in Stored Products* (pp. 29–44). Israel: FTIC Ltd. Publishing.

Diaz, A., K. Okabe, C.J. Eckenrode, M.G. Villani, and B.M. O'Connor. 2000. Biology, ecology and management of the bulb mites of the genus *Rhizoglyphus* (Acari: Acaridae). *Exp. Appl. Acarol.* 24:85–113.

Eaton, M. and S.A. Kells. 2009. Use of vapor pressure deficit to predict humidity and temperature effects on the mortality of mold mites, *Tyrophagus putrescentiae*. *Exp. Appl. Acarol.* 47:201–213.

Evans, G.O., J.G. Sheals, and D. MacFarlane. 1961. *The Terrestrial Acari of the British Isles*, Vol. I. London: British Museum of Natural History.

Fields, P.G. and N.D.G. White. 2002. Alternatives to methyl bromide treatments for stored-product and quarantine insects. *Annu. Rev. Entomol.* 47:331–359.

Hagstrum, D.W. and B. Subramanyam. 2008. *Fundamentals of Stored-Product Entomology*. St. Paul, MN: American Association of Cereal Chemists International and American Phytopathological Society.

Hagstrum, D.W. and B. Subramanyam. 2009. A review of stored-product entomology information sources. *Am. Entomol.* 55(3):174–183.

Heaps, J.W. (ed.) 2006. *Insect Management for Food Storage and Processing*, 2nd ed. St. Paul, MN: American Association of Cereal Chemists International and American Phytopathological Society.

Hughes, A.M. 1976. *The Mites of Stored Food and Houses*, 2nd ed. London: Her Majesty's Stationery Office.

Iatrou, S.A., N.G. Kavallieratos, N.E. Palyvos, C.T. Buchelos, and S. Tomanovic. 2010. Acaricidal effect of different diatomaceous earth formulations against *Tyrophagus putrescentiae* (Astigmata: Acaridae) on stored wheat. *J. Econ. Entomol.* 103:190–196.

Krantz, G.W. 1971. *A Manual of Acarology.* Corvallis: Oregon State University.

Koehler, P.G. 1999. *Pantry and Stored Food Pests*, UF IFAS Publication No. ENY-213. Gainesville: Cooperative Extension Service, University of Florida Institute of Food and Agricultural Sciences (http://edis.ifas.ufl.edu/IG095).

Lesna, I., M. Sabelis, and C. Conijn. 1996. Biological control of the bulb mite, *Rhizoglyphus robini*, by the predatory mite, *Hypoaspis aculeifer*, on lilies: Predator–prey interactions at various spatial scales. *J. Appl. Ecol.* 33:369–376.

Mason, J.J. and C.A. Strait. 1998. Stored product integrated pest management with extreme temperatures. In: G.J. Hallman and D.L. Denlinger (eds.), *Temperature Sensitivity in Insects and Application in Integrated Pest Management* (pp. 141–177). Boulder, CO: Westview Press.

Phillips, T.W. and J.E. Throne. 2010. Biorational approaches to managing stored-product insects. *Annu. Rev. Entomol.* 55:375–397.

Rust, M.K. and D.A. Reierson. 1999. Use of extreme temperatures in urban insect pest management. In: G.J. Hallman and D.L. Denlinger (eds.), *Temperature Sensitivity in Insects and Application in Integrated Pest Management* (pp. 179–200). Boulder, CO: Westview Press.

Schöller, M. 1998. Integration of biological and non-biological methods to control arthropods infesting stored products. *IOBC Bull.* 21:13–25.

Talbot, M. and P. Koehler. 2002. *Pest Management Strategies for Storing Grains in Florida*, UF IFAS Publication No. CIR873. Gainesville: Cooperative Extension Service, University of Florida Institute of Food and Agricultural Sciences (http://edis.ifas.ufl.edu/ae138).

Vincent, C., G. Hallman, B. Panneton, and F. Fleurat-Lessard. 2003. Management of agricultural insects with physical control methods. *Annu. Rev. Entomol.* 48:261–281.

Zhang, Z.-Q. 2003. *Mites of Greenhouses: Identification, Biology, and Control.* Wallingford, U.K.: CAB International.

[illegible]

[illegible] Fourth [illegible]

[illegible]

[illegible]

[illegible]

[illegible] Vincent, C.A., [illegible] (2012) [illegible]

[illegible]

CHAPTER 25

Dust Mites (Pyroglyphidae)

25.1 THE IMPORTANCE OF DUST MITES

Although dust mites are not, strictly speaking, relevant to agriculture, they are of interest to both rural and urban populations around the world where environmental conditions allow their survival and persistence (van Bronswijk 1981).Dust mites were not confirmed until 1968 as sources of allergens that affect humans who are genetically prone to allergies (Colloff 2009). Since then, dust mites have been studied by scientists from multiple disciplines, including acarologists, molecular biologists, ecologists, immunologists, and epidemiologists, because of their significance to human health. At least 100 million people around the world are affected by dust mite allergies, which manifest themselves as asthma, dermatitis, or rhinitis (Colloff 2009). Dust mites produce proteins, present in their feces, that cause the allergic responses. The feces continue to cause symptoms long after the mite that produces them has died. The species of dust mite found varies from area to area and house to house, and their abundance varies based on the availability of food and the temperature and relative humidity in the home. Approximately 300 species of mites are associated with stored foods and homes; however, species of the Pyroglyphidae (Acaridida or Astigmata) are usually considered dust mites. Dust mites live in carpets, bedding, draperies, and furniture, where they feed on human skin scales, which we shed constantly. Dust is not dirt. It contains pet dander, cockroach feces, pollen, molds, and breakdown products from human skin (Denmark and Cromroy 2007). Pet dander, cockroach feces, pollens, and molds in house dust can cause allergic responses in humans, in addition to allergic responses caused by dust mites.

25.2 SPECIES OF DUST MITES

The predominant dust mite species found in houses are *Dermatophagoides pteronyssinus* and *D. farinae*, although at least 13 other species of Pyroglyphidae also may live in house dust (Figure 25.1). In addition, acarid and glycyphagid mites, commonly known as food-storage mites, may be found in homes if molds are present on foods or in organic materials used as floor coverings or bedding (also see Chapter 24). Cheyletid mites in the genus *Cheyletus* also may be present as predators of dust mites; however, these predators may not be present in all homes, and they do not appear to be able to suppress dust mite populations significantly to prevent allergies, so they are not considered adequate biological control agents (Colloff 2009).

Figure 25.1 Scanning electron micrograph of *Dermatophagoides farinae*. This is one of the most common species of dust mites. (Photograph by David Walter, Alberta Royal Museum, Canada.)

25.3 BIOLOGY OF DUST MITES

Dust mites (Pyroglyphidae) are 0.25 to 0.33 mm long and usually invisible to the naked eye. The male is smaller than the female. The mites are whitish in color and can sometimes be seen in dust when it is placed against a dark background. The dust mite life cycle consists of egg, prelarva, larva, protonymph, tritonymph, and male and female adults (Figure 25.2). Note that the Pyroglyphidae has no deutonymphal hypopus, perhaps because dust mites have a relatively stable supply of food in our homes. The acaridids and tyrophagids typically live in less stable environments, and the hypopal stages are effective in dispersal or survival during unfavorable conditions. The duration of the life cycle of dust mites depends on the species, culture medium, temperature, and relative humidity. Ideal conditions are 20 to 25°C and a relative humidity of 70 to 80% (Arlian 1975). Under these conditions, a generation requires about 25 to 30 days. Females of *Dermatophagoides evansi* may lay 35 eggs, but a *D. pteronyssinus* female may deposit 60 to 100 eggs during her lifetime.

Dust mites feed on human skin scales (or at least the bacteria and fungi growing on the scales), fungi, molds, dead insect bodies (carpet beetles, silverfish, clothes moths, cockroaches), pollen grains, bacteria growing on skin scales, and plant material. Dust mites may be found on mold growing on the walls. The fecal pellets they produce contain the proteins that cause allergenic responses (Tovey et al. 1981, Martinez et al. 2000). During digestion (Colloff 2009):

> Cells bud off from the wall of the midgut, engulf food particles and travel along the gut lumen breaking down the food as they go. The products of digestion are absorbed throughout the gut epithelium into the haemolymph. By the time they reach the hindgut, the cells start to dehydrate and die, packaging

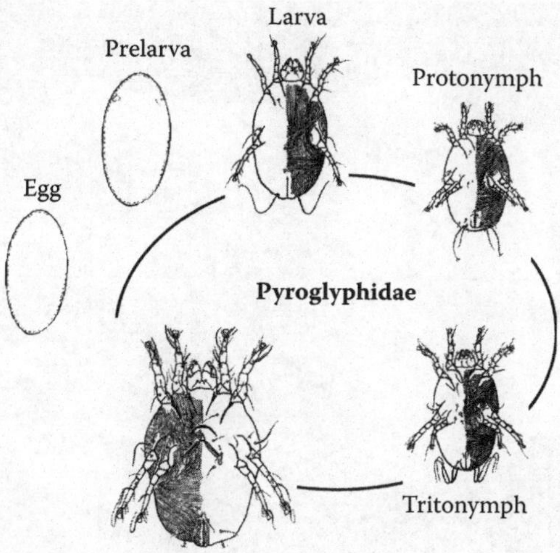

Figure 25.2 The life cycle of the Pyroglyphidae, the family to which dust mite species belong, does not include a deutonymphal stage, perhaps because their environment is relatively stable and they do not have to disperse to new environments. (Adapted from Colloff, M.J., *Dust Mites*, CSIRO Publishing, Collingwood, Australia, and Springer, Dordrecht, 2009. With permission from CSIRO, Australia.)

> themselves into faecal pellets surrounded by a peritrophic membrane that protects the delicate hindgut from damage by abrasion. This mode of digestion results in relatively large quantities of enzymes accumulating in the faecal pellets. The pellets, some 20 to 50 μm in diameter, are egested and accumulate. … The enzymes, being proteins, are immunogenic.

Digestive enzymes known to be allergenic include the Der p 1 cysteine proteinase, chymotrypsins, trypsins, and collagenase (Colloff 2009). Each mite produces about 20 fecal particles a day, and a single gram of house dust can contain as many as 250,000 fecal pellets (Figure 25.3 and Figure 25.4).

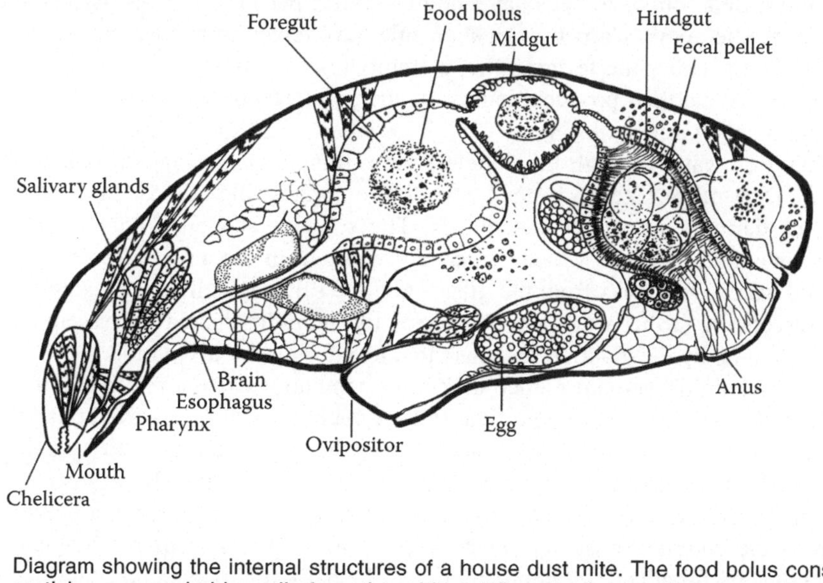

Figure 25.3 Diagram showing the internal structures of a house dust mite. The food bolus consists of food particles surrounded by cells from the midgut. When the food pellet reaches the hindgut, the cells die and the peritrophic membrane that surrounds the hindgut surrounds the fecal pellet. (Adapted from Colloff, M.J., *Dust Mites*, CSIRO Publishing, Collingwood, Australia, and Springer, Dordrecht, 2009. With permission from CSIRO, Australia.)

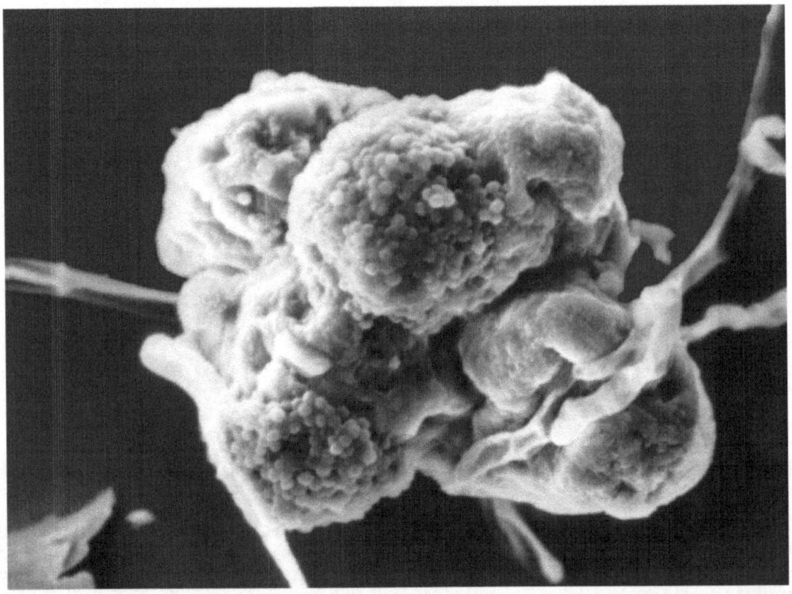

Figure 25.4 Fecal pellets (at least four) of dust mites here are held together by fungal hyphae. The fecal pellets contain allergenic proteins. These pellets can persist for a long time, even after the mites that produced them are dead. (From Colloff, M.J., *Dust Mites*, CSIRO Publishing, Collingwood, Australia, and Springer, Dordrecht, 2009. With permission from CSIRO, Australia.)

In general, humid homes in humid regions have more mites and thus more allergens. Dust mites tolerate fluctuating temperatures in homes, but they do not tolerate relative humidity that is consistently less than 50% (Pike et al. 2005). Dust mites are present where food and shelter are available with an adequate relative humidity. Favorite sites include carpets, mattresses, upholstery, and other fabrics in homes. Bedrooms and beds have the most mites because humans usually spend about 8 hours a day in the bedroom. Carpets with short pile have fewer mites than carpets with long pile. Wood and tile floors harbor the fewest mites, so removing carpets can be a useful cultural control. Airborne dust mites and their products can cause allergenic responses during the changing of bedding (Cunnington and Gregory 1968).

Dust mites lose water through the cuticle, oviposition, anus, and lateral opisthosomal glands (Figure 25.5). Dust mites take up water by feeding on water-filled foods, by uptake through the integument, or through the supracoxal glands. The supracoxal glands contain fluid filled with sodium and potassium chloride, which can absorb water from the air. As the relative humidity is decreased, the water evaporates from the glands and the salts crystallize, blocking the entrance of the gland and reducing further water loss. As the relative humidity increases, the salts dissolve again and water is taken up by the gland to replenish that lost during the dry period.

The use of dehumidifiers can reduce dust mite populations if the relative humidity is maintained sufficiently low over a sufficient time (Olkowski et al. 1991); however, dust mites can survive low relative humidity for awhile by clustering together to restrict water loss, so the low relative humidity must be maintained nearly continuously. Air conditioning, because it lowers relative humidity, can reduce dust mite populations. Arlian et al. (2001) concluded that preventing population growth of the common dust mite *Dermatophagoides farinae* requires maintaining relative humidity below 35% for at least 22 hours per day if the daily relative humidity is 75 to 85% for the remainder of the day.

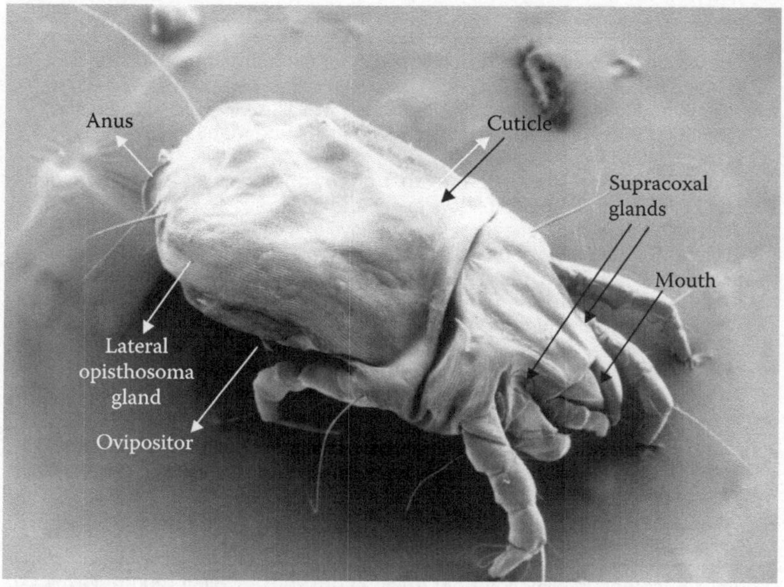

Figure 25.5 Dust mites take up water (black arrows) through the cuticle, the supracoxal glands, and mouth. They lose water (white arrows) through the cuticle, ovipositor, lateral opisthosomal gland, and the anus. Maintaining a consistently low relative humidity in the home (less than 50%) in the microclimate within which the mites live can reduce or eliminate mite populations, thereby reducing the number of allergenic fecal pellets they produce. Sanitation can help to eliminate the fecal pellets. (Adapted from Colloff, M.J., *Dust Mites*, CSIRO Publishing, Collingwood, Australia, and Springer, Dordrecht, 2009. With permission from CSIRO, Australia.)

25.4 INTEGRATED MANAGEMENT OF DUST MITES AND THEIR ALLERGENS

Accurate estimates of population densities (especially in mattresses or carpets) are time consuming and difficult to make. Most estimates are made by vacuuming small areas of the mattress or carpet. Sticky traps also can be employed. The number of mites varies by season, typically being lower in the winter in temperate climates where homes have central heating, which reduces relative humidity. Homes in deserts or in the mountains typically have lower mite densities because the relative humidity experienced in these homes is lower. Dust mites can survive in beds by burrowing into areas of the mattress where moisture may be retained, or they can cluster together to minimize water loss.

Neryl formate was shown to be an airborne aggregation pheromone for both *Dermatophagoides pteronyssinus* and *D. farinae*. Skelton et al. (2010) suggested that synthetic neryl formate has the potential to be used as a lure-and-kill system for dust mite control. Control of dust mites in the home requires multiple cultural tactics to alleviate allergy symptoms (Table 25.1); however, these tactics are not fully effective, because it is important to reduce not only the number of mites in the home but also the fecal pellets that contain the allergens (Colloff 2009). Sanitation is an important control tactic. Frequent washing of bedding can reduce populations of dust mites and their feces. Vacuuming will reduce dust mites and their feces, but removal of both at levels sufficient to eliminate allergy symptoms is difficult. It is better to avoid the buildup of large amounts of mites and their allergenic pellets by other methods described in Table 25.1.

Table 25.1 Cultural Methods for Management of Dust Mites in the Home

1. Encase pillows and mattresses in plastic or other impermeable fabric.
2. Remove carpeting and draperies, especially from bedrooms.
3. Wash bedding weekly at 60°C, or higher; washing at 130°C kills all mites.
4. Washing with cold water removes some mites and many fecal pellets.
5. Dry cleaning kills all mites.
6. Avoid the use of duvets with feather or down fillings. Use synthetic fabrics and pillows, and replace pillows every 2 to 3 years.
7. Put soft toys in the freezer for several days to kill mites.
8. Use a vacuum cleaner daily to reduce dust and dust-mite feces. Vacuum all carpets, under beds, upholstery, curtains, mattresses, and blankets.
9. Apply superheated steam to carpets. Exposing carpets to direct sunlight can kill mites, and hanging rugs in the sun can reduce dust mites. Autoclaving kills dust mites, as does dry heating with electric blankets.
10. Use a dehumidifier to reduce relative humidity in the home to 50% or less on a continuous basis, especially in the microhabitats the dust mites inhabit.
11. Use chemicals to control dust mites, including organochlorides, pyrethroids, and benzyl benzoate. Pyrethroids, an organotin, a phenol, or silver nanoparticles may be added to fabrics used in mattresses, pillows, carpets, and clothing during their manufacture. Boric acid can be used to control ants, fleas, cockroaches, and dust mites, which could reduce the amount of allergenic materials present.

Source: Based on information from Arlian et al. (2001), Colloff (2009), Olkowski et al. (1991), and van Bronswijk (1981).

SELECTED REFERENCES

Arlian, L.G. 1975. Dehydration and survival of the European house dust mite, *Dermatophagoides pteronyssinus*. *J. Med. Entomol.* 12:437–442.

Arlian, L.G., J.S. Neal, M.S. Morgan, D.L. Vyszenski-Moher, C.M. Rapp, and A.K. Alexander. 2001. Reducing relative humidity is a practical way to control dust mites and their allergens in homes in temperate climates. *J. Allergy Clin. Immunol.* 107:99–104.

Colloff, M.J. 2009. *Dust Mites*. Collingwood, Australia: CSIRO Publishing, and Dordrecht: Springer.

Cunnington, A.M. and P.H. Gregory. 1968. Mites in bedroom air. *Nature* 217:1271–1272.

Denmark, H.A. and H.L. Cromroy. 2007. *Featured Creatures: House Dust Mites*. Gainesville: Department of Entomology and Nematology, University of Florida Institute of Food and Agricultural Sciences (http://entnemdept.ifas.ufl.edu/creatures/urban/house_dust_mite.htm).

Martinez, J., E. Eraso, R. Palacios, and J.A. Guisantes. 2000. Cross-reactions between *Dermatophagoides pteronyssinus* and *Dermatophagoides farinae* (Acari: Pyroglyphidae) related to the different growth phases of cultures. *J. Med. Entomol.* 37:35–39.

Olkowski, W., S. Daar, and H. Olkowski. 1991. *Common-Sense Pest Control*. Newtown, CT: Taunton Press.

Pike, A.J., M.J. Cunningham, and P.J. Lester. 2005. Development of *Dermatophagoides pteronyssinus* (Acari: Pyroglyphidae) at constant and simultaneously fluctuating temperature and humidity conditions. *J. Med. Entomol.* 42:266–269.

Skelton, A.C., M.M. Cameron, J.A. Pickett, and M.A. Birkett. 2010. Identification of neryl formate as the airborne aggregation pheromone for the American house dust mite and the European house dust mite (Acari: Epidermoptidae). *J. Med. Entomol.* 47:798–804.

Tovey, E.R., M.D. Chapman, and T.A.E. Platts-Mills. 1981. Mite faeces are a major source of house dust allergens. *Nature* 289:592–593.

van Bronswijk, J.E.M. 1981. *House Dust Biology for Allergists, Acarologists and Mycologists*. Amsterdam: NIB.

PART IX

Some General Conclusions about Integrated Mite Management

Although this book provides examples of crops and their mites that are managed in a manner consistent with all of the principles of integrated pest management (IPM), many agriculturally important mites and ticks are not yet managed using these principles. Chemical control of some mites and ticks will remain a common pest management tool for the foreseeable future, in part because pesticides require less knowledge to deploy than it takes to develop the tools and information appropriate for integrated mite management (IMM) programs. However, social, political, and legislative pressures are increasing to reduce or eliminate the use of broad-spectrum pesticides. The costs of synthetic organic pesticides continue to increase, the propensity of mites and ticks to develop resistance remains constant, and the slowed rate of development and registration of effective new products provide an impetus to develop IMM tools.

In many cases, the lack of basic knowledge of the biology, systematics, and ecology of pest mites and ticks limits the development of IMM programs. Other limiting factors include the lack of effective and rapid sampling methods, a full knowledge of the effectiveness of natural enemies, accurate estimates of the economic injury level in the crop, and adequate methods to deploy what we already know to growers and pest managers. There are insufficient funds for research, limited taxonomic resources to identify mites, and a decrease in the number of scientists trained to work with mites (which this book may help to remedy). How we grow the crop (or farm animals) and deal with all aspects of agricultural production are crucial to developing *appropriate tools* for IMM. What is done to manage a plant disease or a nematode or an insect pest will affect how we manage mites and ticks. Single tactics, or 'silver bullets,' rarely can be found and deployed to solve a mite problem on a sustainable basis. Certainly, the use of broad-spectrum pesticides is not sustainable. Delivery of IPM tools for all crop pests will have to be coordinated to effectively manage them.

If you are a researcher or graduate student, it should be apparent that your career could be devoted to developing improved pest management programs within agricultural systems. Research is urgently needed that will improve our ability to identify mites and ticks accurately and rapidly in the field so we can monitor pests *and* their natural enemies and predict when a pest population is likely to be suppressed by their natural enemies. Creativity will be needed to develop cultural practices or other nonchemical control tactics. The current economic injury levels probably often overestimate the effects of mite feeding and should be modified (or confirmed). We need to increase the use of host-plant resistance and cultural practices (including water and fertilizer management, dust management, crop rotations, and altered planting or harvest dates). Improved biological control programs (augmentative, conservation, and classical) will pay back their investment in research with a high benefit-to-cost ratio.

Mites are fascinating animals worthy of increased basic and applied research. I hope this book will serve as a reference to what we know about these agricultural pests, as well as a stimulus to additional research.

Glossary

A

abiotic Not biotic, not pertaining to life; often involves the effects of temperature, relative humidity, and daylength.

acaricides Pesticides that are effective against mites; these products may control other organisms, as well. Also called *miticides*.

acarology The study of mites and ticks (Acari or Acarina).

Acarus siro Common stored-grain mite (Acaridida) that can feed and reproduce on fungi.

Acarus gracilis Common stored-grain mite (Acaridida) that can feed and reproduce on fungi.

Acaridida See Astigmata.

acarodomatia Tufts of hairs, or invaginations or pits, in the leaf surface of many perennial trees and vines; they are homes for many predatory and mycophagous mites. Also called *domatia*.

accidental acarophagy The unintended consumption of mites.

Aceria aloinis Also known as the aloe wart mite, this pest is found in California, Florida, and southern Africa.

Aceria chondrillae Eriophyoid mite evaluated for the control of skeleton weed (*Chrondrilla juncea*).

Aceria cynodoniensis Eriophyoid mite (bermudagrass mite).

Aceria dianthi This eriophyoid is a leaf vagrant on carnations that causes stunting.

Aceria genistae The broom gall mite is a pest of broom (Leguminosae) in Europe.

Aceria georghioui This eriophyoid attacks carnations, causing distortion.

Aceria guerreronis Eriophyoid mite (coconut mite) that is a pest of coconut palms; it causes scarring and distortion of nuts and premature nut drop.

Aceria lantanae The Lantana gall mite is found in the Caribbean, Florida, and Central and South America. It produces galls and is being evaluated for biological control of *Lantana* where it is considered a weed.

Aceria malherbae Eriophyoid mite that has been evaluated for the control of field bindweed (*Convolvulus arvensis*).

Aceria paradianthi This eriophyoid attacks *Dianthus* species, producing distorted and stunted plants.

Aceria proteae The Protea witches' broom mite is found in South Africa.

Aceria sheldoni Eriophyoid mite (citrus bud mite) that can be a serious pest of citrus, especially lemons; fruits may become distorted or drop off.

Aceria tosichella (A. tritici) Eriophyoid mite associated with high plains disease in the United States. The taxonomy is confusing; it was formerly known as *A. tulipae*.

Aceria (Eriophyes) tulipae Taxonomic problems make the literature confusing with regard to this mite, which attacks bulbs as well as garlic and onion.

Actinedida Also known as the Prostigmata; includes a diverse group of mite families, such as the plant-feeding Tetranychidae, Eriophyoidea, Tarsonemidae, and Tenuipalpidae.

Aculops lycopersici Eriophyoid mite (tomato rust mite) that is a widespread pest of tomatoes, chilies, potatoes, eggplants, tobacco, and petunias.

Aculus cornutus Eriophyoid mite (peach silver mite) that attacks peaches and almonds; may be important alternative prey for phytoseiid predators.

Aculus hyperici This eriophyoid is native to Europe and has been used to control St. John's wort weed (*Hypericum perforatum*) in Australia.

Aculus schlechtendali Eriophyoid mite (apple rust mite) that feeds on flowers, fruits, and leaves of apple; may be a beneficial alternative prey for phytoseiids such as *Metaseiulus occidentalis* in apple orchards.

aedeagus A male intromittent organ, usually sclerotized; found in the Tetranychoidea. Its shape is used to discriminate between species, so males are required to identify spider mites to the species level.

aestival diapause Diapause that allows an organism to survive detrimental summer conditions.

aestivation Diapause that occurs in summer.

allele Diploid organisms have two copies of each type of chromosome, one derived from the mother and one from the father. At a specific location (locus) on the chromosomes, there may be more than one type of DNA coding for slightly different versions of that locus (or gene). The differences may result in a different phenotype (or appearance). The different DNA sequences at that locus are called *alleles*. For example, a gene may be mutated so that it confers resistance to pesticides (resistance allele), but the normal (wild-type) allele does not. Both alleles code for the same basic function, but they are different from each other due to mutation. Multiple alleles of a gene occur in populations because multiple mutations have occurred in the gene in different individuals over time.

allelochemicals Chemicals that function as pheromones, allomones, synomones, or kairomones.

allergens (antigens) Foreign substances that result in the production of antibodies by the animal. Most antigens are proteins. Hypersensitive reactions include itching, redness, swelling, and rash. A very strong reaction can result in anaphylaxis, which can be fatal.

allomone A chemical of advantage to the organism sending it.

Amblyomma americanum This ixodid tick is also known as the Lone Star tick; it is widely distributed in the United States and can transmit human granulocytic ehrlichiosis and tularemia.

Amblyomma hebrae This three-host ixodid tick infests livestock and wildlife and bites humans; it is also called the bont tick. It can transmit *Ehrlichia ruminantium*, the agent of heartwater fever, and is found in the tropics and subtropics.

Amblyomma maculatum This ixodid tick, also known as the Gulf Coast tick, is found in the southeastern United States and Central America.

Amblyomma variegatum This ixodid tick feeds on cattle, sheep, goats, horses, and dogs, as well as humans. It is also known as the tropical bont tick. It, along with *A. hebraeum*, can transmit the causative agent of heartwater fever in ruminants.

Amblyseius barkeri Phytoseiid used in augmentative biological control to suppress populations of *Thrips tabaci* and other thrips species in greenhouses, as well as tarsonemid mites.

Amblyseius (= Neoseiulus) californicus Phytoseiid used in augmentative biological control to suppress populations of spider mites in greenhouses.

Amblyseius (= Neoseiulus) cucumeris Phytoseiid used in augmentative biological control to suppress populations of *Thrips tabaci* and *Frankliniella occidentalis* in greenhouses.

Amblyseius degenerans Phytoseiid used in augmentative biological control to suppress thrips species in greenhouses.

Amblyseius fallacis Phytoseiid used in biological control to suppress populations of spider mites in apple orchards. Pesticide-resistant strains are available.

Amblyseius hibisci Phytoseiid important in suppressing populations of spider mites in orchard and vineyard crops.

Amblyseius swirskii Generalist phytoseiid that is used in augmentative biological control to suppress populations of whiteflies, thrips, spider mites, and tarsonemids; will also feed on pollen.

Anactinotrichida A classification of mites also known as the Parasitiformes; this group of mites has setae that lack actinopilin.

Analgesidae This family of feather mites (Astigmata) has modified legs that aid in their attachment to feathers; these mites may attack chickens, causing dermatitis.

antibiosis In host-plant resistance, it involves traits of a plant that cause decreased survivorship, fecundity, or developmental rate in the pest arthropod. Antibiosis may be caused by heavy pubescence, increased thickness of the cuticle, sticky exudates, secondary plant compounds, or lack of an essential arthropod nutrient.

anus External opening of the hindgut for excretion; typically it is ventral in mites, but in the red-legged earth mite and blue oat mite (Penthaleidae) it is dorsal.

aposematic Having warning coloration, indicating that the organism is distasteful or toxic.

Argas This genus of soft ticks (Argasidae) consists of approximately 57 species and includes the fowl tick (*A. persicus*) and the pigeon tick (*A. reflexus*).

Argasidae The family of soft ticks consists of approximately 183 species.

arrhenotoky Genetic system in which males are haploid and females are diploid. Unmated females deposit haploid eggs that develop into males. In arrhenotokous species, mated females typically produce progeny in which the sex ratio is skewed to female progeny. Most tetranychids are arrhenotokous (also called haplodiploidy).

Astigmata (= Acaridida) Mites lacking stigmatal openings to the tracheal system. Includes stored-products mites, scabies mites, and demodectic mange mites.

augmentative biological control The mass rearing and release of natural enemies to suppress a pest population. Augmentative releases can be *inoculative*, where a small number is released and they are expected to reproduce and ultimately suppress the pest population. Or, they can be *inundative*, which involves releasing sufficiently large numbers of natural enemies that pest population suppression is achieved rapidly. In both situations, release rates, timing, and quality of released natural enemies are of crucial importance in achieving success.

autocidal (genetic) control The use of irradiation or chemosterilants to sterilize pest insects (typically males); if released in adequate numbers, they can reduce pest populations by mating with wild females which do not produce progeny. Autocidal control can be used to suppress pests or to eradicate them. Unfortunately, there are no examples of mites that have been successfully managed, or eradicated, through autocidal control methods. Tests with arrhenotokous spider mites showed that wild females that mated with sterile (haploid) males simply remated with their haploid (fertile) sons.

avermectin Macrocyclic lactone glycoside that is a fermentation product of *Streptomyces avermitilus*, an organism isolated from the soil. This acaricide has translaminar activity and is toxic to spider mites.

azadirachtin Pesticide product (triterpenoid) extracted from the seeds of the neem tree (*Azadirachta indica*). It blocks the action of the molting hormone ecdysone.

B

bait sprays The addition of a toxin to an attractive substrate so pests are killed when they encounter or feed on the bait.

banker plants Plants added to a greenhouse system that aid in the development and dispersal of predators for the control of phytophagous pests; for example, castor bean banker plants may provide pollen and extra-floral nectaries for phytoseiids. Banker plants allow predators to disperse to other plants, suppressing the pests on them for several months. The banker plant system typically includes a predatory mite, the banker plant, and prey or other foods (pollen or nectar).

Berlese funnels Used to extract mites from soil, litter, compost, animal nests, and foliage. Many modifications are known, but the principle is that heat and light from above the sample drive the mites (and other arthropods) down a funnel into a collecting chamber that contains alcohol, or moisture if live specimens are desired.

biological control For mites, this includes the use of predators and pathogens to manage mite populations (only one parasitoid is known for a tick, and it is rare). Biological control can be categorized as *classical*, *augmentative*, or *conservation*.

biological-control treadmill When key pests in a cropping system are under biological control and a new invasive pest arrives, one of the solutions is to attempt to control the invasive pest using classical or augmentative biological control. Without the rapid development of new biological control tools for the invasive pest, the biological-control-based management program can collapse if pesticides are used to suppress the new pest.

biorational control methods Use of mating disruption, mass trapping, and sterile insect techniques, none of which is effective for mites or ticks.

blister mite Eriophyoid mite that causes blister-like growths on the foliage within which it lives.

blue oat mite *Penthaleus major* (Penthaleidae).

bont tick *Amblyomma hebraeum.*

brain The brain of a mite consists of a single, fused mass through which the esophagus passes.

Brevipalpus californicus This tenuipalpid, also known as the citrus flat mite, is a pest of citrus and ornamentals around the world.

Brevipalpus obovatus This tenuipalpid, also known as the privet mite, is a pest of citrus and also transmits leprosis (nailhead rust) in citrus.

Brevipalpus oncidii This tenuipalpid, also known as the oncidium mite, is a pest of orchids.

Brevipalpus phoenicis This tenuipalpid, also known as the red and black flat mite, is a pest of citrus and also transmits leprosis (nailhead rust) in citrus. It is also a pest of palms, privet, and many other plants.

broad mite *Polyphagotarsonemus latus* is a small plant-feeding mite in the family Tarsonemidae that is found on a variety of crops in greenhouses or in tropical and subtropical areas.

brooming Damage to host plants caused by the feeding of some eriophyoid mites; it is characterized by the formation of many small branches.

Bryobia praetiosa The clover mite is found around the world and may be a species complex; it is a pest of grains, grasses, ornamentals, and alfalfa.

Bryobia rubrioculus The brown mite, or brown almond mite, is a pest of stone and pome fruit trees.

bud mite Many species of eriophyoid mites that live within the buds of plants, often causing distortion of subsequent growth of branches or fruits and economic damage.

C

camerostome Cavity into which the mouthparts of some mites can be withdrawn.

cannibalism Killing and eating of one's own species.

capitulum (gnathosoma) Mouthparts of ticks and mites; see also gnathosoma.

Cardinium Microorganism that is an intracellular symbiont of arthropods and is associated with modification of sex ratios and fitness; often located in the ovaries and is transovarially transmitted.

Carios This genus of soft ticks (Argasidae) consists of approximately 87 species and includes parasites of bats and birds.

cattle ticks *Rhipicephalus* (*Boophilus*) *microplus* and *R.* (*B.*) *annulatus.*

Cecidophyes rouhollahi This eriophyoid mite has been released in Canada to control weeds (*Galium aparine* and *G. spurium*).

chaetotaxy The arrangement of setae on the exoskeleton, often used as taxonomic characters in mites.

chelate–dentate Chelicerae that are pincer-like and have teeth.

chelicera(e) Mouthparts of chelicerate arthropods, including mites and ticks. Spider mite chelicerae are modified into long, stylet-like structures used to pierce plant cells. Chelicerae of some male mites may be modified to transfer sperm (spermatodactyl).

Cheyletiella Mites in the Cheyletiellidae are parasites of cats, rabbits, and dogs. These fur mites do not burrow into the skin but feed on lymph by piercing the epidermis with their chelicerae, causing a mild dermatitis.

Cheyletiella blakei This fur mite (Actinedida: Cheyletiellidae) is found on the faces of cats, and heavy infestations can result in loss of hair and intense itching.

Cheyletiella parasitivorax This fur mite (Actinedida: Cheyletiellidae) is found on domestic rabbits around the world.

Cheyletiella yasguri This fur mite (Actinedida: Cheyletiellidae) is found on dogs but rarely causes problems, perhaps because flea-control measures also control this mite.

chiggers Larval mites in the family Trombiculidae. The larvae are parasitic on vertebrates, including humans; some species in Southeast Asia are able to transmit the rickettsia that causes scrub typhus. Nymphs and adults are predatory.

chitin Polysaccharide constituent of arthropod cuticle.

chlorophyll Green pigment of plant cells necessary for photosynthesis.

chloroplast Membrane-bound organelle in plant cells in which chlorophyll is contained; the site of photosynthesis.

Chorioptes Genus of mange mites (Chorioptidae) that attack horses, sheep, goats, rabbits, and llamas, causing chorioptes mange. It is not as serious as psoroptic mange.

Chorioptes bovis This mange mite (Astigmata: Chorioptidae) causes foot mange and itchy leg or chorioptic mange in horses, cows, sheep, goats, rabbits, and llamas. The damage usually is restricted to the legs. These mites feed on dead epidermal tissues, and the effect is less severe than with psoroptic mange.

chorioptic mange Damage caused by *Chorioptes* mites (Astigmata: Chorioptidae) in horses, cows, sheep, goats, rabbits, and llamas. These mites feed on dead epidermal tissues, so they do not cause severe reactions, and infestations may be asymptomatic.

Chrysopidae A family of neuropterans that are generalist predators of small arthropods, including mites. Some species, such as *Chrysoperla* species, are mass reared and released for augmentative biological control.

circadian rhythm Cycle with a natural duration of about 24 hours; most organisms have a biological clock with which they measure time.

classical biological control Primarily conducted to suppress invasive (exotic) species. It assumes that endemic natural enemies are not always adapted to the invasive pest and are unable to suppress it sufficiently. It also assumes that natural enemies in the country of origin may be better adapted and can be imported, evaluated, released, and established permanently in the environment to provide long-term suppression of the pest. Ideally, prey- or host-specific natural enemies are imported so there is less likelihood of nontarget effects.

clofentezine Growth regulator of mites such as tetranychids; it kills eggs but not the active stages.

concentration–response curve Graphical representation of the increase in mortality with an increase in concentration of a toxicant, such as a pesticide. By graphing the responses of populations using a log scale, a line can be generated that allows one to estimate the LC_{50}, the concentration lethal to 50% of the population under the specific test conditions. The LC_{95} or LC_{99} values can also be calculated for the population using these data.

Coniopterygidae A family of neuropterans that are generalist predators of small arthropods, including mites. They are also known as mealy-wings, dusky-wings or dusty-wings.

conservation of natural enemies or **conservation biological control** Involves modifying crop production practices to favor the survival and activity of natural enemies. Often, it involves the modification of pesticide applications, but it may also involve providing refuges, alternative foods, or modifying cultural practices.

conserved Natural enemies should be maintained (conservation biological control) in crops using a variety of tactics such as modified spray practices, provision of alternative prey or other foods, and provision of overwintering sites.

control tactics A variety of methods used to reduce pest populations, including chemical control; physical controls, such as heat or freezing; cultural controls, such as modifying the time of planting, harvesting, or weed control; host-plant resistance; biological and biorational controls, such as the use of mating disruption or mass trapping; and genetic control, which involves the use of sterile insect release (SIR) or sterile insect technique (SIT).

coprophage Feeding on feces.

Cosetacus cameliae This eriophyid mite is a pest of *Camellia japonica* in California.

cover crops Crops, such as clover, planted between the rows of other crops. Cover crops can reduce dust, increase relative humidity in the main crop, and possibly provide pollen, nectar, or alternative prey for natural enemies. Negative effects can include the increased cost of planting the cover crop, the need for additional water and fertilizer, and the possibility that it will harbor pests of the primary crop.

coxa Leg segment closest to the body of the mite.

coxal glands Found in many mites, they appear to be involved in osmoregulation or excretion and may have an endocrine function.

crop pest-control consultants Specialists in pest management, with expertise in agronomy, horticulture, entomology, plant pathology, weed science, nematology, and soil science. They serve as consultants to growers regarding how to manage pests and diseases in crops, with the goal of improving pest management and reducing production costs.

crop rotation A cultural practice that involves growing different crops in a specific area during different growing seasons as a method of reducing pest problems in the crop. Soil fertility can be enhanced when a legume can be planted as one of the crops in a rotation.

cross-resistance Resistance to one pesticide may confer resistance to a second pesticide, even though that pesticide has not been used on the population. Cross-resistances may occur with pesticides within the same class or with pesticides in other classes if the mechanisms of toxicity are similar.

cryptic species Valid species that look so much alike that they cannot be discriminated readily using morphological characters.

Cryptostigmata (= Oribatida) Soil or beetle mites; typical stigmata are absent, but the tracheal system is found at the base of the first and third pairs of legs. These mites are abundant in soil litter and play a role in soil fertility through their activities. Some species can transmit nematode parasites to vertebrates.

cultural controls All the modifications in agronomic practices that tend to reduce pest damage. These may include changing the time of planting or harvest, cultivating weeds or other pests in the soil, managing temperature and relative humidity in greenhouses, dust management, providing adequate water and fertilizer, avoiding overfertilization, and planting cover crops.

cuticular lobes Outgrowths of the cuticle; the shape of cuticular lobes may be used as a taxonomic character. Spider mites have well-developed cuticular lobes during active development, but females in diapause develop without these lobes, perhaps to reduce the loss of water during diapause.

cyclamen mite *Phytonemus* (= *Tarsonemus* or *Steneotarsonemus*) *pallidus*, a plant-feeding mite in the family Tarsonemidae that is found on a variety of ornamental plants, especially those grown under high relative humidity.

D

delayed numerical response The increase in the number of predators (or parasitoids) as a result of an increase in the density of their prey (or hosts). Because an increase in response may require time for development, predator population increases typically lag behind the increase in prey populations.

delusory acariasis The belief that humans are being attacked by mites, when, in fact, they are not.

delusory parasitosis When people become convinced that insects or mites infest them when they are, in fact, not infested. Symptoms may include rash, redness of the skin, and secondary infections caused by scratching or the application of chemicals to eliminate the supposed infestation.

Demodex Genus of mites (Demodicidae) that burrow in the skin of humans and many animals. Currently, 86 *Demodex* species live in 11 of 18 orders of placental mammals. Humans have two species (*D. brevis* and *D. folliculorum*) that do not normally cause disease, but *Demodex* mites in animals (dogs, cats, goats, horses, cattle) can cause demodectic mange.

Demodex bovis This follicle mite (Demodicidae) is found on cattle; it usually is not a serious pest, but large populations can cause damage to hair and hides.

Demodex canis This follicle mite (Prostigmata: Demodicidae) is a parasite of dogs, only occasionally causing disease when the dog has an immature or impaired immune system; however, infestations with *D. canis* can result in red mange, a serious disease caused by bacteria that enter the damaged follicles.

Demodex caprae This follicle mite (Prostigmata: Demodicidae) is a pest of goats, especially young or pregnant animals; it produces bumps in the skin that may contain many mites.

Demodex cati A follicle mite (Demodicidae) that may be found in cats with impaired immune systems.

Demodex equi A follicle mite (Demodicidae) that may be found on horses.

Demodex gatoi A follicle mite (Demodicidae) that may be found on cats with impaired immune systems.

Demodex tauri A follicle mite (Demodicidae) that may be found in hair follicles and sebaceous glands of the eyelids of cattle.

density Number of individuals per unit of measure.

density dependence Predators that respond to changes the density of the prey population behave in a density-dependent manner.

Dermacentor albipictus Winter tick.

Dermacentor andersoni Rocky Mountain wood tick.

Dermacentor occidentalis Pacific Coast tick.

Dermacentor variabilis American dog tick.

dermal LC_{50} or LD_{50} When a toxicant is introduced through the skin or integument, it is a dermal application. LC_{50} is the concentration to which the arthropod was exposed that caused 50% mortality of the treated population, and LD_{50} is the actual amount applied to the exoskeleton that caused 50% mortality.

Dermanyssus gallinae Poultry red mite or roost mite (Dermanyssidae).

deutogyne A structurally different form of female of an eriophyoid species that is found in the winter. Females of the same species in the summer (protogynes) may have been described as different species, causing taxonomic confusion.

deutonymph A developmental stage of mites; the deutonymph is the second nymphal stage (having four pairs of legs). Often it is the last stage before molting to the adult. Some deutonymphs (Acaridida) can be highly modified as hypopi, allowing the mites to survive unfavorable conditions and, perhaps, to disperse to more favorable environments via phoresy. Hypopi typically lack mouthparts and a functional digestive system and may have setae, suckers, or other structures that allow them to attach themselves to another organism to move to a more favorable environment.

diapause A state of arrested development that is genetically determined and occurs in advance of deleterious environmental conditions. Aestival diapause occurs in the summer and hibernal diapause occurs in the winter. In mites, diapause is typically found in eggs or adult females (reproductive diapause). Temperature, day length, and host-plant quality typically serve as cues for diapause induction.

diel periodicity A 24-hour periodicity in physiological or behavioral responses to various environmental cues.

dimorphic Two different forms; for example, males are often a different size and shape compared to females of the same species. Sometimes, there may be two different forms of both males and females.

diplodiploidy Males and females are both diploid (having two sets of chromosomes, with one set derived from the mother and the other from the father).

direct mortality Mortality of predators from a pesticide applied at the label rate.

direct pests Pests that cause damage to a crop or domestic animal (without transmitting a disease).

direct sperm transfer Males and females mate and the male mite transfers sperm into the seminal receptacle of the female by using the spermatodactyl or an aedeagus.

dispersal Movement of individuals away from each other.

dispersion Pattern of spacing of individuals of a population; dispersion patterns may be random, clumped, or hyperdispersed.

diverticula Pouches located off the midgut.

domatia Structures produced by perennial woody trees and vines that are often inhabited by mites. In some cases, the relationship is mutualistic. Domatia often are found on the lower surface of leaves near the midribs and along the veins. They are usually depressions that may be enclosed by hairs. Tydeids and phytoseiids often inhabit domatia where the relative humidity is greater. A correlation appears to exist between cultivars with domatia and the presence of fewer pest mites and fungi. Also called *acarodomatia*.

dominant In genetics, a single copy of the gene in a diploid organism is sufficient to produce the phenotype.

dormant sprays Sprays applied when the plant is not growing actively, usually during the winter.

dorsum Upper side of the body.

dose The actual amount of toxic product to which the arthropod is exposed. If it is eaten, it is an *oral dose*; if it is introduced through the skin or integument, it is a *dermal dose*; and, if it is introduced through the respiratory system, it is an *inhalation dose*.

dose–response curve Represents the relative toxicity of a pesticide; based on data obtained by exposing the test species to increasing doses of a toxicant.

drone brood trapping This is a method for managing *Varroa* mites in honey bee hives. *Varroa* prefers drones, so combs of the drone brood can be used to attract, trap, and remove mites.

dust management In Mediterranean climates or other areas where little rain occurs in the summer, dust can enhance outbreaks of spider mites, perhaps by interfering with their predators or otherwise making the crop more suitable for spider mite reproduction. Reducing the amount of dust by watering or paving roads can reduce spider mite populations.

dust mites Mites in the Pyroglyphidae (Acaridida) that are associated with house dust. They live in carpets, beds, furniture, and draperies and feed on skin scales and other detritus. Their feces contain proteins that cause an allergic response in humans. Common dust mite species are *Dermatophagoides pteronyssinus* and *D. farinae*.

E

ecdysis Molting of the exoskeleton in arthropods.
eclosion Hatching of the egg.
economic damage Amount of damage done by a pest that causes economic loss.
economic injury level Lowest level of pest damage that will justify control costs.
economic threshold Level of the pest at which control measures should be applied to prevent pest levels from reaching the economic injury level.
ecosystem All interacting parts of the biological and physical worlds; a community and its abiotic environment.
empodium Structure between the two true claws of arthropods. Empodia can be pad like or claw like, and their structure may be used as a taxonomic character.
endosymbiont A symbiont that lives within the cells of another organism. *Wolbachia* is a common endosymbiont of arthropods that is transmitted transovarially and is found in the ovaries and testes.
Eotetranychus Genus of Tetranychidae that includes *E. willamettei, E. hicoriae, E. lewisi, E. sexmaculatus*, and *E. yumensis.*
Eotetranychus hicoriae This tetranychid is a pest of pecan, hickory, oak, and chestnuts in the eastern United States.
Eotetranychus lewisi This tetranychid is found around the world on a variety of host plants.
Eotetranychus sexmaculatus Also known as the six-spotted mite, this tetranychid is a pest of citrus, avocados, maples, pyracanthas, azaleas, and camphor.
Eotetranychus willamettei The Willamette mite is a tetranychid species found in vineyards in California and elsewhere; it serves as important prey for the phytoseiid *Metaseiulus occidentalis.*
Eotetranychus yumensis This tetranychid is a pest of citrus, sorghum, grapes, and other plants in the southwestern United States and Mexico. It produces copious webbing and aestivates during the summer.
epicuticle The outermost layer of the exoskeleton; consists of waxes and lipids that waterproof the arthropod.
epidermis In acarines, the layer of cells that secrete the cuticular portion of the exoskeleton.
Epitrimerus alinae Also known as the chrysanthemum rust mite, this eriophyid mite is a pest of ornamentals.
epizootics Large-scale outbreaks of diseases in animals.
erinea Modified leaf structures, typically having large numbers of leaf hairs, that are induced by the feeding of eriophyoid mites. Erinea are thought to provide a favorable site for feeding by these tiny mites, resulting in higher relative humidity and some protection from predation.
Eriophyes lowi Also known as the lilac bud mite, this eriophyid is a pest in Europe.
Eriophyes spiraeae This eriophyid, also known as the bridal-wreath gall mite, is a pest in North America and Europe.
eriophyoid Belonging to the Eriophyoidea, a group of highly specialized plant-feeding mites with only two pairs of legs. Some eriophyoid mites vector viral diseases; they also cause abnormalities in plants ranging from alterations in buds (bud mites), alterations in leaf structure (gall or erinea mites), or modifications of stems (witches' brooms).
esophagus Connects the muscular pharynx to the stomach or ventriculus in mites.
essential oils Oils that are extracted from plants that may have acaricidal activity; they may have a strong odor and are also used in aromatherapy.
Eutetranychus Genus of Tetranychidae that includes *E. banksi* and *E. orientalis.*

Eutetranychus banksi Also known as the Texas citrus mite, this tetranychid is found on citrus, almonds, figs, and castor beans.

Eutetranychus orientalis This tetranychid is the Oriental red mite, a pest of citrus in Asia, the Middle East, and South Africa.

excretion Elimination of liquid and solid waste products of metabolism.

exoskeleton External skeleton of arthropods consisting of hard cuticle; muscles are attached to the inner side of the exoskeleton.

exsanguination Extensive loss of blood; large numbers of ticks feeding on a host can cause exsanguination and even death.

extrafloral nectaries Glandular structures produced by some plants that produce a sugary exudate that may be used by natural enemies, including some phytoseiid species, as a source of supplementary food.

extrinsic muscles Dorsoventral, oblique, rotator, and levator muscles that originate on the body wall; used to modify turgor pressure and cause extension of the chelicerae, palps, and legs.

exuviae The cast exoskeleton left behind after a molt.

F

facultative Not obligatory; may refer to predation or to diapause.

false spider mite A member of the family Tenuipalpidae; these slow-moving mites do not produce silk, and only a few are agricultural pests.

feather mites Members of several mite superfamilies, including Dermanyssoidea and Analgoidea.

fecundity Average number of eggs laid; rate at which females produce progeny.

femur Leg segment of a mite located between the trochanter and the genu.

field failure Occurs when a pesticide fails to control a pest population. Reasons for such failure include resistance in the pests; however, inadequate application methods, improper mixing or concentrations, and other logistical problems can cause field failures without resistance in the pest population.

fitness cost A reduction in the ability of an organism to survive and reproduce.

flexors Muscles that bend a joint or limb.

Floracarus perrepae Eriophyoid mite evaluated for control of the Old World climbing fern (*Lygodium microphyllum*).

formamide Pesticide such as chlorodimeform, amitraz, or formetanate that is thought to inhibit monoamine oxidase, which results in accumulation of biogenic amines.

functional response Changes in the rate of prey attacked by predators in response to changes in prey density. At low prey densities, predators spend a great deal of time searching for prey. At high prey densities, they spend less time searching and more time handling their prey. There are three types of functional response. Type I is typical of passive predators that spend time waiting for prey to arrive and is dependent on prey density. Type II is most typical of predators having a constant search rate and prey mortality declining with prey density; predators of this type cause maximum mortality at low prey density. Type III includes predators that increase their search with increasing prey density, perhaps responding to chemical cues (kairomones) produced by the prey.

fungivorous Feeding on fungi.

G

gall mite Eriophyoid mite that causes the formation of galls through its feeding.

Gamasida See Mesostigmata.

gastric caecae A blind-ending sac or tube off the midgut.

gene DNA sequence that codes for a particular function; many genes code for enzymes, but others code for structural proteins or for RNAs that have cellular functions (such as transfer RNAs, ribosomal RNAs, micro RNAs).

genetic improvement Modification of the genome of an organism to make it more effective under the desired environmental conditions. *Metaseiulus occidentalis*, for example, was subjected to genetic selection for resistance to carbaryl and organophosphorus pesticides in the laboratory. The new strain was considered to be genetically improved because it performed well in almond orchards, where it survived and controlled *Tetranychus* spider mites.

genital shield A sclerotized region surrounding the genital opening.

genital suckers Extensions of the exoskeleton located near the genital opening.

genu Latin word for 'knee.' In the Acari, the genu is the third segment of legs and palps counting from the distal end.

Glycyphagus domesticus Common stored-product mite (Acaridida) that is also found in nests, bat roosts, beehives, and human houses.

gnathosoma The most anterior region of the mite body containing the mouth, chelicerae, and palps. *The Acari have no head.* Mites have two main body regions: the gnathosoma and the idiosoma.

gravid Containing developing young or fertilized eggs.

guanine One of the main nitrogenous compounds excreted by mites.

Gulf coast tick *Amblyomma maculatum.*

H

Haller's organ A sensory field on the tarsus of legs I of ticks that contains olfactory receptors, along with heat and humidity receptors.

Halotydeus destructor Red-legged earth mite (Penthaleidae).

haplodiploidy Reproduction in which the males are haploid and females are diploid; also called *arrhenotoky.* Males contain only the set of chromosomes derived from their mother. Unmated arrhenotokous females can deposit haploid eggs, all of which are males.

hay-itch mite Species of Pyemotidae, including *Pyemotes ventricosus* and *P. triciti,* are pests when they bite humans handling hay or grain. These mites are parasites of grain-infesting insects (and thus are beneficial until they bite farm workers and others). Their bites can cause severe skin reactions.

Hemerobiidae A family of neuropterans, also known as brown lacewings, that are generalist predators of small arthropods, including mites.

Hemitarsonemus tepidariorum This tarsonemid, also known as the fern mite, is a pest of new growth on ferns in warm, humid greenhouses.

hemocoel Main body cavity of arthropods in which the blood moves.

herbivore Organism that eats plants.

herbivore-induced plant volatiles (HIPVs) Substances produced by many plants in response to damage from plant-feeding insect and mite pests. These volatiles vary by herbivore involved, plant species, and cultivar. Methyl salicylate, (*Z*)-3-hexenyl acetate, and other chemicals may serve as attractants to predatory insects and mites. The use of HIPVs to attract specific natural enemies into crops, especially early in the season, is being studied as a possible pest management tool.

hexythiazox Growth regulator of spider mites that kills eggs but not active stages.

hibernal diapause Diapause that allows an organism to survive winter.

high-dose model A model for delaying resistance to pesticides by applying such a high dose of pesticide that the arthropod cannot be selected for resistance. It assumes that total coverage is possible and that the dose is sufficiently high that all pests die, but it does not take into account the environmental damage that may be done by such high doses.

hindgut The region of the digestive system in which most digestion occurs; waste products exit through the anus.

HIPVs See herbivore-induced plant volatiles.

honey bee tracheal mite *Acarapis woodi*, a tarsonemid parasite of honey bees that lives within the tracheae; also known as the *Isle of Wight disease.*

honeydew Watery, sugar-containing fluid excreted from aphids and other hemipterans.

hormoligosis General response by many organisms to low concentrations of toxins, resulting in a physiological stimulation of the organism. In spider mites, hormoligosis may be caused by low concentrations of certain pesticides, resulting in an increased reproductive rate.

host Plant on which an arthropod feeds, often restricted to a plant on which developmentoccurs. May also refer to an arthropod on which a pathogen develops or on which a parasite develops.

host-plant resistance Quality of plants that can be utilized in IMM and IPM to reduce or inhibit pest populations. It is often categorized into three types: antibiosis, tolerance, and nonpreference. *Antibiosis* is achieved when chemical or physical attributes of the plant result in reduced fitness of the organism (e.g., fecundity, longevity). *Nonpreference* occurs when physical or chemical attributes (hairiness, spines, secondary plant compounds) of the plant inhibit or repel the organisms so they do not attack it. *Tolerance* by the plant means it is able to repair or tolerate damage.

house dust mites See dust mites.

Hoyer's mounting medium The most commonly used mounting medium for mites includes gum arabic, chloral hydrate, and glycerin. Mites stored in this medium are not preserved indefinitely due to formation of crystals or air bubbles.

Hyalomma marginatum This ixodid tick may be a species complex and is a vector of bovine tropical theilerosis.

hygienic behavior (honey bee) Bees remove dead brood from their colony by uncapping and removing them. Hygienic behavior may also result in the removal of *Varroa* mites from the colony. Grooming may be an important component of some *Varroa*-resistant lines of bees.

hypersensitive response A type of host-plant resistance in which the plant produces a strong response to damage by plant-feeding arthropods or to plant pathogens. In the case of Bartlett pear trees, a single mite feeding on a leaf can cause defoliation and leaf scorch.

Hypoaspis aculeifer Predator commercially reared and released augmentively to control sciarid flies and *Rhizoglyphus echinopus* in greenhouses.

Hypoaspis miles Also known as *Geolaelaps miles* or *Stratiolaelaps miles*; used to control sciarid flies and *Rhizoglyphus echinopus* in greenhouse releases.

hypopus A nonfeeding deutonymph in the Astigmata adapted for dispersal and avoiding adverse environmental conditions; many hypopi disperse by phoresy.

hypostome A barbed piercing organ in ticks that contains retrorse teeth for attaching to the host.

hysterosoma The region of the body posterior to the second pair of legs in acarines.

I

idiosoma Portion of acarine body without the gnathosoma (all of the body but the mouthparts).

IMM See integrated mite management.

indicator plants Plants grown with a crop that are more attractive to the pest than the crop. By monitoring the indicator plants, it may be possible to identify pest problems before they become serious.

indirect mortality Mortality of obligate predators caused by a lack of prey due to a short-term loss of prey by a pesticide application.

indirect sperm transfer Occurs when males deposit spermatophores on the substrate and the females pick them up; no copulation takes place.

induced host-plant resistance Inoculation of bacteria, fungi, or viruses can induce resistance in plants against subsequent attacks. Attacks by herbivores, including spider mites, also can induce chemical and physical changes in the host plants that make them less suitable for subsequent mite populations.

inoculation In augmentative biological control, the release of natural enemies that will persist, reproduce, and provide control of the pest population over time.

inoculum Any stage or part of a pathogen (spores or virus particles) that can infect a host.

inorganic Not containing carbon; for example, sulfur is an inorganic acaricide and fungicide.

instar The growth stage between two successive molts.

integrated mite management (IMM) Strategy for managing mites using a variety of tactics (host-plant resistance, biological control, cultural controls, monitoring, biorational) in addition to or instead of chemical control. Effective management of the soil, water, and fertilizer enhances these IMM tools. All tactics should be complementary and result in a management program that takes into account environmental, social, and human health issues. Pesticides that do not harm natural enemies and are environmentally friendly should be used whenever possible.

integrated pest management (IPM) Strategy for managing pests that uses a variety of tactics in addition to or instead of chemical control. All tactics employed should be complementary and result in a program that takes into account environmental, social, and human health issues. Pesticides that do not harm natural enemies should be used whenever possible.

intraguild predation Killing and eating of potential competitors. Intraguild predation occurs when a phytoseiid kills and eats the eggs and immatures of another phytoseiid; if it eats the eggs and immatures of its own species, then it is cannibalism.

intrinsic muscles Muscles that extend over the joints of leg segments and are flexors.

intrinsic rate of increase (r_m) Rate of population growth under specific environmental conditions based on age-specific survival and reproductive rates.

intrinsically selective Refers to a pesticide that is more toxic to pest arthropods than to beneficial arthropods (predators and parasitoids). It is an important attribute for integrating the use of chemical control into IMM.

inundation In augmentative biological control, the release of large numbers of natural enemies into the crop in order to obtain rapid control of a pest.

IPM See integrated pest management.

Ixodes pacificus Western black-legged tick, which transmits Lyme disease and anaplasmosis to humans.

Ixodes ricinus Castor bean or sheep tick, which can transmit several pathogens to humans, horses, sheep, dogs, cats, and cattle.

Ixodes scapularis Black-legged tick, which transmits *Borrelia burgdorferi*, the causative agent of Lyme disease.

Ixodida Group of mites that includes ticks; all ticks are mites, but not all mites are ticks. Also known as *Metastigmata*.

Ixodidae The family of hard ticks is the largest family in the Ixodida (Metastigmata), with approximately 683 species. This family includes the Prostriata and the Metastriata.

K

kairomones Chemicals emitted by one organism as a specific signal that is beneficial to another species. Predators often detect chemical or physical cues of their prey and use them to improve their searching.

key pest Recurring, severe pest that causes serious problems in a crop.

Knemidocoptes gallinae This parasitic mite (Astigmata: Knemidocoptidae) is found in the epidermis of chickens, resulting in inflammation so feathers break off easily or are pulled out by the birds.

Knemidocoptes laevis Also known as the depluming mite of chickens, geese, pheasants, and pigeons (Astigmata: Knemidocoptidae), these mites burrow in the epidermis of the feet and legs, which causes displacement and loosening of the scales on the feet and legs, as well as inflammation.

Knemidocoptes mutans Also known as scaly-leg or beak mites, these mites (Astigmata: Knemidocoptidae) are parasites of chickens, turkeys, and pheasants.

L

Laminosioptes cysticola Fowl cyst mite (Laminosioptidae), a parasite of domestic poultry and pigeons that invades subcutaneous tissue and forms small nodules.

larva Stage in which mites and ticks (except the Eriophyoidea) have only three pairs of legs.

LC_{50} Concentration necessary to kill 50% of the population; it is the amount applied to the substrate but not necessarily the amount actually encountered by the test mite or tick. The concentrations that kill 90 or 95% are referred to as the LC_{90} or LC_{95}, respectively.

LD_{50} Dose necessary to kill 50% of the population; it is the actual amount applied to the mite or tick. The doses that kill 90 or 95% are referred to as the LD_{90} or LD_{95}, respectively.

leaf-dip or leaf-spray bioassays Used to assay for resistance of tetranychids, phytoseiids, and other mites to pesticides. The exposure is somewhat similar to field conditions and may be useful to evaluate not only survival but also fecundity and the ability of progeny to develop on the residues.

leaf vagrants Eriophyoid mites that live on the surface of plants and do not form galls; however, surface damage (russeting) may appear on foliage and fruits due to their feeding.

leprosis This is a plant disease caused by *Brevipalpus obovatus* and *B. phoenicis* (tenuipalpids).

life table An analysis (under field or laboratory conditions) of the life stages of an arthropod with a cumulative record of mortality (age-specific)and survival. Results obtained depend on the environmental conditions (biotic and abiotic) at the time of the analysis.

Listrophoridae This family of parasitic fur mites (Astigmata) has the first two pair of legs modified for clasping hairs. These mites parasitize mammals. One species attacks rabbits and another infests cats.

lone star tick *Amblyomma americanum.*

longevity Length of life.

Lyme disease Disease of humans transmitted by ticks, especially *Ixodes scapularis.*

M

Macronyssidae Family of gamasid mites that is parasitic on birds and mammals.

Malpighian tubules Tubules that collect and store nitrogenous wastes, usually as guanine or uric acid.

mating disruption Pest management tool that involves confusing males (usually moths) by releasing a synthetic, species-specific sex pheromone that prevents the males from locating females, thereby resulting in a reduced rate of reproduction by the pest population. This method works best if it is used over large areas (areawide management). Mating disruption of the codling moth in Washington apple orchards has been widely adopted since the mid-1990s.

Mesostigmata (= Gamasida) Suborder of mites in the order Parasitiformes; the stigmatal openings of these mites are in the middle of the body between legs III and IV.

metamorphosis Process in which the body structure of an animal changes during development; it often involves changes in behavior and habitat, as well.

Metastigmata Ticks. Also known as *Ixodida.*

Metastriata Hard ticks (Ixodidae) are divided into two groups, the Metastriata and the Prostriata. The Metastriata have an anal groove posterior to the anus.

microhabitat The immediate habitat in which an organism lives; its local environment.

mite-brushing machine Machine that brushes plant-feeding mites (and predators) off foliage onto a glass plate in a uniform manner so population densities can be estimated in monitoring programs.

mite day Represents one spider mite feeding for one day; used in estimating the damage caused to a crop from spider mites.

mite pockets Some species of lizards in five families have special pockets in which chiggers commonly are found. The pockets provide a protected, warm, and humid site. It is likely that the pockets are a form of damage control, maintaining a level of damage from which the lizards can recover rapidly.

miticides Pesticides that kill mites; also called *acaricides.*

mixtures model Genetic model that has been used as a method for delaying the development of pesticide resistances. It assumes that pesticides in different classes do not confer cross-resistance and that it is difficult to select simultaneously for resistances to two different products (in different classes and with different mechanisms of activity).

mode of inheritance How a trait is inherited; for example, a trait can be genetically determined by a dominant gene, by a recessive gene, by multiple genes (polygenic mode of inheritance), or by genes located on the sex-determining chromosomes (sex-linked gene).

monitoring In acarology or entomology, the recording of various measures, such as the population density of pests or natural enemies.

monoculture Cultivation of a single crop in successive years in a specific site.

Mononychellus This tetranychid genus includes the cassava green mite.

mosaics model Model for delaying the development of resistance in arthropods in which patches of crops or plants will be treated differently and arthropods exposed to these will be subjected to selection for resistance differently. It also assumes that susceptible insects in the untreated patches will interbreed with any resistant insects in the treated plants, thereby delaying resistance (if the resistance is recessive).

multiple resistances Resistance to multiple pesticides due to selection for multiple resistance mechanisms (genes).

multivoltine Having more than one generation a year.

mycetophage Feeding on fungi.

N

natural enemies Living organisms that kill arthropods, weaken them, or reduce their reproductive potential. Mites and ticks primarily have predators or pathogens as their natural enemies.

Nearctic Ecozone that includes temperate North America.

Neoseiulus californicus See *Amblyseius californicus.*

Neoseiulus cucumeris See *Amblyseius cucumeris.*

niche The precise position of an organism in a community; its behavior (e.g., predator, phytophage, pathogen) and location in that habitat.

nongravid Females that do not have fertilized eggs ready to be deposited.

nonpreference In host-plant resistance, the behavior or biology of the arthropod is affected due to chemical or physical traits of the plant that repel or cause an arthropod to prefer to feed or oviposit on another plant.

northern fowl mite *Ornithonyssus sylviarium* is an important pest of chickens and other birds.

numerical response Changes in reproductive rate in response to changes in prey density by predators.

nymph Immature stages between the larva and the adult in the acarine life cycle. The first nymphal stage is the protonymph, the second is the deutonymph, and the third (if present) is the tritonymph.

nymphochrysalis Inactive nymphal stage just prior to molting to the next instar.

O

obligate Able to survive in a single environment; without alternatives. An obligatory predator is unable to feed, develop, or reproduce on nonprey foods.

ocelli Small, simple eye with a simple lens and receptor cells. Some mites have ocelli-like eyes; no mites have compound eyes.

oils Oils may be used as pesticides to control insects and mites. They can be organic (derived from plants such as soybeans or rosemary) or inorganic (derived from petroleum). Petroleum oils vary in their purity and weight; heavier oils can be used in winter when they are less likely to cause phytotoxicity.

Oligonychus A genus of Tetranychidae that includes *O. punicae* (avocado brown mite) and *O. pratensis* (Banks grass mite).

Oligonychus pratensis Also known as the Banks grass mite, this tetranychid is a serious pest of many grasses, dates, and sugarcane.

Oligonychus punicae Also known as the avocado brown mite, this tetranychid is a pest of avocados, grapes, and pomegranates.

one-host tick Some ixodid ticks do not leave the original host animal until after adults emerge, feed, and mate.

oocyte Immature egg within the ovariole.

opisthosoma The portion of the acarine body behind the last pair of legs.

oral LC_{50} or LD_{50} The amount of toxicant taken up orally that will kill 50% of the test population. LC is the concentration to which the arthropod was exposed, and LD is the amount of toxicant taken in.

organosulfurs Acaricide class containing tetradifon, chlorfenson, and propargite; used to control plant-feeding mites.

organotins Acaricide class containing cyhexatin, azocyclotin, and fenbutatin oxide; used to control plant-feeding mites.

oribatid mite See Cryptostigmata.

Oribatida See Cryptostigmata.

Ornithodoros Soft ticks (Argasidae) that attack reptiles, birds, and mammals around the world; *Ornithodoros hermsi* can transmit relapsing fever.

Ornithonyssus bursa Tropical fowl mite (Macronyssidae).

Ornithonyssus sylviarum European or northern fowl mite (Macronyssidae).

osmoregulation Regulation of water balance so an organism's fluids are not too dilute or too concentrated in solutes.

osmoregulatory organ A structure, such as coxal glands, that regulates water balance.

osmosis Movement of water from one solution to another through a membrane.

otoacariasis Invasion of the ear canal by mites or ticks.

Otobius These soft ticks (Argasidae) include *O. megnini* (the spinose ear tick) and *O. lagophilus. Otobius megnini* feeds in the ears of cattle, horses, and domestic ruminants, causing damage and secondary infections (otoacariasis).

Otobius megnini Spinose ear tick, which infests cattle, horses, and domestic ruminants, causing damage to the ear canal and secondary infections (otoacariasis).

Otodectes cynotis Ear mites (Psoroptidae), which attack dogs, cats, and ferrets. The mites live in the ear canal and cause intense itching.

ovary Structure in the female that consists of ovarioles and is the source of oocytes.

ovicide Pesticide used to kill eggs of a pest arthropod.

oviduct Duct or tube through which eggs pass from the ovary to the outside of the body.

oviposition Deposition of eggs.

ovoviviparity Retention of the egg with the female's body until the egg is ready to hatch.

P

Palearctic Ecozone that includes Europe, Africa north of the Sahara, and Asia north of the Himalayas.

palp Second pair of appendages of the gnathosoma, used in sensing and handling food; see also pedipalp.

Panonychus Genus of Tetranychidae, including *P. ulmi* (European red mite) and *P. citri* (citrus red mite).

Panonychus citri This tetranychid is known as the citrus red mite and is a pest of citrus around the world; it also occurs on almonds, pears, and other plants.

Panonychus ulmi This tetranychid has several names, including *Metatetranychus ulmi*. It is a pest on apples, pears, plums, almonds, peaches, and grapes. Its origin is unknown because humans have transported it around the world.

parahaploidy Genetic system (in the Phytoseiidae) in which all eggs are fertilized followed by the loss of paternally derived chromosomes during embryogenesis in males only. Adult phytoseiid males are haploid even through they begin life as diploid organisms. Females remain diploid. Because males are haploid, new mutations can be selected for each generation. Also called *pseudoarrhenotoky*.

parthenogenesis Reproduction without fertilization of the egg; this may occur in arrhenotokous species with haploid males and in thelytokous species that consist of only females.

pathogens Infectious agents of disease; viruses, microsporidia, fungi, and bacteria may infect and kill pest mites and ticks.

peacock mite Mites in the family Tuckerellidae (Actinedida or Prostigmata) that have elaborate setal ornaments that are thought to deter predators. These beautiful mites are most often found in subtropical and tropical regions and usually are not serious plant pests.

pedipalp (palp) In the Acari, the pedipalps of mites and ticks vary depending on the species' feeding mode. Pedipalps are segmented and may be leg like or chelate, resembling a second pair of chelicerae.

Penthaleus major Blue oat mite or winter grain mite (Penthaleidae).

peritreme That part of the exoskeleton surrounding the tracheal opening in mites. The structure of the peritremes (i.e., gutter-like or tube-like) may be used as taxonomic traits. Location, size, and other structural aspects are used in taxonomic descriptions.

pest An organism that is undesirable and is of economic or aesthetic importance.

pesticide Chemical used to suppress pest populations. Pesticides can be stomach, contact, systemic, or fumigant poisons. They may be organic (containing carbon atoms) or inorganic (such as sulfur or arsenic). Some organic pesticides are botanicals, and some are derived from microbes. Synthetic organic acaricides include organophosphates, carbamates, pyrethroids, pyrazoles, quinazolines, methyacrylates, naphthoquinones, tetronic acids, tetrazines, oxazoles, carbazates, benzoylacetonitriles, and trifluoromethanesulfonanilides. Pesticides can be taken up by mites through the stomach, by contact, by inhalation, or as systemic products within a plant host.

pesticide resistance Genetically determined change in the ability of a population to survive applications of pesticides due to selection. The mode of inheritance can be determined by single genes (dominant or recessive), multiple genes (polygenic inheritance), or sex-linked genes. Multiple resistance mechanisms (detoxification, lack of penetration, target-site insensitivity, behavioral changes such as avoidance) are known.

pesticide selectivity Allows pesticides to be deployed in a manner so they have a reduced effect on nontarget species (especially natural enemies). Pesticides can be *intrinsically selective*, where their chemistry is naturally more toxic to the pest than to the natural enemy, or *extrinsically selective*, where they are used in a manner that targets the pest and preserves the natural enemies, perhaps due to application method, application site, or timing or to its systemic mode of action, which is more likely to affect the plant-feeding pests than the natural enemy.

pesticide tolerance Innate ability of an organism to survive particular pesticides even though they have not been selected previously with that product.

pesticide treadmill Term coined to indicate that, when farmers begin to use pesticides as their primary control tool, they tend to continue on this pathway, but broad-spectrum pesticides kill both beneficial and pest insects. This can result in resistance in the pests and in secondary-pest outbreaks, because formerly nonpest arthropods are no longer suppressed by their natural enemies, leading to yet more pesticide use to suppress the secondary pests.

Petrobia A genus of Tetranychidae that includes *latens* (the brown wheat mite).

Petrobia latens Also known as the brown wheat or onion mite, this tetranychid attacks small grains around the world and is also a pest of vegetables and ornamentals.

pharate The immature or adult mite when still inside the cuticle of the previous developmental stage (prior to molting).

pharynx Portion of the alimentary canal with muscular walls, extending from the mouth to the esophagus.

phenology Timing of life cycle events (hatching of eggs, dispersal, diapause) that are influenced by seasonal weather.

phenotype The appearance of an organism that is determined by genes.

pheromones Chemicals emitted by an organism as a specific signal to another member of the same species—for example, sex pheromones.

phoresy Transport of one organism by another, but not involving parasitism; mites may be phoretic on other organisms. Phoresy may occur during the hypopal (modified deutonymphal) stage of some mites.

photoperiod The period of light in the daily cycle of 24 hours; a 12/12 photoperiod involves 12 hours of light and 12 hours of darkness. Most arthropods can perceive daylength, which is often an important cue for species to enter diapause, to disperse, or to mate.

Phyllocoptruta oleivora Citrus rust mite (Eriophyoidea), which is a serious pest of citrus in many regions of the world.

physogastry In pyemotid mites, a condition in which the idiosoma becomes greatly distended due to the growth of ovaries and developing embryos.

Phytonemus (Steneotarsonemus) pallidus Also known as the cyclamen or strawberry mite, this tarsonemid is a pest of strawberries and ornamentals around the world.

phytophagous Able to feed on plants.

phytoseiid Species in the family Phytoseiidae are in the Gamasida (or Mesostigmata) and may be important predators of pest mites in agriculture.

Phytoseiulus persimilis Phytoseiid that is an excellent predator of spider mites, including *Tetranychus urticae* (two-spotted spider mite). This predator is an obligatory predator with a high rate of reproduction and rapid development. It is used in augmentative biological control of pests in greenhouses and strawberries.

phytotoxic Causing damage to plants; phytotoxicity can be a problem with some pesticide oils or sulfur if applied during hot weather to water-stressed plants. Other pesticides can be phytotoxic, as well.

plant disease In its broadest sense, it means any injury or abnormality caused to plants, including damage caused by plant-feeding mites.

podosoma Region of the acarine body containing the legs.

podospermy Occurs when sperm is transferred by the spermatodactyl on the chelicerae of phytoseiid males to openings near coxae III of the female.

polycropping Planting crop mixtures; traditionally occurs in tropical and subtropical areas. Polycropping, in theory, reduces pest damage because detection of the presence of the host plant is reduced if the crops that are planted do not have the same pests. It is unknown whether polycropping is useful in managing plant-feeding mites; however, it is possible that natural enemies may be increased in one crop and are able to move into an adjacent crop to suppress the pests.

Polyphagotarsonemus latus Also known as the broad mite, this tarsonemid mite is a pest of citrus in greenhouses and on lemons and limes in the field, as well as many other crops around the world.

population A group of individuals of the same species occurring within a given space and time.

poultry mites Red poultry mite (*Dermanyssus gallinae*), northern fowl mite (*Ornithonyssus sylviarum*), tropical fowl mite (*Ornithonyssus bursa*), and scaly-leg mite (*Knemidocoptes mutans*).

predators Organisms that will consume multiple individuals during their development and reproductive life. Predators of mites include insects, spiders, and other mites. Predators of ticks include vertebrates, such as birds, or ants.

presence–absence sampling A method of sampling insects or mites (and sometimes their natural enemies) in the field that allows decisions to be made as to whether to spray with a pesticide. Leaves are examined for the presence or absence of the pest (and natural enemies, perhaps); the number of leaves sampled will depend on the population density present. If pest populations are high (and few natural enemies are present), a rapid decision can be made to treat. If populations are at an intermediate level relative to the economic injury level, then sampling may be continued in time until a decision can be made. If pest densities are very low, fewer samples may be required to decide not to treat; future samples on later dates may result in a decision to treat. The densities for each crop that trigger a treat or no-treat decision are determined from detailed predator and prey dynamics data that are collected on a leaf-by-leaf basis.

prey An organism consumed or killed by a predator.

propodosoma The region of the body containing the anterior two pairs of legs in a mite.

prosoma Part of the body consisting of the gnathosoma and the podosoma.

Prostigmata (= Actinedida) Suborder of Acariformes; see also Actinedida.

Prostriata Ixodid ticks are divided into Prostriata and Metastriata. Prostriata ticks have an anal groove anterior to the anus and enclosing the anus.

protogyne Morphologically distinctive adult eriophyoid female that occurs in the summer. The winter form (deutogyne) of the same species may have been given a different species name.

protonymph The first stage in the life cycle in which the mite or tick has four pairs of legs, coming after the larval stage in which only three pairs of legs are present (except in the Eriophyoidea).

pseudoarrhenotoky See parahaploidy.

Psoroptes Genus of mange mites (Psoroptidae) that cause mange in sheep, cattle, wild ruminants, goats, and rabbits.

Psoroptes bovis A scab mite (Astigmata: Psoroptidae) that attacks cattle and may cause psoroptic mange.

Psoroptes equi A scab mite (Astigmata: Psoroptidae) that attacks horses. These mites spend their lives on their hosts and cause inflammation and possible subsequent bacterial infections. Psoroptic mange can spread to the entire body.

Psoroptes ovis A scab mite (Astigmata: Psoroptidae) that attacks sheep, causing psoroptic mange. These mites are able to spread rapidly by direct transfer from other animals or from contaminated fences.

psoroptic mange A condition caused by infestations with psoroptic mites (Astigmata: Psoroptidae). These parasitic mites live on the surface of the skin but feed on lymph blood and cause inflammation, hair loss, crusting, and scabs, as well as itching.

Psoroptidae This family (Astigmata) includes skin parasites and the life cycle never involves a hypopial life stage; they may be called scab mites. All developmental stages live on the skin surface, causing inflammation, hair loss, crusting and scabs. These mites may attack cattle, sheep, goats, horses, and rabbits, as well as cats and dogs.

Pyemotes tritici Species of Pyemotidae that is an ectoparasite of grain-infesting Coleoptera. *P. tritici* can cause severe dermatitis to humans. Also called *straw-itch mite*.

Pyemotes ventricosus Hay-itch or straw-itch mite (Pyemotidae), which parasitizes insects but will attack humans, as well, causing severe dermatitis. The mites inject a toxin that is able to paralyze their insect hosts. Their biology is unusual, as the female bears live young that are sexually mature from her swollen body (physogastry).

pyrethroid An organic synthetic insecticide/acaricide with a structure based on pyrethrum, a botanical insecticide derived from chrysanthemum flowers.

pyrroles A pesticide, such as pyridaben, that works as a mitochondrial electron transport inhibitor to block cellular respiration.

Q

quarantine Local or national in scope, a quarantine involves attempts to prevent the immigration of pests into a particular crop or region. Typically, international trade is subject to regulations and inspections. Locally, monitoring of plants to ensure that they are free of pests and diseases before introducing them into greenhouses or the field can help reduce pest problems.

questing tick A tick that is in the process of seeking a host. Ticks that quest climb grass stems or bushes and wave their anterior legs until a host appears. When the host brushes the grass or shrub it will move onto the host. Carbon dioxide, heat, and movement serve as stimuli for tick host-seeking behavior.

quiescence State of inactivity that is induced directly by adverse environmental conditions in mites and ticks. Typically, normal activity is restored immediately upon a return to favorable environmental conditions. See diapause, in which the onset occurs prior to the onset of adverse conditions.

R

recessive In genetics, refers to two copies of an allele (one on each of the paired chromosomes) being required for the expression of the phenotype in a diploid organism. In haploid male spider mites, which only have one set of chromosomes, only one copy of the gene is present so the phenotype will be expressed even if it is a recessive gene. A recessive allele that confers pesticide resistance will result in a resistant male (haploid) spider mite. Two copies of the recessive allele that confers resistance to a pesticide are required for a diploid female spider mite to be resistant.

red velvet mite See velvet mite.

reservoir host Any person, animal, plant, soil, or substance in which an infectious agent normally lives and multiplies. The reservoir typically harbors the infectious agent without injury to itself and serves as a source from which other individuals can be infected. The infectious agent primarily depends on the reservoir for its survival.

resistance A genetically selected change in the population's response to the effects of pesticides, diseases, or other stressors which allows the population to survive under an otherwise lethal condition.

retrorse teeth Turned, or bent backward; this type of tooth is found on the hypostome of ticks and makes it difficult to remove the tick during feeding.

Rhipicephalus (Boophilus) annulatus Cattle tick that can transmit bovine babesiosis.

Rhipicephalus appendiculatus The brown ear tick is an ixodid and a pest of cattle, buffalo, and antelope, attacking the ears. It can cause anemia and damage to the ears, and it may transmit several pathogens.

Rhipicephalus (Boophilus) microplus Tropical fever tick or southern cattle tick, considered the most important tick parasite of cattle in the world.

Rhipicephalus sanguineus The brown dog tick is an ixodid and a pest of mammals. It is unusual among ticks because it can complete its life cycle indoors. It can transmit canine ehrlichiosis and canine babesia.

Rhizoglyphus echinopus Acarid mite that is a pest of stored bulbs. There is taxonomic confusion as to the number of species that are actually bulb mites. These mites transmit fungal diseases and thrive in high relative humidity.

Rhizoglyphus robini Acarid mite that is a pest of stored bulbs. There is taxonomic confusion as to the number of species that are actually bulb mites.

Rickettsia tsutsugamushi Causal agent of scrub typhus, transmitted to humans by chiggers in parts of Southeast Asia.

rotation of pesticide classes model In an effort to delay pesticide resistances in pests, growers are often told to rotate the class of chemical used; however, effective rotation requires that a number of genetic assumptions are met and that the biology of the arthropod is appropriate. If all of the assumptions are not met, then rotations may not delay resistance development.

russeting Brown, roughened areas that develop on the surface of foliage or fruits as a result of feeding damage from eriophyoid mites.

rust mites Eriophyoid mites that cause russeting (a brown rough area on foliage or fruit) through their feeding.

S

saprophagous Refers to feeding on decaying organic matter.

Sarcoptes Genus of mites (Sarcoptidae) referred to as scabies, mange, or itch mites; they infest mammals, including humans. Each host has its own strain. The mites burrow in the skin, feeding on tissue fluids. In humans, *S. scabiei* damages the skin while burrowing.

sarcoptic mange A skin disorder caused by parasitic mites in the Sarcoptidae (Astigmata) that burrow in the epidermis of their hosts. These mites cause intense itching, lesions, anemia, and weight loss, and secondary bacterial infections may occur. In some hosts, sarcoptic mange can even lead to death.

Sarcoptidae This family (Astigmata) includes skin parasites that are known as scabies, mange, or itch mites; the life cycle never involves a hypopial life stage. These mites may attack mammals in 17 families and 7 orders but their species status is unclear. Females burrow into the epidermis of their hosts, causing dermatitis and mange.

scaly-leg mites Mites (Knemidocoptidae) causing damage to the legs of birds.

secondary metabolite A substance produced by a plant that may defend against herbivory.

secondary pest outbreak The term used when the application of a pesticide applied to control one pest disrupts the natural enemies of another organism in the crop, leading to a new pest problem. It is often the case that pesticides are more toxic to natural enemies than to the pests.

segmentation Arthropods are segmented animals, but mites typically have lost their primary body segmentation (although not the segmentation of legs and palps) so few true segments are seen.

selective pesticides The use of products that are intrinsically more toxic to the target pest than to beneficial species (parasitoids, predators, honey bees). Sometimes, products can be made selective by altering the manner in which they are applied (e.g., alternate-row sprays, application only to the outer portion of trees) or the concentration at which they are applied. Application of a lower concentration may suppress pest populations (but not kill all), while preserving most of the natural enemies. See also pesticide selectivity.

semidominant In genetics, refers to when a phenotype (such as pesticide resistance) is expressed in the F_1 progeny in crosses between a line with and a line without the trait in question, and the progeny are nearly as resistant as the homogeneous resistant population.

semiochemical Any chemical used in communicating within or between species.

sensilla A sense organ.

sequential sampling Sampling program in which the number of samples is not fixed in advance; it is based on the dispersion pattern of the pest and economic decision levels, allowing the pest population to be characterized as "pest" or "not pest." Sequential sampling usually involves a decision table that allows the pest manager to decide whether to spray or not (or release natural enemies or not). Some sequential sampling programs include monitoring of natural enemies as well as the pests.

setae Arthropod bristles; setae have many functions in mites, including chemoreception, mechanoreception, and the detection of relative humidity, among others. Setae take many forms, from simple hairs to highly branched structures. Setae are often named and used in the taxonomy of mites; unfortunately, the names of setae differ among the various mite groups.

sex ratio Ratio between males and females. Diplodiploid species commonly have a 1:1 sex ratio, but arrhenotokous species may have a ratio of 2.5 to 3.0 females for every one male. Thelytokous species produce only females.

sibling species Closely related species that may be difficult to distinguish.

silk Substance produced by some species of spider mites (Tetranychidae), although not all; silk is important in their biology, providing benefits such as protecting them from rain, predation, and desiccation. Silk can, however, serve as a cue to some phytoseiid predators that prey is nearby, in which case the predators use the silk as a kairomone.

silking Refers to the production of silk that allows spider mites to disperse aerially; also called ballooning.

SIR or SIT method The sterile insect release or sterile insect technique. Most often, sterilized males that are reared in a laboratory are released at a high rate to mate with wild females. If sufficient sterile males are released (often at a ratio of 100 sterile males to one wild male) over time, suppression or eradication of the pest can be achieved.

slide-dip bioassay A method for detecting pesticide tolerance or resistances, often used with spider mites or phytoseiids. Healthy young adults are mounted on double-stick tape on a microscope slide, and the slides are then dipped into a series of concentrations of the pesticide. Mortality is determined after 24 to 48 hours.

slow-release bags (or sachets) These bags (or sachets) contain grain, grain mites, and predatory mites that can feed on the grain mites. By releasing phytoseiids in these sachets, the predators can feed, multiply, and disperse from them to provide longer-term control in augmentative releases. These release methods do not work for predators such as *Phytoseiulus persimilis* or *Metaseiulus occidentalis*, phytoseiids that do not feed on grain mites.

smoking and dropping mites Bee hives can be smoked to cause *Varroa* mites to drop off their honey bee hosts. This method works for small apiaries.

somite A segment of the body.

spermatheca An organ in the female that receives and stores sperm received from the male.

spermatodactyl A structure on the chelicerae of some gamasid males that is involved in sperm transfer (podospermy). The structure of the spermatodactyl can be used to identify species. Typically, podospermy involves the male taking up a bundle of sperm from the genital opening on his venter and inserting the sperm packet into a structure in the female located between legs three and four.

spermatophore A bundle of sperm produced by males that is transferred into the female. Transfer can be direct when the male transfers it into the spermatheca of the female or indirect when the male deposits the spermatophore onto a surface, where the female can pick it up later.

spider mite Mite in the family Tetranychidae. Some tetranychids produce silk, giving the family the common name of spider mite. Many species are plant pests of economic importance.

spinneret An enlarged, hollow seta on the distal palptarsus used to spin silk webbing by some spider mite species.

spot treatment Pesticides can be applied to limited areas of the crop where pest populations tend to increase early in the growing season. By doing so, the pests may not spread, and subsequent treatments may not be required if the natural enemies present can suppress them.

Steneotarsonemus laticeps Also known as the bulb scale mite, this tarsonemid is a pest of ornamental bulbs.

stigmata or stigma The opening to the tracheal system in acarines.

stipe Some mite eggs have a small stalk (stipe) on them that may be used by females to attach the egg to the substrate.

stippling Localized damage to foliage characterized by numerous pale dots or points where the chlorophyll has been removed by mites during feeding.

straw-itch mites *Pyemotes ventricosus* and *P. tritici* (Pyemotidae) parasitize insects but will attack humans, as well, causing severe dermatitis. The mites inject a toxin that is able to paralyze their insect hosts. The biology is unusual, as the female bears live young that are sexually mature from her swollen body (physogastry).

stunting Abnormal reduction in plant size.

stylet-like chelicerae Mites in the Tetranychidae and Eriophyoidea have highly modified chelicerae that are like stylets. Those of spider mites are longer than those of the Eriophyoidea.

stylostome Tube-like structure produced when a chigger (Trombiculidae) feeds on a host. It is produced by the interaction between the chigger's saliva and the host, causing intense itching.

sulfur An inorganic acaricide. Sulfur resistance has developed in some populations of spider mites and in some phytoseiid populations; however, it can be a highly effective product for control of most plant-feeding mites (tetranychids, eriophyoids, tarsonemids, tenuipalpids).

summer oils Oils that are applied to control pests during the growing season.

symbiont An organism that has a close relationship with another species. Many symbionts of arthropods are microorganisms (viruses, bacteria, fungi, yeasts). A mutualistic symbiosis is one in which both partners benefit, but the definition (by some) includes parasitism and pathogenesis.

symbiosis A long-standing and close relationship between organisms of two different species; may include mutualism, parasitism, and commensalism. Many mites contain microbial symbionts.

synergist A chemical that results in greater total effect when used in conjunction with another agent, sometimes giving a result greater than the sum of their separate effects.

synganglion A fused mass of nervous tissue that surrounds the esophagus; considered to be the brain of acarines.

synonyms In taxonomy, it means that two names have been given to the same taxonomic entity.

systemic pesticides Pesticides that are absorbed into a plant or animal and protect the plant or animal from pests.

T

tactile Touch receptors.

tagmata Body regions of a mite; mites have two main tagmata: *gnathosoma* and *idiosoma*. These regions are segmented, but the primary segmentation is obscured because fusion has occurred so the primitive segments are no longer visible.

tarsae A leg segment that is commonly segmented and attached to the tibia.

tarsonemid This family has diverse feeding habits, including mites that are pests of plants. Others may feed on fungi or algae or parasitize insects. The tracheal mite of honey bees (*Acarapis woodi*) is a tarsonemid. They may be referred to as thread-footed mites, because the adult female has a fourth pair of legs that are thread like.

Tarsonemus confusus This tarsonemid is a minor pest of ornamentals in greenhouses. It is also a pest of fungal cultures and has broad food habits.

Tegolophus perseaflorae A pest of avocado (Eriophyoidea) that recently invaded Florida.

tenuipalpid Mites in this plant-feeding family are often referred to as false spider mites, although they do not produce silk.

Tenuipalpus pacificus Also known as the phalaenopsis mite, this tenuipalpid is a pest of orchids.

Tetranychus Genus of Tetranychidae that includes numerous web-spinning mite pests, including *T. urticae* (or *T. telarius*, the two-spotted spider mite), *T. ludeni*, *T. evansi*, and *T. pacificus*.

Tetranychus desertorum This spider mite was accidentally introduced into Australia, where it colonized prickly pear cactus, a weed, and provided some suppression of the weed.

Tetranychus evansi This tetranychid is also known as the tomato spider mite and is a pest in North and South America; it also attacks other solanceous plants.

Tetranychus lintearius The gorse spider mite was introduced from Europe into New Zealand to provide biological control of gorse (*Ulex europaeus*).

Tetranychus ludeni Known as the dark-red spider mite, this tetranychid is a pest of vegetables in warm regions of the world.

Tetranychus mcdanieli McDaniel spider mite, an important pest of apples in Washington State.

Tetranychus pacificus This is the Pacific mite, which is a pest of grapes, cotton, and other crops in the western United States.

Tetranychus urticae Two-spotted spider mite, which is also known as *T. telarius*, among other names; may be a species complex.

thelytoky Parthenogenetic reproduction in which females produce only female progeny; males are rarely or never produced. Thelytoky can sometimes be induced by microbial endosymbionts, such as *Wolbachia* or *Cardinium*. Ridding the females of the microbial symbiont with heat or antibiotics can restore her ability to produce male progeny.

three-host ticks Most ixodid ticks use different host animals at the larval, nymphal, and adult stages. Each stage feeds once on each host, then drops off to molt. Larvae and nymphs usually attach to small vertebrates, while adults usually attach to larger mammals.

tibia The leg segment between the genu and tarsus of mites.

tick Mites in the Ixodida (Metastigmata) that are obligate parasites of vertebrates, including reptiles, birds, mammals, and amphibians. These mites feed on blood and can be vectors of disease agents due to their ability to transmit microorganisms *transovarially* and *transstadially*. Ticks may have a single host, two hosts, or three hosts. There are two major families: Ixodidae (hard ticks) and Argasidae (soft ticks).

tick paralysis Paralysis of muscles induced by tick bites of humans, cattle, dogs, and other animals. It is apparently a severe type of sensitization reaction to the salivary secretions or toxins in tick saliva.

tolerance A natural lack of susceptibility of an organism to pesticides, stress, or diseases. In host-plant resistance, it refers to the ability of a plant to repair or survive damage by a pest arthropod.

tracheae Tubular airway that leads from a spiracle or stigmata (in some mites) to tracheoles at the end of the respiratory system. Tracheae branch through the body to allow water, oxygen, and carbon dioxide to diffuse. Some smaller mites lack tracheae.

translaminar A pesticide that can be taken up through the leaves of the plant and translocated to other parts of the plant.

transovarial transmission Transmission of a microorganism into the egg stage (vertical transmission); relevant to the transmission of microbial symbionts of arthropods and to disease-causing microorganisms transmitted by ticks and chiggers.

transstadial transmission Transmission of a microorganism through different life stages, with the microbe being maintained after a molt. This is important with regard to the ability of ticks and other mites to vector disease-causing microorganisms.

tritonymph The third nymphal stage that occurs in some mite species.

trochanter The leg segment located between the coxa and femur.

tropical fowl mite *Ornithonyssus bursa* (Gamasida: Macronyssidae) is a pest of birds but can become a pest of humans if nests are located near homes.

tropical rat mite *Ornithonyssus bacoti* can be a significant pest of humans, although it typically attacks rats and only migrates to humans if their rat hosts are eliminated.

turgor The pressure within the cell or organism resulting from the movement of water into the cell or by contraction of body muscles.

two-host ticks Some ixodid ticks drop off the initial host animal after development to the nymphal stage and require a second host after they have molted to the adult.

tydeid mite A prostigmatid mite in the family Tydeidae that is found on plants, in soil litter, or in stored products or nests. Their biology is variable, with some species feeding on fungi, pollen, or nectar, while others are predators. It is not clear if any are plant pests.

Typhlodromus pyri A phytoseiid species used to control *Panonychus ulmi*, *Tetranychus urticae*, *Eriophyes vitis*, and *Epitrimerus vitis*.

Tyrophagus longior This acarid mite is a pest of stored products and occasionally is a pest of greenhouse cucumbers; it also transmits fungal spores.

Tyrophagus neiswanderi This acarid mite is in stored products and nests but can sometimes be found in European greenhouses on a variety of plants. It is also called the pollen-cap mite because it can feed on pollen of cymbidium orchids.

Tyrophagus putrescentiae A common stored-grain mite (Acaridida) that can transmit fungal spores throughout stored products. This mite can feed and reproduce on fungi.

Tyrophagus similis This acarid mite can become a pest of greenhouse plants.

U

univoltine Having only one generation a year.

V

vector Agent by which a pathogen is transmitted from one host to another. Acarine vectors include both mites and ticks that are able to transmit disease-causing agents to bees, plants, humans, and vertebrates, including humans.

velvet mite or red velvet mite A brilliantly colored red mite that is large enough to have a common name. The Trombidiidae often are found running rapidly over the surface of soil. Adults of some species can live several years. Larvae are parasitic on small arthropods (especially orthopteran insects), and nymphs and adults are predators of small arthropods.

venom A toxic fluid injected into prey that causes death, paralysis, or pain.

ventral Bottom portion of the body.

ventriculus The portion of the digestive tract of a mite in which digestion occurs.

viviparity The egg is retained within the mother until it hatches.

viviparous Refers to when an organism gives birth to living young. This is unusual for mites, but *Pyemotes* species (Actinedida: Pyemotidae) retain their young until they are adults.

voltinism The number of generations per year; see multivoltine or univoltine.

W

wasteful killing The propensity of spiders to capture and kill more prey than they can consume.

western orchard predatory mite A phytoseiid mite, *Metaseiulus* (=*Galendromus* or *Typhlodromus*) *occidentalis*, found in the western United States on deciduous orchard and vineyard crops. It is an effective predator of several pest mite species in a variety of orchard and vineyard crops.

wettable powder A formulation of pesticides applied as a powder suspended in water.

witches' broom Feeding damage caused by some eriophyoid mites, resulting in stunting and excessive branching of the plant.

Wolbachia A microorganism that is a very common microbial endosymbiont of arthropods. Some are associated with modifications of the sex ratio, with cytoplasmic incompatibility, and with fitness costs. Usually, *Wolbachia* is found in the reproductive tracts (ovaries) and transovarially transmitted to progeny. The effects of *Wolbachia* often remain unknown, however.

Z

Zetzellia mali A stigmaeid mite that may be a useful predator of the apple rust mite (*Aculus schlechtendali*) in apple orchards.

Index

A

B

C

D

E

F

G

I

J

K

L

T